Iroductory Botany

Pis, People, and the Environment

Linda R. Berg, Ph.D.
St. Petersburg Junior College

SAUNDERS COLLEGE PUBLISHING
Harcourt Brace College Publishers

Philadelphia   San Diego   New York   Orlando   Austin   San Antonio

Montreal   Toronto   London   Sydney   Tokyo

About the Cover

A pink lady's slipper (*Cypripedium acaule*) blooms in late spring beside a shady rock in the Hudson Valley, New York. Native to eastern North America, this member of the orchid family grows best in acidic soil that is rich in leaves and other decayed organic matter.

Text Typeface: Clearface
Composition: York Graphic Services, Inc.
Acquisition Editors: Julie Levin Alexander; Edith Beard Brady
Development Editors: Lee Marcott; Jane Tufts
Senior Project Editor: Sally Kusch
Copy Editor: Mary Patton
Managing Editor: Carol Field
Manager of Art and Design: Carol Bleistine
Art Director: Caroline McGowan
Illustration Supervisor: Sue Kinney
Art and Design Coordinator: Kathleen Flanagan
Text Designers: Caroline McGowan; Ruth A. Hoover
Cover Designer: Ruth A. Hoover
Text Artwork: Rolin Graphics; Carlyn Iverson
Photo Editor: Robin Bonner
Senior Vice President, Director of EDP: Tim Frelick
Manager of Production: Joanne Cassetti
Vice President/Director of Marketing: Marjorie Waldron
Senior Product Manager: Sue Westmoreland

Cover photograph: ©Carr Clifton/Minden Pictures

Part Opening photo credits: Part 1: Doug Locke/Dembinsky Photo Associates; Part 2: Skip Moody/ Dembinsky Photo Associates; Part 3: Edgar E. Webber; Part 4: Adam Jones/Dembinsky Photo Associates; Part 5: David Cavagnaro/Peter Arnold, Inc.

Printed in the United States of America

Introductory Botany: Plants, People, and the Environment

ISBN 0-03-075453-4

Library of Congress Catalog Card Number: 96-70013

789012345   032   3210   987654321

This book is dedicated with much love and appreciation to my parents, Eileene and Donald Horine.

PREFACE

Introductory Botany: Plants, People, and the Environment is intended primarily as an introductory text for undergraduate students, both non-science and science majors. It is written so that it will be easily understood by students with little or no background in plant biology. As much as possible, the scientific and common names used in this book conform to those listed in *Hortus Third,* New York, MacMillan Publishing Company, 1976.

One of my principal goals in developing this text has been to share with beginning botany students an appreciation of the diverse organisms we call plants, including their remarkable adaptations to the environment and their evolutionary and ecological relationships. Special emphasis has been placed on ecology, particularly the role of plants in the biosphere, and appropriate environmental issues have been integrated into most chapters.

Botanical knowledge is important in its own right, and learning about plants can make your life richer and more rewarding. But we live in a complex society, with science and technology affecting every aspect of our lives every single day. For that reason, botany as a scientific process has also been stressed so that students will understand how scientists think, how they approach problems, and how they obtain scientific knowledge about our world.

TOOLS FOR LEARNING

Learning any science is a challenging endeavor, and botany is no exception. A variety of learning aids are included in the text to help the student achieve mastery of the concepts presented.

1. **Chapter Openers** focus on specific plants relevant to the subjects in the chapters. These short essays promote an awareness and appreciation of plants and motivate the student to read further.
2. **Learning Objectives** indicate, in behavioral terms, what students must be able to do to demonstrate mastery of material covered in the chapter.
3. **Concept-statement heads** introduce each section, previewing for the student the key idea discussed in that section.

4. **Three categories of focus boxes** highlight the environment, the human importance of plants, and topics of general interest. These boxes are written to spark student interest, familiarize students with current environmental issues, and show the relevance of plants to people.
5. **New terms are boldfaced,** permitting easy identification and providing emphasis.
6. **Study Outlines** at the end of each chapter provide quick reviews of material presented.
7. **Selected Key Terms** at the end of each chapter list important new words alphabetically with page references to their definitions.
8. **Review Questions** focus on important chapter concepts and applications.
9. **Thought Questions** can be used for essay assignments or class discussion. They challenge students to apply knowledge learned in the chapter to new situations.
10. **Suggested Readings** provide references for further learning.

Wildflower in the daisy family. (© *John Arnaldi 1995*)

11. The **Glossary** contains definitions of all Selected Key Terms, appropriately cross-listed.
12. **Appendices** include metric conversions, major periods in Earth's history, and answers to review questions.

THE ORGANIZATION OF *INTRODUCTORY BOTANY: PLANTS, PEOPLE, AND THE ENVIRONMENT*

Chapter 1, An Introduction to the Science of Botany, introduces important biological concepts and lays the foundation for understanding how science works. The remaining text is organized into five main parts.

Part 1 The Plant Cell

Chapter 2 provides the basic chemistry needed to understand botany and introduces the biological molecules that interact in the construction and maintenance of plants. Chapter 3 focuses on the structure and functions of cells and their membranes. Chapter 4 discusses the energy transactions involved in life processes; both photosynthesis and cellular respiration are introduced with simplified overviews, followed by a more detailed explanation for those courses that require more.

Part 2 Plant Structure and Life Processes

Chapters 5 through 8 discuss the structural adaptations of roots, stems, and leaves. Chapter 9 describes reproduction in flowering plants, including discussions of flowers, fruits, and seeds. Chapter 10 focuses on mineral nutrition and the mechanisms of transport in xylem and phloem. Chapter 11 presents plant hormones and responses and includes such up-to-date information as the role of phytochrome in competition for sunlight among shade-avoiding plants.

Part 3 The Continuity of Plant Life

The molecular basis of inheritance is discussed in Chapter 12; both transcription and translation are introduced with simplified overviews, followed by a more detailed explanation for those courses that require more. The processes of mitosis and meiosis are explained and contrasted in Chapter 13. Chapter 14 presents patterns of inheritance, and Chapter 15 presents concepts on the cutting edge of genetic research. Chapter 16 includes natural selection and the scientific evidence that supports evolution. The evolution within populations and of new species is discussed in Chapter 17.

Part 4 Diversity

Chapter 18 describes how biologists classify plants and other organisms and includes a rationale for the six-kingdom clas-

sification scheme. Student-accessible material on cladistics is also included in Chapter 18. The various groups of organisms traditionally studied in botany are presented in an evolutionary framework in Chapters 19 through 25: archaebacteria and eubacteria in Chapter 19; protists in Chapter 20; fungi in Chapter 21; bryophytes in Chapter 22; seedless vascular plants in Chapter 23; gymnosperms in Chapter 24; and flowering plants in Chapter 25. Current material includes a new fossil discovery that adds significantly to what we know about bryophyte evolution.

Part 5 Plant Ecology

Chapter 26 presents the foundations of ecology, such as how organisms interact with one another, how matter cycles through ecosystems, and how energy flows through ecosystems. Chapter 27 describes nine biomes and examines two ways that humans are altering the ecosphere: global climate change and stratospheric ozone depletion.

SUPPLEMENTS

To further facilitate learning and teaching, a supplements package has been carefully designed for the student and instructor. It includes:

1. *An Instructor's Manual with Test Bank* was prepared by Maralyn Renner, The College of the Redwoods. It contains materials to help the instructor and 50 questions per chapter that address the chapter learning objectives.
2. **A Computerized Test Bank** enables instructors to edit, revise, add to, or delete from the printed Test Bank. Available in IBM™ 3 1/2, Macintosh™, and Windows™ formats.
3. **Overhead Transparencies** feature 100 full-color acetates from the text, using labels with large type for easy classroom viewing.
4. *A Laboratory Manual for General Botany*, Seventh Edition, by Margaret Balbach, Illinois State University, and Lawrence C. Bliss, University of Washington, is available for use with this text. The manual contains 32 experiments and is lavishly illustrated with line drawings.

Saunders College Publishing may provide complimentary instructional aids and supplements or supplement packages to those adopters qualified under our adoption policy. Please contact your sales representative for more information. If as an adopter or potential user you receive supplements you do not need, please return them to your sales representative or send them to

Attn: Returns Department
Troy Warehouse
465 South Lincoln Drive
Troy, MO 63379

ACKNOWLEDGMENTS

The development and production of *Introductory Botany: Plants, People, and the Environment* was a complex process involving interaction and cooperation among the author, editors, and reviewers. I appreciate the valuable input and support from my editors, colleagues, and students. I thank my family and friends for their understanding, support, and encouragement as I worked through many revisions and deadlines.

The Editorial Environment

Preparing a book of this complexity is challenging and requires a great deal of time and effort. I could not have produced it without the help and support of the outstanding editorial and production staff at Saunders. I thank Publisher Barry Fetterolf and Executive Editor for Biology Edith Beard Brady for their support and help. I am grateful to Developmental Editors Jane Tufts and Lee Marcott who provided valuable input in every aspect of the project. Art Developmental Editor Carlyn Iverson made a unique contribution by sharing her wonderful artistic talents. I thank Photo Editor Robin Bonner for working hard to find exceptional photographs to enhance the text. I also recognize the support and help of Elizabeth Widdicombe and Julie Alexander.

I greatly appreciate the help of Project Editor Sally Kusch who guided the project through the complexities of production. I also thank Art Director Caroline McGowan for creating the book design and coordinating the art program with the capable help of Illustration Supervisor Sue Kinney. I appreciate the efforts of Sue Westmoreland, Senior Product Manager, in marketing the book. All of these dedicated professionals and many others at Saunders made important contributions to the production of *Introductory Botany: Plants, People, and the Environment.* I thank them for their help and support throughout this project.

My colleagues and students at both the University of Maryland and St. Petersburg Junior College have provided valuable input. I thank them and ask for their comments and suggestions as they use this text. I can be reached through my editors at Saunders College Publishing.

The Professional Environment—Reviewers

I very much appreciate the input of the many professors and scientific workers who have reviewed the manuscript during various stages of its preparation and provided me with valuable suggestions for improving it. Their work contributed greatly to the final product. Reviewers include:

Rolf W. Benseler, *California State University—Hayward*
Lawrence C. Bliss, *University of Washington*
Frank Bowers, *University of Wisconsin—Stevens Point*
Maynard Bowers, *Northern Michigan State University*
Richard C. Bowmer, *Idaho State University*
Richard Churchill, *Southern Maine Technical College*
Curtis Clark, *California State Polytechnic University—Pomona*
W. Dennis Clark, *Arizona State University*
James T. Dawson, *Pittsburg State University*
Darleen DeMason, *University of California—Riverside*
Roger el Moral, *University of Washington*
Michael Farris, *Hamline University*
Daniel Flisser, *Cazenovia College*
Doug Friend, *University of Hawaii—Manoa*
Stephen W. Fuller, *Mary Washington College*
Michael Gardiner, *University of Puget Sound*
John H. Green, *Nicholls State University*
William M. Harris, *University of Arkansas*
Hon H. Ho, *SUNY, The College at New Paltz*
William Jensen, *Ohio State University*
J. Morris Johnson, *Western Oregon State College*
Robert J. Lebowitz, *Mankato State University*
Marion Blois Lobstein, *Northern Virginia Community College—Manassas Campus*
George H. Manaker, *Temple University*
Michael Howard Marcovitz, *Midland Lutheran College*
James E. Marler, *Louisiana State University—Alexandria*
Bill Mathena, *Kaskaskia College*
W.D. McBryde, *Central Texas College*
John McMillon, *Lee College*
Robert S. Mellor, *University of Arizona*
H. Gordon Morris, *University of Tennessee at Martin*
Jeanette C. Oliver, *Flathead Valley Community College*
John Olsen, *Rhodes College*
Lee R. Parker, *California Polytechnic State University—San Luis Obispo*
Pat Pendse, *California Polytechnic State University—San Luis Obispo*
William J. Pietraface, *SUNY, College at Oneonta*
James A. Raines, *North Harris College*
James A. Rasmussen, *Southern Arkansas University*
Maralyn A. Renner, *College of the Redwoods*
Robert Rupp, *Ohio State University, Agricultural Technical Institute*
Barbara Schumacher, *San Jacinto College Central*
Richard Sims, *Jones County Junior College*
Ann Spencer, *Charles County Community College*
Michael P. Timko, *University of Virginia*
B. Dwain Vance, *University of North Texas*
Judith Verbeke, *University of Arizona*
David Williams, *Anne Arundel Community College*
James Winsor, *Pennsylvania State University, Altoona Campus*

Linda R. Berg
September 1996

Contents Overview

CONTENTS

Purple cone flowers (*Echinacea* sp). (*Doug Locke/Dembinsky Photo Associates*)

Fringed orchid (*Habenaria carnea*). (© *John Arnaldi, 1995*)

Prickly pear (*Opuntia* sp). (© *John Arnaldi, 1995*)

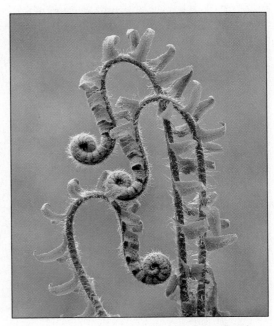

Christmas fern (*Polystichum acrostichoides*). (*Skip Moody/Dembinsky Photo Associates*)

Aspen (*Populus* sp) and spruce (*Picea* sp).
(*© John Arnaldi, 1995*)

Lady's slipper (*Phragmipedium caricinum*).
(© *John Arnaldi, 1995*)

Common sunflower (*Helianthus annuus*). (*Edgar Webber*)

Spleenwort fern (*Asplenium verecundum*).
(© *John Arnaldi, 1995*)

Rose moss (*Portulaca grandiflora*). (© *John Arnaldi, 1995*)

Texas paintbrush (*Castilleja* sp) and blue-bonnets (*Lupinus subcarnosus*). (*Adam Jones/Dembinsky Photo Associates*)

Yellow patches mushrooms (*Amanita flavoconia*). (*Skip Moody/Dembinsky Photo Associates*)

Sori on rabbit's foot fern (*Polypodium aureum*).
(© *John Arnaldi, 1995*)

White fir (*Abies concolor*). (© David Cavagnaro/
Peter Arnold, Inc.)

FOCUS BOXES

Moth orchid (*Phalaenopsis* sp). (© *John Arnaldi, 1995*)

CHAPTER 1

AN INTRODUCTION TO THE SCIENCE OF BOTANY

Cartoon artists often depict the sole survivor of a shipwreck stranded on a tiny tropical island that sports a single coconut palm (*Cocos nucifera*). If such a situation were to occur in real life, the shipwreck survivor would be truly fortunate, because this versatile plant provides almost everything a person needs to survive and be comfortable.

A single coconut palm tree usually produces 50 to 100 fruits (called "nuts") each year, and each nut contains a single large seed—the largest seed in the world. The coconuts that we buy in stores have had most of the outer layers of fruit removed (much like a peach pit without the rest of the peach). The outer husks of the coconut fruit can be separated from the nut, softened by soaking for several months in salty water, and shredded into soft fibers that can be spun into yarn and twisted to make ropes or woven to make floor mats.

An immature green coconut seed has a highly nutritious liquid center called endosperm. As the seed matures, its endosperm solidifies to form coconut meat. Copra, or dried coconut meat, can be eaten just as it is, or it can be pressed to extract its oil. The oil can then be used to make soap and other

A coconut palm (*Cocos nucifera*) on the Clipperton Atoll in the tropical eastern Pacific. The coconut palm provides food, shelter, and a variety of useful products. (*SharkSong/M. Kazmers/Dembinsky Photo Associates*)

products. Coconut oil is found in dozens of everyday products, such as margarine, nondairy whipped topping, nondairy creamer, cosmetics, suntan lotions, and hand lotions, among others. Although it is still found in some commercial baked goods, coconut oil is no longer considered desirable in foods because it is highly saturated. (Chapter 2 discusses the chemical explanation for saturated fats and their effects on health.)

The coconut shell that surrounds the seed can be used as an eating utensil, and the trunk of the coconut palm can provide wood for a shelter. Its leaves can be used to thatch a roof or to weave baskets and hats. An adventurous shipwreck survivor could also hollow out the coconut palm's trunk, make a canoe, and leave the island for more populated shores!

The coconut palm may be the world's most useful tree. Small wonder, then, that natives of the tropics call it the "tree of life." The coconut palm is an appropriate introduction to the study of botany because it demonstrates how important plants are to us. Indeed, humans depend on plants for food, shelter, fibers, fuel, and countless other products. Life would be impossible without plants.

After reading this chapter, you should be able to:
1. *Briefly describe the field of botany and give short definitions of at least five subdisciplines of plant biology.*
2. *Summarize and discuss the features of plants and other organisms that distinguish them from nonliving things.*
3. *Distinguish among the six kingdoms of organisms recognized by many biologists and give representative organisms for each of them.*
4. *Summarize the main steps in the scientific method and explain how science differs from many other human endeavors.*

THE HUMAN POPULATION THREATENS MANY PLANTS AND OTHER ORGANISMS

In 1996 the population of the world exceeded 5.8 billion individuals—approximately 100 million more humans than the Earth supported in 1995. The 1995 world population, in turn, was approximately 95 million greater than the 1994 population. Although our numbers continue to grow, the global *rate* of population increase is slowing. The United Nations periodically publishes population projections for the 21st century. Their latest available figures forecast that the human population will increase to between 7.8 billion (their low projection) and 12.5 billion (their high projection) by the year 2050.

As our numbers continue to increase during the next century, environmental deterioration, hunger, persistent poverty, and health issues will become more critical. The need for food for the increasing numbers of people living in arid regions around the world has already led to overuse of the land for grazing and crop production. Partly as a result of such overuse, many of these formerly productive lands have been degraded into unproductive deserts.

Although the human population is increasing, not all countries have the same rate of population increase. **Highly developed countries**, which include industrialized countries such as the United States, Canada, Japan, and European countries, have low rates of population growth. **Less developed countries**, such as the relatively unindustrialized countries of Laos, Bangladesh, Nigeria, and Ethiopia, have high rates of population growth. **Moderately developed countries**, such as Mexico, Turkey, Thailand, and most of South America, fall in the middle, with population growth rates that are higher than in highly developed countries but lower than in less developed countries.

The relationships among population growth, natural resource consumption, and environmental degradation are complex, but two generalizations are useful:
1. In many less developed and moderately developed countries, individual resource use is small. However, a rapidly increasing *number* of people tends to overwhelm and deplete these countries' soils, forests, and other natural resources.
2. In highly developed countries, individual resource demands are large, far above the minimum requirements for survival. To satisfy their desires rather than their basic needs, people in more affluent countries exhaust natural resources and degrade the global environment through extravagant *consumption* and "throwaway" lifestyles. A single child born in a country such as the United States has a greater impact on the environment and on resource use than a dozen or more children born in a country such as Nigeria.

Thus, the disproportionately great consumption of resources by people in highly developed countries affects natural resources and the environment as much as, or more than, the population explosion in the developing world (Figure 1-1). In this book, references to population issues refer to both too many people *and* too much consumption per person.

Why is a botany textbook starting with a discussion of human population? For one thing, the increase in human numbers affects Earth's organisms, including its plants. A dramatic reduction in biological diversity—that is, the number and kinds of organisms—is occurring worldwide, largely as a result of human endeavors. Many human activities that harm plants and other wildlife—clearing forests to grow crops or graze livestock, for example—are the direct outcomes of population growth.

The seriousness of the current decline in biological diversity often goes unrecognized, even by some well-educated people. The Earth's plants and other organisms contribute to a sustainable environment, one that allows humans as well as other organisms to survive long-term. For example, humans and other animals depend on plants to produce food by the process of photosynthesis. Bacteria and fungi perform the essential role of decomposition, which releases essential minerals for plants to reuse.

In addition to their essential roles in the environment, plants and other organisms provide us with a variety of products. For example, plants supply us with such diverse prod-

(a)

(b)

Figure 1-1   People and natural resources. (a) A large family in Tanzania. The rapidly increasing numbers of people in less developed and moderately developed countries overwhelm their natural resources, even though individual resource requirements may be low. **(b)** A small family in the United States. People in highly developed countries consume a disproportionate share of the world's natural resources. (a, *Food and Agriculture Organization of the United Nations;* b, *The Stock Market/Michael Keller*)

ucts as oils and lubricants, perfumes and fragrances, dyes, paper, lumber, waxes, rubber and other elastic latexes, resins, poisons, cork, fibers, and medicines (see Plants and People: Taxol). Yet, although there are more than 260,000 different kinds of plants, 225,000 of them have never been evaluated with respect to their potential usefulness! As long as biological diversity remains high, it represents an important natural resource for future uses and benefits. A decline in biological diversity results in lost opportunities and lost solutions to current and future problems.

It is now time to embark on a journey of discovery. During the coming weeks, you will learn much about the biology of plants, a fascinating group of organisms very much like you and yet very different. In doing so, you will not only broaden your own intellectual horizons but develop an awareness of many environmental issues that tie plants and people together.

BOTANY IS THE BRANCH OF BIOLOGY THAT DEALS WITH PLANTS

Botany, or **plant biology**, is the scientific study of plants. Its scope is extensive. Its time frame extends from the present back to almost 3.5 billion years ago, the age of the earliest fossilized cells. The study of plant biology encompasses the origin, diversity, structure, and internal processes of plants as well as their relationships with other organisms and with the nonliving physical environment. Some plant biologists study how global climate affects plants, whereas others examine the molecules that make up plant cells.

Today is arguably the most exciting time to be studying plant biology. During the past decade or so, scientists have developed new research tools that are helping us unravel some of the most fundamental mysteries of life. Moreover, some of

Taxol

Plants produce many organic chemicals that are vitally important to the pharmaceutical industry. Plants produce the active ingredients in a wide variety of medicines—laxatives, diuretics, heart medicines, ulcer treatments, analgesics, hormones, cancer treatments, and anesthetics, to name a few. In fact, one of every four prescription drugs contains ingredients originally derived from plants.

One of the most exciting new drugs to be used against cancer is taxol, a complex organic compound first obtained from the bark of the Pacific yew (*Taxus brevifolia*) (see figure). This tree is found primarily in the old-growth forests of the Pacific Northwest.

Taxol, which was first tested as a potential anticancer drug in the mid-1960s, inhibits the growth of cancer cells. After extensive clinical trials, it has been approved for use in treating advanced stages of ovarian cancer, which kills 12,000 women in the United States each year. In clinical trials, taxol caused tumors to shrink in up to 50 percent of patients with advanced ovarian cancer. Taxol has also shown promise in treating breast cancer (40,000 deaths from breast cancer occur in the United States each year); cancers of the cervix, colon, lung, prostate, and skin; and other cancers.

Bark from the Pacific yew (*Taxus brevifolia*), found primarily in old-growth forests of the Pacific Northwest, is the original source of taxol, an anticancer drug. (*Visuals Unlimited/Peter K. Ziminski*)

To meet the demand for taxol, drug manufacturers planted yew trees, and several biotechnology firms are growing the taxol-producing yew cells in culture. In 1994 chemists at Scripps Research Institute and Florida State University independently succeeded in synthesizing taxol. Because taxol is such a complicated molecule, almost 30 steps are required in its synthesis, which makes it unlikely that drug companies will manufacture it synthetically in the near future.

The synthesis of taxol is significant, however, because it allows chemists to use the steps they have developed to synthesize similar compounds and determine whether they are also effective in treating cancer. Sometimes the discovery of a single plant compound leads to the development of a whole family of drugs.

Taxol is not the only plant chemical under current study. In the early 1990s, the National Cancer Institute screened thousands of plants and discovered three, in Malaysia, Cameroon, and Samoa, that produce chemicals that inhibit the AIDS virus.

The plant kingdom has the potential to provide us with other medicines. Of the 235,000 or so species of flowering plants on Earth, only about 5000 have been evaluated in laboratories for their pharmaceutical potential. The ability of the world's plants to provide us with useful medicines is a strong case for their preservation. After all, if we destroy forests and natural areas before studying and evaluating the organisms that live there, we've permanently lost a valuable potential resource.

our most pressing contemporary human problems have a botanical component. Some plant biologists confront the challenge of producing enough food to support the ever-expanding world population, while others identify plants that might be important sources of future drugs to treat diseases such as cancer or AIDS. Certain botanists choose to study plants that are in danger of extinction, with the ultimate goal of keeping these organisms from disappearing forever from our planet. The effects of human-produced pollution on plants is another focus for many plant biologists. Some of these pollution problems are regional or global in scope—such as acid precipitation, global warming, and the destruction of the ozone layer in the upper atmosphere.

Botany Comprises Many Disciplines

Because of its immense scope, plant biology is composed of a number of specialties. Botanists who specialize in **plant molecular biology** study the structures and functions of important biological molecules such as proteins and nucleic acids.

Plant biochemistry is the study of the chemical interactions within plants, including the variety of chemicals that plants produce. **Plant cell biology** encompasses the structures, functions, and life processes of plant cells; **plant anatomy** is microscopic plant structure (cells and tissues); and **plant morphology** is the structures of plant parts such as leaves, roots, and stems, including their evolution and development. Plant biologists who specialize in **plant physiology** study such processes as photosynthesis and mineral nutrition to understand how plants function. Plant heredity and variation make up the specialty of **plant genetics**. **Plant ecology** is the study of the interrelationships among plants and between plants and their environment. **Plant systematics** encompasses the evolutionary relationships among different plant groups. **Plant taxonomy**, a subdiscipline of systematics, deals with the description, naming, and classification of plants. **Paleobotany** is the study of plants in the geological past.

In addition to the specialties of plant biology just described, many botanists specialize in particular types of plants. For example, **bryology** is the study of mosses and similar plants. Also, many areas of applied plant biology exist, including **agronomy** (field crops and soils), **horticulture** (ornamental plants and fruit and vegetable crops), **economic botany** (plants with commercial importance), and **forestry** (forest conservation and forest products such as lumber).

PLANTS AND OTHER ORGANISMS SHARE CERTAIN ATTRIBUTES

Although plants are a dominant part of our landscape, they are easy to overlook or take for granted because they appear so passive. Plants do not appear to "live" in the sense that animals live. Plants do not run or swim or slither or fly; they do not eat other plants or animal prey; nor do they reproduce by an obvious coupling of two partners. Plants have adapted to life on land in ways that seem completely different from the adaptations of humans and other animals. Despite the perceived differences between plants and animals, plants are complex living organisms that share many important characteristics with other organisms. In this section we examine in detail these general characteristics of life.

Plants Are Highly Organized

Like all living things, plants have a complex organization. Although a palm tree looks very different from a tulip, both are composed of basic building blocks called cells. The **cell**, which is microscopic, is the smallest unit able to perform all of the activities associated with life. Some organisms—diatoms, for example—are single-celled, whereas trees are composed of trillions of cells (Figure 1-2). Those trillions of cells interact in a coordinated way to carry out the life processes required by the multicellular organism.

The biological world is organized on more than just a cellular level (Figure 1-3). The elements (basic chemical substances) of which cells are composed constitute the simplest level of organization in the biological world. An **atom** is the smallest particle of an element that possesses the properties of that element. An atom of carbon, for example, is the smallest quantity of carbon possible. Atoms combine chemically by forming bonds to produce **molecules**. A molecule can be composed of two or more atoms of a single element (for example, O_2, molecular oxygen) or different elements (for example, CO_2, carbon dioxide). Molecules, in turn, may be organized into **macromolecules**, large biological molecules such as pro-

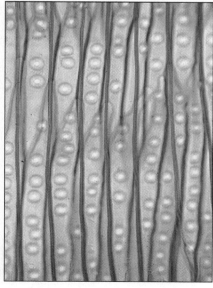

Figure 1-2 Organisms are composed of cells. (a) Diatoms (*Navicula* sp) are single-celled organisms that live in aquatic environments. **(b)** A cluster of water-conducting cells (tracheids) in a pine (*Pinus* sp) stem. Plants and other complex multicellular organisms are composed of billions or even trillions of cells. (a, *Ed Reschke;* b, *Visuals Unlimited/John D. Cunningham*)

(a)   (b)

6

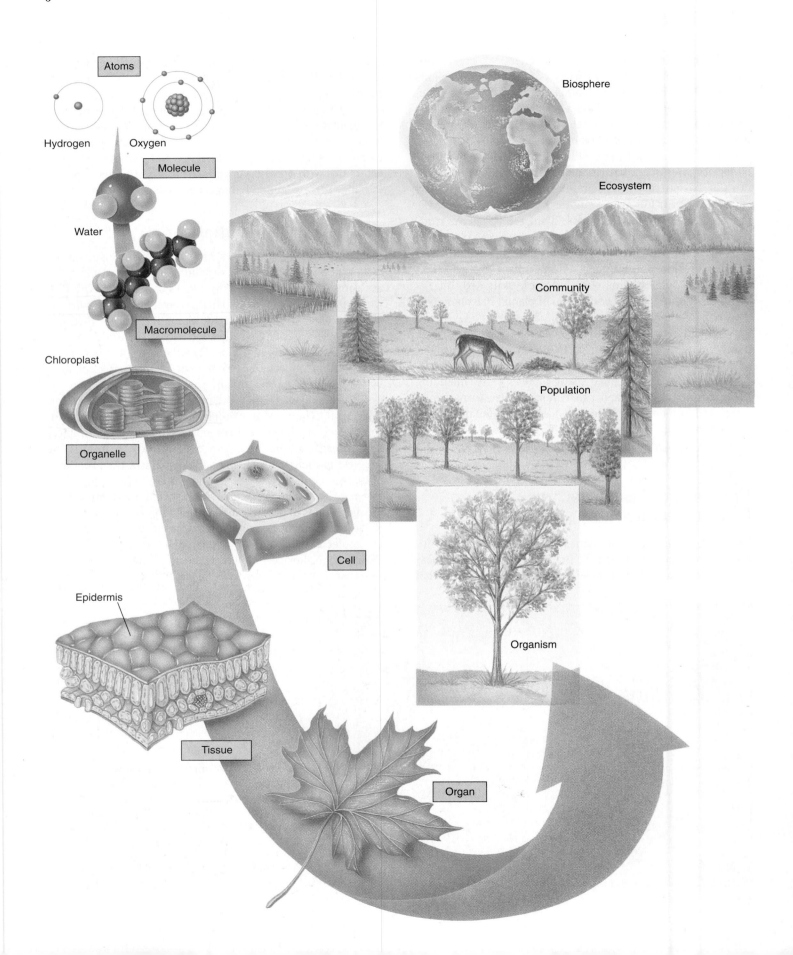

Atoms

Hydrogen   Oxygen

Molecule

Water

Macromolecule

Chloroplast

Organelle

Cell

Epidermis

Tissue

Organ

Organism

Population

Community

Ecosystem

Biosphere

teins and nucleic acids. Macromolecules associate with one another to form structures called **organelles** within cells. A cell's nucleus, which contains the hereditary material of the cell, is an example of an organelle. Organelles associate to form the cell.

In most multicellular organisms, cells are organized into **tissues**, associations of cells that perform specific functions. Examples of plant tissues include epidermis, the protective tissue that covers the plant's surface, and xylem, the water-conducting tissue. Tissues, in turn, are organized into **organs**, complex functional units that perform specific roles. Leaves and roots are examples of plant organs. Several types of organs, functioning with great precision, make up a multicellular plant.

Several more complex levels of biological organization occur at the ecological level, above the level of the organism. Organisms are arranged into **populations**, groups of members of the same species that live together in the same area at the same time, such as a population of daisies in a field. Populations, in turn, are organized into communities. A **community** consists of all the populations of different organisms that live and interact within an area. A pond and a forest are two examples of communities; each might contain hundreds of types of organisms. "Ecosystem" is a more inclusive term than "community," because an **ecosystem** is a community together with its nonliving environment. Thus, an ecosystem encompasses not only all interactions among organisms but also the interactions between organisms and their physical environment. All of Earth's communities are collectively called the **biosphere**. The organisms of the biosphere depend on one another and on the Earth's physical environment—its water (liquid and frozen, fresh and salty), its crust (soil and rock), and its gaseous atmosphere.

Plants Take In and Use Energy

A continual input of energy from the sun enables life to exist on Earth. All organisms require energy for their activities, which include growth, repair, and maintenance. The two most important energy-related activities in the living world are photosynthesis and cellular respiration.

◀ **Figure 1-3   Levels of biological organization.** Starting at the simplest level, atoms are organized into molecules, which are organized into macromolecules. Macromolecules are organized into organelles, organelles into cells, cells into tissues, tissues into organs, and organs into individual multicellular organisms. A group of individuals of the same species is a population. Populations of many species interact to form communities. A community and its physical environment are an ecosystem. The highest level of organization is the biosphere, which consists of all communities of organisms on Earth.

Figure 1-4   Sunlight filters through trees to a field of blue lupines (*Lupinus hirsutus*). Lupines and other plants absorb radiant energy for photosynthesis. (*Skip Moody/Dembinsky Photo Associates*)

Plants, algae, and a few bacteria perform photosynthesis, which is essential for almost all life. In **photosynthesis**, radiant (light) energy from the environment is converted to chemical energy that is stored in molecules such as the sugar glucose (Figure 1-4). Photosynthesis provides plants, algae, and photosynthetic bacteria with a ready supply of energy (in molecules of glucose) that they can use as the need arises. The energy can also be transferred from one organism to another—for example, from plants to animals that eat plants, and to animals that eat animals that eat plants. Thus, animals receive their energy from the sun indirectly by eating plants. Photosynthesis also produces oxygen, which most organisms require when they break down food.

The chemical energy that plants store in food molecules is released within the cells of plants, animals, and other organisms through **cellular respiration**. In cellular respiration, food molecules such as glucose are broken down, usually in the presence of oxygen. Cellular respiration makes the chemical energy stored in food molecules available to the cell for biological work. All organisms obtain energy via cellular respiration.

Plants Respond to Stimuli

Plants and other organisms respond to **stimuli**, changes in their environment. Stimuli to which plants respond include changes in the direction, color, or intensity of light; in temperature or in the orientation toward gravity; and in the chemical composition of the surrounding soil, air, or water.

The responses of plants to environmental stimuli are often not as obvious as those of animals. For example, when a coyote charges a rabbit, the rabbit responds by darting away. In contrast, plants often respond to stimuli by the differential growth of parts of their bodies. Roots, for example, normally grow in the direction of gravity. When a potted plant is placed on its side, its root tips respond by changing their direction of growth so that the root is again growing downward (Figure 1-5). Stems normally grow upward, so when the plant is placed on its side, its stem tip (the site of growth in a stem) bends upward.

Some plants respond to stimuli in a dramatic fashion. The Venus flytrap, with its hinged leaves that resemble tiny bear traps, is exceptionally sensitive to touch and can catch insects (Figure 1-6). Tiny hairs on the leaf detect when an insect alights, and the leaf rapidly folds shut, preventing the escape of the prey. The leaf then secretes enzymes to digest the insect. Venus flytraps and similar plants usually live in soil that is deficient in nitrogen; they obtain part of the nitrogen they require by capturing and "eating" insects.

Plants Grow and Develop

Growth is an increase in the size and weight of an organism. In plants, growth results from both an increase in the *number* of cells and an increase in the *size* of cells. Many plants—most trees, for example—continue to grow larger throughout their life. In contrast, growth ceases when humans and many other animals reach adulthood.

Plants and other organisms develop as well as grow. **Development** includes all the changes in an organism from the start of its life through its immature (juvenile) stage, through its mature (adult) stage, to its death. An oak tree begins life as a fertilized egg, which grows and develops into a multicellular embryo—with an embryonic root, stem, and leaves—within a seed (the acorn). Development continues when the seed germinates and the young plant (called a seedling) emerges from the acorn. The seedling eventually develops into an adult oak tree.

Plants Reproduce

One of the basic concepts of biology is that living things come from previously existing organisms. **Reproduction**, the formation of a new individual by sexual or asexual means, is the most distinctive characteristic of life. Although some organisms exist for a long time—individual bristlecone pines have

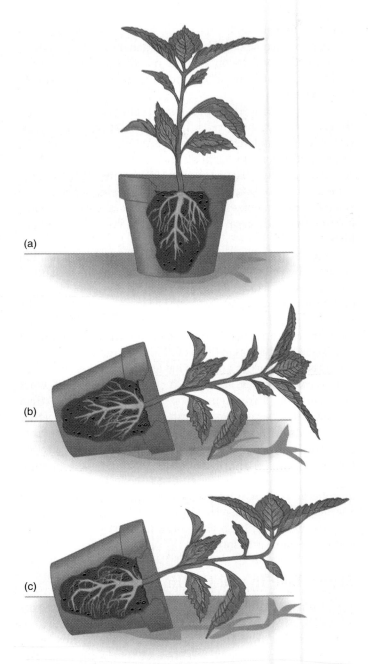

Figure 1-5   Root growth in the direction of gravity. When a potted plant **(a)** is placed on its side **(b)**, making the roots horizontal, the root tips change the direction of their growth so that they are again growing downward **(c)**. The roots other than the tips do not change their direction, because they are no longer growing; the root tips are the only parts that increase in length.

lived for thousands of years—all organisms eventually die. Reproduction enables an organism to perpetuate itself.

In many organisms, reproduction is **asexual** and does not involve the sexual process of the union of gametes (reproductive cells). In asexual reproduction, one parent gives rise to

(a)

(b)

Figure 1-6   The modified leaves of a Venus flytrap (*Dionaea muscipula*) snap shut on its prey. Venus flytraps grow in bogs in North and South Carolina. **(a)** A Venus flytrap with open traps. **(b)** A closed trap and insect prey. After enzymes produced by the plant digest as much of the insect as is digestible, the trap will reopen. (a, *Patti Murray;* b, *Ed Reschke*)

offspring that are virtually identical to it (Figure 1-7). In most plants, **sexual reproduction** involves the union of gametes that often come from two separate parents. The gametes—an egg cell and a sperm cell—unite to form a fertilized egg, which develops into the new offspring. In sexual reproduction, the offspring are not exact copies of a single parent but are products of the unique combination of traits inherited from both parents.

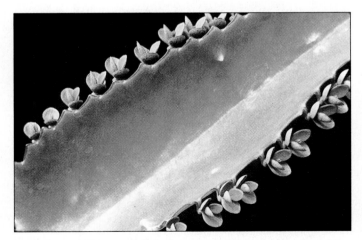

Figure 1-7   The mother-in-law plant (*Kalanchoe pinnata*) produces young plants asexually along the margins of the leaves. When the young plants attain a certain size, they drop off and root in the ground. Each young plant is genetically identical to its parent. (*Dennis Drenner*)

Plants Use DNA To Transmit Information from One Generation to the Next

The characteristics of an organism are encoded in its **genes**, which are the units of hereditary information. The genes of a peach tree provide information about all aspects of the plant, such as how tall the tree grows, what color its flowers are, whether it can tolerate full sun or shade, and the number of seeds that each fruit produces. Genes are composed of **DNA,** the organic molecule that stores and carries important genetic information in cells. DNA, which stands for *deoxyribonucleic acid*, is the molecule of inheritance for all organisms, from beans to giraffes to bacteria.

Information encoded in genes—that is, in the DNA that composes genes—is transmitted from one generation to the next during reproduction. Genes ensure that a bean plant produces seeds that grow into bean plants, not into roses or cucumbers.

Plant Populations Undergo Genetic Changes Over Time

The ability of a plant population to adapt to its environment enables it to exist in a changing world. **Adaptations** are characteristics that enable an organism to better survive in a certain environment. Adaptations may involve changes in structure, form, or function. The thick, succulent (fleshy, water-storing) stem of a cactus is an adaptation that enables it to live in dry regions where precipitation is low (Figure 1-8). Likewise, the prickly spines covering the surface of the cactus stem are an adaptation that discourages desert animals from eating the succulent stem tissue.

Figure 1-8 The cactus (a species of *Melocactus* is shown) possesses adaptations that enable it to survive in dry climates. The stem of a cactus functions for both photosynthesis and water storage. The leaves of cacti are modified into spines for protection. (*James Mauseth, University of Texas*)

Every organism contains many interdependent adaptations that enable it to survive in the particular environment to which it is adapted. **Evolution,** the process by which organisms adapt to their environment over time, is the genetic change in a population of organisms from generation to generation. Evolutionary processes typically require long periods of time and occur over many generations.

Biologists consider evolution the unifying theory of biology because it encompasses all the changes that have transformed life from the first primitive cells into the millions of different organisms that exist today. Evolution helps us understand why organisms are the way they are and how they are related, both to each other and to earlier forms of life.

MANY KINDS OF PLANTS AND OTHER ORGANISMS HAVE EVOLVED

We do not know exactly how many kinds of organisms exist, but most biologists estimate that there are at least 5 to 10 million species. So far, approximately 235,000 flowering plant species, 150,000 lower plant species, and more than 1 million animal species have been identified. ("Lower plants" refers to all nonflowering plants as well as algae and fungi, both of which were traditionally classified as plants.) The variation among organisms staggers the imagination and has required the development of a system to organize this diversity.

Many Biologists Assign Organisms into Six Kingdoms

For hundreds of years, biologists viewed organisms as belonging to two broad groups, the plant and animal kingdoms.

As organisms underwent closer study, it became increasingly obvious that many did not fit very well into either the plant or animal kingdom. For example, bacteria, single-celled organisms that lack cell nuclei, were originally considered plants. The fact that all bacteria lack nuclei is now considered a fundamental difference between them and other organisms—a difference more significant than the differences between plants and animals. With our present knowledge, it is difficult to consider bacteria as plants.

In addition, certain organisms—*Euglena,* for example—seem to possess characteristics of plants *and* animals (Figure 1-9). In fact, because many single-celled organisms seem to have more in common with one another than with either plants or animals, biologists assigned them to a separate kingdom.

These and other considerations have led to the six-kingdom system of classification used by most biologists today. The kingdoms currently recognized are Archaebacteria, Eubacteria, Protista, Fungi, Plantae, and Animalia (Figure 1-10).

Bacteria are now divided into two groups on the basis of significant biochemical differences. Members of the **kingdom Archaebacteria** are often adapted to harsh conditions and frequently live in oxygen-deficient environments, including hot springs, salt ponds, and hydrothermal vents in the ocean depths. The thousands of kinds of other bacteria are classified in the **kingdom Eubacteria**.

Kingdom Protista contains protozoa, algae, water molds, and slime molds. These organisms are single-celled or simple multicellular organisms. **Kingdom Fungi** consists of molds, yeasts, and mushrooms, which obtain their nutrients by secreting digestive enzymes into food and then absorbing the

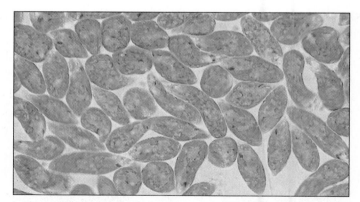

Figure 1-9 Numerous euglenoids crowded into a drop of water. *Euglena* has both plantlike and animal-like traits. This single-celled organism is plantlike because it can photosynthesize and is animal-like because it can move about from place to place. In the past, *Euglena* was classified in the plant kingdom (when algae were considered plants) *and* in the animal kingdom (when protozoa were considered animals). Today most biologists place *Euglena* in the kingdom Protista. (*Dwight R. Kuhn*)

KINGDOMS

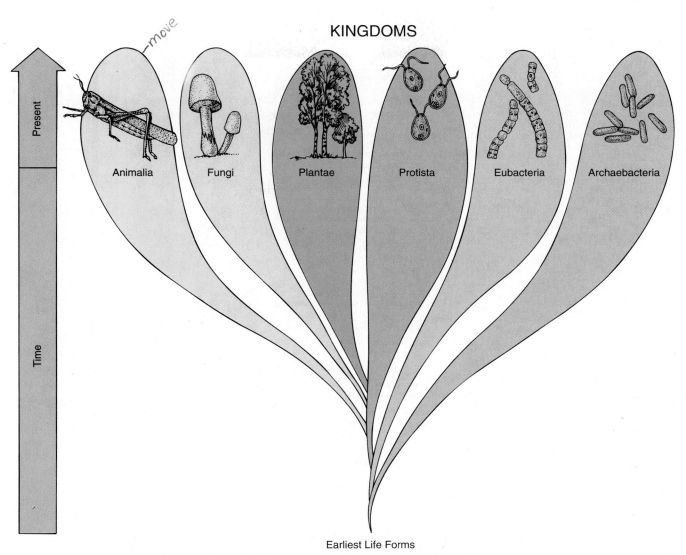

Figure 1-10   **The six-kingdom system of classification includes the archaebacteria, eubacteria, protists, plants, animals, and fungi.**

predigested nutrients. Animals belong to the **kingdom Animalia**. Animals are complex multicellular organisms that must eat other organisms to obtain nourishment. They possess muscular systems, which enable them to move, and nervous systems, including the brain, spinal cord, and sense organs such as eyes.

Plants, complex multicellular organisms that almost always photosynthesize, are placed in the **kingdom Plantae**. Plants possess a **cuticle** (a waxy covering over their outer parts that reduces water loss), **stomata** (tiny openings in leaves and stems for gas exchange),[1] and multicellular **gametangia** (reproductive organs that protect gametes). Plants are usually rooted in the ground and, unlike animals, do not possess a nervous system. The kingdom Plantae includes vascular plants—such as ferns, conifers, and flowering plants—which possess tissues that conduct food and water throughout the plant body, and nonvascular plants—such as mosses—which lack these tissues.

Classification Is Hierarchical

The broadest classification category is the kingdom, and the narrowest is the species. Between them is a range of categories that forms a hierarchy (Figure 1-11). A **species** is a group of similar organisms that interbreed in their natural environment but do not interbreed with other species. Similar species are assigned to the same **genus** (pl. *genera*), and similar genera are grouped together in the same **family**. Families are grouped into **orders**, orders into **classes**, classes into **phyla** (sing. *phylum*), and phyla into **kingdoms**. Plants and fungi were traditionally classified into *divisions* rather than phyla. The International Botanical Congress, however, approved the phylum designation for plants and fungi in 1993.

---

[1]Liverworts, the simplest of plants, lack stomata.

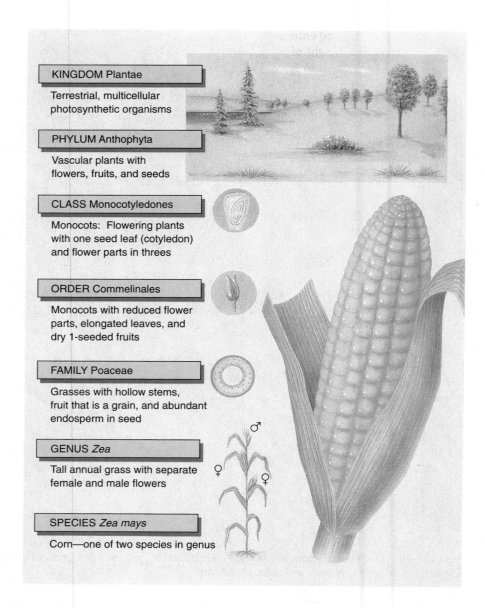

Figure 1-11 The classification of corn (*Zea mays*) using the hierarchical system from kingdom to species.

Each Species Is Named Using a Binomial System

In the 18th century, Carolus Linnaeus, a Swedish botanist, simplified the naming of organisms. In Linnaeus' system, called the **binomial system of nomenclature**, each species receives a two-part name. The first word designates the genus to which the organism is assigned, and the second word is a specific epithet, that is, a descriptive word that characterizes the organism. The specific epithet is always used with the full or abbreviated generic name preceding it. For example, the white oak (*Quercus alba,* sometimes abbreviated *Q. alba*) and the red oak (*Quercus rubra*) belong to the same genus, *Quercus.* The American beech (*Fagus grandifolia*) belongs to a different genus. The genus is capitalized and the specific epithet is usually not capitalized; both names are italicized.

SCIENTISTS USE THE SCIENTIFIC METHOD

Most people think of science as a body of knowledge—a collection of facts about the natural world. However, science is also a dynamic *process*. It is a particular way to investigate nature in an attempt to understand the Universe. Science seeks to reduce the apparent complexity of our world to general principles, which can then be used to solve problems or provide new insights. In other words, science is devoted to discovering the general principles that govern the operation of the natural world. The information collected by scientists is called **data** (sing. *datum*). Data are collected by observation and experimentation and then analyzed or interpreted. Scientific conclusions are inferred from the available data and are not based on faith, emotion, or intuition.

Science is an ongoing enterprise, and scientific concepts must be reevaluated in light of newly discovered data. Thus, scientists can never claim to know the "final answer" about anything, because scientific knowledge changes.

The process that scientists use to answer questions or solve problems is called the **scientific method** (see Focus On: Using the Scientific Method). Although there are many versions of the scientific method, it basically involves five steps:

1. Recognize a question or an unexplained occurrence in the natural world. Science is based on knowledge accumulated previously. Therefore, after one recognizes a problem, one determines what is already known about it.
2. Develop a **hypothesis**, or educated guess, to explain the problem. A good hypothesis makes a prediction that can be tested. The same factual evidence can be used to formulate several alternative hypotheses, each of which must be tested.
3. Design and perform an experiment to test the hypothesis. An experiment involves collection of data by making observations and measurements.
4. Analyze and interpret the data to reach a conclusion. Does the evidence match the prediction stated in the hypothesis? In other words, do the data support or refute the hypothesis?
5. Share new knowledge with the scientific community. The scientist does this by publishing articles in scientific journals or books and by presenting the information at scientific meetings. Sharing new knowledge with the scientific community permits other scientists to design new experiments to verify or refute the work.

An Experiment Must Have a Control

Most often, the question we wish to answer is influenced by many factors, which are called **variables**. To test a particular hypothesis about *one* variable, it is necessary to hold all other variables constant so that we are not misguided by them. This is done by performing two forms of the experiment. In the main experiment, the conditions are altered in one way (one variable is adjusted). The second experiment, called the **control,** is identical to the main experiment in all respects except that conditions are *not* altered. By designing an experiment with a control, the researcher ensures that any difference in the outcomes between the main experiment and its control must be the result of the variable.

Scientists Use Inductive and Deductive Reasoning

Discovery of general concepts by the careful examination of specific cases is called **inductive reasoning**. The scientist begins by organizing data into manageable categories and asking the question "What does this information have in common?" He

or she continues by seeking a unifying explanation for the data. Inductive reasoning is the basis of modern experimental science.

As an example of inductive reasoning, consider the following.

Datum: Poppies obtain their energy by photosynthesis.
Datum: Daisies obtain their energy by photosynthesis.
Datum: Roses obtain their energy by photosynthesis.
Conclusion: All flowering plants obtain their energy by photosynthesis.

Even if inductive reasoning makes use of correct data, the conclusion may be either true or false. As new data come to light, they may show that the generalization arrived at through inductive reasoning is false. Science has shown, for example, that a flowering plant called Indian pipe (*Monotropa uniflora*) is nonphotosynthetic and relies on fungi living in the soil for nutrient energy (Figure 1-12); the fungi, in turn, are usually associated with the roots of nearby trees, from which they obtain nutrients. When one adds this information to the preceding list, one must formulate a different conclusion. Inductive reasoning, then, produces new knowledge but is error-prone.

Science also makes use of **deductive reasoning**, which proceeds from generalities to specifics. Deductive reasoning

Figure 1-12   The Indian pipe (*Monotropa uniflora*) does not contain chlorophyll or photosynthesize, and it lives underground except for its flowers (*shown*). (*Carlyn Iverson*)

In this chapter I stated that roots grow downward, in the direction of gravity. Such a response of plants to the Earth's gravitational pull is known as **gravitropism**, that is, growing in response to gravity. Roots are considered to be positively gravitropic, and some stems are negatively gravitropic because they grow upward, opposite the direction of gravity.

Suppose you wished to conduct an experiment to determine whether the young roots of germinating (sprouting) corn kernels exhibit this downward growth. How would you begin?

(1) Recognize a question. Do the roots of germinating corn kernels exhibit positive gravitropism?

After stating the question, you would do some library research related to the problem. You might determine, for example, that each corn kernel has a top and a bottom; the young root emerges from the bottom of the germinating kernel, and the young shoot (stem and leaves) from the top. You would have to learn the environmental requirements for seed germination. For example, seeds must absorb water before they begin growing. You would also have to investigate ways to grow seeds so that their roots could be observed. Normally, of course, the roots are hidden from view in the soil.

(2) Develop a hypothesis (in the form of a simple declarative sentence):

Regardless of the direction in which a corn kernel is planted, the emerging root always grows downward.

(3) Design and perform an experiment to test the hypothesis. There are several ways of designing an experiment to test this hypothesis. One way is to place the corn kernels on moistened blotting paper between two plates of glass (see figure).

Every well-designed experiment has a control. In this experiment, the control would be a corn kernel oriented in the normal direction, that is, with its "bottom" side downward. The variables in this experiment would be other kernel positions, which are oriented in various directions.

After setting up the experiment, you would collect data, probably by observing the experiment over a period of several days. You would confirm that the emerging root always grows out of the bottom of the kernel regardless of its orientation. A few days after germination, you would notice that the roots of all the kernels were growing downward, even those coming from an upside-down kernel.

It is helpful to present the data in a table.

(4) Analyze and interpret the data to reach a conclusion. Examine the data to determine whether the evidence matches the prediction stated in the hypothesis. What is your conclusion?

(5) Share new knowledge with the scientific community. Since you are a student in a college class, you might write a laboratory report. Most laboratory reports have a format similar to papers that scientists submit for publication in scientific journals.

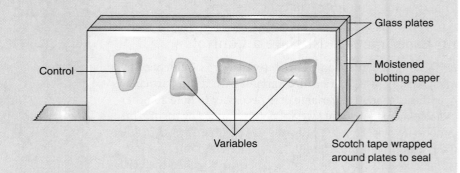

Kernel No.	Initial Orientation of Kernel	Direction of Root, Day 1*	Direction of Root, Day 2*	Direction of Root, Day 3*	Direction of Root, Day 4*
1 (control)					
2					
3					
4					

*After germination.

adds nothing new to knowledge, but it can make relationships among data more apparent. For example:

General rule: A plant's aerial parts (those exposed to air) are covered by a waxy cuticle that helps reduce water loss.

Specific example: Corn is a plant.

Conclusion: Corn plants possess a cuticle.

This is a valid argument. The conclusion that corn plants possess a cuticle follows inevitably from the information given. Scientists use deductive reasoning to determine the type of experiment or observations that are necessary to test a hypothesis.

Scientific Theories Are Supported by Many Observations and Experiments

A hypothesis supported by a large body of observations and experiments becomes a **theory**. A good theory relates many data that previously appeared to be unrelated, grows as additional information becomes known, and predicts new data and suggests new relationships among phenomena.

By demonstrating the relationships among classes of data, a theory simplifies and clarifies our understanding of the natural world. Theories are the solid ground of science; they are the explanations of which we are most sure. This definition contrasts sharply with the general public's usage of the word "theory," implying *lack* of knowledge, or a guess (as in "I have a theory about the assassination of J.F.K."). In this book, we *always* use the word "theory" in its scientific sense, to refer to a broadly conceived, logically coherent, and very well-supported explanation.

Some theories—for example, the cell theory that every organism is composed of one or more cells, and all cells come from preexisting cells—are so strongly supported that the likelihood of their being rejected in the future is very small. Because these theories have withstood repeated testing and are the strongest statements we can make about the natural world, the scientific community promotes them to a higher level of confidence and calls them scientific **principles**.

Yet there is no absolute truth in science—only varying degrees of uncertainty. The possibility always remains that future evidence will require the revision of a hypothesis, a theory, or even a principle. A scientist's acceptance of a hypothesis, theory, or principle is always provisional.

S T U D Y   O U T L I N E

I. Botany, or plant biology, is the scientific study of plants and comprises many areas of specialization.

II. Plants and other organisms share certain characteristics.
   A. Plants are highly organized.
     1. Cells are the basic building blocks of plants.
     2. The hierarchical levels of biological organization are: atoms → molecules → macromolecules → organelles → cells → tissues → organs → organisms → populations → communities (and ecosystems) → biosphere.
   B. Plants take in and use energy.
   C. Plants respond to stimuli in their environment.
   D. Plants grow and develop.
     1. Growth involves increases in the size and number of cells.
     2. Development involves all the changes in an organism from the start of its life to its death.
   E. Plants reproduce and form new individuals by asexual or sexual means.
     1. In asexual reproduction, one parent gives rise to offspring that are virtually identical to it.
     2. In sexual reproduction, the union of gametes, often from separate parents, results in variation among the offspring.
   F. DNA is the molecule that transmits genetic information from one generation to the next.
   G. In the process of evolution, populations change, or adapt, in order to survive in changing environments.

III. Biologists estimate there are at least 5 to 10 million species of living organisms.
   A. Many biologists classify organisms into six kingdoms: Archaebacteria, Eubacteria, Protista (protozoa, algae, water molds, and slime molds), Fungi (molds and yeasts), Animalia, and Plantae.
   B. Taxonomic classification is hierarchical. The levels are kingdom (most general), phylum, class, order, family, genus, and species (most precise).
   C. In the binomial system of nomenclature, each species is assigned a scientific name with two parts: genus and specific epithet.

IV. Science is a systematic process of investigating the natural world.
   A. The scientific method encompasses five steps.
     1. Define a problem or an unanswered question.
     2. Develop a hypothesis to explain the problem.
     3. Design and perform an experiment to test the hypothesis.
     4. Analyze and interpret the data to reach a conclusion.
     5. Share new knowledge with the scientific community.
   B. A well-designed experiment has two parts, a control and a variable (the part of the experiment that is identical to the control in all ways except one). Any difference in outcome between the control and the variable must be the result of the variable.
   C. Both inductive and deductive reasoning are used in the scientific method.
     1. Inductive reasoning provides new knowledge but is error-prone.
     2. Deductive reasoning adds nothing new to knowledge, but it can make relationships among data more apparent.
   D. A hypothesis is an educated guess; a theory is a higher level of scientific interpretation that is well supported by scientific evidence; a principle is a theory that is almost universally accepted.

SELECTED KEY TERMS

adaptation, p. 9
asexual reproduction, p. 8
binomial system of
  nomenclature, p. 12
botany, p. 3
cell, p. 5
cellular respiration, p. 7

control, p. 13
deductive reasoning, p. 13
development, p. 8
DNA, p. 9
ecosystem, p. 7
evolution, p. 10

gene, p. 9
growth, p. 8
hypothesis, p. 13
inductive reasoning, p. 13
photosynthesis, p. 7
plant biology, p. 3

principle, p. 15
scientific method, p. 13
sexual reproduction, p. 9
species, p. 11
stimulus, p. 8
theory, p. 15

REVIEW QUESTIONS

1. How is the conspicuous consumption by people in highly developed countries related to environmental problems?
2. What features distinguish plants from nonliving things?
3. Outline the hierarchical levels of biological organization, from atoms to the biosphere.
4. A cactus has a thick, succulent stem and spines that are modified leaves. Explain why these adaptations may have evolved over time.
5. List the six kingdoms of life, and briefly describe the distinguishing features of the kingdom Plantae.
6. Outline the hierarchical levels of classification, from species to kingdom.

7. Two plants are classified in the same species, and two other plants are classified in the same family. Which pair of plants is more similar? Why?
8. Define the binomial system used to name species, and give an example.
9. How is science different from other human endeavors?
10. Contrast inductive and deductive reasoning. Contrast a hypothesis, a theory, and a principle.
11. What is a control? Why is it an important part of an experiment?

THOUGHT QUESTIONS

1. Make a list of the natural resources you use in a single day. How would this list compare to a similar list made by a poor person in a less developed country?
2. What parts of the scientific method involve inductive reasoning? Deductive reasoning?
3. The cell theory, which states that organisms are composed of cells, was arrived at by what type of reasoning? List at least five observations that might have been used by the scientists who developed the cell theory.

4. Explain the following statement: In science a hypothesis is hypothetical, but a theory is *not* theoretical.
5. Design a suitably controlled experiment to test the effect of low temperature on a bean seedling's rate of growth. What is the hypothesis? What is the control? The variable?
6. Why could photosynthesis be considered the single most important biological process in the living world?

SUGGESTED READINGS

Capon, B., *Plant Survival: Adapting to a Hostile World,* Portland, Oreg., Timber Press, 1994. How plants survive and thrive in adverse environments.

Careers in Biology, prepared by the Botanical Society of America, c/o Manager of Publications, Botanical Society of America, 1735 Neil Avenue, Columbus, OH 43210-1293. Provides information on careers in plant biology and discusses the many specialties within the field.

Daily, G.C., and P.R. Ehrlich, "Population, Sustainability, and Earth's Carrying Capacity," *BioScience* Vol. 42, No. 10, November 1992. Explores the question of how many people and what kinds of human lifestyles the Earth can support indefinitely.

Dasgupta, P.S., "Population, Poverty, and the Local Environment," *Scientific American,* February 1995. The three issues mentioned in the title of this article interact with one another in complex ways.

Durning, A.T., "Long on Things, Short on Time," *Sierra,* January/February 1993. Arguments for a low-consumption, sustainable lifestyle.

Joyce, C., "Taxol: search for a cancer drug," *BioScience* Vol. 33, No. 3, March 1993. This effective anticancer drug was originally obtained from the bark of the yew tree.

Moore, J.A., *Science as a Way of Knowing: The Foundations of Modern Biology,* Cambridge, Mass., Harvard University Press, 1993. An account of scientific thought as it is related to the history of modern biology.

"World Population," a special report of *The Christian Science Monitor,* July 8, 1992. Several articles in this special supplement report on progress in slowing the increase in human population.

PART

1

The Plant Cell

CHAPTER 2

THE CHEMICAL COMPOSITION OF CELLS

Among the novelties early Spanish explorers brought to Europe from South America were bouncing balls made from the heated, solidified sap of certain trees. Although this material, later named rubber, was considered an interesting curiosity, its potential uses were not appreciated for about 300 years. Rubber was used in the early 19th century to coat and waterproof clothing, but it had certain drawbacks: it got sticky at warmer temperatures and brittle at cooler temperatures. In the mid-19th century Charles Goodyear developed vulcanization, a process that combined rubber with sulfur and eliminated these undesirable properties.

Rubber came into its own in the early 20th century with the invention and mass production of the automobile, which required rubber for its tires, hoses, gaskets, and so on. Demand outstripped supply, and large rubber plantations were established in Asia. When World War II disrupted the supply of rubber, chemists in the United States developed a synthetic rubber from petroleum. Today, rubber tires contain predominantly synthetic rubber, mainly because it costs less to produce synthetic rubber than to grow and harvest natural rubber. However, synthetic rubber does not have all of the desirable properties of natural rubber. For example, natural rubber is more

Latex is tapped from a rubber tree (*Hevea brasiliensis*) in Ecuador. (*James L. Castner*)

resilient that synthetic rubber; that is, it returns its original shape after being stretched—a characteristic required in high-performance radial tires. As a result, natural rubber makes up a large proportion of the rubber in radial tires.

Although many plants produce rubber, most of our natural supply comes from the rubber tree (*Hevea brasiliensis*), which is native to South America. The bark of its stem contains a network of canals that hold latex, a milky substance that is a mixture of many chemical compounds, some of which have elastic properties. Latex is tapped from the tree and used to make rubber, but its role in the rubber tree is not known. Some evidence suggests that latex discourages animals from eating the tissues—latex compounds taste bad, and many of them are poisonous. Other biologists think that latex compounds are waste products that are simply stored in the canals.

Because plants produce a substantial number of different chemical compounds, they are sometimes referred to as nature's "chemical factories." Although many of these chemicals, like latex, have unknown roles in the plants that produce them, humans often make use of them for products ranging from perfumes to medicines.

After reading this chapter, you should be able to:

1. *Diagram the basic structure of an atom, showing its protons, neutrons, and electrons. Explain chemical formulas, diagrams of chemical structures, chemical equations, and ionic and covalent bonds.*
2. *Discuss the properties of water, and explain its importance to life.*
3. *Distinguish between acids and bases, and describe the pH scale.*
4. *Describe the chemical compositions and functions of carbohydrates, lipids, proteins, and nucleic acids.*
5. *Discuss the role of enzymes in cells.*
6. *State the first and second laws of thermodynamics and describe how each applies to plants and other organisms.*

PLANTS ARE COMPOSED OF CHEMICAL ELEMENTS

The Earth contains an astonishing variety of materials in its water, rocks, minerals, soil, atmosphere, and organisms, but there is an underlying order to this complex profusion. All materials are made of matter, and all matter, living and non-living, is composed of chemical **elements**, substances that cannot be broken down into simpler substances by chemical changes. There are 92 naturally occurring elements, ranging from hydrogen (the lightest) to uranium (the heaviest). In addition to the naturally occurring elements, 19 elements heavier than uranium have been made artificially.

Instead of writing out the name of each element, chemists use a system of abbreviations called **chemical symbols**. The symbol for an element is usually the first one or two letters of the element's English or Latin name. For example, O is the symbol for oxygen, C for carbon, Cl for chlorine, and Fe for iron (the Latin name for iron is *ferrum*). Table 2-1 gives the

chemical symbols and functions of some common elements found in plants.

Atoms Are the Smallest Particles of Elements

If you take a piece of an element—for example, a bit of iron—and divide it into smaller and smaller pieces, eventually you will get to a particle of iron that cannot be divided further. The smallest possible particle of an element that still retains the properties of that element is called an **atom** (Greek *atomos*, "indivisible").

Protons, neutrons, and electrons are constituents of atoms

During the 19th century, numerous experiments led scientists to conclude that atoms themselves are composed of units called subatomic particles (Figure 2-1). A **proton** is a subatomic particle that has a positive electrical charge and a small amount of mass (mass is a measure of the quantity of matter in an object); a **neutron** is an uncharged particle with about the same mass as a proton; and an **electron** is a particle with a negative electrical charge and an extremely small mass (each electron has about 1/1800 the mass of a proton). Protons and neutrons, which are located in the center of each atom (the atomic nucleus), make up almost all the mass of an atom. An

TABLE 2-1   Some Common Elements Found in Plants

Name	Symbol	Functions
Carbon	C	Backbone of organic molecules
Oxygen	O	Present in most organic molecules; required for aerobic respiration
Hydrogen	H	Present in most organic molecules
Nitrogen	N	Present in all proteins and nucleic acids; present in chlorophyll
Phosphorus	P	Present in nucleic acids and energy transfer molecules such as ATP
Potassium	K	Helps provide ionic balance in cells
Magnesium	Mg	Present in chlorophyll
Iron	Fe	Component of certain enzymes
Calcium	Ca	Structural component in cell walls

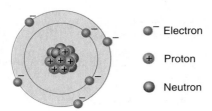

Figure 2-1   Model of a carbon atom. Positively charged protons and electrically neutral neutrons are situated in the center (the nucleus) of the atom. Negatively charged electrons move about the nucleus. (This simple representation of an atom is easy to visualize and is accurate enough for our discussion of plant biology, but it does not accurately reflect the locations of electrons.)

atom's electrons spin about in the space surrounding the nucleus.

Each atom has an atomic number and a mass number

Any given atom of each kind of element has a fixed number of protons in its atomic nucleus. The number of protons in an atom is its **atomic number**. Because a single atom is electrically neutral (has the same number of electrons as protons), the atomic number also designates the number of electrons in an atom. The atomic number is usually written as a subscript to the left of the chemical symbol. Thus, hydrogen, with an atomic number of 1 (indicating one proton in its atomic nucleus), is designated $_1H$. The atomic number determines the chemical identity of an element; that is, the only element with the atomic number 1 is hydrogen, the only element with the atomic number 6 is carbon, and so on.

Because electrons have so little mass, almost the entire mass of an atom is in the protons and neutrons that make up its nucleus. The total number of protons plus neutrons in the nucleus is known as the **mass number**, and it is usually written as a superscript to the left of the chemical symbol. The common form of carbon, for example, is designated $^{12}_{6}C$, indicating that it has 6 protons and 6 neutrons ($6 + 6 = 12$).

Although atoms of the same element always possess the same number of protons, they can vary in the number of neutrons. Atoms of the same element that contain different numbers of neutrons and thus have different mass numbers are called **isotopes** (Figure 2-2). A sample of an element taken from nature is often a mixture of its isotopes, although one isotope usually predominates.

Electrons move around the nucleus, but their exact positions are uncertain

Although the precise location of an electron at any given time is uncertain, its approximate location—that is, where it will probably be found 90 percent of the time—can be described. An atom's **electron configuration** is the arrangement of electrons around its nucleus. Because electrons are negatively charged, they are attracted to the positively charged nucleus (opposite electrical charges attract). At the same time, electrons repel one another. These considerations help determine an atom's electron configuration.

An atom may have several energy levels in which electrons are located. The first energy level is the one closest to the nucleus; only two electrons can occupy this energy level. The second energy level can hold a maximum of eight electrons. The third and fourth energy levels can each hold more than eight electrons, but they are most stable when only eight are present. Thus, the first energy level is considered complete when it contains two electrons, and every other energy level is complete when it contains eight electrons. Although the simple diagrams of electron configuration shown in Figure 2-3 help us understand atomic structure, they are highly oversimplified. Electrons move around the nucleus in an unpredictable way, first closer and then farther.

An electron's distance from the nucleus depends on its energy level. When an electron is provided with more energy—in the form of light, for example—it can move farther from the nucleus (such a move requires energy). When an electron gives up energy, it drops back to a lower energy level closer to the nucleus. We say more about electrons absorbing light energy to move from lower to higher energy levels in the discussion of photosynthesis in Chapter 4.

Atoms Can Combine Chemically

In addition to existing as single atoms, elements can unite in fixed ratios to form chemical **compounds**. Examples of compounds include water (H_2O), which is composed of the elements hydrogen and oxygen in a ratio of $2:1$, and glucose ($C_6H_{12}O_6$), composed of 6 carbons, 12 hydrogens, and 6 oxygens. Although there is a fixed number of elements (111), they can combine to form an infinite number of different compounds.

Atoms are Held Together by Chemical Bonds

A **chemical bond** is the attractive force that holds two or more atoms together in a compound. The atoms of each element can form only a particular number of chemical bonds, a number determined by the number and arrangement of the atom's electrons in the outermost energy level. When an atom's outer energy level contains fewer than eight electrons, it tends to gain, lose, or share electrons to achieve an outer energy level

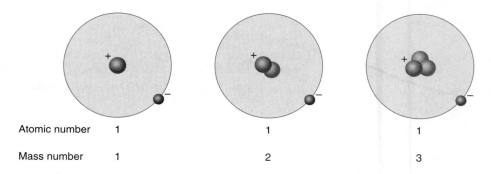

Figure 2-2   Isotopes of hydrogen.
Isotopes differ in the number of neutrons per atom; the numbers of protons and electrons are the same. These isotopes are all atoms of the element hydrogen, and their chemical properties are identical.

Atomic number	1	1	1
Mass number	1	2	3

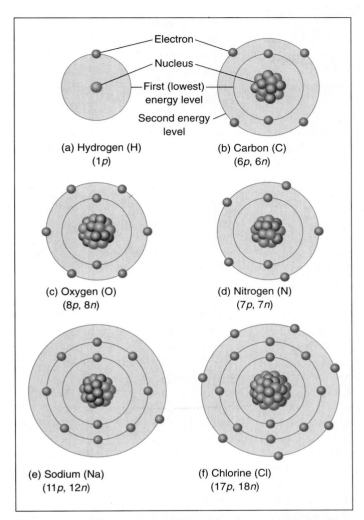

Figure 2-3   Electron configurations of some biologically important atoms. Each circle represents an energy level in which electrons occur; p = proton and n = neutron.

of 8. (The exceptions are hydrogen and helium, the only two elements with a single energy level, which holds a maximum of two electrons.) To complete its outer energy level, an atom can form one of two types of chemical bonds: ionic and covalent.

Atoms gain or lose electrons to form ionic bonds

In attempting to fill their outer energy levels, some atoms are able to completely pull one or more electrons away from another atom. Because the number of protons in the nuclei of the two atoms remains the same, the gain or loss of electrons results in an **ion**, an atom with a negative or positive electrical charge. An atom that has one, two, or three electrons in its outer energy level may donate those electrons to other atoms and become a positively charged ion in the process (because it has more positively charged protons than negatively charged electrons). An atom with five, six, or seven electrons in its outer energy level may gain electrons from other atoms, which makes it a negatively charged ion (because it has more electrons than protons). The force of attraction between two oppositely charged ions is called an **ionic bond**.

Sodium chloride (table salt) is an example of a compound formed by ionic bonding. The sodium atom donates its outermost electron to chlorine. As a result, the outer energy levels of both sodium and chlorine are complete. The positively charged sodium ion is attracted to the negatively charged chloride ion to form an ionic bond. Ionic compounds are not composed of individual molecules. A crystal of sodium chloride, for example, is composed of millions of sodium and chloride ions in a 1:1 ratio. Each sodium ion is surrounded by a number of chloride ions and is equally attracted to each of them. Likewise, each chloride ion is surrounded by and attracted to several sodium ions. Thus, a single sodium ion is not bonded to a single chloride ion to form a molecule.

Atoms share electrons to form covalent bonds

A **covalent bond** forms when two atoms share a pair of electrons to complete their outermost energy levels. When two or more atoms join to one another by covalent bonding, a molecule forms. A **molecule** is the smallest unit of a covalent compound.

The simplest example of a covalent bond is the one that joins two hydrogen atoms together to form a molecule of hydrogen gas (Figure 2-4). Each atom of hydrogen contains a single electron in its first energy level, which needs two electrons to be complete. The two hydrogen atoms share their single electrons so that each of the two electrons is

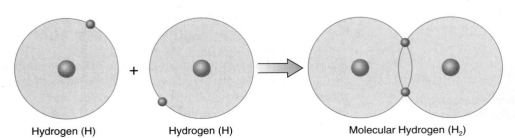

Figure 2-4   Covalent bonding between two hydrogen atoms. The two hydrogens share a pair of electrons between them, thereby becoming a molecule of hydrogen.

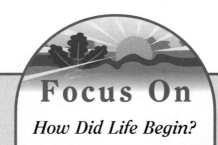
Scientists generally accept the hypothesis that the first living cells developed from nonliving matter. This development process, called **chemical evolution**, probably involved several stages. First, small organic molecules formed spontaneously and accumulated over time. Then organic macromolecules such as proteins and nucleic acids assembled from smaller molecules and gradually accumulated. The macromolecules interacted with one another, combining into more complicated structures that could eventually carry on chemical reactions and duplicate themselves. These macromolecular assemblages developed into cell-like structures that ultimately became the first true cells. It is thought that life originated only once and that this occurred under environmental conditions quite different from those we experience today.

Although we will never be certain of the exact conditions on Earth when life arose, scientific evidence from a number of sources provides us with valuable clues (see figure). The atmosphere of early Earth, which contained little or no molecular (or free) oxygen, included carbon dioxide (CO_2), water vapor (H_2O), carbon monoxide (CO), hydrogen (H_2), and nitrogen (N_2). The early atmosphere may have also contained some ammonia (NH_3), hydrogen sulfide (H_2S), and methane (CH_4), although these molecules may have been rapidly broken down by ultraviolet radiation from the sun. As the temperature of the Earth cooled, water vapor may have condensed, causing torrential rains to fall. This may have been the origin of the ocean. The falling rain would have eroded the Earth's surface, adding minerals to the ocean and making it salty.

There are four requirements for chemical evolution to have occurred: no free oxygen, a source of energy, the availability of chemical building blocks, and time. First, life could have begun only in the absence of free oxygen because oxygen is very reactive and would have broken down the organic molecules that were the basis of life's origin. Evidence indicates that the Earth's early atmosphere was strongly reducing; that is, any free oxygen would have reacted with other elements so that the oxygen would be tied up in chemical compounds.

A second requirement for life to have evolved from nonliving chemicals was energy. Early Earth was a place of high energy, with violent thunderstorms, volcanoes, and intense radiation from the sun. The young sun probably produced more ultraviolet radiation than it does today, and the Earth had no protective ozone layer to block much of this radiation. The radiation provided the energy needed to form organic molecules from simple inorganic compounds.

Third, the chemicals that would be the building blocks for chemical evolution must have been present. These included water, dissolved inorganic minerals (present as ions), and the gases present in the early atmosphere.

The fourth requirement was time—time for molecules to accumulate and react. The age of the Earth provides adequate time for chemical evolution. The Earth is approximately 4.6 billion years old, and there is geological evidence that simple organisms appeared about 3.5 billion years ago. Once the first cells originated, they evolved over millions of years into the rich biological diversity on Earth today.

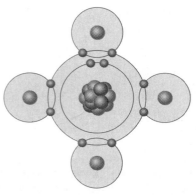

Figure 2-5   Covalent bonding between a carbon atom and four hydrogen atoms results in the formation of a molecule of methane.

attracted simultaneously to the two hydrogen nuclei. Because the two electrons are under the influence of both atomic nuclei, the two hydrogen atoms are joined, or bonded, together.

The carbon atom has four electrons in its outer energy level, but it would be most stable if it possessed eight (recall that atoms tend to gain, lose, or share electrons to achieve an outer energy level of eight). Each of the four electrons in carbon's outer energy level is available to form a covalent bond. If each of carbon's outer electrons forms a covalent bond with a hydrogen atom, a molecule of methane (CH_4) is formed (Figure 2-5). Each atom in a covalently bonded molecule shares only the electrons in its outer energy level. In methane, this sharing results in a full first energy level (two electrons) for each hydrogen and a full second energy level (eight electrons) for carbon. Each line in the following structural for-

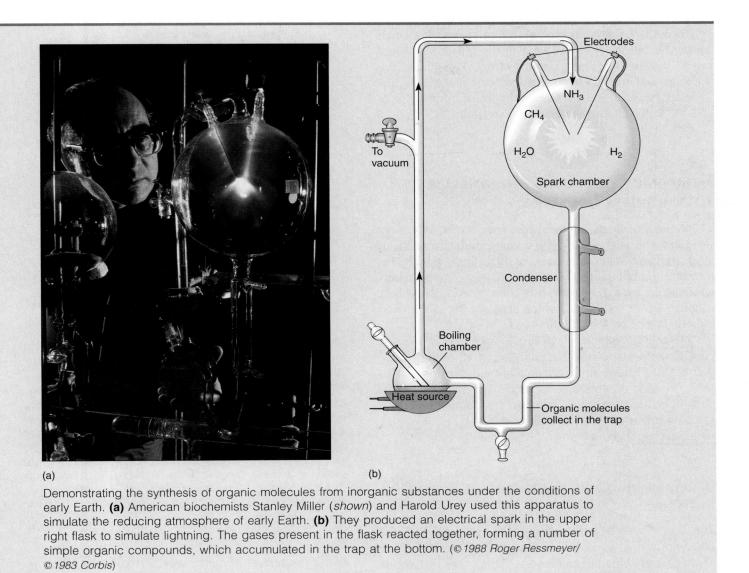

(a)   (b)

Demonstrating the synthesis of organic molecules from inorganic substances under the conditions of early Earth. **(a)** American biochemists Stanley Miller (*shown*) and Harold Urey used this apparatus to simulate the reducing atmosphere of early Earth. **(b)** They produced an electrical spark in the upper right flask to simulate lightning. The gases present in the flask reacted together, forming a number of simple organic compounds, which accumulated in the trap at the bottom. (*© 1988 Roger Ressmeyer/ © 1983 Corbis*)

mula designates a covalent bond formed by a single pair of shared electrons.

$$
\begin{array}{c}
H \\
| \\
H-C-H \\
| \\
H
\end{array}
$$

We refer to the sharing of a single pair of electrons as a **single bond**. Some atoms can form **double bonds** by sharing *two* pairs of electrons. The ability of carbon to form double bonds with other carbon atoms is evident in many biologically important molecules, such as unsaturated fatty acids (discussed shortly). In structural formulas, two lines between carbon atoms ($C=C$) designate a double bond.

Covalent bonds in which the electrons are equally shared, such as in the hydrogen molecule (H_2), are called **nonpolar covalent bonds**. However, some atoms, such as oxygen and nitrogen, have a stronger attraction for the shared electrons in a covalent bond so that the electrons are unequally shared between the bonded atoms. Such a bond is known as a **polar covalent bond**. In such a bond, electrons are much closer to one atom than to the other, and as a result of this unequal sharing, different regions of the molecule are slightly charged.

Hydrogen Bonds Are Attractions Between Adjacent Molecules

A **hydrogen bond** is an attraction between a positively charged hydrogen atom in one polar molecule and a negatively charged oxygen or nitrogen atom in another polar molecule. Hydrogen bonds, which are weak compared to covalent bonds, form and break readily. Although a single hydrogen bond is only about one-tenth or one-fifteenth as strong as a single covalent bond, the collective force of many hydrogen bonds can be significant. In organisms, hydrogen bonding affects the shape and function of protein and nucleic acid molecules (discussed shortly) and is important in determining the properties of water (discussed in the next section).

INORGANIC COMPOUNDS ARE SMALL AND LACK CARBON BACKBONES

Chemical compounds can be divided into two broad groups, inorganic and organic. **Inorganic compounds** are relatively small compounds of elements other than carbon. Among the biologically important groups of inorganic compounds are water, acids, and bases. (For a discussion of how life may have originated from inorganic compounds, see Focus On: How did Life Begin? on the preceding pages).

Organic compounds are generally large and complex and contain carbon and usually hydrogen. In an organic compound, many carbon atoms are usually bonded to one another to form a kind of molecular backbone that may consist of a straight chain, a branched chain, or rings. Examples of biologically important organic compounds are carbohydrates, lipids, and proteins.

Water, an Inorganic Compound, Is Essential to Plants

The view of planet Earth from outer space reveals that it is different from other planets in the Solar System. Earth is a predominantly blue planet because of the water that covers three-fourths of its surface. Water, which can exist as a solid (ice), liquid, or vapor (steam), has a great effect on our planet: it helps shape the continents, moderates our climate, and allows organisms to evolve and survive. All life forms, from simple bacteria to complex multicellular plants and animals, contain water.

Water is vital to plants and other organisms because it carries dissolved nutrients and other important materials to cells. Solutions of water and dissolved materials surround every living cell and are present inside every cell. Water is also essential to life because almost all of the chemical reactions needed to sustain life occur in water; many of these chemical reactions involve water directly. Water also affects how the biological molecules that compose cells are organized. For example, the interaction of lipids (fats and oils) with water shapes cell membranes. Because water influences plants and other organisms so profoundly, we now examine some of its properties.

(a) Polar nature of water molecule

(b) Hydrogen bonding of water molecules due to their polarity

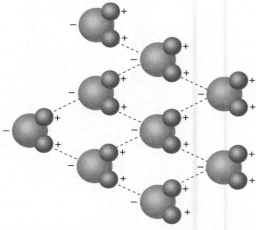

Figure 2-6   Many of the properties of water are results of its chemical structure. (a) Water, which consists of two hydrogen atoms and one oxygen atom, is a polar molecule with positively and negatively charged areas. **(b)** The polarity of water molecules causes hydrogen bonds (*represented by dashed lines*) to form between the positive area of one water molecule and the negative areas of others. Each water molecule can form up to four hydrogen bonds with other water molecules.

Many properties of water are caused by its polarity

Water is a polar molecule; that is, the electrical charge is unevenly distributed. The negative area (the oxygen part) of one water molecule is attracted to the positive area (the hydrogen part) of another water molecule, forming a hydrogen bond between the two molecules (Figure 2-6).

The hydrogen bonds in water are the basis for a number of its physical properties, including its high melting/freezing point (0°C, 32°F) and high boiling point (100°C, 212°F). Because most of the Earth's surface has a temperature between 0°C and 100°C, most water exists as the liquid on which organisms depend.

Because of hydrogen bonding, water must absorb a lot of heat (to break hydrogen bonds) before it vaporizes, or changes from liquid to vapor. When it does evaporate, water vapor carries the heat with it. Thus, evaporating water has a cooling effect. That is why evaporation of perspiration from your skin cools your body. Leaves are cooled in a similar way as water evaporates from their surfaces.

Water is sometimes called the "universal solvent," and although this is an exaggeration, many materials do dissolve in water as a result of its polarity. In nature, water is never completely pure, because it contains dissolved gases from the atmosphere and dissolved mineral salts from the Earth. Seawater, for example, is a solution of water and a variety of dissolved salts, including sodium chloride, magnesium chloride, magnesium sulfate, calcium sulfate, and potassium chloride. Water's dissolving ability has one major drawback, however: many pollutants (such as fertilizers, road salt, and acids) also dissolve in water.

Two properties of water, cohesion and adhesion, are particularly important in the transport of water and dissolved minerals in plants. The tendency of *like* molecules to adhere or stick together is known as **cohesion**. Water is very cohesive, and this strong attraction of water molecules to one another results from the hydrogen bonds among them.

Adhesion is the tendency of *unlike* molecules to adhere to one another. Water is strongly adhesive with many other materials, particularly those that are polar—that is, have positive or negative charges. The ability of water to make things wet is the result of adhesion.

ACIDS AND BASES ARE IMPORTANT COMPOUNDS IN PLANTS

The chemical makeup of a plant's internal environment and the external environment in which it lives affect its overall health and well-being. One of the most important aspects of either environment is how acidic or basic (alkaline) it is. The degree of acidity or alkalinity inside a plant cell must remain fairly stable, for example, or important chemical reactions will not take place. Likewise, certain plants decline and may even die if the acidity of the soil is too high.

Acids and Bases Dissociate When Dissolved in Water

An **acid** is a compound that dissociates, or breaks up, in a solution of water to form hydrogen ions (H^+, or protons) and negatively charged ions. Hydrochloric acid (HCl) is an example of an acid.

$$HCl \xrightarrow{\text{(in water)}} H^+ + Cl^-$$

Hydrochloric acid   Hydrogen ion   Chloride ion

Some acids are known as strong acids because they dissociate almost completely in water. Hydrochloric acid is a very strong acid because most of its molecules dissociate, producing hydrogen ions and chloride ions. Other acids, called weak acids, dissociate only slightly. Vinegar, which is a dilute solution of acetic acid, is an example of a weak acid.

A compound that dissociates in water to produce negatively charged hydroxide ions (OH^-) and positively charged ions is known as a **base**. Sodium hydroxide (NaOH) is an example of a base.

$$NaOH \xrightarrow{\text{(in water)}} Na^+ + OH^-$$

Sodium hydroxide   Sodium ion   Hydroxide ion

Acids and bases react with each other, in the process neutralizing the chemical effect each group had originally. Each hydrogen ion of an acid combines with a hydroxide ion of a base to form a water molecule:

$$H^+ + OH^- \longrightarrow H_2O$$

A Solution's Acidity or Alkalinity Can Be Expressed in Terms of pH

The **pH scale** measures the relative concentrations of hydrogen ions and hydroxide ions in a solution. The pH scale extends from 0, the pH of a strong acid such as HCl, to 14, the pH of a strong base such as NaOH. Pure water, which is neutral (neither acidic nor basic), has a pH of 7.

A solution that contains a greater concentration of OH^- ions than H^+ ions has a pH greater than 7 and is basic. A solution that contains a greater concentration of H^+ ions than OH^- ions has a pH less than 7 and is acidic.

For purposes of comparison, pure water has a pH of 7, tomato juice has a pH of 4, vinegar has a pH of 3, and lemon juice has a pH of 2. Normally, rainfall, with a pH of 5 to 6, is slightly acidic because carbon dioxide and other materials in the air dissolve in rainwater, forming dilute acids. However, the pH of precipitation in the northeastern United States and Canada averages 4 and is often 3 or even lower. Such acidic precipitation is commonly called **acid rain** (see Plants and the Environment: Acid Rain).

CARBON FORMS LARGE, COMPLEX MOLECULES ESSENTIAL TO LIFE

Organic compounds are the foundations on which the structures of plants and other organisms are built. Within cells, organic compounds participate in thousands of chemical reactions, including those that provide energy to sustain life. Four groups of organic compounds are essential for all organisms: carbohydrates, lipids, proteins, and nucleic acids.

Sugars, Starches, and Cellulose Are Examples of Carbohydrates

Carbohydrates are organic compounds, such as sugars, starches, and cellulose, that plants use as fuel molecules, as constituents

PLANTS AND THE Environment

Acid Rain

During the past few decades, acid rain has received considerable media coverage. Although the extent and acidity of acid rain varies widely from one region to another, rain in certain parts of eastern North America, Europe, and Japan-China-Korea is sometimes as acidic as vinegar! Acid rain corrodes materials such as metal and stone and poses a serious threat to the environment. It kills aquatic organisms such as fish and has been linked to widespread damage to forests, perhaps by changing soil chemistry.

Scientists are studying how acid rain changes the chemistry of soils and interferes with the uptake of essential minerals by plant roots. Some essential minerals, such as calcium and magnesium, wash out of acidic soil. Other minerals, including heavy metals such as aluminum, accumulate in acidic soil and may then be absorbed by plants in toxic amounts.

Acid rain can result from natural events such as volcanic activity, which adds large amounts of sulfuric acid to the atmosphere. In recent years, industrial activity and auto-mobile exhaust have become impor-tant contributors to acid rain by releasing nitrogen and sulfur oxides into the atmosphere. Motor vehicles are a major source of nitrogen oxides. Electrical power plants and heavy industrial equipment, both of which burn coal, are the main sources of sulfur dioxide emissions and produce substantial amounts of nitrogen oxides, as well.

During their stay in the atmos-phere, sulfur oxides and nitrogen oxides combine with water to produce dilute solutions of sulfuric acid and nitric acid. Precipitation such as rain or snow returns these acids to the ground, bathing plants in acid and causing the soil to become acidic.

A practice that has increased and complicated the problem of acid rain is the construction of tall smokestacks to release the sulfur-rich smoke high up in the atmos-phere. Unfortunately, the wind blows the pollutants long distances. For example, tall smokestacks allow the midwestern United States to "export" its acid emissions hundreds of miles away, to New England and Canada.

The pollutants in smokestack emissions can be captured and removed, but this requires the instal-lation of pollution control devices and thus makes prevention of acid rain expensive. The estimated cost to install and maintain the necessary scrubbers in smokestacks is about $4 to $5 billion per year in the United States. The Clean Air act of 1990 requires coal-burning power plants in the United States to reduce their sulfur and nitrogen oxide emissions. Such clean-air legislation is beginning to have an effect, and the acidity of precipitation is being reduced in some places.

---

of other important compounds such as nucleic acids, and as structural components of cells.

Carbohydrates are composed of carbon (C), hydrogen (H), and oxygen (O) atoms in an approximate ratio of 1C:2H:1O. The general equation for carbohydrates is $(CH_2O)_n$, where n refers to any number from 3 to several thousand. Three kinds of carbohydrates occur: monosaccharides, disaccharides, and polysaccharides.

Monosaccharides, or simple sugars, usually contain three to six carbon atoms (Figure 2-7). **Glucose** ($C_6H_{12}O_6$), a six-carbon sugar, is commonly called blood sugar because it is the form of sugar that is transported in the bloodstreams of humans and many other animals. Plants produce glucose by photosynthesis, but most of the glucose is immediately con-verted to other compounds. Another six-carbon sugar, **fruc-tose** ($C_6H_{12}O_6$), is commonly called fruit sugar because it is often found in fruits, along with other sugars. Glucose, fruc-tose, and other monosaccharides are fuel molecules that cells break down to obtain energy for cellular activities. Although glucose and fructose have the same chemical formula,

$C_6H_{12}O_6$, their atoms are arranged differently and, as a result, they have different properties. Fructose is sweeter than glu-cose, for example.

A **disaccharide**, or double sugar, consists of two mono-saccharide units bonded together. The disaccharide **sucrose** ($C_{12}H_{22}O_{11}$, common table sugar) is the carbohydrate stored in sugar cane (*Saccharum officinarum*) and sugar beets (*Beta vulgaris*). Sucrose is also the form of sugar transported in a plant's vascular system (conducting tissue). Sucrose consists of a molecule of glucose combined with a molecule of fruc-tose (Figure 2-8).

$$C_6H_{12}O_6 + C_6H_{12}O_6 \longrightarrow C_{12}H_{22}O_{11} + H_2O$$
Glucose   Fructose   Sucrose   Water

Note that the formation of sucrose from glucose and fructose involves the removal of a molecule of water. Such a reaction, in which two molecules are joined and a molecule of water is removed, is called a **condensation reaction**.

Starches and cellulose are the most important **polysac-charides**, carbohydrates composed of many sugar units. Each

(a) Glucose $C_6H_{12}O_6$ (b) Fructose $C_6H_{12}O_6$ (c) Ribose $C_5H_{10}O_5$ (d) Deoxyribose $C_5H_{10}O_4$

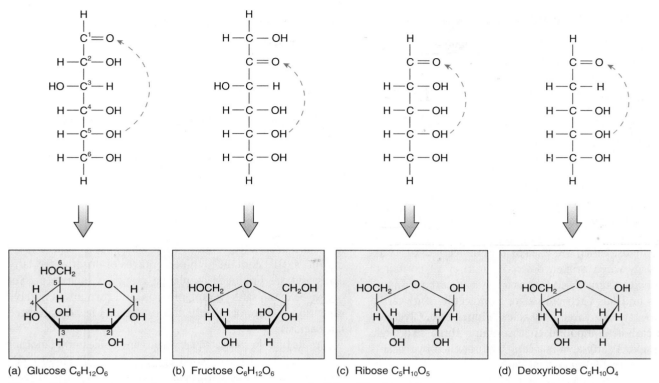

Figure 2-7 Structures of some common monosaccharides. The linear form (*top row*) of each is shown because it is easier to visualize bonds and count the number of carbon, hydrogen, and oxygen atoms. In solution, these molecules have a ring form (*bottom row*). The six-carbon monosaccharides, glucose **(a)** and fructose **(b),** are important fuel molecules, whereas the five-carbon monosaccharides, ribose **(c)** and deoxyribose **(d),** are components of nucleic acids.

polysaccharide molecule consists of a long chain, either branched or unbranched, of thousands of monosaccharide units. **Starches,** enormous polysaccharide molecules composed of thousands of glucose units, are the main storage carbohydrates in plants. Plants build up their energy reserves by storing starch. Many of the foods humans consume, such as potatoes (*Solanum tuberosum*), corn (*Zea mays*), and rice (*Oryza sativa*) contain abundant starch.

Cellulose, a major component of plant cell walls, is an important structural polysaccharide. Cellulose, the most abundant carbohydrate on Earth, accounts for about 50 percent by weight of the organic compounds in plants. Although cellulose, like starch, is composed of glucose units, the bonds that join the glucose units in cellulose form differently from those in starch. Although many organisms (including humans) can digest starch, few can digest cellulose. Plant fiber, which consists mainly of cellulose, is an important part of the human diet. Although it cannot be digested to provide nutrients, cellulose adds bulk to the materials moving through the intestines and aids in bowel function.

Glucose Fructose Sucrose

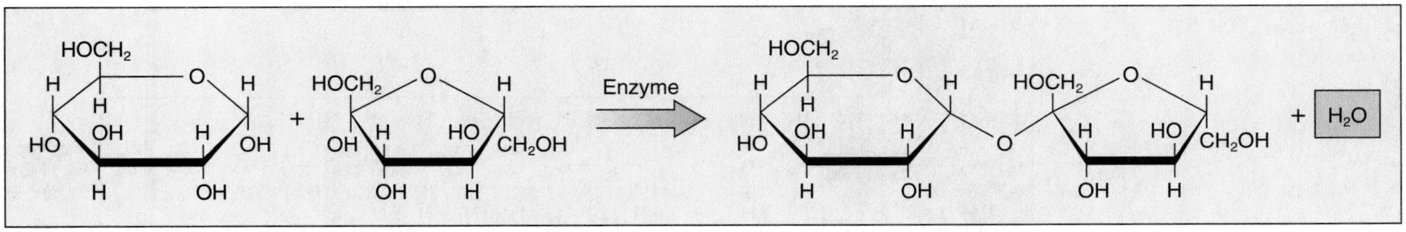

Figure 2-8 Synthesis of sucrose. The monosaccharides glucose and fructose combine covalently by a condensation reaction to form the disaccharide sucrose and a molecule of water.

Fats and Oils Are Examples of Lipids

Lipids are organic compounds that have a greasy or oily consistency and do not readily dissolve in water. Lipids are composed primarily of carbon and hydrogen, although they also contain some oxygen. They function in all cells as fuel molecules and as essential components of cell membranes. In plants, lipids also function as waterproof coverings over the plant body and as light-gathering molecules for photosynthesis. Lipids include neutral fats and oils, phospholipids, steroids, certain pigments, and waxes. We confine our discussion here to neutral fats and oils; other sections of the text discuss phospholipids and other types of lipids.

Neutral fats and oils are easily distinguished: a fat is solid at room temperature, and an oil is liquid at room temperature. These lipids, which are used as fuel molecules, provide a lot of energy when broken down: a gram of fat contains more than twice as much energy as a gram of carbohydrate. Neutral fats and oils each consist of a molecule of glycerol joined to one, two, or three fatty acids (Figure 2-9). **Glycerol** is a three-carbon compound that contains three hydroxyl (—OH) groups. A **fatty acid** is a long, unbranched hydrocarbon (composed of hydrogen and carbon) chain with a carboxyl (—COOH) group at one end. Fatty acids are typically composed of even numbers of carbon atoms and range in length from 4 to 20 carbons or even longer. The most common fatty acid, oleic acid, contains 18 carbon atoms.

A fatty acid is either saturated or unsaturated. **Saturated fatty acids** contain the maximum number of hydrogen atoms possible. For example, the 12-carbon fatty acid at the top of the next column is saturated.

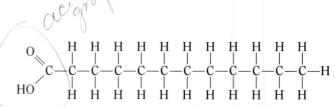

If a fatty acid contains one or more carbon-carbon double bonds, the molecule is **unsaturated**, or not fully saturated with hydrogens. The following 12-carbon fatty acid is unsaturated.

Fats, which contain high percentages of saturated fatty acids, are generally produced by animals; butter and lard are examples. Oils contain high percentages of unsaturated fatty acids and are produced by plants; examples include peanut oil, soybean oil, safflower oil, olive oil, and corn oil. (A notable exception is coconut oil, which contains very few unsaturated fatty acids.)

Diets high in saturated fatty acids and cholesterol (another lipid) tend to raise the blood cholesterol level. These lipids have been associated with heart disease, particularly atherosclerosis, a progressive disease in which the arteries become blocked with fatty material. On the other hand, ingestion of unsaturated fats, particularly polyunsaturated fats,[1] tends to

---

[1]"*Poly*unsaturated" refers to the presence of three or more carbon-carbon double bonds in a fatty acid.

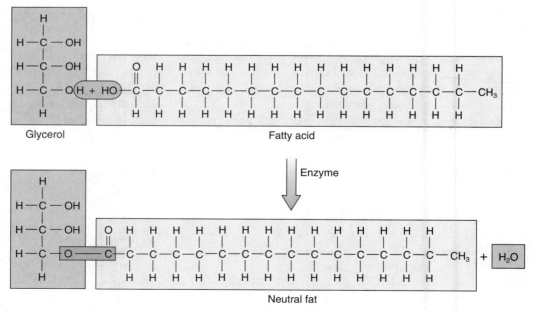

Figure 2-9   A neutral fat or oil is formed when glycerol combines with one (*shown*), two, or three fatty acids in a condensation reaction. The green boxed area of the neutral fat indicates where the fatty acid and glycerol combined.

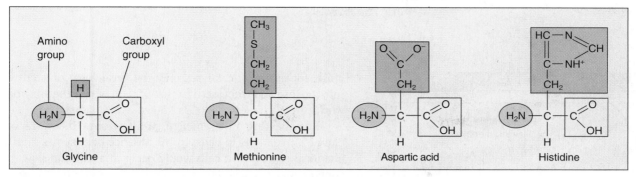

Figure 2-10 Four representative amino acids, which differ in their R groups (*green boxed areas*). Note that each amino acid contains an amino group ($-NH_2$) and a carboxyl group ($-COOH$).

decrease the blood cholesterol level. For this reason, many people now cook with vegetable oils rather than with butter and lard, drink skim milk rather than whole milk, and eat frozen yogurt or ice milk instead of ice cream.

Proteins Are Large Organic Molecules Composed of Amino Acids

Proteins are macromolecules composed of carbon, oxygen, hydrogen, nitrogen, and usually sulfur that serve as structural components of cells and tissues. Proteins also regulate biochemical processes in plants and other organisms. Every chemical reaction that occurs in living cells is controlled by its own specific enzyme, a protein that affects the rate at which a chemical reaction occurs (discussed shortly). The importance of enzymes cannot be overstated: *enzymes are crucial for an organism's survival*.

Structural proteins are particularly important in animals. Skin, hair, nails, muscles, tendons, and cartilage are some of the animal tissues composed largely of protein.

The types of proteins found in a cell determine to a large extent what type of cell it is—how it looks and how it functions. Proteins vary slightly from one species to another, and this fact partly explains differences among species. A rose is a rose because it produces proteins characteristic of roses. Protein composition also reflects evolutionary relationships: the proteins of closely related species have more similarities than those of species that are only distantly related. Information on the kinds of proteins that organisms possess is encoded in their hereditary material and passed from one generation to the next.

Proteins are composed of hundreds of units called **amino acids**. Each amino acid contains a carbon atom bonded to an amino group ($-NH_2$), a carboxyl group ($-COOH$), and a side chain, designated R. The general formula for amino acids is

$$\begin{array}{c} H \quad\quad R \quad\quad O \\ \diagdown \quad | \quad \diagup\diagup \\ N-C-C \\ \diagup \quad | \quad \diagdown \\ H \quad\quad H \quad\quad OH \end{array}$$

About 20 different amino acids are found in proteins; they differ in their R groups (Figure 2–10). Most plants can synthesize the various amino acids that they need from simpler substances. Animals, including humans, can manufacture some amino acids, but they must obtain other amino acids, referred to as **essential amino acids**, in their diets.

Amino acids are not randomly linked together to form proteins. Just as the order of letters determines the meanings of words, the order of amino acids in a protein determines its structure and function. The amino acid chains of many proteins fold up on themselves to form the complex three-dimensional shapes that determine their functions.

The bond that links one amino acid to another forms between the carboxyl carbon of one amino acid and the amino nitrogen of another (Figure 2-11). This covalent bond, called a **peptide bond**, forms as a result of a condensation reaction.

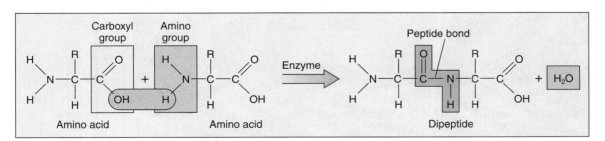

Figure 2-11 Two amino acids combine by forming a peptide bond between the carboxyl group of one amino acid and the amino group of another. The reaction is a condensation reaction.

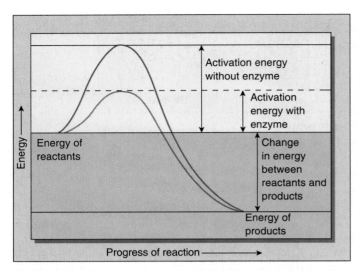

Figure 2-12   An enzyme speeds up a reaction by lowering its activation energy. A reaction catalyzed by an enzyme (*red curve*) proceeds faster than an uncatalyzed reaction (*blue curve*) because it requires less activation energy to initiate.

Each additional amino acid added to the growing chain likewise forms a peptide bond between itself and the **polypeptide** chain. Some proteins are composed of a single polypeptide chain, whereas other proteins contain two or more polypeptide chains.

Enzymes lower the energy needed to initiate chemical reactions

Enzymes are protein molecules that function as catalysts; that is, they increase the rate at which chemical reactions occur but are not used up in the reactions. Most enzymes are highly specific and catalyze only a single chemical reaction or a few closely related chemical reactions. The enzyme su-

crase, for example, catalyzes only the breakdown of sucrose into glucose and fructose. The material on which the enzyme works, in this case sucrose, is known as the **substrate**.

Enzymes are very efficient at increasing the rates of chemical reactions in cells. In the absence of enzymes, most chemical reactions in cells would occur at a rate too slow to support life. The disaccharide sucrose, for example, is a very stable molecule that, in the absence of sucrase, would take *years* to break down into glucose and fructose. However, if a tiny amount of sucrase is present, the reaction occurs very quickly. Each molecule of sucrase catalyzes the breakdown of *thousands* of sucrose molecules *per second!*

How do sucrase and other enzymes work? Before new chemical bonds can form during a chemical reaction, old ones must be broken. The process of breaking old bonds requires an input of **activation energy**. An enzyme works by lowering the activation energy needed to initiate a chemical reaction (Figure 2-12).

Each enzyme has one or more **active sites**, areas that fit particular substrates and enable the enzyme to form a temporary *complex* with them. During the course of a chemical reaction, the enzyme exerts pressure on existing chemical bonds so that they break more readily. Once the chemical reaction has occurred, the products separate from the enzyme, and the enzyme is free to bind to other substrate molecules (Figure 2-13).

Enzyme + Substrate(s) $\longrightarrow$ Enzyme-substrate complex
$\longrightarrow$ Enzyme + Product(s)

DNA and RNA Are Nucleic Acids

Nucleic acids are macromolecules composed of carbon, oxygen, hydrogen, nitrogen, and phosphorus. There are only two types of nucleic acids: **deoxyribonucleic acid (DNA)** and

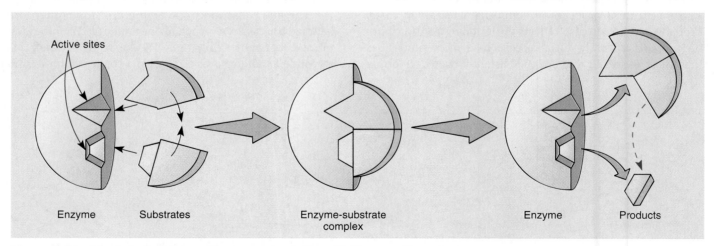

Figure 2-13   A substrate forms a temporary complex with the specific enzyme that catalyzes its reaction. When the products separate from the enzyme, the enzyme can bind to additional substrate molecules.

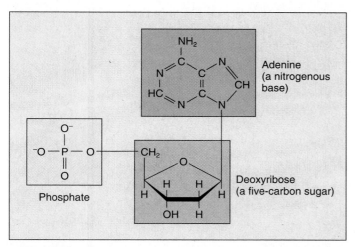

Figure 2-14   A DNA nucleotide. Each nucleotide has three parts: a nitrogenous base, a five-carbon sugar, and a phosphate group.

ENERGY IS NEEDED FOR BIOLOGICAL WORK

Energy is the capacity or ability to do work. In organisms, the biological work that requires energy includes processes such as growing, moving, reproducing, and repairing damaged tissues.

Energy exists in several forms: heat, radiant energy from the sun, chemical energy in the chemical bonds of molecules, mechanical energy, and electrical energy. Energy can exist as stored energy—called **potential energy**—or as **kinetic energy**, the energy of motion. You can think of potential energy as an arrow on a drawn bow (Figure 2-15). When the string is released and the arrow shoots through the air, the potential energy is converted to kinetic energy. Thus, energy can change from one form to another.

ribonucleic acid (RNA). Together, these two nucleic acids control all the life processes of an organism and are involved in the transmission of hereditary information from one generation to the next.

DNA contains the instructions for making all the proteins needed by an organism. These instructions are encoded in **genes**, units of hereditary information that consist of DNA and are part of the chromosomes. RNA functions in the process of protein synthesis.

Nucleic acids are composed of repeating units called **nucleotides**. Each nucleotide molecule is composed of three parts: (1) a nitrogenous base, (2) a five-carbon sugar, and (3) a phosphoric acid (phosphate) molecule (Figure 2-14). In a nucleic acid molecule, the phosphate portion of one nucleotide is attached to the sugar of the next nucleotide. The specific information encoded in a nucleic acid molecule is determined by the sequence of nucleotides in the nucleotide chain.

Energy Is Temporarily Stored in ATP

In addition to being the building blocks of nucleic acids, some nucleotides serve other important functions in cells. **Adenosine triphosphate (ATP)**, for example, is a modified nucleotide compound composed of the base adenine, the sugar ribose, and three phosphate molecules. ATP is present in *all* living cells as their "energy currency." Unstable bonds join the two terminal phosphate groups in the ATP molecule to each other. The biologically usable energy in these bonds is transferred to other molecules when the bonds are broken. Much of the chemical energy of a cell is temporarily stored in the bonds of ATP, ready to be released to do biological work when a phosphate group is transferred to another molecule.

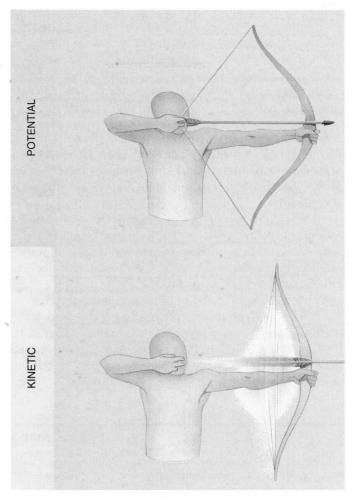

Figure 2-15   Potential and kinetic energy can be represented by a bow and arrow. Potential energy is stored in the drawn bow and is converted to kinetic energy as the arrow speeds toward its target.

The Laws of Thermodynamics Govern Energy Transformations

The study of energy and its transformations is called **thermodynamics**. There are two laws about energy that apply to all things in the Universe: the first and second laws of thermodynamics.

According to the **first law of thermodynamics**, energy cannot be created or destroyed, although it can change from one form to another. As far as we know, the energy present in the Universe at its formation, approximately 15 billion years ago, equals the energy present in the Universe today and represents all the energy that can ever be present in the Universe. Similarly, the energy of any object and its surroundings is constant. An object may absorb energy from its surroundings, but the total energy content of that object and its surroundings is always the same.

As stipulated by the first law of thermodynamics, then, plants and other organisms cannot create the energy that they require to live. Instead, they must capture energy from the environment and use it to do biological work, a process that involves transforming energy from one form to another. Through the process of photosynthesis, for example, plants absorb the radiant energy of the Sun and convert it to the chemical energy contained in the bonds of fuel molecules such as carbohydrates.

As each energy transformation occurs, some of the energy changes to heat energy that is then given off into the surroundings. This energy can never again be used by the organism for biological work, but because of the first law of thermodynamics, it is not "gone"; it still exists in the surroundings.

The **second law of thermodynamics** can be stated as follows: when energy is converted from one form to another, some usable energy—that is, energy available to do work—degrades into a less usable form, usually heat, that disperses into the surroundings. As a result, the amount of usable energy available to do work in the Universe decreases over time.

It is important to understand that the second law of thermodynamics is consistent with the first law. That is, the total amount of energy in the Universe is *not* decreasing with time. However, the energy available to do work in the Universe decreases over time.

Figure 2-16   Maintaining order requires energy. A white-footed mouse eats swamp red currants (*Ribes trista*). Animals must eat to replace the energy they continuously give up to the surroundings as they carry on life processes. (*Dwight R. Kuhn*)

Less-usable energy is more dilute, or disorganized. **Entropy** is a measure of this disorder, or randomness; organized, usable energy such as radiant, electrical, and chemical energy has a low entropy, whereas disorganized energy such as heat has a high entropy. Entropy is continuously increasing in the Universe in all natural processes. It may be that at some time billions of years from now, all energy will exist as heat uniformly distributed throughout the Universe. If that happens, the Universe will cease to operate because no work will be possible. Another way to explain the second law of thermodynamics, then, is that entropy, or disorder, in a system tends to increase over time.

Living things are highly organized and at first glance appear to refute the second law of thermodynamics. That is, as plants and other organisms grow and develop, they maintain a high level of order and do not appear to become more disorganized. However, organisms can maintain their degree of order over time only with the constant input of energy. That is why plants must photosynthesize (to obtain radiant energy) and animals must eat (to obtain chemical energy); (Figure 2-16).

S T U D Y   O U T L I N E

I. All matter is composed of elements. An atom is the smallest possible particle of an element.
   A. Atoms are composed of protons, neutrons, and electrons. A proton has a positive electrical charge and a small amount of mass. A neutron is uncharged and has about the same mass as a proton. An electron is negatively charged and has an extremely small mass.
   B. The number of protons in an atom is its atomic number. The total number of protons plus neutrons is an atom's mass number.
   C. Electrons move around the nucleus in different energy levels.
II. Elements combine chemically in fixed ratios to form chemical compounds. A molecule is the smallest particle of a compound.
   A. Some atoms tend to gain or lose electrons, forming ions.

Ionic bonds are attractions between two oppositely charged ions.

B. Some atoms share electrons with other atoms. A covalent bond involves the sharing of a pair of electrons.

C. Electrons are equally shared in nonpolar covalent bonds, whereas electrons are unequally shared in polar covalent bonds.

D. The attraction between the slightly positively charged hydrogen of one polar molecule and the slightly negatively charged oxygen or nitrogen of another is called a hydrogen bond.

III. Biologically important inorganic compounds include water, acids, and bases.

A. Water has a strong dissolving ability and cohesive and adhesive properties. Many of water's traits are the result of its polarity and hydrogen bonding.

B. Acids dissociate in water to form hydrogen ions (protons, H^+); bases dissociate in water to yield negatively charged hydroxide ions (OH^-). A solution's acidity or alkalinity can be expressed in terms of the pH scale, a measure of the relative concentrations of H^+ and OH^- in a solution.

IV. Carbohydrates, lipids, proteins, and nucleic acids are the most important biological molecules.

A. Carbohydrates include sugars, starches, and cellulose.

1. Carbohydrates are important as fuel molecules, as constituents of compounds such as nucleic acids, and as a component of cell walls.

2. Simple sugars are monosaccharides; disaccharides are composed of two monosaccharide units; polysaccharides are composed of many monosaccharide units.

B. Lipids have a greasy consistency and do not readily dissolve in water.

1. Lipids are important as fuel molecules, as components of cell membranes, as waterproof coverings over plant surfaces, and as light-gathering molecules for photosynthesis in plants.

2. A neutral fat or oil molecule is composed of a molecule of glycerol plus one, two, or three fatty acids.

3. A saturated fatty acid has no carbon-carbon double bonds; an unsaturated fatty acid contains one or more carbon-carbon double bonds.

C. A protein is a macromolecule composed of amino acid units.

1. Enzymes are proteins that increase the rate of chemical reactions. Other proteins are important as structural molecules.

2. There are about 20 different amino acids. The order of amino acids determines the structure and function of a protein molecule.

D. DNA and RNA are the two types of nucleic acids.

1. Nucleic acids control the cell's life processes because they contain instructions for making the cell's proteins. DNA transmits hereditary information from one generation to the next. RNA is involved in protein synthesis.

2. Nucleic acids are composed of repeating units called nucleotides. In turn, each nucleotide consists of a five-carbon sugar, a nitrogenous base, and a phosphate group.

3. The order of nucleotides in a nucleic acid chain determines the specific information it encodes.

E. ATP is a modified nucleotide compound composed of adenine, ribose, and three phosphate groups. ATP is important in energy transfers in cells.

V. An enzyme speeds up a chemical reaction by lowering its energy of activation (the energy needed to initiate the reaction).

A. A substrate forms a temporary complex with the specific enzyme that catalyzes its reaction.

B. When the products separate from the enzyme, it can bind to additional substrate molecules.

VI. Energy, the ability to do work, can exist as stored energy (potential energy) or as the energy of motion (kinetic energy).

A. The first law of thermodynamics states that energy can be converted from one form to another, but it can be neither created nor destroyed.

B. The second law of thermodynamics states that entropy continually increases in the Universe as usable energy is converted to a lower-quality, less usable form—usually heat.

SELECTED KEY TERMS

activation energy, p. 30
adenosine triphosphate (ATP), p. 31
amino acid, p. 29
carbohydrate, p. 25
condensation reaction, p. 26
covalent bond, p. 21
deoxyribonucleic acid (DNA), p. 30

disaccharide, p. 26
enzyme, p. 30
fatty acid, p. 28
first law of thermodynamics, p. 32
hydrogen bond, p. 23
ionic bond, p. 21
lipid, p. 28

monosaccharide, p. 26
nonpolar covalent bond, p. 23
nucleic acid, p. 30
nucleotide, p. 31
organic compound, p. 24
peptide bond, p. 29
pH scale, p. 25

polar covalent bond, p. 23
polysaccharide, p. 26
protein, p. 29
ribonucleic acid (RNA), p. 31
saturated fatty acid, p. 28
second law of thermodynamics, p. 32
unsaturated fatty acid, p. 28

REVIEW QUESTIONS

1. What is the difference between an element and a compound? An atom and a molecule? An atom and an ion?

2. Distinguish between an element's atomic number and its mass number.

3. For the element phosphorus, $^{31}_{15}P$: (a) How many protons, neutrons, and electrons are present in one atom? (b) What is the atomic number? (c) What is the mass number?

4. Distinguish between a covalent bond and an ionic bond.

5. Describe some of the unique properties of water.
6. Contrast a monosaccharide such as glucose with a disaccharide such as sucrose. Contrast a monosaccharide with a polysaccharide such as starch.
7. What are the molecular components of neutral fats and oils?

8. Draw the structure of an amino acid, and indicate which portions correspond to the amino and carboxyl groups.
9. What are enzymes? Why are they essential to plants and other organisms?

THOUGHT QUESTIONS

1. "When one form of energy is changed into another form, some of the energy is turned into heat." Which law of thermodynamics does this statement describe? Explain your answer.
2. From experience you know that fats and oils are not very soluble in water. Explain that fact based on their chemical structure. (*Hint:* Are fatty acids polar or nonpolar?)

3. What connection is there between a cell's DNA and the enzymes that catalyze its biochemical reactions?
4. How can life exhibit so much order in an increasingly disordered Universe?

SUGGESTED READINGS

Begley, S., "Beyond Vitamins," *Newsweek,* April 25, 1994. Broccoli, citrus fruits, tomatoes, garlic, and many other plants eaten as food are rich in chemicals that help the body fight diseases, including cancer.

Monastersky, R., "Acid Precipitation Drops in United States," *Science News* Vol. 144, July 10, 1993. The Clean Air act is reducing acid rain in the United States.

Nemecek, S., "Not Yet Elemental, My Dear Seaborg," *Scientific*

American, March 1995. A short news article about the recent creation of two new elements with atomic numbers 110 and 111.

Solomon, Berg, Martin, and Villee, *Biology,* 4th ed., Philadelphia, Saunders College Publishing, 1996. Contains a detailed introduction to cell chemistry.

Trefil, J., "However It Began on Earth, Life May Have Been Inevitable," *Smithsonian,* February 1995. An interesting summary of scientific research on the origin of life.

PLANT CELLS

Submerged along the rocky shores of warm seas lives a most remarkable organism, the mermaid's wineglass (*Acetabularia*). This little organism has a body composed of three parts: a delicate base that anchors it to a rock or piece of coral, a long slender stalk, and a cap. Different species of mermaid's wineglass have caps with different shapes. One (*Acetabularia mediterranea*) is smooth and cup-shaped, and another (*Acetabularia crenulata*) is rounded with a series of finger-like projections. Green in color, the mermaid's wineglass is an alga, an aquatic organism that obtains energy by photosynthesis; its green color is due to the presence of chlorophyll, the light-trapping pigment found in most photosynthetic organisms.

Why is the mermaid's wineglass remarkable? This dainty alga, from 2 to 10 centimeters (0.8 to 4 inches) tall, is composed of a single gigantic cell. The mermaid's wineglass has also had a significant role in the development of scientific knowledge, because biologists have used it since the 1930s to answer basic questions about how cells function.

When the cap of *Acetabularia* is cut off at a certain time in the cell's life cycle, the cell base grows a new cap identical in appearance to the original. (The replacement of a lost part by the organism is called *regeneration*.) Biologists wondered what kind of cap would form if they

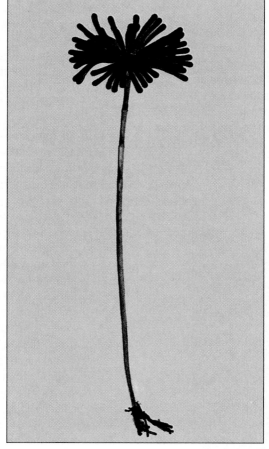

The marine alga *Acetabularia* is a single-celled organism consisting of a base, a stalk, and a cap. (*From* Dasycladales: An Illustrated Monograph of a Fascinating Algal Order, *by Sigrid Berger and Matthias J. Kaever, Georg Theime Publishers, Stuttgart, Germany*)

grafted the base of one species onto the stalk of another. When the base of *A. mediterranea* was grafted onto the stalk of *A. crenulata,* the cap that first regenerated on the *A. crenulata* stalk was an *A. crenulata* cap. If the regenerated cap was removed, however, the second regenerated cap always matched the base (*A. mediterranea*). No matter how many times the regenerated caps were removed, new ones continued to regenerate to match the base.

Because the base contains the cell's nucleus, biologists concluded that the nucleus controls the type of cap the organism makes. They suggested that the nucleus does this by producing some type of temporary messenger substance that is sent into the stalk to tell the stalk what type of cap to grow. Because the newly grafted stalk initially contained some of the messenger substance produced by its old base, the first cap to regenerate resembled the old shape. As the nucleus in the new base began sending messenger substance into the grafted stalk, the stalk regenerated caps with the new shape.

These experiments provided some of the first evidence that the nucleus contains hereditary information—a master plan of everything about the cell. They also indicated that the nucleus sends out a messenger substance, which we know today as messenger RNA (discussed in Chapter 12).

After reading this chapter, you should be able to:

1. *Contrast prokaryotic and eukaryotic cells.*
2. *Describe the functions of the following ten parts of a plant cell: nucleus, plasma membrane, chloroplasts, mitochondria, endoplasmic reticulum, ribosomes, Golgi apparatus, vacuole, cytoskeleton, and cell wall.*
3. *Summarize the differences between plant cells and animal cells.*
4. *Draw a diagram of the fluid mosaic model of a membrane, and label its parts.*
5. *Define the following processes, and explain their importance to the cell: diffusion, osmosis, facilitated diffusion, and active transport.*

THE INVENTION OF THE MICROSCOPE LED TO THE DISCOVERY OF CELLS AND THE DEVELOPMENT OF THE CELL THEORY

The first microscopes, invented in the late 1500s, were not very useful instruments. Although they could magnify specimens, the magnified images were blurry and details were undiscernible. In the 1600s, Anton van Leeuwenhoek (1632–1723) perfected the art of grinding lenses (pieces of glass with curved surfaces) and, by using them in microscopes of his own design, was able to produce clear, magnified images. Leeuwenhoek made many microscopes and peered through them to discover a world of organisms that had previously been unknown.

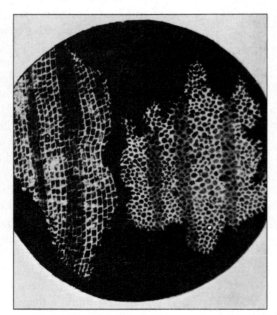

Figure 3-1   Robert Hooke's drawing of a thin slice of cork that he observed through his homemade microscope.
(*From the book* Micrographia, *published in 1665, in which Hooke described many of the objects he had viewed with his compound microscope.*)

Robert Hooke, an English physicist and microscopist of the 1600s, made significant discoveries with the compound microscope (a microscope with two lenses, the eyepiece and the objective). In 1665 Hooke used a microscope to examine a sliver of cork from the bark of a certain tree (Figure 3-1). Hooke saw that cork was composed of tiny boxes or compartments, which he named cells. Hooke recognized that he was looking at *dead* cells, of which all that remained were the cell walls. As Hooke and later biologists observed the contents of living cells, they began to realize that the cell's *interior* was also an important part of the cell.

As microscopes continued to improve during the 19th century, biologists observed tiny structures within cells, which they called **organelles** (little organs). The biologists envisioned these structures carrying out special functions for the cell much as organs, such as the heart and stomach or leaves and roots, carry out specific jobs for a multicellular organism. The cell's nucleus (an organelle that serves as its control center) was first identified and named in 1830 by Robert Brown, a Scottish botanist.

By the late 1830s, enough different kinds of tissues had been examined for biologists to conclude that all organisms are composed of cells. Two German biologists, Matthias Schleiden, a botanist, and Theodor Schwann, a zoologist, published separate papers in 1838 and 1839, respectively, that clearly stated that cells are the structural units of life. This statement has come to be known as the **cell theory.**

Another German scientist, Rudolf Virchow, extended the cell theory in 1855 by stating that all cells come from preexisting cells. That is, cells divide to give rise to new cells. In 1880 August Weismann pointed out that, since cells come from preexisting cells, all cells in existence today can trace their origins back to ancient cells.

Plant Biologists Study Cells Today Using a Variety of Methods

As the preceding material indicates, the knowledge of cells—their existence, structure, and composition—advanced as microscope technology improved. Today scientists use a variety of microscopes in their studies of various organisms. Figure 3-2 shows how two different microscopes work. Many of

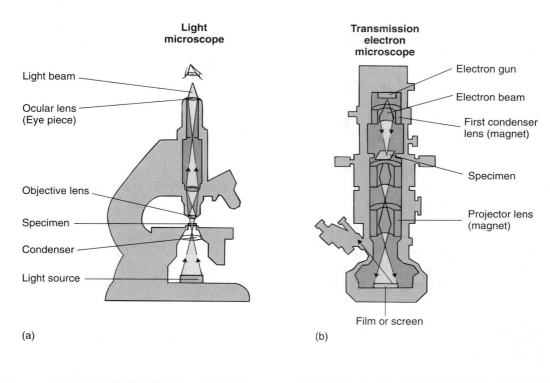

Light microscope

Light beam

Ocular lens (Eye piece)

Objective lens

Specimen

Condenser

Light source

(a)

Transmission electron microscope

Electron gun

Electron beam

First condenser lens (magnet)

Specimen

Projector lens (magnet)

Film or screen

(b)

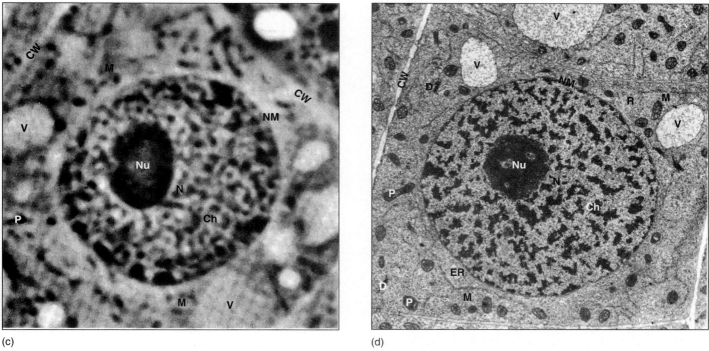

(c)

(d)

Figure 3-2 Microscopes magnify objects that are too small to be seen by the unaided eye. (a) A light microscope focuses a beam of light through the sample. **(b)** An electron microscope directs a beam of electrons through the sample. Lenses in the electron microscope are actually magnets that bend the beam of electrons. **(c,d)** An onion (*Allium cepa*) root cell as viewed under **(c)** a light microscope and **(d)** an electron microscope. Both cells are magnified ×1000, which is near the maximum magnification for the light microscope but a very low magnification for the electron microscope. Note the structural detail of the cell that is viewed with the electron microscope. (N = nucleus, NM = nuclear envelope, Nu = nucleolus, Ch = chromatin, V = vacuole, M = mitochondrion, P = plastid, R = ribosomes, and CW = cell wall) (c,d; *Photos by William A. Jensen, from* Cell Ultrastructure, *by Jensen and Park, Wadsworth, 1967, p. 57*)

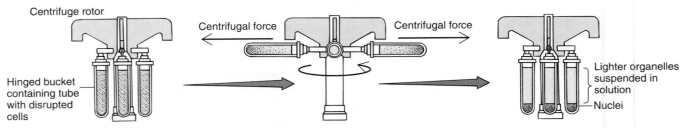

Figure 3-3 Cellular organelles can be separated from one another in a centrifuge, an instrument that spins test tubes. Centrifugal force caused by the spinning tubes forces the heavier organelles to settle at the bottom of the tube. When the speed of rotation is increased, organelles with lighter densities settle out.

the photographs in this book, especially in this chapter, were shot with the assistance of microscopes.

The **light microscope**, which has been much improved since Leeuwenhoek's time, focuses a beam of visible light through a transparent sample (Figure 3-2a). Light microscopes provide **magnification**, an increase in the apparent size of an object, of up to about 1000 times; they also provide **resolving power**, the ability to reveal fine detail, up to 500 times better than the human eye.

The **electron microscope**, which is much more powerful than a light microscope, passes a beam of electrons rather than light through the sample being studied (Figure 3-2b). The electron microscope can magnify an object 250,000 times or more and has resolving power up to 10,000 times better than the human eye.

Cells can also be studied using a variety of chemical and physical methods. For example, cells can be broken apart, placed in tubes, and centrifuged (spun) at high speeds to separate the cellular organelles (Figure 3-3). In centrifuges, centrifugal force (the force that makes rotating bodies move away from the center of rotation) causes larger, heavier organelles to settle out at the bottom of a tube, while smaller, lighter components remain suspended in solution. Once various cellular organelles are separated from one another by centrifugation, they can be studied to determine their chemical compositions and their activities.

THERE ARE TWO DIFFERENT TYPES OF CELLS

Cells consist of a small mass of jelly-like living material called **cytoplasm** surrounded by a **plasma membrane**, the outer boundary of the cell. (A cell wall encloses the plasma membrane of a plant cell, but, for reasons to be discussed shortly, the cell wall is *not* considered the outer boundary of the cell.)

All cells contain genetic material, the deoxyribonucleic acid (DNA) that encodes instructions for the cell's activities. In some cells, called **eukaryotic cells**, the genetic material is located in a special structure called the **nucleus**, which is bounded by a membrane. In cells that lack nuclei, called **prokaryotic cells**, the genetic material is in the cytoplasm.

Prokaryotic cells are generally smaller and simpler than eukaryotic cells. A prokaryotic cell has a simple internal organization: it has no nucleus and lacks certain other membrane-bounded organelles commonly found in eukaryotic cells. The word "prokaryotic" is derived from the Greek *pro*, "before," and *kary*, "nucleus"; evidence indicates that prokaryotic cells evolved before eukaryotic cells. Archaebacteria and eubacteria, introduced in Chapter 1, are the only **prokaryotes**.

Eukaryotic cells are generally larger and have more complex structures than prokaryotic cells. The term "eukaryotic" is derived from the Greek *eu*, "true," and *kary*, "nucleus," reflecting the fact that the genetic material of eukaryotic cells is located in a membrane-bounded nucleus. The cytoplasm of eukaryotic cells also contains membrane-bounded organelles not found in prokaryotic cells. All organisms other than bacteria—including algae, fungi, plants, and animals—are composed of eukaryotic cells. Organisms with a eukaryotic cell structure are sometimes called **eukaryotes**.

EUKARYOTIC CELLS HAVE MANY PARTS

When early biologists examined living cells under a microscope, the contents of the cells' interiors appeared to be a uniform jelly-like fluid. As light microscopes were improved, however, it became clear that a cell's cytoplasm was filled with numerous organelles. With the invention of the electron microscope, the interiors of these organelles, as well as organelles that were too small to be seen with the light microscope, became visible (Figure 3-4).

Each eukaryotic cell is like a tiny state; it possesses a control center, power plants, factories that make products, packaging and transport systems, a communication system, and a waste removal system. The cell's organelles perform these various functions.

Plant Cells Are Surrounded by Cell Walls

Although *every* cell is contained by a plasma membrane, a plant cell also has a firm **cell wall** *outside* the plasma membrane. The cell wall, which is a coating secreted by the cell, supports and protects each plant cell while providing routes for water and dissolved materials to pass to and from the cell.

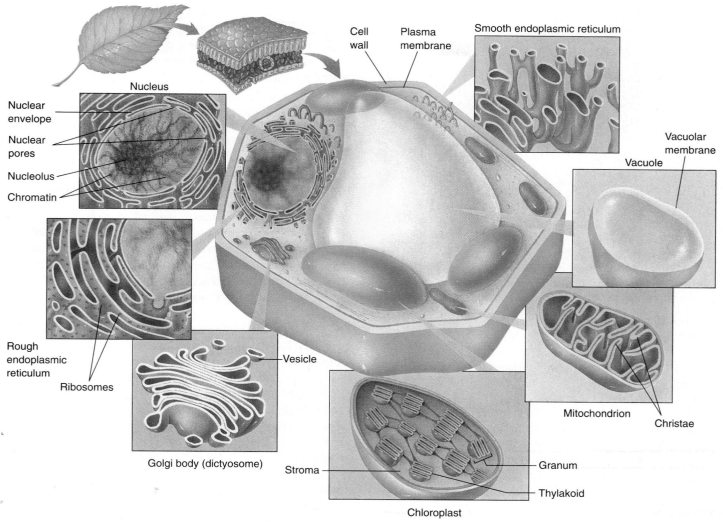

Cell wall
Plasma membrane
Smooth endoplasmic reticulum
Nucleus
Nuclear envelope
Nuclear pores
Nucleolus
Chromatin
Vacuolar membrane
Vacuole
Rough endoplasmic reticulum
Ribosomes
Vesicle
Mitochondrion
Christae
Golgi body (dictyosome)
Stroma
Granum
Thylakoid
Chloroplast

Figure 3-4   Structure of a "typical" plant cell. Some plant cells do not possess all the organelles depicted here. Root cells, for example, do not contain chloroplasts.

Collectively, cell walls provide strength to the entire plant; a massive tree stands tall and doesn't collapse on itself because of the combined strength of its cell walls.

Although animal cells do not have cell walls, the cells of many other organisms—bacteria, algae and some other protists, and fungi—do. The chemical compositions of cell walls in these organisms often differ from those in plants, however. Plant cell walls are composed largely of **cellulose**, a long-stranded polysaccharide that consists of as many as several thousand glucose molecules linked together. Cellulose and other cell-wall components are produced within the cell and transported out of the cytoplasm by the Golgi apparatus (discussed later in this chapter). Cellulose forms bundles of fibers that are held together by other polysaccharides, including **pectin** (the material that thickens jellies). Each layer of cel-

lulose fibers in a plant cell wall runs in a different direction from the adjacent layer, thereby providing strength to the wall (Figure 3-5).

A growing plant cell secretes a thin **primary cell wall**, which stretches and expands as the cell increases in size. After the cell stops growing, additional wall material may be secreted that thickens and solidifies the primary cell wall. Alternatively, after growth ceases, multiple layers of a **secondary cell wall** with a different chemical composition may form between the primary cell wall and the plasma membrane (Figure 3-6). In addition to cellulose, secondary cell walls usually contain **lignin**, a hard substance in which the cellulose fibers become embedded. (Lignin may also be found in primary cell walls.) It is lignin that gives wood, which consists largely of secondary cell walls, many of its distinctive properties.

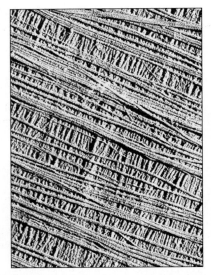

Figure 3-5   Cellulose fibers in a cell wall as viewed under an electron microscope. Each of these fibers consists of many long, thin cellulose molecules. (*Biophoto Associates*)

The Plasma Membrane Is the Outer Boundary of the Cell

In order for a cell to exist, it must be able to keep its contents together and separated from the environment. The plasma membrane is a physical boundary that confines the contents of the cell to an internal compartment. Although a plant cell is surrounded by an outer cell wall, it is the plasma membrane that defines the boundary of a cell, because the plasma membrane regulates the flow of materials into and

out of the cell. This regulation is *selective*: some substances pass through the plasma membrane unimpeded, other materials are allowed passage in a controlled fashion, and some substances are denied entrance or exit. Thus, the plasma membrane is a selective barrier that allows the interior of the cell to have a chemical composition quite different from the outside environment. The importance of the plasma membrane and other cellular membranes is discussed in more detail later in the chapter.

The Nucleus Is the Cell's Control Center

Housed within the nucleus is a complete set of genetic plans—in the form of DNA—for everything about the cell and its activities. This plan is not simply stored in the nucleus for safekeeping but is used continually to direct the activities of the cell. When a cell divides, its DNA is carefully replicated (copied) so that the two new daughter cells contain identical plans.

The nucleus has a remarkably complex structure (Figure 3-7). It is separated from the rest of the cell by a double membrane, the **nuclear envelope**, which contains pores lined with protein molecules. Substances can enter and leave the nucleus through these pores, but the pores are very selective, and only certain materials can pass through the nuclear envelope. The jelly-like interior of the nucleus, called **nucleoplasm**, contains the DNA, which is associated with certain protein molecules to form a threadlike material called **chromatin**. Although chromatin is normally not visible under the light microscope, during cell division it coils and thickens and becomes visible as distinct structures called **chromosomes**.

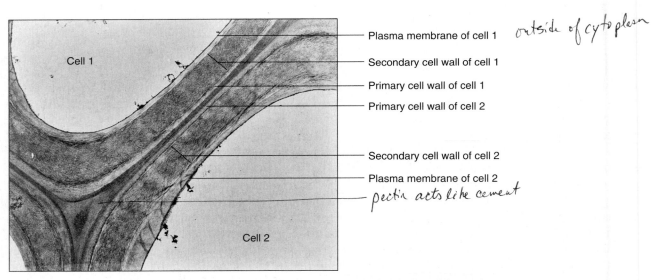

Plasma membrane of cell 1 — *outside of cytoplasm*
Secondary cell wall of cell 1
Primary cell wall of cell 1
Primary cell wall of cell 2
Secondary cell wall of cell 2
Plasma membrane of cell 2 — *pectin acts like cement*

Figure 3-6   Layers of the plant cell wall. The primary cell wall is the first wall formed. When present, the secondary cell wall is laid down between the plasma membrane and the primary cell wall. The thickness of plant cell walls varies greatly. (*Biophoto Associates*)

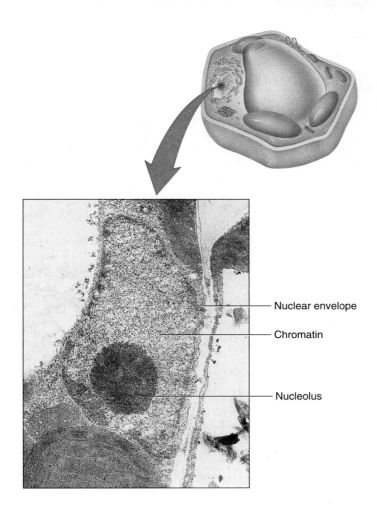

Figure 3-7   The nucleus in a tomato (*Lycopersicon lycopersicum*) leaf cell. Note the nuclear envelope, chromatin, and prominent nucleolus. (*Dwight R. Kuhn*)

— Nuclear envelope

— Chromatin

— Nucleolus

The **nucleoli** (sing. *nucleolus*), visible dark areas within the nucleus, are involved in making and assembling the subunits of ribosomes, important organelles in the cytoplasm (ribosomes are discussed shortly).

Chloroplasts Convert Light Energy to Chemical Energy

Algal and plant cells contain large organelles called **plastids**, each of which is surrounded by a double membrane. Although there are several kinds of plastids, the most common are **chloroplasts**, plastids that have a photosynthetic function and occur in certain leaf and stem cells. Recall from Chapter 1 that during photosynthesis plants use the energy of light to convert carbon dioxide and water to carbohydrates such as glucose. Chloroplasts contain the enzymes necessary for photosynthesis plus the green pigment **chlorophyll**, a molecule with the vital role of absorbing light energy.

Chloroplasts are usually disc-shaped in plant cells but occur in a variety of shapes in algae. Figure 3-8 shows the structure of a chloroplast. In addition to inner and outer

Figure 3-8   The chloroplast. Electron micrograph (*right*) and interpretive drawing (*left*) of a chloroplast from a corn leaf cell. The thylakoids—flat, disclike sacs—are arranged in stacks called grana. The inset on the left is a closeup of a granum. (*E.H. Newcomb and W.P. Wergin, University of Wisconsin/Biological Photo Service*)

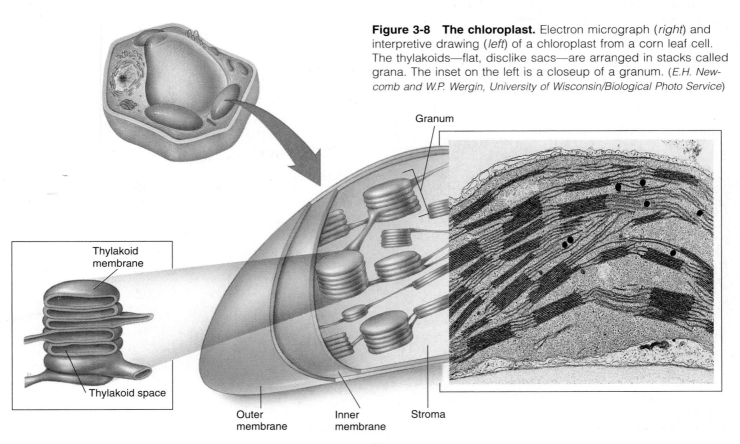

Granum

Thylakoid membrane

Thylakoid space

Outer membrane

Inner membrane

Stroma

41

Focus On

The Evolution of Eukaryotic Cells

According to fossil records, eukaryotic cells appeared approximately 1.9 to 2.1 billion years ago, long after the evolution of prokaryotic cells some 3.5 billion years ago. Fossil evidence suggests that eukaryotic cells evolved from prokaryotic cells.

The **endosymbiont theory** offers an explanation of how eukaryotic cells got organelles such as chloroplasts and mitochondria but does not explain how the genetic material in the nucleus came to be surrounded by a nuclear envelope. According to the endosymbiont theory, organelles such as chloroplasts and mitochondria originated from symbiotic relationships between two prokaryotic organisms (see figure). (A symbiotic relationship is an intimate relationship between two different kinds of organisms.) Chloroplasts are thought

to have evolved from cyanobacteria (a type of photosynthetic bacteria) that lived inside other prokaryotes. Mitochondria are thought to have evolved from aerobic (oxygen-requiring) bacteria that lived inside other prokaryotes.

How did these bacteria come to be **endosymbionts** (organisms that live symbiotically inside a host cell)? Probably they were originally ingested by the host cell but not digested by it. Thus, they survived and reproduced along with the host cell so that subsequent generations of the host also

contained endosymbionts. The two organisms became dependent on each other, and eventually the endosymbiont lost the ability to exist outside its host.

The principal evidence in favor of the endosymbiont theory is that chloroplasts and mitochondria possess their own DNA and their own ribosomes, a heritage received from their free-living ancestors. Both chloroplasts and mitochondria conduct protein synthesis on a limited scale, independently of the nucleus. Additional evidence for the endosymbiont theory comes from endosymbiotic relationships that exist today— for example, certain algae (zooxanthellae) that reside in the bodies of corals and other marine animals (discussed in Chapter 20).

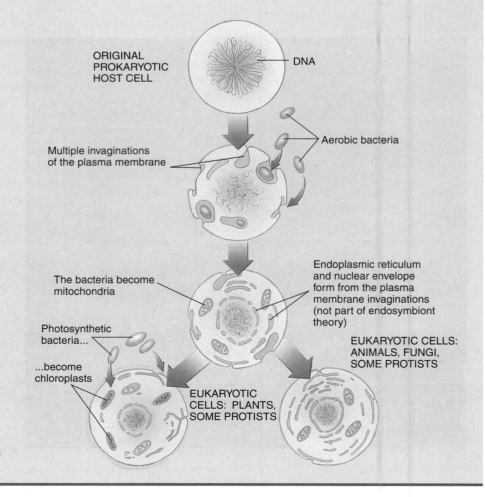

The endosymbiont theory of the origin of eukaryotes.

membranes, the interior of a chloroplast contains membranous stacks of thin, flat, circular plates called **thylakoids**; a stack of thylakoids is called a **granum** (pl., *grana*). The grana are embedded in a jelly-like fluid called the **stroma**, which contains enzymes that catalyze the chemical reactions of photosynthesis that convert carbon dioxide to carbohydrate.

Each chloroplast also contains a small amount of DNA and a few ribosomes. The presence of DNA and ribosomes in chloroplasts is significant because it indicates that they had free-living ancestors (see Focus On: The Evolution of Eukaryotic Cells). Chapter 4 considers chloroplasts and photosynthesis in detail.

Plant cells also contain **leucoplasts**, colorless plastids that form and store starch, oils, or proteins. Leucoplasts are common in seeds and in roots and stems modified for food storage. When leucoplasts are exposed to light, they can synthesize chlorophyll and function as chloroplasts; this happens, for example, when potato tubers are exposed to light.

A third type of plastid, known as a **chromoplast**, contains pigments that provide yellow, orange, and red colors to certain flowers such as marigolds and ripe fruit such as tomatoes and red peppers. Chromoplasts often form from chloroplasts when chlorophyll breaks down; this happens, for example, when green tomatoes ripen and turn red.

Mitochondria Convert the Chemical Energy in Food Molecules to ATP

The powerhouses of eukaryotic cells are the **mitochondria** (sing., *mitochondrion*), tiny organelles bounded by double membranes. Cellular respiration, a series of chemical reactions in which fuel molecules are broken down into carbon dioxide and water with the release of energy, occurs in mitochondria. The released energy is temporarily packaged in the chemical bonds of adenosine triphosphate (ATP) molecules, which are distributed to whatever part of the cell needs chemical energy.

Mitochondria vary in shape but often appear as tiny rods. They are too small to be seen under a student-grade light microscope. Their internal structure is complex. Each mitochondrion is bounded by a double membrane. Because the inner membrane has a larger surface area than the outer membrane, it folds inward, with the folds, called **cristae**, projecting into the interior of the mitochondrion (Figure 3-9). Some of the enzymes needed for cellular respiration are arranged along the cristae; other respiratory enzymes are found in the **matrix**, the jelly-like fluid inside the inner mitochondrial membrane.

Each mitochondrion also contains a small amount of DNA and a few ribosomes. As in chloroplasts, the presence of DNA and ribosomes in mitochondria is significant (see Focus On: The Evolution of Eukaryotic Cells). Chapter 4 considers mitochondria and cellular respiration in detail.

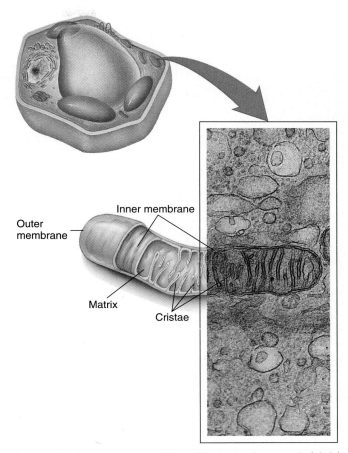

Figure 3-9 The mitochondrion. Electron micrograph (*right*) and interpretive drawing (*left*) of a mitochondrion, showing details of the interior, including the matrix and cristae. (*D.W. Fawcett*)

Ribosomes Are Sites of Protein Synthesis

Ribosomes are small organelles that are protein-manufacturing centers of the cell. Specifically, ribosomes use instructions from DNA in the nucleus to assemble proteins by joining amino acids together in precise sequences. Some of these proteins are used inside the cell; others are exported for use outside the cell.

Each ribosome is composed of two subunits; each subunit, in turn, consists of ribonucleic acid (RNA) and protein molecules. Although they occur in the nucleus, plastids, and mitochondria, ribosomes are most numerous in the cytoplasm, where they are found free—not associated with a particular organelle—or bounded to the endoplasmic reticulum.

The Endoplasmic Reticulum Has Many Functions

The **endoplasmic reticulum (ER)** is an extensive network of parallel membranes that extends throughout the cell's interior (Figure 3-10). The ER appears to be continuous with both

Figure 3-10   Endoplasmic reticulum (ER). The electron micrograph and its paired interpretive drawing show both smooth and rough ER. The smooth ER is more tubular, whereas the rough ER is more flattened. The rough ER is studded with ribosomes. (*Visuals Unlimited/R. Bolender– D. Fawcett*)

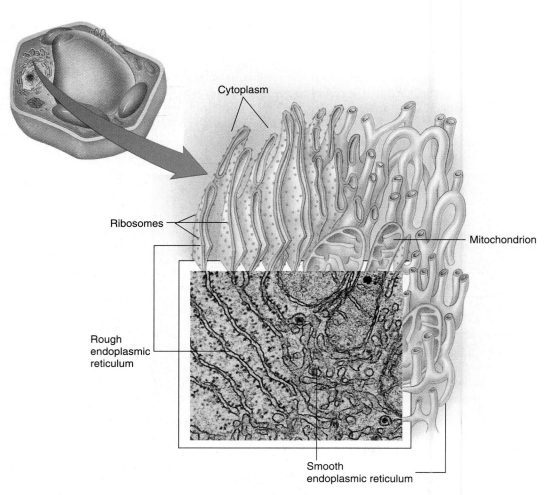

Cytoplasm

Ribosomes

Mitochondrion

Rough endoplasmic reticulum

Smooth endoplasmic reticulum

the plasma membrane and the nuclear envelope surrounding the nucleus.

The ER is one of the major manufacturing centers of the cell. Many enzymes are associated with the ER membranes; these enzymes catalyze chemical reactions that synthesize biologically important molecules. ER that has ribosomes attached to it is called **rough ER** and is the site of protein synthesis; ER without ribosomes is known as **smooth ER** and is associated with lipid synthesis.

ER also synthesizes the membranes for various organelles throughout the cell, including the nuclear envelope (recall that the ER is continuous with the nuclear envelope) and other cellular organelles such as the Golgi apparatus.

The Golgi Apparatus Is the Cell's Collecting and Packaging Center

A **Golgi body**, or **dictyosome**, is a factory for processing and packaging proteins and polysaccharides. It consists of several flattened sacs, each of which is surrounded by a membrane (Figure 3-11). The edges of the sacs often bulge out to form **vesicles**, sacs that contain cellular products. The vesicles

transport materials to the plasma membrane, to the outside of the cell, or to other organelles within the cell. The collective term for all the Golgi bodies (dictyosomes) in a cell is the **Golgi apparatus**.

The Golgi apparatus collects and processes materials that are to be exported from the cell. In plant cells, for example, the Golgi apparatus produces and transports some of the polysaccharides that make up the cell wall. The Golgi apparatus also collects materials that are stored inside large, membrane-bounded sacs called vacuoles.

The manufacture and export of materials from the cell is a complex process that involves several cellular organelles. For example, consider the secretion of proteins. Proteins that are manufactured by the ribosomes along the rough ER are sealed in tiny membrane-bounded sacs and transported to a Golgi body (dictyosome). The sac from the ER then fuses with the membrane of the Golgi body (dictyosome) and deposits its contents inside. Once inside, the proteins are modified in various ways. The finished products are then sealed inside a vesicle that pinches off the edge of the Golgi body (dictyosome). The vesicle then migrates to the cell's plasma membrane, fuses with it, and deposits its contents outside the cell.

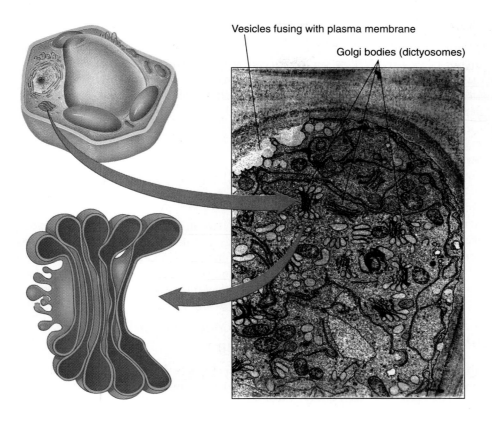

Vesicles fusing with plasma membrane

Golgi bodies (dictyosomes)

Figure 3-11   Golgi bodies (dictyosomes).
(*Right*) Electron micrograph of Golgi bodies (dictyosomes) in a mucilage-secreting plant cell. (*Left*) Three-dimensional view of a Golgi body (dictyosome). Note the small vesicles pinching off the flattened sacs. (*Courtesy of H. Mollenhauer, Texas A&M University*)

Vacuoles Are Large, Membrane-Bounded Sacs

Cell sap

A **vacuole** is a membrane-bounded sac filled with a liquid that contains a variety of materials in addition to water—dissolved salts, ions, pigments, and waste products. Vacuoles are present in many types of cells but are most common in plant cells and the cells of certain protists. In certain mature plant cells, the vacuole may occupy more than 90 percent of the volume of the cell (Figure 3-12).

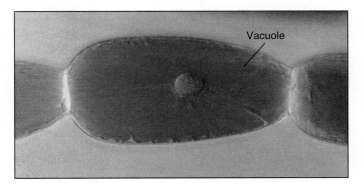

Vacuole

Figure 3-12   The vacuole. A pigmented vacuole of a stamen hair of spiderwort (*Tradescantia virginiana*) dramatically conveys the size of the vacuole. The nucleus and cytoplasm are restricted to the edges of the cell. (The nucleus looks as though it is in the middle of the vacuole because it is lying in the cytoplasm on *top* of the vacuole.) (*Phil Gates, University of Durham/Biological Photo Service*)

The vacuole performs several important functions for a plant cell. One of the most significant is that the vacuole helps the cell maintain its shape by making it turgid (from the Latin *turg,* "swollen"). A **turgid** cell is one that is swollen or firm because it contains water under pressure. The large concentration of ions and other materials dissolved in the vacuole causes water to accumulate. The vacuole swells and presses against the cytoplasm, which in turn presses against the cell's plasma membrane and cell wall. Collectively, vacuoles provide strength for nonwoody plants. These plants are erect because the vacuoles pressing against the cell walls give them rigidity. Without the collective strength of their turgid vacuoles, plants would wilt.

The vacuole also serves as a temporary storage area; excess materials such as calcium ions can be stored in the vacuole until the cell needs additional calcium. Water-soluble pigments are often stored in the vacuole; for example, a red pigment is stored in the vacuoles of red onion cells, giving them their characteristic color. Waste products, malformed proteins, and the like also enter the vacuole, where they may be disassembled so that their component parts can be used again. Some wastes accumulate and form small crystals.

The Cytoskeleton Is Composed of Protein Fibers

The **cytoskeleton** is a complex network of fibers that extends throughout the cytoplasm and provides structure to a eukaryotic cell. The cytoskeleton, which is also important in cell movement, includes two types of fibers, microtubules and microfilaments.

Microtubules are cytoskeletal elements that make up the spindle, a special structure that moves chromosomes during cell division. Microtubules are also a part of flagella and cilia, hairlike extensions of certain cells that aid in locomotion; flagella are longer than cilia and occur in smaller numbers.

Microfilaments, which are much thinner than microtubules, can contract and are responsible for **cytoplasmic streaming**, the movement of cytoplasm within the cell. Cytoplasmic streaming has a variety of purposes; for example, the movement of cytoplasm in leaf cells helps orient the chloroplasts for optimal exposure to light, which strikes the leaf cells at different angles during the day as the sun crosses the sky.

PLANT CELLS AND ANIMAL CELLS ARE MORE ALIKE THAN THEY ARE DIFFERENT

Because plants and animals are eukaryotes, their cell structures are fundamentally the same. Both plant and animal cells are bounded by plasma membranes; both have nuclei, mito-chondria, ribosomes, ER, the Golgi apparatus, and a cytoskeleton. Plant cells differ from animal cells in several respects, however: plant cells have cell walls, plastids, and conspicuous vacuoles, whereas animal cells do not. In addition, animal cells contain centrioles that function in cell division and lysosomes that are involved in digestion; plant cells lack both of these organelles.

CELLS CANNOT SURVIVE WITHOUT MEMBRANES

To exist, every cell must have an outer boundary that separates it from its surroundings and defines its limits. The plasma membrane serves this function in all cells. Most of a eukaryotic cell's organelles also possess specialized membranes that are intimately connected with their structures and functions.

The Fluid Mosaic Model Describes Membrane Structure

The **fluid mosaic model** characterizes the plasma membrane and other cell membranes as consisting of a double layer, or bilayer, of lipid molecules (Figure 3-13). A number of proteins

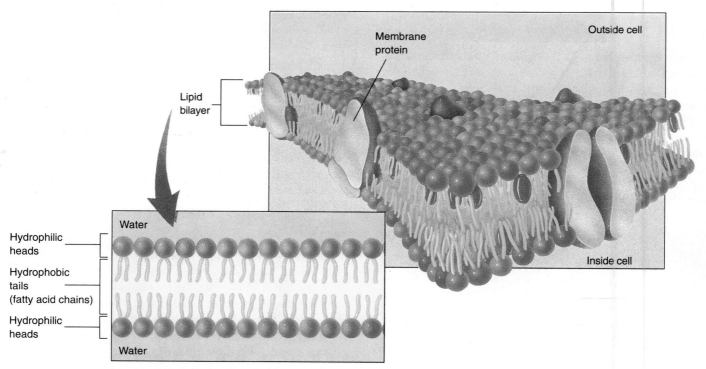

Figure 3-13 Structure of a membrane. The fluid mosaic model describes membranes as consisting of a lipid bilayer in which are embedded a number of proteins. (*Inset*) A phospholipid bilayer forms in watery surroundings. The hydrophilic heads point outward toward the watery surroundings, and the hydrophobic tails point inward so that they are not exposed to the water. In diagrams of phospholipid bilayers, the hydrophilic heads are usually represented by circles and the hydrophobic tails by two lines.

are embedded in the lipid bilayer in a way that resembles a *mosaic* pattern. The membrane structure is *fluid* rather than motionless, and the lipids (and protein molecules, to a lesser extent) can move laterally (sideways) within the membrane.

One of the important lipid components of membranes is **phospholipid**, composed of a glycerol molecule to which are attached two fatty acids and a molecule containing a phosphate group. The phosphate end of the phospholipid molecule is polar—that is, slightly charged—whereas the fatty acid chains are nonpolar (Figure 3-13, inset). The polar "head" is **hydrophilic** (from the Greek *hydro,* "water," and *phil,* "love"); that is, it has an affinity for water. The nonpolar "tail" is **hydrophobic** (from the Greek *hydro,* "water," and *phobos,* "fear"): that is, it has an aversion to water.

Because the cell and its surroundings are composed largely of water, phospholipids and other lipid components of membranes spontaneously arrange themselves in a double layer. The hydrophilic heads are positioned on the outer edges of each side of the layer—toward the watery surroundings inside *and* outside the cell or organelle. The hydrophobic tails form the inside of the double layer (away from the water). No chemical reactions occur to hold the membrane molecules together, nor do covalent bonds connect adjacent molecules. The forces of hydrophilic portions being attracted to water and of hydrophobic regions being repelled by water are strong enough to form and maintain the membrane's structure.

Membranes Perform Many Functions

Membranes are not just passive boundaries around cells and their organelles; they play a number of important roles in cellular function.

First, membranes regulate the passage of materials because they are **selectively permeable**; that is, they prevent the entrance or exit of certain materials while permitting—and even helping—the entrance or exit of other materials. For example, the lipid bilayer is impermeable to ions and polar molecules. Thus, charged ions such as Na$^+$ and Cl$^-$ are not allowed to pass into or out of a cell on their own. The membrane's regulation of the passage of materials enables the cell to maintain **homeostasis**, a relatively constant set of internal conditions. All materials that the cell's cytoplasm requires for survival must pass through the plasma membrane. Nutrient molecules

and oxygen must enter the cell, and wastes such as carbon dioxide must exit through the plasma membrane.

Second, membranes—particularly the plasma membrane—receive information from their surroundings, including other cells. Chemical messengers such as hormones often bind to special molecules in a membrane and set off some type of response in the cell. Thus, membranes help the cell to respond to its environment.

Some of the proteins in membranes are involved in important energy and enzymatic functions. Mitochondria, for example, cannot store energy in ATP without intact, functioning membranes, nor can chloroplasts use the sun's energy without thylakoid membranes. Membranes of other organelles, such as the ER, are the sites of enzymatic activity.

MATERIALS ARE TRANSPORTED ACROSS MEMBRANES IN A VARIETY OF WAYS

Some materials move passively through membranes by physical processes such as diffusion and osmosis. Other materials are moved in and out of the cell by processes such as active transport, which requires the cell to expend energy.

Diffusion Is the Movement of a Substance from a Region of Higher Concentration to a Region of Lower Concentration

Some materials pass into and out of cells by simple **diffusion**, the net movement of atoms or molecules of a particular substance from a region of higher concentration to a region of lower concentration (Figure 3-14). During diffusion, atoms and molecules move along a **concentration gradient**—that is, from where they are more concentrated to where they are less concentrated.

Diffusion occurs because atoms and molecules are in constant random motion that makes them collide with other particles. As a particle diffuses, it moves in a straight line until it collides with some other particle. This collision causes the original particle to rebound in another direction. Eventually the particles are uniformly distributed throughout a particular space. Even though the particles continue to move

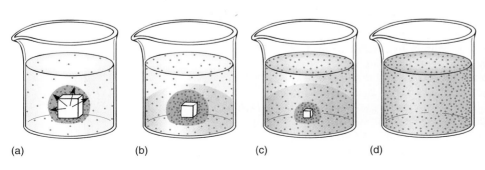

(a)   (b)   (c)   (d)

Figure 3-14 Diffusion. (a) When a lump of sugar is placed in a beaker of water, it begins to dissolve and diffuse into the water. **(b,c)** The sugar molecules continue to diffuse. **(d)** Eventually, the sugar is evenly distributed throughout the water.

and collide, they remain uniformly distributed, because as fast as some particles move in one direction, other particles travel in the opposite direction.

Diffusion is important to cellular function because it is responsible for the movement of many materials throughout the cytoplasm and into and out of cells. Oxygen, carbon dioxide, and water, for example, diffuse readily into and out of cells. Other materials, however, cannot pass through a cell's plasma membrane by diffusion because they are either too large or too polar.

Osmosis Is the Diffusion of Water Across a Selectively Permeable Membrane

Osmosis, a special kind of diffusion, is the movement of water through a selectively permeable membrane from a solution with a higher concentration of water to a solution with a lower concentration of water. In biological systems a **solution** is a mixture in which salts, sugars, and other materials are dissolved in water. The substances that are dissolved in water are referred to as **solutes**, and the water is referred to as the **solvent**.

A cell's plasma membrane is relatively impermeable to sugars and salts, but water moves across the membrane freely in either direction. When a cell is placed in a solution with a solute concentration equal to that inside the cell, water molecules diffuse through the plasma membrane equally in both directions. Such solutions are said to be **isotonic** (from the Greek *iso,* "equal")—that is, of a solute concentration equal to that in the cell.

When a cell is placed in a solution with a solute concentration higher than that within the cell, the solution is said

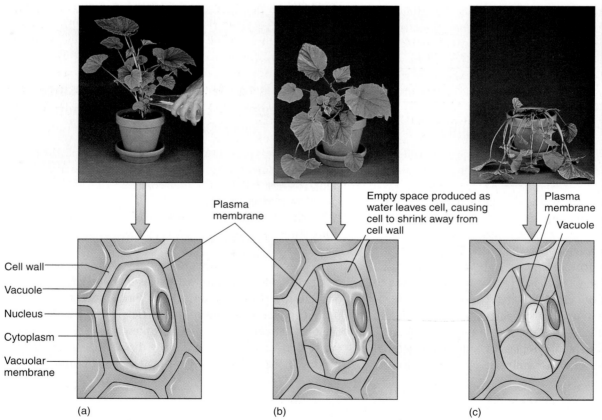

Figure 3-15   The soil water surrounding root cells is hypotonic to root cells. Consequently, water moves into the cells, generating turgor pressure, which provides rigidity to the nonwoody plant body. **(a)** Photograph of a begonia and drawing of a single normal cell in hypotonic surroundings. Note that a salt solution is being added to the pot. **(b,c)** Exposing begonia to a salt solution causes it to wilt and eventually die, due to the hypertonic environment. When cells are placed in a hypertonic solution, water moves out of them and their cytoplasm shrinks. Such cells have lost their turgor pressure. (*Dennis Drenner*)

to be **hypertonic** (from the Greek *hyper,* "over") to the cell. In such a situation, water flows out of the cell and into the surrounding solution. Suppose, for example, that cells are placed in a concentrated sugar solution. Because the sugar solution has more solute molecules than the cell, it has proportionately *fewer* water molecules than the cell's cytoplasm. Water passes freely through the plasma membrane from the cell to the surrounding sugar solution, but the sugar does not move from the solution into the cell, because it cannot get through the membrane. Because water always moves from an area of higher concentration (of water) to an area of lower concentration (of water), water moves out of the cell, and the cytoplasm shrinks.

When a cell is placed in a solution with a solute concentration lower than that within the cell, the solution is said to be **hypotonic** (from the Greek *hypo,* "under") to the cell. In such a situation, water flows into the cell from the surrounding solution. For example, suppose cells are placed in distilled water—that is, 100 percent water with no solute molecules. In this situation, the cell contains more solute molecules and has proportionately *fewer* water molecules than the water it has been placed in. Because water passes freely through the plasma membrane and always moves from an area of higher concentration (of water) to an area of lower concentration (of water), water moves into the cell, and the cell swells.

Normally, the roots of plants are exposed to soil water that is actually a very dilute solution of mineral salts (Figure 3-15). Because soil water has a lower concentration of solutes than the cytoplasm of a root cell, the soil-water solution that bathes plant roots is hypotonic to the root cells. As water moves into the root cells, their cell walls enable them to withstand the building pressure caused by the incoming water. As the internal pressure, known as **turgor pressure**, increases, an equilibrium is reached in which the turgor pressure forces water molecules out of the cell in numbers equal to that of the molecules coming in by osmosis.

Some Substances Cross Membranes by Facilitated Diffusion and by Active Transport

In **facilitated diffusion**, materials diffuse from a region of higher concentration to a region of lower concentration through special passageways in the membrane (Figure 3-16). These passageways are actually membrane proteins called **carrier proteins**, which function as conveyor belts. Facilitated diffusion enhances the diffusion of certain materials such as ions through cell membranes, but *always* in the direction of the concentration gradient—that is, from high to low concentration of that material. Without facilitated diffusion, these materials would not be able to move through the plasma membrane.

Active transport is the assisted movement of a substance from a *lower* concentration to a *higher* concentration of that substance (Figure 3-16). In other words, during active transport substances move *against* the concentration gradient.

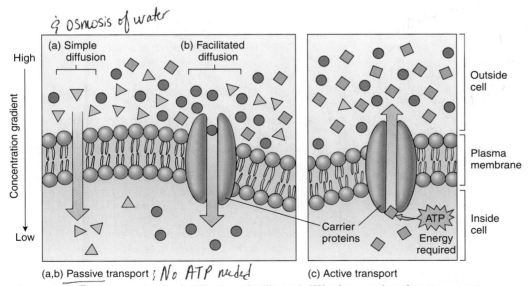

Figure 3-16 A comparison of diffusion, facilitated diffusion, and active transport. (a) In simple diffusion, materials move directly through the lipid bilayer from high to low concentration. **(b,c)** Carrier proteins are involved in transporting materials across a membrane in both facilitated diffusion and active transport. **(b)** The movement of materials in facilitated diffusion is along the concentration gradient, from high to low concentration. **(c)** The movement of materials in active transport is *against* the concentration gradient, from low to high concentration. The cell expends energy for active transport.

Working against the concentration gradient requires a direct expenditure of energy, usually supplied by ATP, the cell's energy-carrying molecule. Active transport occurs with the assistance of carrier proteins in the membrane that move the substance from one side of the membrane to the other.

Why does the cell expend energy for active transport? Cells require some materials—potassium, K^+, for example— in greater concentrations than are found in their surroundings. Active transport allows these materials to build up inside the cell. Other materials, such as Na^+ and H^+, are found in the surroundings in greater concentrations than cells can tolerate. They move into the cell by diffusion or facilitated diffusion and are then pumped out of the cell by active transport.

STUDY OUTLINE

I. Organisms are composed of cells. The cell is considered the basic unit of life because it is the smallest self-sufficient unit of living material.

II. Two fundamentally different types of cells exist, prokaryotic and eukaryotic.
  A. A prokaryotic cell has a simple internal organization; it lacks a nucleus and other membrane-bounded organelles.
  B. A eukaryotic cell possesses many cellular organelles, including a nucleus.

III. A eukaryotic cell has a complex structure with specialized organelles that perform specific functions.
  A. The plasma membrane separates the cell from its surroundings and controls what materials enter and leave the cell.
  B. The nucleus is the cell's control center. It is bounded by a double-layered nuclear envelope that encloses one or more nucleoli (where ribosomes are assembled) and genetic material (DNA) associated with protein, to form chromatin.
  C. Plastids are found in algal and plant cells.
    1. Chloroplasts contain chlorophyll and are the sites of photosynthesis, the process in which plants use the energy of light to convert carbon dioxide and water to carbohydrates such as glucose.
    2. Leucoplasts are colorless plastids that form and store starch, oils, or proteins.
    3. Chromoplasts contain yellow, orange, and red pigments and provide color to certain flower petals and ripened fruits.
  D. Mitochondria are the cell's powerhouses; they perform cellular respiration, in which the chemical energy in fuel molecules is transferred to ATP.
  E. Ribosomes are sites of protein synthesis.
  F. The endoplasmic reticulum (ER) is a system of internal membranes that is the site of enzymatic activity and that synthesizes membranes such as the nuclear envelope and the Golgi apparatus.
    1. Smooth ER is the site of lipid synthesis.
    2. Rough ER is studded with ribosomes and is associated with protein synthesis.
  G. The Golgi apparatus collects, modifies, and packages material for export from the cell.
  H. Vacuoles are large, membrane-bounded sacs that contain dissolved salts, ions, pigments, and waste products.

  I. Microtubules and microfilaments form the cell's cytoskeleton; they maintain the shape of the cell and have a role in cellular movement.
  J. The cell wall, composed largely of cellulose, supports and protects the cell.

IV. Plant and animal cells have the following structures in common: plasma membrane, nucleus, mitochondria, ribosomes, ER, Golgi apparatus, and cytoskeleton.
  A. In addition, plant cells possess plastids, cell walls, and large vacuoles.
  B. Animal cells have centrioles and lysosomes, which plant cells lack.

V. The fluid mosaic model explains membrane structure. According to this model, each membrane is composed of a phospholipid bilayer in which varying proteins are embedded.
  A. The nonpolar, hydrophobic fatty acid chains of the phospholipids project into the interior of the double-layered membrane.
  B. The polar, hydrophilic heads are located on the two surfaces of the double-layered membrane.

VI. Membranes are selectively permeable and allow the passage of some materials but not others.
  A. Diffusion is the net movement of a substance from an area of higher concentration to an area of lower concentration, along the concentration gradient.
  B. Osmosis is the diffusion of water across a selectively permeable membrane.
    1. An isotonic solution has a solute concentration identical to that of the solution with which it is compared.
    2. A hypertonic solution has a solute concentration greater than that of the solution with which it is compared.
    3. A hypotonic solution has a solute concentration less than that of the solution with which it is compared.
  C. In facilitated diffusion, a carrier protein helps move a material across a membrane in the direction of the concentration gradient, from high to low concentration.
  D. In active transport, energy is expended to move a material across a membrane against the concentration gradient, from low to high concentration.

SELECTED KEY TERMS

active transport, p. 49
cell wall, p. 38
chloroplast, p. 41
concentration gradient, p. 47
cytoplasm, p. 38
cytoskeleton, p. 45
diffusion, p. 47

endoplasmic reticulum (ER),
   p. 43
endosymbiont theory, p. 42
eukaryotic cell, p. 38
fluid mosaic model, p. 46
Golgi apparatus, p. 44
hydrophilic, p. 47

hydrophobic, p. 47
hypertonic, p. 49
hypotonic, p. 49
isotonic, p. 48
mitochondrion, p. 43
nucleolus, p. 41

nucleus, p. 38
osmosis, p. 48
plasma membrane, p. 38
prokaryotic cell, p. 38
ribosome, p. 43
vacuole, p. 45

REVIEW QUESTIONS

1. Distinguish between a microscope's magnification and its resolving power.
2. How are mitochondria and chloroplasts alike? How are they different?
3. A plant cell secretes a protein. Trace the protein's production, packaging, and release from the cell.
4. The animal body obtains support and strength from its skeletal system; plants lack skeletons and yet can attain great size. Explain this phenomenon for both woody and nonwoody plants.
5. Describe the fluid mosaic model of membrane structure. Make sure you use the following terms: hydrophilic, hydrophobic, phospholipid, protein, bilayer, fluid, mosaic.
6. Describe what happens to a plant cell placed in a hypertonic solution, and why.
7. Carrier proteins are used in both facilitated diffusion and active transport, yet these processes are very different. Explain their differences.
8. Label the following diagram of a plant cell. Check your answers against Figure 3-4.

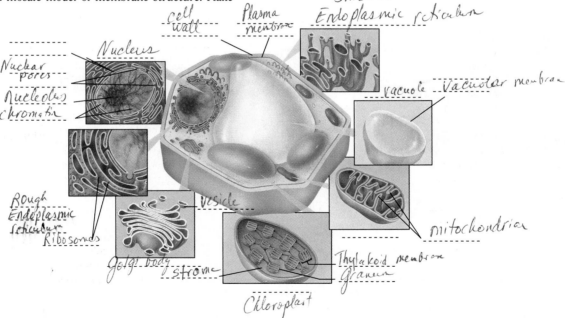

Smooth
Endoplasmic reticulum

Cell wall

Plasma membrane

Nucleus

Nuclear pores

Nucleolus

chromatin

Vacuole   Vacuolar membrane

mitochondria

Rough Endoplasmic reticulum

Ribosomes

Vesicle

Golgi body

stroma

Thylakoid membrane

Granum

Chloroplast

9. Label the following diagram of a membrane. Check your answers against Figure 3-13.

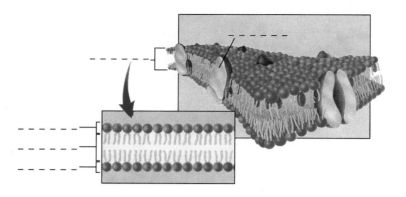

THOUGHT QUESTIONS

1. In a related experiment involving *Acetabularia* (discussed in the chapter introduction), the caps are removed from two species, and the bases are joined together; they subsequently regenerate a common cap.

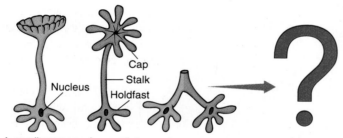

A. mediterranea   A. crenulata

Predict the appearance of the cap that regenerates. State your answer in the form of a hypothesis.

2. What would be the fate of a mutant cell that lacked mitochondria? Why?
3. What would be the fate of a plant cell that possessed a plasma membrane but not a cell wall? Why?
4. What would be the fate of a plant cell that possessed a cell wall but not a plasma membrane? Why?
5. Membranes have been described in a number of ways. Explain how each of the following analogies is both accurate and inaccurate.
   a. Protein icebergs floating in a lipid sea
   b. A block of Jello with fruit chunks
   c. A piece of cake with frosting layered on the top and bottom
6. Why do carrot and celery sticks become limp after a period of time? How could they be made crisp once more? Explain your answers in terms of turgor pressure.

SUGGESTED READINGS

de Duve, C., *A Guided Tour of the Living Cell,* New York, Scientific American Library, 1986. An illustrated tour of the cell in the form of journeys through different membrane and organelle systems.

Goodsell, D.S., "A Look Inside the Living Cell," *American Scientist* Vol. 80, No. 5, 1992. A fascinating effort to visualize how biological molecules actually look in their natural environment, the cell.

Solomon, Berg, Martin, and Villee, *Biology,* 4th ed., Philadelphia, Saunders College Publishing, 1996. Chapter 4 contains a detailed focus box on *Acetabularia.*

Wayne, R., "Excitability in Plant Cells," *American Scientist* Vol. 81, March–April 1993. A discussion of how modern technology, including the patch clamp technique, is being used to study ion movements in plants.

METABOLISM IN CELLS

R ice (*Oryza sativa*) is one of the most important human foods. It is a staple crop for half the world's population, particularly people in the humid tropics and subtropics of Asia, Africa, and Latin America. Rice, which is usually dehusked and boiled before it is eaten, feeds more people than any other crop.

A rice plant grows 3 to 5 feet tall and produces a foot-long terminal cluster containing multiple seeds, or grains. Like wheat and corn, two other very important human foods, rice is a cereal crop. Cereals are grasses that are cultivated for their hard, dry "seeds." (Although rice kernels and other cereal grains are called seeds, botanically they are fruits.) Like other cereals, rice is an annual; that is, it grows, flowers, sets seed, and dies in one growing season. Rice must therefore be planted each year.

Rice has been cultivated for thousands of years, but the exact location of its domestication from wild rice is unknown. It is thought

Rice heads ready to be harvested. Rice is one of the world's most important food crops. (*The Stock Market/Thomas del Brase*)

that rice cultivation began in Southeast Asia (Thailand), China, or India. Ancient Chinese writings some 5000 years old describe the importance of rice as a food crop, as do ancient Hindu texts in India.

Rice is unusual among important food crops in that it is generally grown submerged in 4 to 6 inches of water.

Most crops would die under those conditions, because their roots would not get enough oxygen in the waterlogged soil. In contrast, rice roots thrive under water. Rice grown in watery paddies is known as lowland rice to distinguish it from upland rice, which is cultivated "dry" like other crops. (The terms "lowland" and "upland" have nothing to do with elevation.) Upland rice crops do not yield as much rice per acre as lowland crops, and more than 80 percent of the world's rice is lowland rice.

Although rice is nutritious, it contains a relatively small amount of protein (about 10 percent). As a result, a balanced diet with rice as the principal food must be supplemented with meat, fish, or a protein-rich plant such as beans. Modern milling procedures, which strip off the protein- and vitamin-rich bran and germ (embryo), thereby converting brown rice to white rice, further reduce the nutritional value of rice. Although many people prefer the taste and appearance of white rice over those of brown rice, white rice is far less nutritious. White rice consists almost entirely of carbohydrates, which provide the body with necessary calories but lack essential nutrients.

In this chapter we examine photosynthesis, the complex chemical process by which rice and other plants capture energy from their environment and use it to manufacture carbohydrates. By one estimate, photosynthesis worldwide produces more than 100 times the food required by humans.

This "food" supports not only humans but almost the entire biosphere—its plants, animals, fungi, protists, and bacteria. Chapter 4 also considers cellular respiration, the essential process by which plants and other organisms release energy from carbohydrates and other fuel molecules.

LEARNING OBJECTIVES

After reading this chapter, you should be able to:

1. *Describe how photosynthesis is important not only to plants but to the entire web of life on planet Earth.*
2. *Write a summary reaction for photosynthesis, explaining the origin and fate of each substance involved.*
3. *Summarize the events of the light-dependent reactions of photosynthesis, including the role of light in the activation of chlorophyll. Describe how a proton gradient allows the formation of ATP according to the chemiosmotic model.*
4. *Summarize the events of the light-independent reactions of photosynthesis.*
5. *Write a summary reaction for aerobic respiration, giving the origin and fate of each substance involved.*
6. *List and give a brief overview of the four stages of aerobic respiration.*

METABOLISM INCLUDES ANABOLIC REACTIONS AND CATABOLIC REACTIONS

The chemical processes that occur in a cell are collectively referred to as its **metabolism**. Two kinds of chemical reactions occur in a cell's metabolism. **Anabolic reactions** (also called **anabolism**) are chemical reactions in which energy is stored in complex molecules. Anabolism is a building-up process in which large, complex molecules are synthesized from simpler molecules. Photosynthesis, in which carbon dioxide (CO_2) and water are used to synthesize glucose molecules, is an example of anabolism. **Catabolic reactions** (also called **catabolism**) are chemical reactions in which energy is released from molecules. Catabolism is a breaking-down process in which complex molecules are split apart, or degraded, into simpler ones to release energy. The cellular respiration of glucose, in which glucose is broken down into CO_2 and water with the release of biologically useful energy, is an example of catabolism.

Oxidation-Reduction Reactions Occur in Metabolism

The processing of energy by cells involves a transfer of energy through the flow of electrons. **Oxidation** is a chemical process in which a substance *loses* electrons, and **reduction** is a chemical process in which a substance *gains* electrons. An oxidation reaction is always accompanied by a reduction reaction because there are no free electrons in living cells—all electrons are in atoms. If a substance gains an electron in a

reduction reaction, that electron must come from another substance that has lost an electron.

As quickly as electrons are released, they are accepted by another atom or molecule. *At the same time as the oxidized molecule gives up electrons, it also gives up energy; the reduced molecule receives energy when it gains electrons.* Oxidation-reduction reactions are characteristic of many cellular processes including photosynthesis and cellular respiration. Generally, a sequence of oxidation-reduction reactions takes place as electrons associated with hydrogen are transferred from one compound to another.

Electrons Associated with Hydrogen are Transferred to Electron Acceptor Molecules

You may recall from Chapter 2 that when atoms share a pair of electrons, a covalent bond forms. Because such bonds are stable, it is difficult to remove electrons from covalent compounds unless an entire atom is removed. In a cell, oxidation usually involves the removal of a hydrogen atom and its single electron from a compound. Reduction usually involves a gain of a hydrogen atom (and thus a gain in an electron).

When hydrogen atoms are removed from an organic compound, they take with them some of the energy that was stored in their electrons. The electrons are transferred to an **electron acceptor molecule**, which temporarily accepts them until they move along to the next acceptor molecule. As we will see shortly, oxidation-reduction reactions and electron acceptor molecules are of crucial importance to both photosynthesis and cellular respiration.

PHOTOSYNTHESIS CAPTURES RADIANT ENERGY THAT SUSTAINS THE BIOSPHERE

For billions of years, almost all life in the biosphere has run on solar energy. Of all organisms, however, only plants, algae, and certain bacteria are capable of absorbing and converting light energy from the sun to chemical energy through the process of **photosynthesis** (Figure 4-1). The end products of photosynthesis are carbohydrates (formed from the simple raw materials water and carbon dioxide) and oxygen (O_2).

Thus, plants are of central importance to life. Without plants and other photosynthetic organisms, there would be no energy for animals or other nonphotosynthetic organisms. Animals, humans included, depend on plants for food, which provides energy, and for oxygen.

Humans also burn **biomass**, which is any organic material used as fuel (see Plants and the Environment: The Environmental Effects of Using Biomass for Fuel). Biomass, which includes wood, agricultural wastes, and fast-growing plants, contains chemical energy, the source of which can be traced back to energy from the sun, which photosynthetic organisms used to form the organic molecules of biomass.

LIGHT EXHIBITS PROPERTIES OF BOTH WAVES AND PARTICLES

To understand how plants convert light energy to chemical energy in photosynthesis, it is necessary to know a little about the nature of light. Light is a very small portion of the electromagnetic spectrum, which is a vast, continuous range of

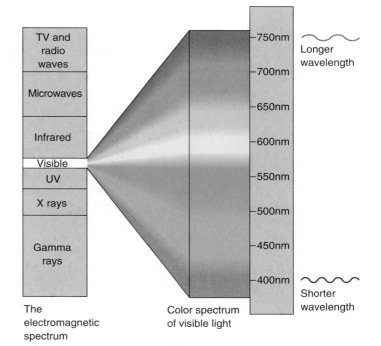

Figure 4-2 The electromagnetic spectrum. Visible light is only a portion of the electromagnetic spectrum and consists of a mixture of wavelengths (380 to 760 nm). The energy of visible light is used in photosynthesis.

electromagnetic radiations propagated through space and matter (Figure 4-2). All radiations in the electromagnetic spectrum behave as though they travel in waves. A **wavelength** is the distance from one wave peak to the next. At one end of the spectrum are gamma rays with extremely short wavelengths, measured in nanometers (nm), billionths of a meter.

 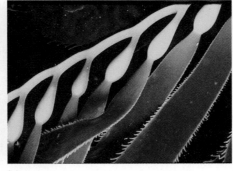 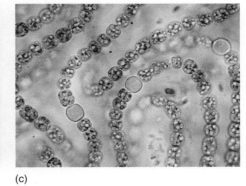

(a)   (b)   (c)

Figure 4-1 Photosynthetic organisms include (a) plants, such as the prayer plant (*Maranta* sp): (b) algae, such as this kelp (*Macrocystis pyrifera*); and (c) *cyanobacteria* (*Nostoc* sp). Plants are primarily terrestrial, whereas algae are primarily aquatic. Algae range in size and complexity from microscopic single-celled organisms to large seaweeds that make up underwater "forests." Cyanobacteria are prokaryotic organisms that photosynthesize as plants and algae do. (a, *James L. Castner*; b, *Visuals Unlimited/D. Gotshall*; c, *Visuals Unlimited/R. Calentine*)

PLANTS
AND THE
Environment

The Environmental Effects of Using Biomass for Fuel

Humans burn biomass fuel, which can be a solid, liquid, or gas, to release its energy. Solid biomass, such as wood, is burned directly for energy. Biomass can also be converted to liquid fuels such as methyl alcohol and ethyl alcohol, which can then be burned in internal combustion engines. However, a major disadvantage of alcohol fuels is that 30 to 40 percent of the energy in the starting material is lost during the conversion to alcohol. Some experts have calculated that alcohol production from a crop such as corn consumes *more* energy than it yields, when all energy costs associated with growing and harvesting a crop for alcohol production are assessed.

Biomass—particularly firewood, charcoal (wood that has been turned into coal by partial burning), animal dung, and peat (partly decayed plant matter found in bogs and swamps)—meets a substantial portion of world-wide energy needs (see figure). At least half of the world's people rely on biomass as their main source of energy. In developing countries, for example, wood is the primary fuel for cooking.

In many areas of the world, wood is burned faster than it can be replaced. Intensive use of wood has resulted in severe damage to the environment, including soil erosion, deforestation (loss of forests), and desertification (expansion of deserts). (See Chapters 6 and 7 for discussions of deforestation and desertification.)

Air pollution, particularly CO_2 emissions, is a problem with the use of biomass as a fuel. (See Chapters 8 and 27 for discussions of air pollution and CO_2 emissions.) However, burning biomass produces less sulfur and ash than burning certain other fuels, such as coal. Tree planting can offset the CO_2 released into the atmosphere by biomass combustion. As trees photosynthesize, they absorb atmospheric CO_2 and lock it up in organic molecules that make up the body of the tree. Thus, if biomass is regenerated to replace the biomass burned as a fuel, there is no net contribution of CO_2 to the atmosphere.

Biomass production requires land and water. Because the use of agricultural land for energy crops competes with the cultivation of food crops, shifting the balance toward energy production might decrease food production, leading to higher food prices. For this reason, some scientists are interested in the commercial development of certain desert shrubs that produce oils that could be used for fuel. Such shrubs do not require prime agricultural land, although the growers would have to take care to ensure that the desert soils were not degraded or eroded by overuse.

Women in Rwanda gather firewood, a form of biomass. Firewood is the major energy source for most of the developing world. (*Robert E. Ford/Biological Photo Service*)

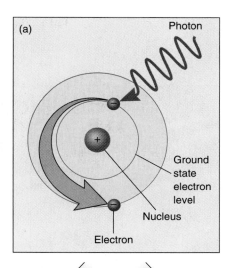

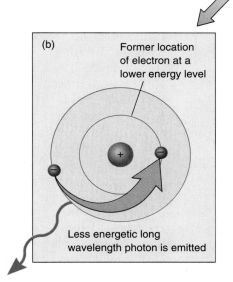

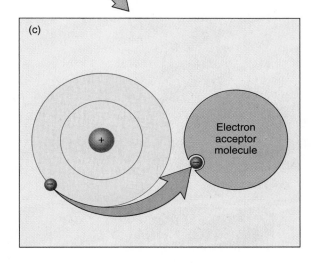

Figure 4-3 **Light excites certain types of biological molecules, moving electrons into higher energy levels.** **(a)** When a photon of light energy strikes an atom or the molecule of which the atom is a part, the energy of the photon may push an electron to a higher energy level (that is, to an orbit farther from the nucleus). **(b)** If the electron returns to the next lower energy level, a less energetic photon is emitted. **(c)** Alternatively, the electron may be accepted by an electron acceptor molecule. In photosynthesis, an electron acceptor molecule captures the energetic electron and passes it along an electron transport chain.

At the other end of the spectrum are radio waves with wavelengths so long that they are measured in full meters.

Within the spectrum of visible light (wavelengths between 380 and 760 nanometers), violet light has the shortest wavelength and red light the longest wavelength. Ultraviolet radiation, which is invisible to the human eye, has a shorter range of wavelengths than visible light, and infrared radiation, also invisible, has a longer range.

Light behaves as though it is composed not only of waves but also of discrete packets or particles of energy, which are called **photons**. The amount of energy in a photon depends on the wavelength of light. The shorter the wavelength, the more energy per photon; the longer the wavelength, the less energy per photon.

Photosynthesis depends on *visible light* rather than some other wavelength of radiation. This may be so because most of the radiation reaching our planet from the sun is within this portion of the electromagnetic spectrum. Only radiation in the visible-light portion of the spectrum excites certain

types of biological molecules, causing their electrons to jump into higher energy levels. Wavelengths of radiation that are longer than visible light (for example, microwaves, television waves, and radio waves) do not possess enough energy to excite biological molecules. Wavelengths shorter than visible light (for example, ultraviolet radiation, x-rays, and gamma rays) possess so much energy that they disrupt biological molecules by breaking chemical bonds.

The interaction between photons and atoms depends on the arrangement of electrons in the atoms. Recall from Chapter 2 that an atom consists of an atomic nucleus (protons and neutrons) surrounded by electrons positioned in one or more energy levels. The lowest energy state an electron possesses is called the **ground state**, but energy can be added to an electron to boost it to a higher energy level. When an electron is raised to a higher energy level than its ground state, the electron is said to be **excited**.

When an electron is raised to a higher energy level by absorbing light (Figure 4-3), one of two things can happen.

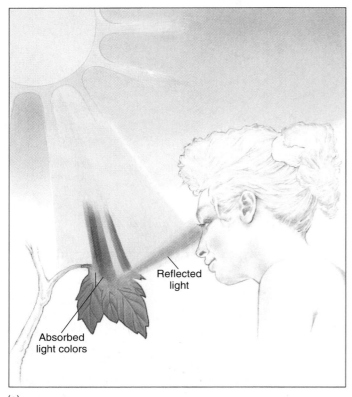

(a)

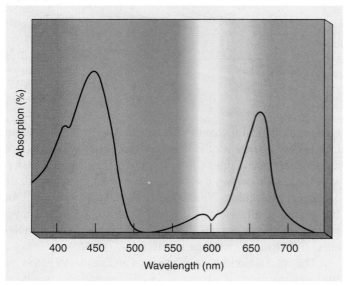

(b)

Figure 4-4   Why chlorophyll is green. (a) Leaves appear green because when light strikes a leaf, most of the colors composing the light are absorbed by chlorophyll. Since chlorophyll does not absorb most green light, the green light is reflected off the leaf to our eyes. **(b)** The absorption spectrum of chlorophyll *a*. This graph was obtained by exposing chlorophyll *a* to different wavelengths of light and measuring how much light it absorbed at each wavelength.

(1) The electron can return to its ground state. When it does, the "extra" energy is released as heat or as light of a longer wavelength (that is, light with less energy). (2) Instead of returning to its ground state, an excited electron may leave the atom. In this instance, the electron may be accepted by an electron acceptor molecule, which is reduced in the process. As we will see shortly, this is what occurs in photosynthesis.

In Plants and Algae, Photosynthesis Takes Place in Chloroplasts

In plants and algae, photosynthesis occurs in **chloroplasts**, cellular organelles that contain molecules of **chlorophyll**, which absorb light. Like other pigments (colored substances that absorb light), chlorophyll does not absorb all wavelengths (colors) of light in the same amount. Whatever wavelengths are not absorbed by the pigment may be reflected. Chlorophyll is a *green* pigment, which means that it absorbs and uses light primarily in the violet, blue, and red regions of the visible spectrum and reflects most of the green light that strikes it. As a result, chlorophyll molecules and the plant cells that contain them (primarily in a plant's leaves and stems) appear green in color (Figure 4-4).

There are several kinds of chlorophyll, the most common of which are chlorophyll *a* and chlorophyll *b*. Plants also have other photosynthetic pigments, such as the yellow and orange **carotenoids**, which absorb wavelengths of light different from those absorbed by chlorophyll. The large quantity of chlorophyll in most leaves usually masks the presence of carotenoids in spring and summer; in autumn, when the chlorophyll breaks down, other pigments, including carotenoids, become visible.

In Photosynthesis, Plants Convert Light Energy to the Chemical Energy Stored in Sugar Molecules[1]

The principal raw materials for photosynthesis are water and carbon dioxide. The energy that chlorophyll molecules absorb from sunlight is expended to split water, which releases oxygen (Figure 4-5) and hydrogen (that is, the electrons and protons of hydrogens). Hydrogen joins with carbon dioxide to form the sugar glucose, a six-carbon carbohydrate. Although photosynthesis is a complex process composed of many steps, it can be summarized as follows:

---

[1]This section provides a very brief and general description of the reactions of photosynthesis. For a more detailed approach, see the section beginning on page 59.

$$6CO_2 + 12H_2O \xrightarrow{\text{Light energy, chlorophyll}} C_6H_{12}O_6 + 6O_2 + 6H_2O$$

Carbon dioxide   Water   Glucose   Oxygen   Water

This equation describes *what* happens during photosynthesis but not *how* it happens. The reactions of photosynthesis occur in two stages: the light-dependent reactions and the light-independent reactions.

The Light-Dependent Reactions Capture Energy

The light-dependent reactions, which take place only in the presence of light, occur in the thylakoids of the chloroplast. During this phase of photosynthesis, several important events occur.

1. Chlorophyll absorbs light energy, which triggers a flow of energized, excited electrons from the chlorophyll molecule.
2. Some of the energy of the excited electrons is transformed into chemical energy and used to make **adenosine triphosphate (ATP)**, a temporary energy storage molecule.
3. Some of the light energy trapped by the chlorophyll is used to split water, with oxygen from the water being released as molecular oxygen (O_2). Some of the oxygen may be used by the plant for cellular respiration, in which fuel molecules are broken down to release biologically useful energy, but most of it is released into the atmosphere.
4. The electrons and protons of hydrogens (from the water) combine with **nicotinamide adenine nucleotide phosphate (NADP$^+$)**, forming **NADPH** (reduced NADP$^+$).[2] Again, as in step 2, some of the energy of the excited electrons is converted to chemical energy and stored temporarily in NADPH.

Thus, *in the light-dependent reactions, the energy from sunlight is used to make ATP and NADPH.* Some of the captured energy of sunlight is temporarily stored within these two compounds. Although ATP and NADPH are sources of quick energy for various metabolic functions, neither is useful for the long-term storage of chemical energy. Cells possess very limited amounts of the materials from which ATP and NADPH are made, so neither molecule can accumulate in large enough quantities to ensure a continual, long-term supply of energy.

Note that carbon dioxide is not used in the light-dependent reactions of photosynthesis, nor is glucose produced. These steps take place in the next phase of photosynthesis.

Figure 4-5   On sunny days, the oxygen (O_2) released by aquatic plants is sometimes visible as bubbles in the water. This plant (*Elodea*) is actively carrying on photosynthesis, as evidenced by the oxygen bubbles. (*Visuals Unlimited/Bernard Wittich*)

The Light-Independent Reactions Fix Carbon

During the light-independent reactions of photosynthesis, glucose is produced from carbon dioxide. The bonds that hold glucose molecules together make it a source of energy for the cell that, unlike ATP and NADPH, can be produced in large quantities and stored for future use.

The light-independent reactions depend on the products of the light-dependent reactions. The energy stored in ATP and NADPH during the light-dependent phase of photosynthesis is used to form carbohydrate molecules from CO_2 in a process called **CO_2 fixation**. In other words, during the light-independent reactions of photosynthesis, the energy in the chemical bonds of ATP and NADPH is transferred to the chemical bonds of sugar molecules.

The reactions of the light-independent phase of photosynthesis are known as the **Calvin cycle**, named after Melvin Calvin, who first described it (Figure 4-6). Calvin was awarded a Nobel Prize in 1961 for this significant scientific contribution. The enzymes for the Calvin cycle are in the stroma of the chloroplast.

PHOTOSYNTHESIS IS A COMPLEX PROCESS

The brief overview of photosynthesis in the preceding section is sufficient for some introductory botany courses. The following material provides greater detail.

Photosystems I and II Are Light-Harvesting Units of the Light-Dependent Reactions

According to current scientific evidence, chlorophyll molecules and associated electron acceptor molecules are physically organized in thylakoid membranes into units called **photo-**

---

[2]Although the correct way to write the reduced form of NADP$^+$ is NADPH + H$^+$, for simplicity's sake we present the reduced form as NADPH throughout the chapter.

Figure 4-6 Calvin's classic experiment that demonstrated the steps in the light-independent reactions of photosynthesis, also known as the Calvin cycle. Melvin Calvin and his colleague Andrew Benson grew algae in the green "lollipop." Radioactively labeled $^{14}CO_2$ was bubbled through the algae, which were periodically killed by dumping the "lollipop" contents into a beaker of boiling alcohol. By identifying which compounds contained the radioactive ^{14}C at different times, Calvin determined the steps of CO_2 fixation in photosynthesis. (*Melvin Calvin, University of California, Berkeley*)

systems, of which there are two types. *Photosystem I* contains a reactive pigment (probably a special form of chlorophyll *a*) that is known as P700 because it absorbs far-red light with a wavelength of 700 nm very strongly. *Photosystem II* uses a reactive pigment, P680 (also probably a special form of chlorophyll *a*), which strongly absorbs red light at a wavelength of 680 nm. Photosystems I and II are joined together by an electron transport chain—that is, a chain of alternately oxidized and reduced compounds.

All the chlorophyll molecules of a photosystem apparently serve as antennae that gather solar energy (Figure 4-7). Once absorbed, the energy is passed from one chlorophyll molecule to another within the photosystem until it reaches a P700 or P680 pigment molecule, referred to as the **reaction center**. Only the reaction center can give up energized electrons to an electron acceptor molecule.

As you read the following details of the light-dependent reactions, refer to the diagram in Figure 4-8. The energized

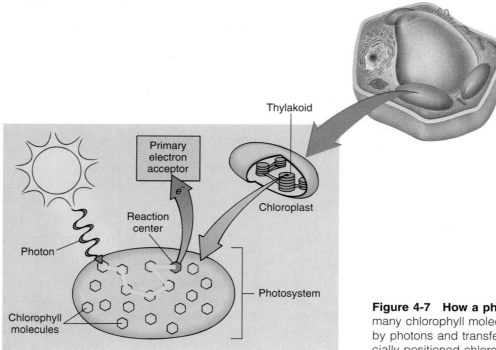

Figure 4-7 How a photosystem traps light energy. The many chlorophyll molecules in the photosystem are excited by photons and transfer their excitation energy to the specially positioned chlorophyll molecule at the reaction center.

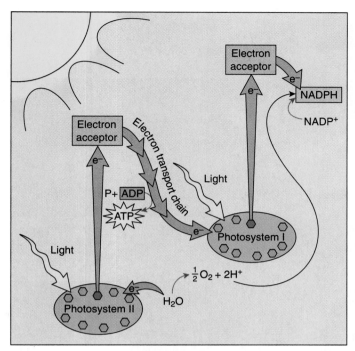

Figure 4-8 Light-dependent reactions (noncyclic photophosphorylation). When photosystem I is activated by absorbing photons, electrons are passed along an electron transport chain and eventually donated to NADP$^+$. Photosystem II, also activated by light energy, passes energized electrons along a series of acceptor molecules to replace the ones given up by photosystem I. Photosystem II is also linked to the splitting of water and the production of O$_2$. By following the blue arrows starting at H$_2$O, you can see that the flow of energized electrons is in one direction, from the electrons of hydrogen (when water is split) to the formation of NADPH. ATP formation is also connected to the electron flow, as the energy released from the energized electrons is used to establish a proton gradient that produces ATP.

electrons that leave photosystem I are transferred from one electron acceptor molecule to another through an electron transport chain until they are finally accepted by NADP$^+$. When NADP$^+$ accepts the electrons, they unite with protons (H$^+$) in the chloroplast to form hydrogen, resulting in the formation of NADPH.

Like photosystem I, photosystem II is activated by photons, and the chlorophyll molecule in its reaction center passes electrons along from one electron acceptor molecule to another through an electron transport chain. As they are transferred along this chain, the electrons lose some of their energy. Some of the energy released in this way is used to synthesize ATP (see next section). Electrons emitted from photosystem II are eventually donated to photosystem I.

The electrons that leave photosystem II are replaced by electrons from the hydrogen atoms in water. When P680 in the reaction center of photosystem II absorbs light energy, it becomes positively charged and exerts a strong pull on the

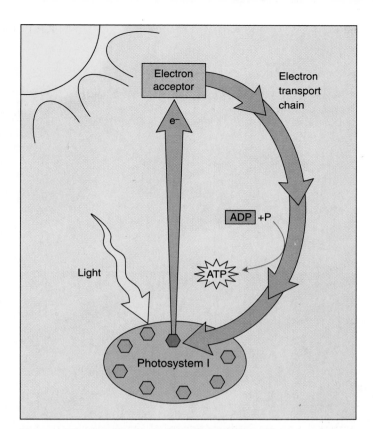

Figure 4-9 Cyclic photophosphorylation. Note that only photosystem I is involved. The energized electrons that leave photosystem I pass along a series of electron acceptor molecules back to the reaction center of photosystem I (*follow the blue arrows*). The energy that is released as the electrons pass along the electron transport chain is used for ATP synthesis. Water is not split in cyclic photophosphorylation, and neither O$_2$ nor NADPH is produced.

electrons in water molecules. Water is split into its components: protons (H$^+$), electrons, and oxygen. Most of the oxygen split from the water is released into the atmosphere as molecular O$_2$.

The entire process just described, which produces NADPH and ATP, is known as **noncyclic photophosphorylation**. It is called photophosphorylation because electrons obtain energy from *photons* of light and then contribute that energy to the *phosphorylation* (addition of phosphate) of adenosine diphosphate (ADP), producing ATP. In other words, in photophosphorylation the energy of light is used to synthesize ATP. It is noncyclic because there is a one-way flow of electrons from water to NADP$^+$—that is, H$_2$O $\rightarrow$ photosystem II $\rightarrow$ photosystem I $\rightarrow$ NADP$^+$.

The light-dependent reactions of photosynthesis also include **cyclic photophosphorylation**, a "shortcut" version of noncyclic photophosphorylation that produces ATP. Only photosystem I is involved in cyclic photophosphorylation. In this pathway, the excited electrons in photosystem I do not go to NADP$^+$ but cycle back to photosystem I. In cyclic photophosphorylation, the energy from light is used to synthesize ATP, but water is not split, and as a result, NADPH is not produced and O$_2$ does not form (Figure 4-9). The biological significance of cyclic photophosphorylation in chloroplasts is unclear.

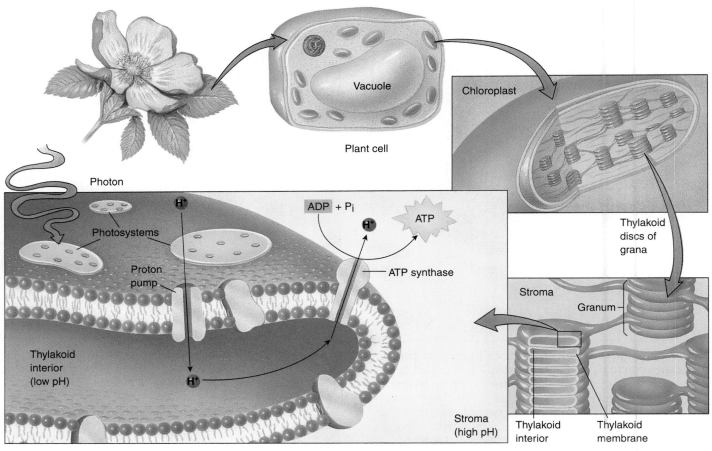

Figure 4-10   The production of ATP by chemiosmosis occurs in the thylakoid membrane. The high concentration of protons in the interior of the thylakoid provides energy (as those protons move across the membrane), which is used by ATP synthase to phosphorylate ADP.

The Chemiosmotic Model Explains ATP Synthesis During the Light-Dependent Reactions

Photosystems I and II, that is, the pigment molecules and electron transport chains involved in cyclic and noncyclic photophosphorylation, are embedded in the thylakoid membrane. Some of the energy released from electrons as they travel through the electron transport chain is used to pump protons (that is, H^+ ions) from the stroma of the chloroplast across the thylakoid membrane and into the center of the thylakoid. Thus, protons accumulate within the thylakoids, lowering the pH of the thylakoid interior and making it more acidic. The difference in concentration of protons between the thylakoid and the stroma, referred to as a **proton gradient**, represents potential energy that can be used to form ATP.

ATP synthase, an enzyme complex embedded in the thylakoid membrane, forms a channel through which the accumulated protons can leave the thylakoid. As the protons pass through ATP synthase, some of their energy is used to form ATP from ADP and inorganic phosphate. The synthesis

of ATP by first producing a proton gradient, using the energy released during electron transport, is called **chemiosmosis**[3] (Figure 4-10).

Fixation of Carbon Dioxide Occurs During Light-Independent Reactions

As you read the following details of the Calvin cycle, refer to the reactions illustrated in Figure 4-11. Let us begin with *ribulose phosphate (RP)*, a five-carbon sugar that has been activated by the addition of a phosphate group. In the first chemical reaction, ATP from the light-dependent reactions is expended to add a second phosphate to ribulose phosphate. This reaction converts ribulose phosphate to *ribulose bisphosphate (RuBP)*.

---

[3]The term "chemiosmosis" is somewhat misleading in that chemiosmosis has nothing to do with osmosis.

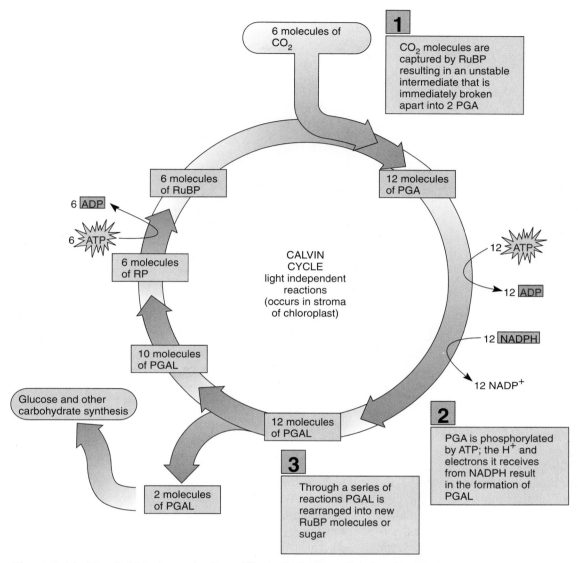

Figure 4-11   The light-independent reactions of photosynthesis, also known as the Calvin cycle. This diagram depicts six turns of the cycle. For every six turns, six molecules of CO_2 are converted to one molecule of a six-carbon sugar such as glucose. The energy that drives the Calvin cycle comes from the products of the light-dependent reactions (that is, from ATP and NADPH).

A key enzyme, **rubisco**,[4] then combines a molecule of CO_2 with RuBP. Instantly, this six-carbon molecule splits into two three-carbon molecules called *phosphoglycerate (PGA)*. Because the first detectable molecules to be formed contain three carbon atoms, the Calvin cycle is also referred to as the **C_3 pathway**. The PGA molecules are then converted to *phosphoglyceraldehyde (PGAL)* using NADPH and more ATP. (Keep in mind that the ATP and NADPH used throughout the Calvin cycle formed during the light-independent reactions.)

For every six turns of the Calvin cycle, 2 of the 12 PGAL molecules leave the pathway to be used in carbohydrate synthesis. Each three-carbon molecule of PGAL is essentially half of a six-carbon sugar molecule; PGAL molecules are joined in pairs to produce glucose or fructose.

In some plants, glucose and fructose then combine to form the disaccharide sucrose, or common table sugar. This we harvest from sugarcane, sugar beets, or maple sap. Plants also use glucose to produce the polysaccharides, cellulose (a constituent of plant cell walls) and starch (the principal storage compound in plants).

Note that although 2 of the 12 PGAL molecules formed during six turns of the cycle were removed from the cycle, 10

---

[4]Rubisco is an acronym for the enzyme ribulose bisphosphate carboxylase/oxygenase.

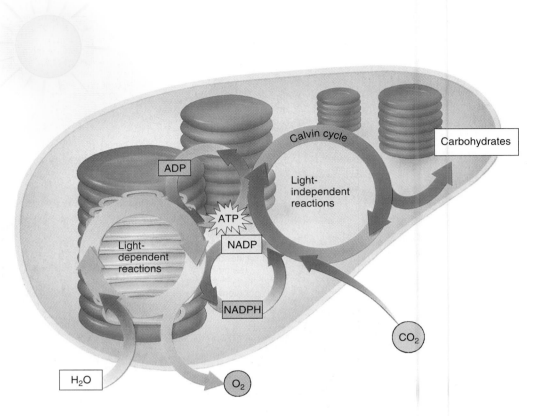

Figure 4-12   A summary of photosynthesis. Light energy is used to form ATP and NADPH during the light-dependent reactions of photosynthesis. These reactions also result in the splitting of water and the subsequent release of O_2. In the light-independent reactions of photosynthesis (that is, the Calvin cycle), CO_2 is fixed into glucose and other carbohydrate molecules. The energy required for the Calvin cycle is supplied by ATP and NADPH that were produced in the light-dependent reactions.

PGALs remain, containing 30 carbon atoms in all. Through a complex series of reactions, these 30 carbons and their associated atoms are rearranged into six molecules of the five-carbon compound ribulose phosphate, with which the Calvin cycle started. Each of these ribulose phosphate molecules is in a position to begin the process of CO_2 fixation and sugar production once again.

In summary, the inputs required for the light-independent reactions are CO_2, NADPH, and ATP (Figure 4-12). The light-dependent reactions of photosynthesis provide the ATP and NADPH. In the end, the sugar glucose is manufactured.

The C_3 Pathway Is Not Efficient, Because C_3 Plants Also Have Photorespiration

Many plants, including certain agriculturally important crops such as soybeans, wheat, and potatoes, do not yield as much carbohydrate from photosynthesis as might be expected. This reduced yield is especially significant during very hot spells in summer, which cause water stress in plants. Water-stressed plants lose more water by evaporation from their surfaces than the roots can replace by drawing from the soil. To con-

serve water, plants close the tiny pores (*stomata*) in the leaf surfaces through which CO_2 enters and O_2 leaves during photosynthesis. Once the stomata close, photosynthesis in the chloroplasts rapidly uses up the CO_2 remaining in the leaf, and the O_2 produced in photosynthesis accumulates in the chloroplasts.

When these conditions exist, a series of reactions involving ribulose bisphosphate (RuBP) occurs in which rubisco (the enzyme responsible for attaching CO_2 to RuBP in the Calvin cycle) binds RuBP to O_2 instead of to CO_2. When this occurs, some of the intermediate molecules involved in the Calvin cycle are degraded to CO_2 and H_2O instead of being synthesized into carbohydrates. This degradation process, called **photorespiration**, occurs during the daylight (hence the prefix "photo") and, like cellular respiration, requires oxygen and produces CO_2 and H_2O. Unlike respiration, however, photorespiration does *not* produce biologically useful energy.

Photorespiration reduces photosynthetic efficiency because it removes some of the molecules used in the Calvin cycle, thus reducing CO_2 fixation and carbohydrate formation. The reason plants carry out photorespiration is unknown. From a human viewpoint, the process is wasteful, particularly because it reduces the yields of important crop plants.

Many Plants with Tropical Origins Fix Carbon Using the C_4 Pathway

Not all plants have reduced photosynthetic efficiency due to photorespiration. Many plants with tropical origins, including crabgrass, corn, and sugarcane, have evolved ways to bypass photorespiration. One example is the **C_4 pathway**. The C_4 plants get their name from the fact that the first detectable molecule formed by CO_2 fixation is a four-carbon compound (produced by the joining of CO_2 to a three-carbon compound), rather than the three-carbon compound formed in the Calvin (C_3) cycle. This four-carbon compound is then rapidly transported to special cells, called **bundle sheath cells**, that surround veins (conducting tissue) of the leaf (Figure 4-13). In the bundle sheath cells, the CO_2 is removed from the four-carbon molecule and fixed into sugar by the regular C_3 pathway. The C_4 pathway in effect concentrates the amount

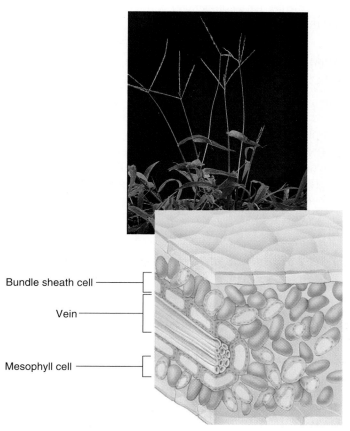

Bundle sheath cell ———

Vein ———

Mesophyll cell ———

Figure 4-13   Arrangement of cells in a C_4 leaf. In C_4 plants, such as crabgrass (*Digitaria* sp) reactions that fix CO_2 into four-carbon compounds take place in the chloroplast-containing mesophyll cells. Then the Calvin cycle occurs in the chloroplast-containing bundle sheath cells that surround the veins of the leaf. (*Dennis Drenner*)

of CO_2 in the bundle sheath cells, making it greater than it could possibly be as a result of diffusion of CO_2 from the atmosphere.

Photorespiration rarely occurs in C_4 plants, because the concentration of CO_2 in bundle sheath cells (where the enzyme rubisco works) is always high. Some scientists are attempting to transfer the C_4 pathway to crops such as soybeans and wheat. If they succeeded, these plants would be able to produce a lot more carbohydrates during hot weather, resulting in increased crop yields.

Many Desert Plants Fix Carbon Using the CAM Pathway

Plants living in almost continuously dry conditions have many special adaptations that enable them to survive. For example, their stomata may open during the cool night hours and close during the hot daylight hours to reduce water loss. (This feature is in contrast with most plants, which have stomata that open during the day and close at night.) Desert plants that close their stomata during the day, however, cannot exchange gases (CO_2 and O_2) during photosynthesis. (Most plants fix CO_2 during the day when sunlight is available to produce ATP and NADPH.)

Many desert plants have a special photosynthetic pathway called **crassulacean acid metabolism (CAM)** that in effect solves this dilemma. The name comes from the stonecrop plant family, the Crassulaceae, although it is only one of the more than 25 plant families that possess the CAM pathway.

CAM plants fix CO_2 during the night, when stomata are open for gas exchange, by combining it with a three-carbon compound to form a four-carbon compound. This compound is temporarily stored in the vacuoles of leaf cells. During the day, when stomata are closed and gas exchange cannot occur between the plant and the atmosphere, CO_2 is removed from the four-carbon compound. Thus, the CO_2 is available from the *leaf itself* to be fixed into sugar by the usual photosynthetic pathway, the C_3 pathway.

While CAM photosynthesis may sound familiar to you because it is similar to the C_4 pathway, there are important differences. The C_4 plants initially fix CO_2 into four-carbon compounds in leaf *mesophyll* cells. The four-carbon compounds are then transported to the bundle sheath cells, where the CO_2 is removed and fixed by the C_3 pathway. Thus, the C_4 and C_3 pathways occur in *different locations* within the leaf of a C_4 plant.

In CAM plants, the initial fixation of CO_2 occurs at night. Removal of CO_2 from the four-carbon compound and subsequent production of sugar from the CO_2 by the C_3 pathway occur during the day. In other words, the C_4 and C_3 pathways occur at *different times* within the same cell of a CAM plant.

The CAM pathway is a very successful adaptation to desert conditions. CAM plants can carry out gas exchange for photosynthesis *and* significantly reduce water loss during hot daylight hours. Plants with CAM photosynthesis can survive in deserts where neither C_3 nor C_4 plants can.

ENERGY CAN BE RELEASED VIA AEROBIC OR ANAEROBIC PATHWAYS

Every organism must extract energy from the organic fuel molecules that it either manufactures (for example, when plants photosynthesize) or captures from the environment (for example, when an animal eats another organism). These fuel molecules are transported to all the cells of a complex organism, where they can be broken down to provide the energy for cellular work.

Thus, every plant cell must extract energy from its long-term storage molecules—the organic fuel molecules it manufactures by photosynthesis. Within each plant cell, glucose and other fuel molecules are broken down by the process of **cellular respiration**, a series of chemical reactions that break apart fuel molecules and transfer the energy stored in their bonds to **adenosine triphosphate (ATP)** for use in cellular work.

Cells use three different catabolic pathways to extract energy from the fuel molecules they manufacture or ingest: aerobic respiration, anaerobic respiration, and fermentation (another type of anaerobic pathway). The type of environment a cell inhabits may determine which catabolic pathway it uses to break down fuel molecules. Cells that live in environments where oxygen is plentiful use an **aerobic** pathway, one that requires O_2. Cells that inhabit waterlogged soil or polluted water where oxygen is absent must use **anaerobic** pathways (either anaerobic respiration or fermentation) that do not require O_2.

IN AEROBIC RESPIRATION, CELLS BREAK DOWN FUEL MOLECULES INTO CARBON DIOXIDE AND WATER[5]

Plant cells and most other cells extract energy from fuel molecules, such as glucose, fatty acids, and other organic compounds, by using **aerobic respiration**. This process involves a long sequence of 30 or more chemical reactions, each regulated by a specific enzyme. During aerobic respiration, energy is released as fuel molecules are catabolized to CO_2 and water. One of the most common pathways of aerobic respiration involves the breakdown of the nutrient glucose. The overall reaction for the aerobic respiration of glucose can be summarized as follows:

$$C_6H_{12}O_6 + 6O_2 + 6H_2O \longrightarrow 6CO_2 + 12H_2O + Energy$$
Glucose   Oxygen   Water   Carbon dioxide   Water

(See Table 4-1 for a comparison of photosynthesis and aerobic respiration.)

Aerobic Respiration Occurs in Four Stages

The chemical reactions of aerobic respiration are grouped into four stages: (1) glycolysis, (2) formation of acetyl coenzyme A, (3) the citric acid cycle, and (4) electron transport and chemiosmosis (Figure 4-14).

1. **Glycolysis.** The term "glycolysis" is derived from two Greek words that, taken together, mean "splitting sugar." During glycolysis, the six-carbon glucose molecule is split into two three-carbon molecules of *pyruvic acid*. During this

---

[5]This section provides a very brief and general description of aerobic respiration. For a more detailed approach, see the section beginning on page 67.

TABLE 4-1   A Comparison of Photosynthesis and Aerobic Respiration

	Photosynthesis	Aerobic Respiration
Raw materials	CO_2, H_2O	$C_6H_{12}O_6$, O_2
End products	$C_6H_{12}O_6$, O_2	CO_2, H_2O
Plant cells that have these processes	Plant cells that contain chlorophyll	All plant cells
Organelles involved	Chloroplast	Cytoplasm, mitochondrion
Pathway of energy	Light energy $\rightarrow$ NADPH/ATP $\rightarrow$ energy stored in carbohydrate molecules	Energy stored in fuel molecules $\rightarrow$ NADH/ATP $\rightarrow$ energy for work in cell

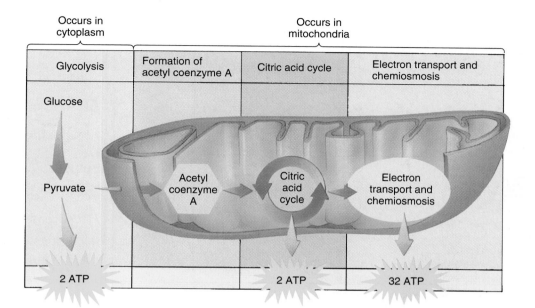

Figure 4-14   The four stages in aerobic respiration are glycolysis, formation of acetyl coenzyme A, the citric acid cycle, and the electron transport system with its associated chemiosmosis. Most ATP is synthesized by chemiosmosis during electron transport.

sequence of reactions, the electrons and protons of hydrogens removed from glucose combine with **nicotinamide adenine dinucleotide (NAD⁺)** to form NADH; a small amount of ATP is also produced. Glycolysis, which takes place in the cytoplasm, does not require O_2 and so can proceed under aerobic or anaerobic conditions.

2. **Formation of acetyl coenzyme A.** The second stage of aerobic respiration, the formation of acetyl coenzyme A, links glycolysis to the citric acid cycle. Each molecule of pyruvic acid produced during glycolysis passes from the cytoplasm into the mitochondrion (the "power plant" organelle), where it is degraded to a two-carbon fuel molecule (an acetyl group) that combines with coenzyme A (a carrier molecule), forming *acetyl coenzyme A (acetyl CoA).* The carbon removed from pyruvic acid is released as CO_2, and the electrons and protons of the hydrogens released combine with NAD⁺ to produce more NADH. The formation of acetyl CoA is an important intermediate step in the aerobic respiration of energy-rich fuel molecules because it prepares the fuel molecule to enter the citric acid cycle (as acetyl CoA).

3. **Citric acid cycle.** In this cycle, which occurs in the mitochondrion, each acetyl CoA combines with a four-carbon molecule (oxaloacetic acid) to form a six-carbon molecule (citric acid). Citric acid eventually degrades into oxaloacetic acid plus two molecules of CO_2. A small amount of ATP is produced, but most importantly, the remaining hydrogens are removed from the fuel molecule and picked

up by NAD⁺ and FAD (flavin adenine dinucleotide), forming NADH and FADH₂.

4. **Electron transport and chemiosmosis.** The NADH and FADH₂ produced in the preceding three stages transfer the electrons that they have accepted to a chain of electron acceptor molecules within the inner membrane of the mitochondrion. As electrons pass along the electron transport chain, energy is released. This energy is used to pump protons (H⁺) across the inner mitochondrial membrane, from the interior of the mitochondrion to the space between the inner and outer mitochondrial membranes. The difference in concentration of protons across the inner mitochondrial membrane, referred to as a **proton gradient**, represents potential energy much as water held behind a dam does. During **chemiosmosis** the protons diffuse across the membrane, and their energy is used to produce ATP, much as the energy of water spilling over a dam is used to generate electricity. Chemiosmosis in mitochondria is essentially the same process as occurs in chloroplasts.

AEROBIC RESPIRATION IS A COMPLEX PROCESS

The brief overview of aerobic respiration in the preceding section is sufficient for some introductory botany courses. The following material and Figure 4-15 provide greater detail.

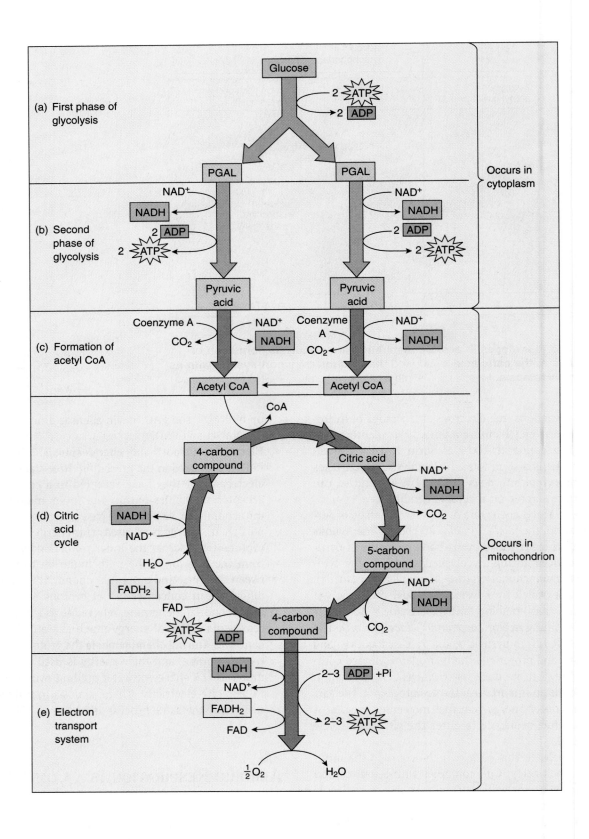

◀ **Figures 4-15   A detailed summary of aerobic respiration.**
The reactions of glycolysis can be divided into two phases.
(a) In the first phase, ATP is used to split glucose into two
three-carbon molecules of PGAL; **(b)** In the second phase,
PGAL is converted to pyruvic acid with the formation of ATP
and NADH. **(c)** Acetyl CoA is formed from each pyruvic acid
molecule with the removal of CO_2 and the attachment of the
remaining two-carbon fragment to coenzyme A; NADH forms
in the process. **(d)** In the citric acid cycle, a two-carbon
acetyl group combines with a four-carbon molecule to form
six-carbon citric acid. Two molecules of CO_2 are removed,
and the four-carbon starting molecule is ultimately regenerated to begin the cycle again. The two CO_2 molecules
account for the two carbons that entered the cycle as part of
one acetyl CoA molecule. Each turn of the citric acid cycle
produces one ATP, three NADH, and one $FADH_2$. **(e)** The
NADH and $FADH_2$ formed in the previous three stages of
aerobic respiration enter the electron transport system. As
the electrons of their hydrogens pass down the electron
transport chain, ATPs are produced by chemiosmosis.

In Glycolysis, Glucose Is Converted to Pyruvic Acid

The reactions of glycolysis can be divided into two phases.
The first phase requires the input of energy and phosphates
from two molecules of ATP (Figure 4-15a). The two phosphates
are added to glucose. (Recall that the addition of one or more
phosphate groups to a molecule is called **phosphorylation.**)
The phosphorylated sugar molecule is then split, forming two
molecules of the three-carbon compound *phosphoglyceraldehyde (PGAL).*

In the second phase of glycolysis (Figure 4-15b), each
PGAL molecule is oxidized with the removal of two hydrogens, and certain other atoms are rearranged so that each
molecule of PGAL is transformed into a molecule of pyruvic
acid. During these reactions, enough chemical energy is
released from the original sugar molecule to produce four
ATP molecules. Because the first phase of glycolysis required
two molecules of ATP, but the second phase produced four
molecules of ATP, glycolysis yields a *net* energy profit of two
ATPs per molecule of glucose.

The two hydrogens removed from each PGAL immediately combine with NAD^+, forming NADH. The fate of these
hydrogen atoms is discussed in the section on the electron
transport system.

Pyruvic Acid Is Converted to Acetyl CoA

Pyruvic acid, the end product of glycolysis, contains most of
the energy that was present in the original glucose molecule.
Pyruvic acid molecules move into the mitochondrion, where
all subsequent reactions of aerobic respiration take place.

Once a pyruvic acid molecule enters the mitochondrion,
it is converted in a multistep, enzyme-controlled reaction to
the two-carbon compound acetyl coenzyme A, or simply acetyl
CoA (Figure 4-15c). First a carbon is removed and released
as CO_2. Then the two-carbon fragment remaining is oxidized;
the hydrogens removed during this oxidation are accepted by
NAD^+, forming NADH. Finally the oxidized fragment, an
acetyl group, is attached to coenzyme A.

Note that the original six-carbon glucose molecule has
now been oxidized to two acetyl groups and two CO_2 molecules. Hydrogens have been removed and accepted by NAD^+,
reducing it to NADH. Two NADH molecules have been formed
during glycolysis, and two more during the oxidation of pyruvic acid.

The Citric Acid Cycle Oxidizes Acetyl CoA

The citric acid cycle (Figure 4-15d) is also known as the
Krebs cycle after Sir Hans Krebs, a British biochemist who
ascertained the pathway in the 1930s, for which he received
a Nobel Prize. This cycle is the common pathway for the final
oxidation reactions of the cell's fuel molecules—glucose, fatty
acids, and the carbon chains of amino acids—with the carbons
being released as CO_2. The citric acid cycle also takes place
in the mitochondrion and consists of eight reactions, each of
which is catalyzed by a specific enzyme.

The first reaction of the citric acid cycle occurs when
acetyl CoA transfers its two-carbon acetyl group to the four-
carbon compound *oxaloacetic acid,* forming *citric acid,* a six-
carbon compound. The citric acid then goes through a series
of chemical transformations, losing first one, then another
carboxyl (—COOH) group as CO_2. The two CO_2 molecules
produced thus account for the two carbon atoms that entered
the citric acid cycle as acetyl CoA. Eventually the original
starting molecule, oxaloacetic acid, is reformed; hence, the
reactions are part of a *cycle.*

Most of the energy made available by the oxidation steps
of the citric acid cycle is transferred as energy-rich electrons
to NAD^+. For each acetyl group that enters the citric acid
cycle, three molecules of NAD^+ are reduced to NADH. In
addition, electrons are also transferred to FAD, forming one
molecule of $FADH_2$ for each acetyl group entering the cycle.

Because two acetyl CoA molecules are produced from
each glucose molecule, the citric acid cycle must turn twice
to process each glucose. At the end of each turn of the cycle,
a four-carbon oxaloacetic acid is all that is left, and the cycle
is ready for another turn.

Only one molecule of ATP is produced directly with each
turn of the citric acid cycle. Thus, at this point in aerobic
respiration, the energy of one glucose molecule has resulted
in the formation of only four ATPs (two ATPs from glycolysis
and two ATPs from two turns of the citric acid cycle). To maintain their highly ordered state, most cells need to expend
much more energy than these four ATPs can provide. How,
then, is most of the ATP produced?

The Electron Transport System Produces Most of the ATP

Now we consider the fate of all the hydrogens removed from the fuel molecule during glycolysis, acetyl CoA formation, and the citric acid cycle. Recall that these hydrogens (and the energy of their electrons) were transferred to NAD^+ and FAD, forming NADH and $FADH_2$. What becomes of the hydrogens?

The electron transport chain accepts hydrogens or their electrons from the previous three stages

The electron transport system is a chain of electron acceptor molecules embedded in the inner membrane of the mitochondrion (Figure 4-15e). The electrons associated with the hydrogens released from NADH (and $FADH_2$) are passed along the chain of acceptors in a series of oxidation-reduction reactions. Electrons pass from NADH to the first electron acceptor molecule in the chain. That acceptor molecule then transfers electrons to another acceptor molecule, which passes them on to yet another.

The electrons entering the electron transport system have a relatively high energy content. As they pass along the chain of electron acceptor molecules, they lose much of their energy. Some of this energy is used to make ATP (discussed in the next section.) Finally, the last electron acceptor molecule in the electron transport chain passes the two (now low-energy) electrons on to oxygen. Simultaneously, the electrons reunite with protons (H^+) in the surrounding environment to form hydrogen. The hydrogen and oxygen combine chemically to produce water. It is because oxygen is the final electron acceptor in the electron transport system that O_2 is required for aerobic respiration.

The passage of each pair of electrons down the electron transport chain from NADH to oxygen is thought to yield enough energy to produce a maximum of three ATPs. Because the phosphorylation of ADP to form ATP is connected to the oxidation of electron transport components, ATP synthesis in the electron transport system is referred to as **oxidative phosphorylation**.

In chemiosmosis, the energy of a proton gradient is used to make ATP

It had long been known that the transfer of electrons from NADH to oxygen resulted in the production of ATP. Just *how* ATPs were synthesized remained a mystery until 1961, when Peter Mitchell proposed the **chemiosmotic model**, which states that the link between the electron transport chain and ATP synthesis is a proton gradient established across the inner mitochondrial membrane. Mitchell was awarded a Nobel Prize for this scientific contribution.

As hydrogens released from NADH and $FADH_2$ are transferred from one acceptor molecule to another in the electron transport chain, the protons (H^+) become separated from their electrons and are released into the surrounding environment. Some of the energy released by the electron transport chain is used to pump the protons from the interior of the mitochondrion across its inner membrane to the intermembrane space (the space between the inner and outer mitochondrial membranes). The proton pumps result in the establishment of a proton gradient between the intermembrane space and the interior of the mitochondrion. The difference in concentration of protons across the inner mitochondrial membrane represents potential energy that is used to synthesize ATP.

Because the inner mitochondrial membrane is impermeable to protons, the protons can flow back to the interior of the mitochondrion only through special protein channels in the membrane. As in photosynthesis, the protein channels occur within the enzyme complex **ATP synthase**. The protons move down the energy gradient—that is, through ATP synthase from the area where they are highly concentrated (the space between the inner and outer mitochondrial membranes) to the matrix (interior of the mitochondrion), where they are present in a lower concentration. The energy released by the flow of protons is used by ATP synthase to produce ATP.

Just how much ATP can be produced from the complete breakdown of one molecule of glucose is currently under study. The complete aerobic respiration of one molecule of glucose may produce a *maximum* total of 36 to 38 ATPs. Recently, some scientists have challenged this number and have suggested that the actual number of ATPs produced is substantially lower.

MANY CELLS USE ANAEROBIC PATHWAYS IN THE ABSENCE OF OXYGEN

Bacteria and some other types of organisms that inhabit waterlogged soil or stagnant ponds where oxygen is absent must engage in anaerobic respiration. In **anaerobic respiration**, energy is released from glucose and other fuel molecules without O_2; that is, oxygen is not the final electron acceptor in the electron transport chain. Instead, an inorganic compound, such as nitrate (NO_3^-) or sulfate (SO_4^{2-}), serves as the final acceptor of electrons.

Fermentation, another anaerobic pathway, also degrades glucose and other organic molecules without oxygen. Like aerobic respiration, fermentation depends on the reactions of glycolysis (Figure 4-16). Unlike the final acceptor in anaerobic respiration (just described), the final acceptor of hydrogen

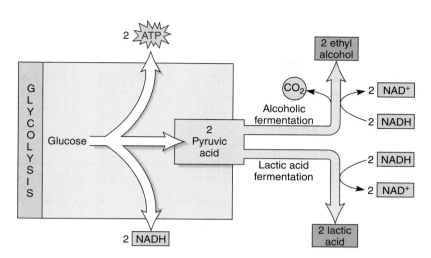

Figure 4-16   Glycolysis is part of fermentation reactions. In alcoholic fermentation, CO_2 is split off, and the two-carbon ethyl alcohol is the end product. In lactic acid fermentation, the final product is the three-carbon compound lactic acid. In both alcoholic and lactic acid fermentation, there is a net gain of only two ATPs per molecule of glucose.

in fermentation is an *organic* molecule. Two common types of fermentation are alcoholic fermentation and lactic acid fermentation.

Yeasts (unicellular fungi) can carry on a type of fermentation known as **alcoholic fermentation** (Figure 4-17). First they degrade glucose to pyruvic acid through the process of glycolysis. When deprived of O_2, yeasts split CO_2 off from pyruvic acid, eventually forming *ethyl alcohol*. Alcoholic fermentation is the basis for the production of beer, wine, and other alcoholic beverages. Yeasts are also used in baking to produce the CO_2 that causes the dough to rise (the alcohol evaporates during baking).

Certain fungi and bacteria carry on **lactic acid fermentation**. In this pathway, pyruvic acid produced during glycolysis is converted to lactic acid. Lactic acid fermentation occurs when bacteria cause milk to go sour or ferment cabbage to form sauerkraut. Under conditions of insufficient oxygen, human muscle cells can also use lactic acid fermentation to obtain limited amounts of energy. Lactic acid buildup may cause the fatigue and muscle cramps people sometimes experience during heavy exercise.

Fermentation Is Inefficient

Fermentation is a very inefficient method of extracting energy from fuel molecules in comparison with aerobic respiration, because the fuel is only partially oxidized. Alcohol, the end product of fermentation by yeast cells, can be burned and can even be used as automobile fuel. Obviously, it contains a great deal of energy that the yeast cells were unable to extract anaerobically. Lactic acid, a three-carbon compound, contains even more energy than the two-carbon alcohol. During aerobic respiration, *all* available energy is removed because fuel molecules are completely oxidized. Fermentation can produce a

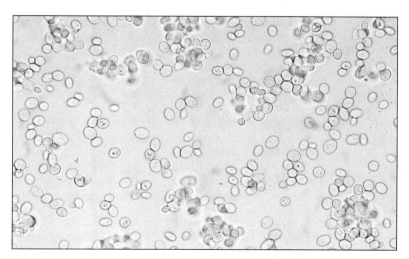

Figure 4-17   Live yeast cells as seen under a light microscope. Yeast cells possess mitochondria and carry on aerobic respiration when O_2 is present. In the absence of O_2, yeasts carry on alcoholic fermentation. (*Dwight R. Kuhn*)

net profit of only two ATPs from one molecule of glucose, compared with an estimated maximum of 36 to 38 ATPs when oxygen is available.

The inefficiency of fermentation necessitates a large supply of fuel. By rapidly degrading many fuel molecules, a cell can compensate somewhat for the small amount of energy that can be gained from each molecule. To perform the same amount of work, an anaerobic cell must consume up to 20 times as much glucose or other carbohydrate as a cell metabolizing aerobically.

STUDY OUTLINE

I. The chemical processes that occur in a cell are collectively referred to as its metabolism.
   A. Anabolism is a building-up process in which large, complex molecules are synthesized from simpler materials.
   B. Catabolism is a breaking-down process in which complex molecules are degraded into simpler ones to release energy.
II. During photosynthesis, light energy is captured by chlorophyll and used to chemically combine the hydrogen from water with CO_2 to produce carbohydrates. Oxygen is released as a by-product of photosynthesis.
III. The two phases of photosynthesis are the light-dependent reactions (in the chloroplast thylakoids) and the light-independent reactions (in the chloroplast stroma).
   A. During the light-dependent reactions of photosynthesis, chlorophyll absorbs light and becomes energized. Some of this energy is used to make ATP, and some is used to split water. Hydrogen from the water is transferred to $NADP^+$, forming NADPH.
      1. In noncyclic photophosphorylation, ATP and NADPH are formed using the energy of light; water is split.
      2. In cyclic photophosphorylation, ATP is formed; water is not split, and NADPH is not formed.
      3. ATP is synthesized by chemiosmosis in both noncyclic and cyclic photophosphorylation.
         a. The flow of energized electrons along an electron transport chain provides energy to establish a proton gradient across the thylakoid membrane.
         b. Protons flow back through the membrane through a channel within the enzyme ATP synthase; energy released in this process is used to synthesize ATP.
   B. During the light-independent reactions, energy stored within ATP and NADPH during the light-dependent reactions is used to chemically fix CO_2.
      1. The light-independent reactions proceed in most plants via the Calvin cycle, also known as the C_3 pathway.
      2. In the Calvin cycle, CO_2 is combined with ribulose bisphosphate, a five-carbon sugar. With each turn of the cycle, one carbon atom enters the cycle.
      3. Six turns of the cycle result in the synthesis of two molecules of a three-carbon compound (PGAL), which combine to produce a molecule of glucose.
IV. To extract energy from fuel molecules such as glucose, cells use three different catabolic pathways: aerobic respiration, anaerobic respiration, and fermentation.
   A. Aerobic respiration requires O_2. Anaerobic respiration and fermentation do not require O_2.
   B. During aerobic respiration, a fuel molecule such as glucose is oxidized, forming CO_2 and water with the release of energy.

V. The chemical reactions of aerobic respiration occur in four stages: glycolysis, formation of acetyl CoA, the citric acid cycle, and the electron transport system with its associated chemiosmosis. Glycolysis occurs in the cytoplasm; all other parts of aerobic respiration occur in the mitochondria.
   A. During glycolysis, a molecule of glucose is degraded, forming two molecules of pyruvic acid.
      1. A small amount of ATP is produced during glycolysis.
      2. Hydrogen atoms are removed from the fuel molecule and transferred to NAD^+, forming NADH.
   B. The two pyruvic acid molecules each lose a molecule of CO_2, and the remaining acetyl groups combine with coenzyme A, producing acetyl CoA. One NADH is formed as each pyruvic acid is converted to acetyl CoA.
   C. Each acetyl CoA enters the citric acid cycle by combining with a four-carbon compound to form citric acid, a six-carbon compound.
      1. With two turns of the citric acid cycle, the two molecules of acetyl CoA are completely degraded, and four CO_2 molecules are released.
      2. Hydrogens are transferred to NAD^+ and FAD, forming NADH and $FADH_2$. Only one ATP is produced directly with each turn of the citric acid cycle.
   D. Hydrogen atoms (or their electrons) removed from fuel molecules are transferred from one electron acceptor molecule to another along the electron transport system.
      1. The final acceptor in the chain is O_2, which combines with the hydrogen to form water.
      2. According to the chemiosmotic model, energy liberated in the electron transport chain is used to establish a proton gradient across the inner mitochondrial membrane.
      3. Protons flow back through the membrane through a channel within the enzyme ATP synthase; energy released in this process is used to synthesize ATP.
VI. Organisms that inhabit oxygen-poor environments use anaerobic pathways for capturing energy.
   A. In anaerobic respiration, fuel molecules are broken down in the absence of O_2; an inorganic compound serves as the final electron acceptor.
   B. Fermentation is an anaerobic pathway that initially begins with glycolysis. Fermentation does not release as much energy per glucose molecule as aerobic respiration.
      1. Yeasts carry on alcoholic fermentation in which ethyl alcohol and CO_2 are the final products.
      2. Certain fungi and bacteria, as well as animal muscle cells, carry on lactic acid fermentation, in which lactic acid is the final product.

SELECTED KEY TERMS

aerobic, p. 66
aerobic respiration, p. 54
anabolism, p. 54
anaerobic, p. 64
ATP synthase, p. 62
Calvin cycle (C₃ pathway),
   p. 59

catabolism, p. 54
chemiosmosis, pp. 62, 67
citric acid cycle, p. 67
electron acceptor molecule,
   p. 54
electron transport, p. 67

fermentation, p. 70
glycolysis, p. 66
metabolism, p. 54
oxidation, p. 54
oxidative phosphorylation,
   p. 70

photon, p. 57
photophosphorylation,
   p. 61
photosynthesis, p. 55
reduction, p. 54
rubisco, p. 63

REVIEW QUESTIONS

1. Explain the roles of light and chlorophyll in photosynthesis.
2. How are almost all organisms on planet Earth dependent on the process of photosynthesis?
3. Write the overall equation for photosynthesis.
4. Summarize the light-dependent reactions of photosynthesis. Summarize the events of the Calvin cycle.
5. Explain how ATP is produced according to the chemiosmotic model. How does a proton gradient contribute to ATP synthesis?
6. Justify referring to mitochondria as the "power plants" of the cell. Be specific, using information you have acquired from this chapter.

7. Write the overall equation for aerobic respiration.
8. Trace the fate of hydrogens removed from the fuel molecule during glycolysis when O₂ is present.
9. What is the specific role of O₂ in cells? What happens when such cells are deprived of O₂?
10. Label the following overview of photosynthesis. Check your answers against Figure 4-12.

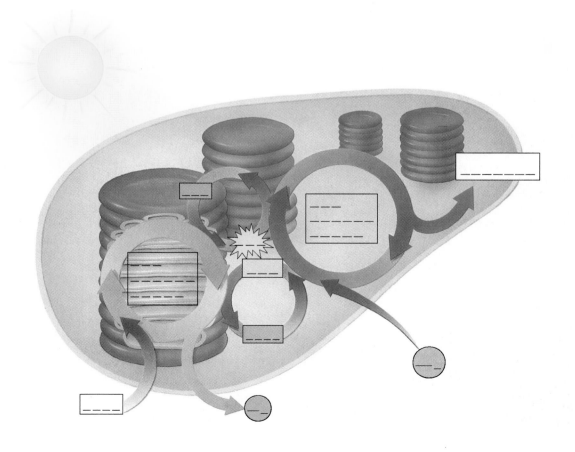

THOUGHT QUESTIONS

1. Would placing a plant under green light increase or decrease its rate of photosynthesis? (The plant was previously growing in sunlight.) Why?

2. The chemiosmotic model (generation of ATP using the energy of a proton gradient) has been compared to hydroelectric power (generation of electricity using the energy of water flowing over a dam). Explain the analogy.

3. Kentucky bluegrass thrives in eastern lawns during the spring months but is usually overrun by crabgrass during the hot summer months. Based on what you know about the C_3 and C_4 pathways, which plant would you suppose possesses the C_3 pathway? The C_4 pathway? Explain your reasoning.

4. The poison cyanide works by blocking the electron transport system. On the basis of what you have learned in this chapter, why is cyanide poisoning fatal?

5. Louis Pasteur, an important biologist in the 19th century, noted that yeast cells added to a closed container of grape juice broke down the sugar very slowly as long as O_2 was present in the container. Once the O_2 was used up, however, the yeast cells consumed the remainder of the grape sugar very quickly, producing ethyl alcohol in the process. Explain this observation on the basis of what you know about the relative efficiencies of aerobic respiration and alcoholic fermentation.

6. The burning of biomass contributes no net CO_2 to our atmosphere, but the burning of fossil fuels increases the level of CO_2 in our current atmosphere. Explain.

SUGGESTED READINGS

Balzani, V., "Greener Way to Solar Power," *New Scientist* Vol. 12, November 1994. Chemists are imitating the light-gathering photosystems of plants in an attempt to convert solar energy to electricity.

Govindjee and W.J. Coleman, "How Plants Make Oxygen," *Scientific American,* February 1990. Probes the photosynthetic process of using solar energy to split water molecules into oxygen, protons, and electrons.

Hendry, G., "Making, Breaking, and Remaking Chlorophyll," *Natural History,* May 1990. Examines the endless process by which plants make chlorophyll in the spring and dispose of it in the fall.

Solomon, Berg, Martin, and Villee, *Biology,* 4th ed., Philadelphia, Saunders College Publishing, 1996. Chapters 8 and 9 present detailed discussions of photosynthesis and respiration.

PLANT STRUCTURES AND
LIFE PROCESSES

CHAPTER 5

PLANT TISSUES AND THE MULTICELLULAR PLANT BODY

At dawn most of us are still in bed sleeping. Not so the squirrel, which begins gathering acorns, beechnuts, pine seeds, berries, and similar foods at sunrise. Although it usually eats perishable food immediately, the squirrel often stores acorns and seeds in holes in the ground to provide itself with winter food. But it never retrieves many of them, and so a new oak (*Quercus* sp) often begins its life after a squirrel has planted an acorn.

The seed within the acorn first absorbs water from the surrounding soil. Then germination occurs as a root emerges and works its way down into the soil, absorbing additional water and anchoring the young plant. A miniature shoot grows upward and breaks through the soil. At this point the young plant is called a seedling because it has just emerged from the seed and because it still depends on the food supply stored within the seed.

A massive oak grows from a tiny acorn.
(*E.R. Degginger/Bruce Coleman, Inc.*)

at their tips. As it ages, the tree also increases in girth (circumference) by forming additional wood and bark. This growth occurs along the sides of the more mature stems and roots. Thus, the oak progressively accumulates more wood and bark, more stem and root tissues, and more leaves.

Growth and expansion of both the root and shoot systems continue throughout the life of the oak. As in all plants, growth is flexible, enabling the oak to respond to its environment. For example, the young oak may grow very slowly for several years, particularly if it is shaded by mature trees. When conditions change—perhaps when an older tree nearby dies and falls to the ground, thereby permitting direct sunlight to reach the oak—the tiny tree is poised for rapid growth.

Oaks reproduce by forming separate male and female flowers in spring. The minute male flowers are more conspicuous than the female flowers, because they are clustered together on slender, drooping, caterpillar-shaped catkins; in contrast, the tiny female flowers are often solitary. After the wind transports pollen from the male flower to the female flower, a sperm cell fertilizes the egg and an acorn develops, partially enclosed by a scaly cup. First green in color, the acorn fruit turns a deep brown as it matures. Within the acorn is a seed containing the embryo of a tiny miniature oak along with a supply of stored food. Thus, the cycle of life repeats itself, as squirrels busily collect these acorns and cache them for later retrieval.

The stem continues to elongate, and small leaves develop, expand, and begin to photosynthesize. The young plant is now independently established—it is anchored in the ground, it absorbs water and dissolved minerals from the soil, and it uses energy from organic molecules, such as glucose, that it has produced by photosynthesis.

Smaller roots branch off the original root, and the shoot tips produce buds, dormant structures that develop into branches the following spring. The young tree gains height, always by growth at the tips of its branches; the roots likewise elongate

PLANTS EXHIBIT SIMILARITY AND DIVERSITY IN STRUCTURE AND LIFE SPAN

Remarkable variety exists among the 235,000 or so species of flowering plants[1] that live in and are adapted to the many environments offered by our planet. Yet all of these plants—from desert cacti with enormously swollen stems to cattails partly submerged in marshes to orchids growing in the upper-most branches of lush tropical-forest trees—are recognizable as plants. From the tiny floating water-meal (*Wolffia microscopica*), the smallest flowering plant known, to Australian gums (*Eucalyptus* spp), some of Earth's tallest trees, all plants have the same basic body plan.

Plants are either herbaceous or woody. **Herbaceous plants** do not develop persistent woody parts aboveground, whereas **woody plants** (trees and shrubs) do.

Annuals are herbaceous plants (such as corn, geranium, and marigold) that grow, reproduce, and die in 1 year. Other herbaceous plants (such as carrot, cabbage, and beet) are **biennials** and take 2 years to complete their life cycles (see Figure 11–7). During their first season, biennials produce extra carbohydrates, which they store and use during their second year, when they typically form flowers and reproduce.

Perennials are woody or herbaceous plants that live for more than 2 years. In temperate climates, the aerial (above-ground) stems of herbaceous perennials (for example, iris, rhubarb, onion, and asparagus) die back each winter. Their underground parts (roots and underground stems) become dormant during the winter and send out new growth each spring. (In dormancy, an organism reduces its metabolic state to a minimum level to survive unfavorable conditions.) Likewise, in certain tropical climates with pronounced wet and dry seasons, the aerial parts of herbaceous perennials (for example,

many grasses) die back and the underground parts become dormant during the dry season. Other tropical plants, such as orchids, are herbaceous perennials that grow year-round.

All woody plants are perennials, and some of them live for hundreds or even thousands of years (see Focus On: Plant Life-History Strategies). In temperate climates, the above-ground stems of woody plants become dormant during the winter. Most temperate woody perennials shed their leaves before winter and produce new stem tissue with new leaves the following spring. Because they have permanent woody stems that are the starting points for new growth the following season, many trees attain massive sizes.

ROOTS, STEMS, LEAVES, FLOWERS, AND FRUITS MAKE UP THE PLANT BODY

The plant body is organized into a root system and a shoot system (Figure 5-1). The **root system** is generally the below-ground portion. The aboveground portion, the **shoot system**, consists of a vertical stem bearing leaves, flowers, and fruits that contain seeds.

Each plant grows in two different environments: the dark, moist soil and the relatively bright, dry air. A plant must have both roots and shoots because it needs resources from *both* environments. Thus, roots branch extensively through the soil, forming a network that anchors the plant firmly in place and absorbs water and dissolved minerals from the soil. Leaves, the flattened organs of photosynthesis, are attached more or less regularly on the stem to capture the sun's light and absorb the atmospheric CO_2 needed for photosynthesis.

THE PLANT BODY IS COMPOSED OF CELLS AND TISSUES

The cell is the basic structural and functional unit of plants as well as of other organisms. During the course of evolution, plants evolved a diversity of cell types, each specialized for particular functions.

---

[1]About 90 percent of the 262,000 or so species of plants are flowering plants, which are characterized by the presence of flowers and of seeds enclosed within fruits. Because flowering plants are the largest group of plants, they are the focus of much of this and subsequent chapters.

Figure 5-1 The primary plant body (here, a geranium) includes roots, stems, and leaves.

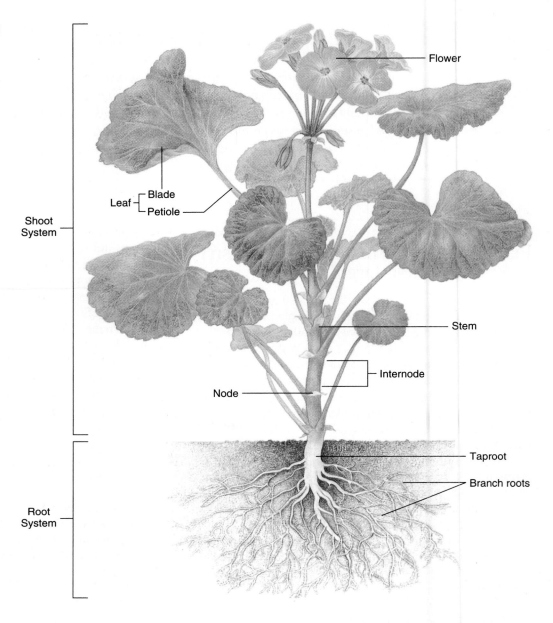

As in animals, plant cells are organized into tissues. A **tissue** is a group of cells that forms a structural and functional unit. Some plant tissues (*simple tissues*) are composed of only one kind of cell, whereas other plant tissues (*complex tissues*) have two or more kinds of cells. In plants, tissues are organized into three tissue systems, each of which extends throughout the plant body. Each tissue system contains two or more kinds of tissues (Table 5-1).

Most of the plant body is composed of the following parts:

1. The **ground tissue system** consists of three tissues that perform a variety of functions, including photosynthesis, storage, and support.
2. The **vascular tissue system**, an intricate plumbing system that extends throughout the plant body, conducts various substances, including water, dissolved minerals, and dissolved food materials (carbohydrates). It also functions in strengthening and supporting the plant.
3. The **dermal tissue system** provides a covering for the plant body.

Roots, stems, and leaves are referred to as **organs** because each is composed of several different tissues. The tissue systems of different plant organs form an interconnected network throughout the plant. For example, the vascular tissue of a leaf is continuous with the vascular tissue of the stem to which the leaf is attached, and the vascular tissue of the stem is continuous with the vascular tissue of the root.

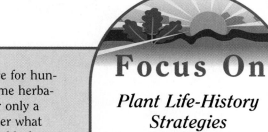

Focus On
Plant Life-History Strategies

Woody perennials often live for hundreds of years, whereas some herbaceous annuals may live for only a few weeks or months. Under what conditions is it more favorable for a species to be long-lived or short-lived? Botanists have considered the relative advantages of each life-history strategy and have concluded that in some environments a longer life span is advantageous, whereas in others a shorter life span increases reproductive success.

When an environment is relatively favorable, it is filled with plants that compete for the available space. Because such an environment is so crowded, it has few open spots where new plants can become established. When a plant dies, the empty area it leaves behind is quickly filled by another plant, but not necessarily of the same species. Thus, an adult

perennial survives well in such a habitat, but young plants, whether perennials or annuals, do not. A plant with a long life span thrives in this type of environment because it can "hold onto" a piece of soil and continue to produce seeds for many years. In a tropical rain forest, for example, such competition prevents most young plants from becoming established; hence, woody perennials predominate.

In a relatively unfavorable environment or one that is frequently disturbed, many growing sites are usually available. This type of environ-

ment is not crowded, and young plants usually don't have to compete against large, fully established plants. Here, smaller, short-lived plants have the advantage. These plants are opportunists—they grow and mature quickly during the brief periods when environmental conditions are most favorable. As a result, all of their resources are directed into producing as many seeds as possible before dying. In deserts, for example, annuals are more prevalent than woody perennials.

Thus, each species has its own characteristic life-history strategy, with some plants adapted to variable environments and others adapted to stable environments. The longer life span conferred on woody perennials is just one of several successful life-history plans.

The Ground Tissue System Is Composed of Three Simple Tissues

The bulk of a herbaceous plant is its ground tissue, which is composed of three tissues: parenchyma, collenchyma, and sclerenchyma. These tissues can be distinguished by their cell wall structures. Recall that a plant cell is surrounded by a cell wall that provides structural support (see Chapter 3). A growing cell secretes a thin **primary cell wall**, which stretches and

in meristem

expands as the cell increases in size. After the cell stops growing, it sometimes secretes a thick, strong **secondary cell wall**, which is deposited *inside* the primary cell wall—that is, between the primary cell wall and the plasma membrane.

Parenchyma cells have thin primary walls

Parenchyma tissue, a simple tissue composed of parenchyma cells, is found throughout the plant body and is the most

TABLE 5-1   Tissue Systems, Tissues, and Cell Types of Flowering Plants

Tissue System	Tissues	Cell Types
Ground tissue system	Parenchyma tissue	Parenchyma cells
	Collenchyma tissue	Collenchyma cells
	Sclerenchyma tissue	Sclerenchyma cells (sclereids, fibers)
Vascular tissue system	Xylem	Tracheids, vessel elements, parenchyma cells, fibers
	Phloem	Sieve-tube members, companion cells, parenchyma cells, fibers
Dermal tissue system	Epidermis	Parenchyma cells, guard cells, trichomes
	Periderm	Cork cells, cork cambium cells, cork parenchyma

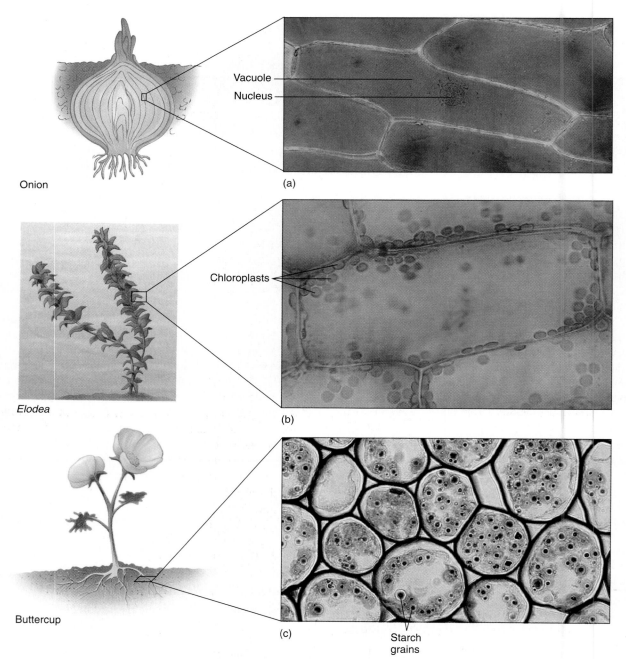

Figure 5-2   Parenchyma are living, actively metabolizing cells. (a) Parenchyma cells from an epidermal peel of red onion (*Allium cepa*). The large vacuole contains pigmented material and occupies most of the cell. The nucleus and cytoplasmic strands are positioned under and on top of the vacuole, between it and the plasma membrane. **(b)** Some parenchyma cells contain chloroplasts, and their primary function is photosynthesis. These parenchyma cells are from a waterweed (*Elodea* sp) leaf. **(c)** Parenchyma cells often function in storage. These parenchyma cells are from a buttercup (*Ranunculus* sp) root. Note the starch grains filling the cells. (a, *Ed Reschke;* b, c, *Dennis Drenner*)

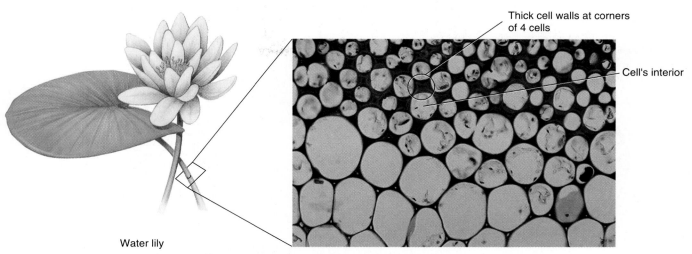

Water lily

Figure 5-3   Collenchyma cells provide flexible support. Note the unevenly thickened cell walls that are especially thick in the corners, making the cell contents appear spherical in cross section. These collenchyma cells are from a water lily (*Nymphaea* sp) petiole. (*James Mauseth, University of Texas*)

common type of cell and tissue (Figure 5-2). The soft parts of a plant, such as the edible part of an apple or a potato, consist largely of parenchyma.

Parenchyma cells perform a number of important functions for plants, such as photosynthesis, storage, and secretion. Parenchyma cells that function in photosynthesis contain chloroplasts, whereas nonphotosynthetic parenchyma cells lack chloroplasts. Materials stored in parenchyma cells include starch grains, oil droplets, water, and salts (sometimes visible as crystals). The various functions of parenchyma require that they be living, metabolizing cells.

Collenchyma cells have unevenly thickened primary walls

Collenchyma tissue, a simple tissue composed of collenchyma cells, is an extremely flexible support tissue that provides much of the support in soft, nonwoody plant organs (Figure 5-3). Support is a crucial function in plants, in part because it allows plants to grow upward, thus enabling them to compete with other plants for available sunlight in a plant-crowded area. Instead of the bony skeletal system that is typical of many animals, individual cells, including collenchyma cells, provide support for the plant body.

Collenchyma cells, which are usually elongated, are alive at maturity. Their primary cell walls are unevenly thickened and are especially thick in the corners. Collenchyma is not found uniformly throughout the plant and often occurs as long strands near stem surfaces and along leaf veins. The "strings" in a celery stalk (petiole), for example, consist of collenchyma.

Sclerenchyma cells have both primary walls and thick secondary walls

A second simple tissue specialized for structural support is **sclerenchyma** tissue, whose cells have both primary and secondary cell walls. The root of the word "sclerenchyma" is the Greek word root (*sclero*) meaning "hard." The secondary cell walls of sclerenchyma cells become strong and hard due to extreme thickening. At functional maturity, when sclerenchyma tissue is providing support for the plant body, its cells are often dead.

Sclerenchyma tissue may occur in several areas of the plant body. Two types of sclerenchyma cells are sclereids and fibers. **Sclereids** are short, more-or-less cuboidal cells common in the shells of nuts and the pits of stone fruits, such as cherries and peaches (Figure 5-4a). Pears owe their slightly gritty texture to the presence of sclereids. **Fibers**, which are long, tapered cells that often occur in groups or clumps, are particularly abundant in the wood and bark of flowering plants (Figure 5-4b).

The Vascular Tissue System Consists of Two Complex Tissues

The vascular tissue system, which is embedded in the ground tissue, transports needed materials throughout the plant via two complex tissues: xylem (pronounced "ZYE-lem") and phloem (pronounced "FLOW-em"). Both xylem and phloem are continuous throughout the plant body. (Chapter 10 discusses the mechanisms of transport in xylem and phloem.)

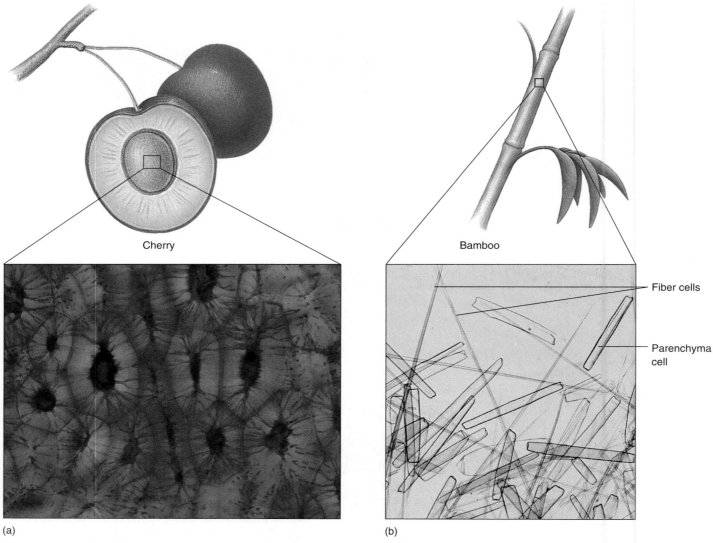

Cherry

Bamboo

Fiber cells

Parenchyma
cell

(a)   (b)

Figure 5-4   Sclerenchyma cells provide support. (a) Sclereids from a cherry (*Prunus
avium*) pit. The cells walls are extremely thick and hard, providing structural support.
(b) Long, tapering fibers and shorter parenchyma cells from a bamboo (*Bambusa* sp)
stem. The stem was treated with acid to separate the cells. (a, *Dennis Drenner;* b,*James
Mauseth, University of Texas*)

Two kinds of conducting cells, tracheids and vessel elements, occur in xylem

Xylem conducts water and dissolved minerals from the roots
to the stems and leaves and provides structural support. In
flowering plants, xylem is a complex tissue composed of four
different cell types: tracheids, vessel elements, parenchyma
cells, and fibers. Two of the four cell types found in xylem—
the **tracheids** and **vessel elements**—actually conduct water
and dissolved minerals (Figure 5-5). In addition to these cells,
xylem contains parenchyma cells that perform storage functions and fibers that provide support.

Tracheids and vessel elements are highly specialized for
conduction of water and minerals. At maturity, both cell types
are dead and therefore hollow; only their cell walls remain.
Tracheids, the chief water-conducting cells in gymnosperms[2]
and seedless vascular plants such as ferns, are long, tapering
cells. Water is conducted upward from roots to shoots, passing from one tracheid into another through **pits**, thin areas

[2]Gymnosperms are seed-bearing plants that do not produce flowers. They
include cone-bearing trees such as pines, hemlocks, and firs as well as
ginkgoes and cycads.

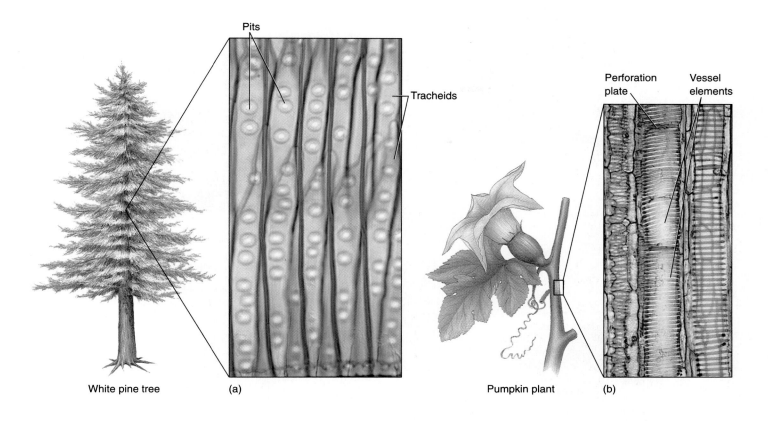

White pine tree   (a)

Pumpkin plant   (b)

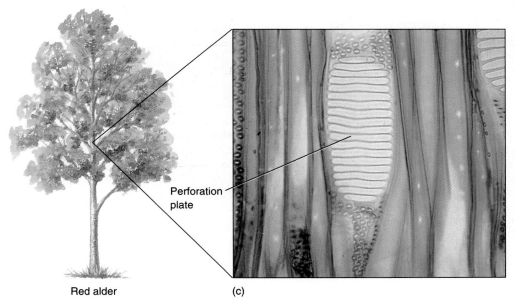

Red alder   (c)

Figure 5-5   Conducting cells in xylem. (a) Tracheids from a white pine (*Pinus strobus*) stem in longitudinal section (that is, cut lengthwise). These cells, which occur in clumps, transport water and dissolved minerals. Water passes readily from tracheid to tracheid through pits, thin places in the cell wall. **(b)** Vessel elements from a pumpkin (*Cucurbita mixta*) stem in longitudinal section. Note how they are stacked end-on-end. **(c)** The end walls of vessel elements, called perforation plates, have large holes. Water passes through the perforation plate from one vessel element to the next. Shown is a perforation plate from a red alder (*Alnus rubra*) stem in longitudinal section. (a, *Visuals Unlimited/John D. Cunningham;* b, *J. Robert Waaland, University of Washington/Biological Photo Service;* c, *Visuals Unlimited/J.D. Litvay*)

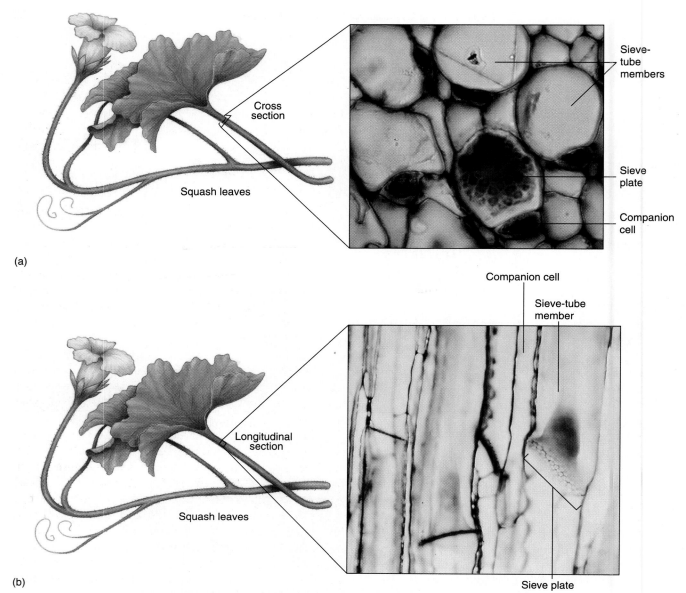

(a)

(b)

Labels in figure:
- Cross section
- Squash leaves
- Sieve-tube members
- Sieve plate
- Companion cell
- Companion cell
- Sieve-tube member
- Longitudinal section
- Squash leaves
- Sieve plate

Figure 5-6   Dissolved food materials are conducted through sieve-tube members of the phloem. (a) Phloem tissue from a squash (*Cucurbita* sp) petiole in cross section. Note the sieve plate, the end wall of the sieve-tube member. Most of the sieve-tube members appear empty because they were sectioned in the middle of the cells rather than at the end walls. The smaller cells are companion cells. **(b)** Phloem tissue from a squash (*Cucurbita* sp) petiole in longitudinal section. (a, *J. Robert Waaland, University of Washington/Biological Photo Service;* b, *James Mauseth, University of Texas*)

in the tracheids' cell walls where a secondary cell wall did not form.

In addition to relatively few tracheids, flowering plants possess very efficient water-conducting cells called vessel elements. The cell diameters of vessel elements are usually greater than those of tracheids. Vessel elements are hollow, but unlike tracheids, they have holes in their end walls known as **perforations**, or the end walls may be entirely dissolved away. Vessel elements are stacked one on top of the other, and water is conducted readily from one vessel element into the next. A stack of vessel elements, called a **vessel**, resembles a miniature water pipe. Vessel elements also have pits, which

permit the lateral transport (sideways movement) of water from one vessel to another.

Sieve-tube members are the conducting cells of phloem

Phloem conducts food materials—that is, carbohydrates formed in photosynthesis—throughout the plant and provides structural support. In flowering plants, phloem is a complex tissue composed of four different cell types: sieve-tube members, companion cells, fibers, and parenchyma cells (Figure 5-6). In the phloem of herbaceous plants, fibers are frequently extensive, providing additional structural support for the plant body.

Food materials are conducted *in solution*—that is, dissolved in water—through the **sieve-tube members**, which are among the most specialized cells in nature. Sieve-tube members are long, thin cells that are stacked end-on-end to form extended sieve tubes. The cell's end walls, called **sieve plates**, are pierced by a series of holes through which cytoplasm extends from one sieve-tube member into the next. Sieve-tube members are alive at maturity, but many of their organelles, including the nucleus, vacuole, mitochondria, and ribosomes, disintegrate as they mature.

Sieve-tube members are among the few eukaryotic cells that can function without nuclei, although they typically live for less than a year. There are, however, notable exceptions: certain palms have sieve-tube members that have remained alive approximately 100 years!

Adjacent to each sieve-tube member is a **companion cell**, which assists in the functioning of the sieve-tube member. The companion cell is a living cell, complete with a nucleus and other organelles. Numerous tiny openings in the cell walls between a companion cell and its sieve-tube member allow cytoplasm to extend from one cell to the other. Although the companion cell does not conduct, it plays an essential role in moving food materials into the sieve-tube members for transport to other parts of the plant.

The Dermal Tissue System Consists of Two Complex Tissues

The dermal tissue system provides a protective covering over plant parts. In herbaceous plants, the dermal tissue system is a single layer of cells called the epidermis. Woody plants initially produce an epidermis, but it splits apart as the plant increases in diameter, due to the production of additional woody tissues. Periderm, a tissue several to many cell layers thick, forms under the epidermis to provide a new protective tissue. Periderm, which replaces the epidermis in the stems and roots of older woody plants, composes the outer bark.

Epidermis is the outermost layer of cells on a herbaceous plant

The **epidermis** is a complex tissue comprising mostly parenchyma cells with scattered guard cells. In most plants, the epidermis consists of a single layer of cells. Epidermal cell walls are somewhat thicker toward the outside of the plant to provide protection. Epidermal parenchyma cells generally contain no chloroplasts and are therefore transparent, allowing light to penetrate into the interior tissues of stems and leaves. In both stems and leaves, photosynthetic tissues lie *beneath* the epidermis.

An important requirement of aerial shoots is the ability to control water loss. Epidermal cells of stems and leaves secrete a waxy layer called a **cuticle** over the surfaces of their walls; this wax greatly restricts the loss of water from plant surfaces.

Although the cuticle is very efficient at preventing most water loss through epidermal cells, it also prevents the carbon dioxide required for photosynthesis from diffusing from the atmosphere into the leaf or stem. **Stomata** (sing. *stoma*) facilitate the diffusion of carbon dioxide. Stomata are tiny pores in the epidermis between two cells called **guard cells** (Figure 5-7). Many gases, including carbon dioxide, oxygen, and water vapor, pass through stomata by diffusion. Stomata are generally open during the day, when photosynthesis is occurring, and the loss of water that also takes place when

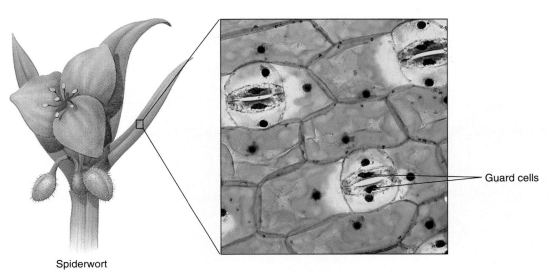

Spiderwort

Guard cells

Figure 5-7   Light microscope view of a peel of spiderwort (*Tradescantia virginiana*) leaf epidermis (×350). Note the stomata, minute pores formed by guard cells (*stained pink*). (*Ed Reschke*)

stomata are open helps to cool the leaf. To conserve water, stomata are usually closed at night, when photosynthesis is not taking place and cooling is not required. In drought conditions, the need to conserve water overrides the need to cool the leaves and exchange gases. Thus, during a drought, the stomata close in the daytime. Chapter 10 discusses stomata in greater detail.

The epidermis may also contain special outgrowths, or hairs, called **trichomes**, which occur in many sizes and shapes and have a variety of functions. **Root hairs** are simple, unbranched trichomes that increase the surface area of the root epidermis, which comes into contact with the soil, for more effective water and mineral absorption. Plants that can tolerate salty environments often have specialized trichomes on their leaves that remove excess salt that has accumulated in the plant. The presence of trichomes on the aerial parts of desert plants may increase the reflection of light off the plants,

thereby keeping the internal tissues cooler and decreasing water loss. Other trichomes have a protective function. For example, the trichomes on stinging nettle leaves and stems contain irritating substances that discourage herbivorous animals from eating the plant.

Epidermis is replaced by periderm in woody plants

As a woody plant begins to increase in girth, its epidermis sloughs off and is replaced by **periderm**. Periderm forms the protective outer bark of older stems and roots (Figure 5-8). It is a complex tissue composed mainly of cork cells and cork parenchyma cells. **Cork cells** are dead at maturity, and their walls are heavily coated with a waterproof substance called *suberin*, which helps reduce water loss. **Cork parenchyma (*phelloderm*) cells** function primarily in storage.

Table 5-2 summarizes selected cell types found in the ground, vascular, and dermal tissue systems.

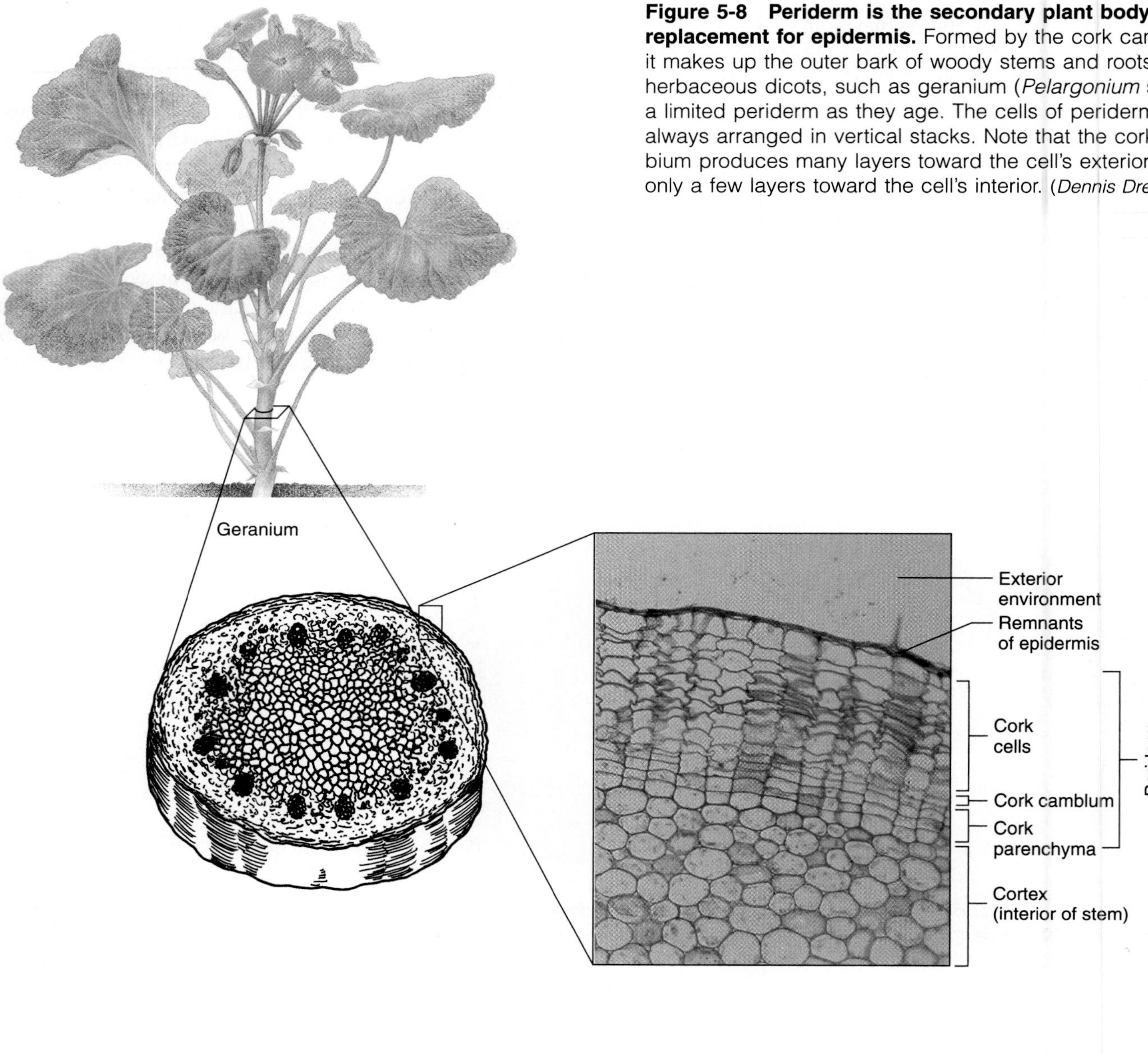

Figure 5-8   Periderm is the secondary plant body replacement for epidermis. Formed by the cork cambium, it makes up the outer bark of woody stems and roots. Some herbaceous dicots, such as geranium (*Pelargonium* sp), form a limited periderm as they age. The cells of periderm are always arranged in vertical stacks. Note that the cork cambium produces many layers toward the cell's exterior, but only a few layers toward the cell's interior. (*Dennis Drenner*)

TABLE 5-2   A Summary of Selected Cell Types

Cell Type	Description	Function	Location
Parenchyma cell	Living; thin primary wall	Secretion, storage, photosynthesis	Throughout the plant body
Collenchyma cell	Living; unevenly thickened primary wall	Support	Just under stem epidermis; along leaf veins
Sclerenchyma cell (fiber)	Dead at maturity; thick secondary wall	Support	Throughout the plant body; common in stems and certain leaves
Tracheid	Dead at maturity; lacks secondary wall at pits	Conduction of water and minerals; support	Xylem
Vessel element	Dead at maturity; end walls have perforations; lacks secondary wall at pits	Conduction of water and minerals; support	Xylem
Sieve-tube member	Living; end walls are sieve plates; lacks nuclei and other organelles at maturity	Conduction of dissolved food materials (carbohydrates)	Phloem
Companion cell	Living; cytoplasmic connections with sieve-tube member	Aids sieve-tube member	Phloem
Guard cells	Living; change shape	Open and close stomatal pore	Epidermis

PLANTS EXHIBIT LOCALIZED GROWTH AT MERISTEMS

Growth is a complex phenomenon involving three different processes: cell division, cell elongation, and cell differentiation. Cell division is an essential part of growth that results in an increase in the number of cells. However, an increase in cell number without a corresponding increase in cell size would contribute little to the overall size increase in a plant. Thus, cell elongation (the expansion of a cell) is an essential part of growth.

Plant cells also **differentiate**, or specialize, into the various cell types just discussed. These cell types compose the mature plant body and perform the various functions required in a complex, multicellular organism. Although differentiation does not contribute to an increase in size, it is considered an important aspect of growth because it is essential for tissue formation.

One difference between plants and animals is the *location* of growth. When a young animal is growing, all parts of its body grow, although not necessarily at the same rate. In contrast, plants grow only in specific areas called **meristems**, which are composed of cells that do not differentiate. Meri-

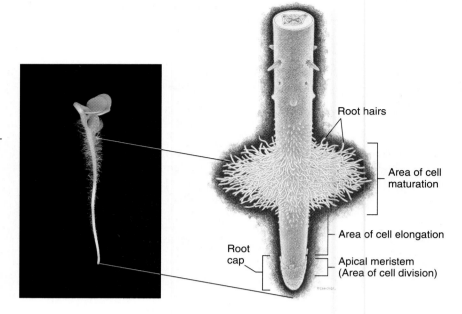

Figure 5-9 The root tip. The root apical meristem (where cells divide and thus increase in number) is protected by a root cap. Farther from the tip is an area of cell elongation, where cells enlarge and begin to differentiate. The area of cell maturation has fully mature, differentiated cells. Note the root hairs in this area and on the root of a young radish (*Raphanus sativus*) seedling (*left*). (*Dennis Drenner*)

Root hairs

Area of cell maturation

Area of cell elongation

Root cap

Apical meristem (Area of cell division)

stematic cells retain the ability to divide (see Chapter 13), a trait that differentiated cells lose. The persistence of meristems means that plants retain the capability for growth throughout their entire life spans.

Two kinds of growth may occur in plants. **Primary growth** is an increase in the length of a plant. All plants have primary growth, and tissues produced by primary growth compose the entire plant body in herbaceous plants and the young, soft shoot tips and root tips of woody trees and shrubs. **Secondary growth** is an increase in the girth of a plant. For the most part, only gymnosperms and woody dicots[3] have extensive secondary growth. Wood and bark, which make up most of the bulk of trees and shrubs, consist of tissues produced by secondary growth. A few annuals (for example, sunflower and geranium) have limited secondary growth but do not have obvious wood and bark tissues.

Primary Growth Takes Place at Apical Meristems

Primary growth occurs as a result of the activity of **apical meristems**, areas at the tips of roots and the buds of stems. Such growth is evident in a root tip (Figure 5-9). The root tip is covered by a protective layer of cells called the root cap. Directly behind the root cap is the root apical meristem, which consists of meristematic cells. Meristematic cells, which are very small and "boxy" in shape, remain small because they are continually dividing.

Farther back from the tip of the root, just behind the area of cell division, is an area of cell elongation where the cells

---

[3]On the basis of a variety of structural features, flowering plants are divided into two groups, informally called dicots and monocots. Oak, sycamore, ash, cherry, apple, and maple are examples of woody dicots, whereas bean, daisy, and snapdragon are examples of herbaceous dicots. Palm, corn, bluegrass, lily, and tulip are examples of monocots.

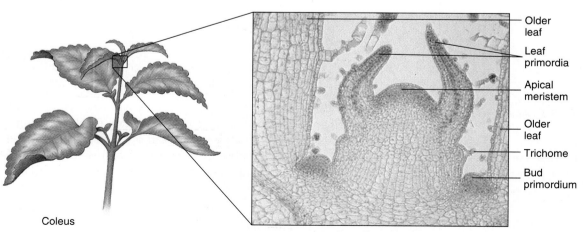

Coleus

Older leaf

Leaf primordia

Apical meristem

Older leaf

Trichome

Bud primordium

Figure 5-10 A longitudinal section through a terminal bud of coleus (*Coleus* sp) showing the stem apical meristem, leaf primordia, and bud primordia. (*James Mauseth, University of Texas*)

are no longer dividing, but instead grow longer, pushing the root tip deeper into the soil. Some differentiation also occurs in the area of cell elongation, and immature tissues become evident. The immature tissues continue to develop and differentiate into the mature tissues of the adult plant. Farther back from the tip, above the area of cell elongation, the cells have completely differentiated and are fully mature. Root hairs are evident in this area.

A stem bud—the terminal bud, for example—is quite different in structure from a root tip (Figure 5-10). Within every bud is a dome of tiny meristematic cells, the stem apical meristem. **Leaf primordia** (embryonic leaves) and **bud primordia** (embryonic buds) emerge from the embryonic stem tip. The leaf primordia tend to cover and protect the stem apical meristem. Farther back from the tip of the stem, the immature cells enlarge and differentiate into the three tissue systems.

Secondary Growth Takes Place at Lateral Meristems

In addition to primary growth (elongation), trees and shrubs have secondary growth, or increase in girth, in areas that are no longer elongating. Secondary growth is due to cell divisions that occur in **lateral meristems**, areas that extend the entire lengths of stems and roots except at the tips. Two lateral meristems are responsible for secondary growth: the vascular cambium and the cork cambium (Figure 5-11).

The **vascular cambium** is a layer of meristematic cells that forms a thin cylinder around the stem and root trunk, between the wood and bark of a woody plant. Cells of the vascular cambium divide, adding more cells to the wood (secondary xylem) and to the inner bark (secondary phloem).

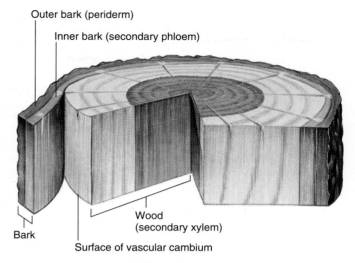

Outer bark (periderm)

Inner bark (secondary phloem)

Wood (secondary xylem)

Bark

Surface of vascular cambium

Figure 5-11   In secondary growth, plants increase in girth as a result of the activity of two lateral meristems. The vascular cambium, a thin layer of cells sandwiched between the wood and bark, produces secondary vascular tissues: the wood, which is secondary xylem, and inner bark, which is secondary phloem. The cork cambium produces the periderm, the outer bark tissues that replace the epidermis in the secondary plant body.

The **cork cambium** is a thin cylinder or irregular arrangement of meristematic cells in the outer bark region. Cells of the cork cambium divide to form the cork cells and cork parenchyma cells that make up the periderm. Chapters 6 and 7 contain a more comprehensive discussion of secondary growth.

STUDY OUTLINE

I. A plant body typically consists of a root system and a shoot system.
   A. The root system is generally underground and obtains water and dissolved minerals for the plant. Roots also anchor the plant firmly in place.
   B. The shoot system is generally aboveground and obtains sunlight and carbon dioxide for the plant.
      1. The shoot system consists of a vertical stem that bears leaves (the main organs of photosynthesis) and flowers and fruits (reproductive structures).
      2. Terminal and lateral buds (undeveloped embryonic shoots) develop on stems.
II. The plant body comprises three tissue systems.
   A. The ground tissue system consists of three tissues.
      1. Parenchyma tissue is composed of living parenchyma cells that possess thin primary cell walls. Functions include photosynthesis, storage, and secretion.
      2. Collenchyma tissue is composed of collenchyma cells

with unevenly thickened primary cell walls. This tissue provides flexible structural support.
      3. Sclerenchyma tissue is composed of sclerenchyma cells that have both primary and secondary cell walls. Sclerenchyma cells are often dead at maturity, but they provide structural support.
   B. The vascular tissue system conducts materials throughout the plant body and provides strength and support.
      1. Xylem is a complex tissue that conducts water and dissolved minerals. The actual conducting cells of xylem are tracheids and vessel elements.
      2. Phloem is a complex tissue that conducts food materials (carbohydrates) in solution. Sieve-tube members are the conducting cells of phloem; they are assisted by companion cells.
   C. The dermal tissue system is the outer protective covering of the plant body.
      1. The epidermis is a complex tissue that covers the herba-

ceous plant body. The epidermis that covers aerial parts secretes a layer of wax, called the cuticle, that reduces water loss. Gas exchange between the interior of the shoot system and the surrounding atmosphere occurs through stomata.

    2. The periderm is a complex tissue that forms the outer bark in woody plants.

  D. Although separate organs (roots, stems, and leaves) exist in the plant, many tissues are integrated throughout the plant body, providing continuity from organ to organ.

III. Growth in plants is localized in specific regions called meristems and involves cell division, cell elongation, and cell differentiation.

  A. Primary growth is an increase in stem and root length.

    1. Primary growth occurs in all plants.

    2. Primary growth is due to the activity of apical meristems at the tips of roots and at the buds of stems.

  B. Secondary growth is an increase in stem and root girth.

    1. In addition to primary growth, woody plants have secondary growth.

    2. Secondary growth is localized, typically as long cylinders of active growth throughout the lengths of older stems and roots.

    3. The two lateral meristems responsible for secondary growth are the vascular cambium and the cork cambium.

SELECTED KEY TERMS

annual, p. 77
apical meristem, p. 88
biennial, p. 77
collenchyma, p. 81
dermal tissue system, p. 78

epidermis, p. 85
ground tissue system, p. 78
lateral meristem, p. 89
parenchyma, p. 79
perennial, p. 77

periderm, p. 86
phloem, p. 84
primary growth, p. 88
root system, p. 77
sclerenchyma, p. 81

secondary growth, p. 88
shoot system, p. 77
tissue, p. 78
vascular tissue system, p. 78
xylem, p. 82

REVIEW QUESTIONS

1. What are some of the functions of roots? Of shoots?
2. What are the three tissue systems in plants? Describe the functions of each.
3. Compare the cellular structures and functions of parenchyma, collenchyma, and sclerenchyma cells.

4. What are the functions of xylem and phloem?
5. Compare and contrast epidermis and periderm.
6. What is the role of meristems in plants?
7. Distinguish between primary and secondary growth. Distinguish between apical and lateral meristems.

THOUGHT QUESTIONS

1. Grasses have special meristems at the bases of the leaves. Relate this information to what you know about mowing the lawn.
2. Why is meristematic tissue so important to a plant? What would happen if all the meristematic regions of a plant were removed?
3. A couple carved a heart with their initials in a tree trunk, 4 feet above ground level; the tree was 25 feet tall at the time. Twenty years later, the tree was 50 feet tall. How far above the ground were the initials? Explain your answer.

4. Sclerenchyma in plants is the functional equivalent of bone in animals (that is, both sclerenchyma and bone provide support). However, sclerenchyma is dead, and bone is living tissue. What are some of the advantages to a plant of having dead support cells? Can you think of any disadvantages?

SUGGESTED READINGS

Bell, A.D., *Plant Form: An Illustrated Guide to Flowering Plant Morphology,* Oxford, Oxford University Press, 1991. This beautifully illustrated book examines the external features of plants.

Petrides, G.A., *Peterson Field Guide to Eastern Trees,* Boston, Houghton Mifflin Company, 1988. Provides general information about the many oak species common in eastern North America. A similar field guide for western trees is also available.

Steele, M., and P. Smallwood, "What Are Squirrels Hiding?" *Natural History,* October 1994. Examines the complex relationship between squirrels and oaks.

PLANT ORGANS: ROOTS

If you chanced upon an American ginseng *(Panax quinquefolius)* plant growing in the eastern woodlands of North America, you would probably not pay much attention to it. This small, herbaceous plant consists of a short, erect stem bearing a whorl of three compound leaves, each divided into five leaflets, and a rounded cluster of tiny, greenish-white flowers that form bright red fruits. Ginseng is attractive and is sometimes grown as a ground cover in cool-temperate woodland gardens. From only a superficial look at the plant, however, it would be difficult to fathom why it has been collected—at one time almost to the point of extinction—since European colonists discovered it in the early 1700s. Furthermore, it would be hard to understand why the American ginseng is cultivated commercially today, with much of it being used in the United States, Russia, Europe, and China.

American ginseng (*Panax quinquefolius*), growing in a woodland in eastern North America. Many people attribute restorative powers to the roots of ginseng. (*Gregory K. Scott*)

two armlike branch roots, two leglike branch roots, and sometimes a branch root between the "legs" that resembles male genitals. The root of a related species of ginseng (*Panax ginseng*) that is native to China and Korea also possesses this remarkable appearance.

Because of its resemblance to the human body, ginseng root is thought by some to have curative powers for many human conditions and illnesses. This idea is based on the *Doctrine of Signatures,* a belief that was popular in the 1500s and 1600s but was later discredited. According to the Doctrine of Signatures, God created plants for human use and gave each a visible sign, or "signature," to indicate its purpose. Ginseng, considered a powerful Doctrine of Signatures plant, has been used as a tonic for a wide variety of conditions, including cancer, rheumatism, diabetes, impotence, sterility,

Although the aboveground portion of American ginseng is unremarkable, the underground part is quite extraordinary. The root of the American ginseng is thick and often forked. To an imaginative mind, it sometimes resembles a tiny human. Depending on how it branches, the root may possess

and aging. Indeed, its scientific name, *Panax*, is Latin for "cure-all." Some people dry ginseng root, grind it into a powder, and use it to brew a tea that is taken as a general tonic or as an aphrodisiac. Although the Chinese have used ginseng root medicinally for centuries and many Americans extol its virtues, the therapeutic value of ginseng has not been scientifically verified.

After reading this chapter, you should be able to:

1. Describe the functions of roots and how their structures correlate with these functions.

2. Label cross sections of a primary dicot root and a monocot root, and describe the functions of each tissue.

3. Trace the pathway of water from the soil through the various root tissues.

4. Describe several roots that are modified to perform unusual functions.

5. Discuss the significance of roots to humans.

THE ROOT SYSTEM GROWS INTO THE SOIL

Because roots are underground and out of sight, people do not always appreciate the important functions they perform for plants. A plant's branching underground root system is often more extensive than its aboveground parts. The roots of a corn plant, for example, may grow to a depth of 2.5 meters (about 8 feet) and spread outward 1.2 meters (4 feet) from the stem. Desert-dwelling tamarisk trees reportedly have roots that grow to a depth of 50 meters (165 feet) to tap underground water. The total root length, not counting root hairs, of a 4-month-old rye plant was measured and found to exceed 500 kilometers (310 miles)! The extents of a plant's root depth and spread vary considerably among species and even among different individuals in the same species. Soil conditions greatly affect the extent of root growth. (See Plants and the Environment: Soil Erosion, and Plants and the Environment:

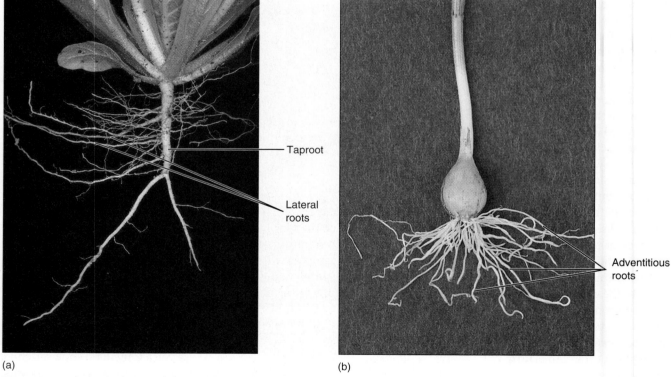

(a)   (b)

Figure 6-1   Root systems in plants. (a) The taproot system of a tobacco (*Nicotiana tabacum*) seedling, a dicot. The taproot developed from the embryonic root in the seed, and all lateral roots branching from the taproot came from the taproot. **(b)** The fibrous root system of an onion (*Allium cepa*), a monocot. All of the fibrous roots shown are adventitious and developed from stem tissue. (a, *James Mauseth, University of Texas;* b, *Marion Lobstein*)

Desertification for discussions of two soil problems that affect root growth.) Chapter 10 discusses soils.

Roots generally grow downward, in the direction of gravity. Two types of root systems—a taproot system and a fibrous root system—may develop from the embryonic root (the *radicle*) in the seed (Figure 6-1). A **taproot system** consists of one main root (formed from the enlarging embryonic root) with many smaller lateral (branch) roots coming out of it. Lateral roots often occur initially in regular rows along the length of the main root. Taproots are generally characteristic of dicots and gymnosperms. A dandelion is a good example of a common herbaceous dicot with a taproot system. A few trees, such as hickory, retain their taproots, which become quite massive as the plants age. Most mature trees, however, do not retain their taproots but have root systems that consist of large, shallow lateral roots from which other roots branch off and grow downward.

A **fibrous root system** has several to many roots of the same size that develop from the end of the stem, with smaller lateral roots branching off of them. Onions, crabgrass, and other monocots have fibrous root systems. Fibrous root systems form in plants in which the embryonic root is short-lived. The roots originate initially from the base of the embryonic root and later from stem tissue. Because fibrous roots do not arise from preexisting roots but from the stem, they are said to be *adventitious*. **Adventitious** organs occur in unusual locations, such as roots that develop on a stem or buds that develop on roots.

ROOTS PROVIDE PLANTS WITH ANCHORAGE, ABSORPTION/CONDUCTION, AND STORAGE

Roots perform three main functions for plants. First, as anyone who has ever pulled weeds can attest, roots anchor a plant securely in the soil. Because a plant remains in one location throughout its life and needs a solid foundation from which to grow, firm anchorage is essential for its survival.

Second, roots absorb water and dissolved mineral salts such as nitrates, phosphates, and sulfates from the soil. These materials are then transported throughout the plant, in the xylem. The two types of root systems, the taproot system and the fibrous root system, are adapted to obtain water in a variety of ways. For example, taproot systems often extend down into the soil to obtain water deep underground, whereas fibrous root systems, which are relatively close to the ground surface, are adapted to obtain rainwater as it drains into the soil.

Storage is the third main function performed by many roots. Surplus carbohydrates produced in the leaves by photosynthesis are transported in the phloem, as sugar, to the roots for storage, usually as sugar or starch, until they are

Figure 6-2   Beets (*Beta vulgaris*) and other storage roots are important sources of human food because of the accumulated sugars and starches stored in the roots.
(*Dennis Drenner*)

needed. Carrot roots, for example, have an extensive phloem for this purpose. Although roots use some of this sugar for their own respiratory needs, most of it is stored and later transported out of the root when the plant needs it. Either type of root system, taproot (beets, carrots, radishes, and turnips) or fibrous root (sweet potatoes and yams), may be modified for storing food (Figure 6-2; also see Plants and People: Important Root Crops). Storage taproots are usually biennials that store their food reserves in the root during the first year's growth and use it to reproduce during the second year's growth. Other plants, particularly those living in very dry regions, possess storage roots adapted to store water.

In certain species, roots are modified for functions other than anchorage, absorption/conduction, and storage. Later in this chapter we will discuss roots that are specialized to perform unusual functions.

PLANTS AND THE Environment

Soil Erosion

Soil, which is composed of mineral particles, organic material, water, and air, is a valuable natural resource on which humans depend for food. Water, wind, ice, and other agents cause **soil erosion**, the wearing away or removal of soil from the land. Water and wind are particularly effective in removing soil: rainfall loosens soil particles that can then be transported away by moving water (see figure), while wind loosens soil and blows it away, particularly if the soil is exposed and dry.

Soil erosion is a national and international problem that does not make the headlines very often. To get a feeling for how serious the problem is, consider that approximately 2.7 billion metric tons (3.0 billion tons) of topsoil are lost from U.S. farmlands as a result of soil erosion *each year*. The U.S. Department of Agriculture estimates that approximately one fifth of U.S. cropland is vulnerable to soil erosion damage. Because erosion reduces the amount of soil in an area, it limits the growth of plants. Erosion also causes a loss in soil fertility, because essential minerals and organic matter that are part of the soil are removed. As a result of these losses, the productivity of eroded agricultural lands declines.

Humans often accelerate soil erosion through poor soil manage-

ment practices. Agriculture is not the only culprit. The removal of natural plant communities during the construction of roads and buildings accelerates erosion. Unsound logging practices such as clear-cutting large forested areas for lumber and pulpwood cause severe erosion.

Soil erosion has a detrimental impact on other natural resources. Sediments that get into streams, rivers, and lakes, for example, affect water quality and fish habitats. If the sediments contain pesticide and fertilizer residues, as they often do, they further pollute the water.

Sufficient plant cover limits soil erosion: leaves and stems cushion the impact of rainfall, and roots help to hold the soil in place. Although soil erosion is a natural process, plant cover makes it negligible in many natural ecosystems.

Wind Has Contributed to Soil Erosion in Grasslands

Semiarid lands such as the Great Plains[1] of North America have low annual precipitation rates and are subject to periodic droughts that may last for extended periods. Prairie grasses, the plants that grow best in

[1]The Great Plains extend from Texas north to Alberta and from the base of the Rocky Mountains east for about 650 kilometers (400 miles).

semiarid lands, are adapted to survive droughts. Although the aboveground portions of the plant may die, the dormant root systems can survive several years of drought. When the rains return, the root systems send up new shoots. Soil erosion is minimal because the living, dormant root systems hold the soil in place against the assault by wind.

Problems arise, however, when large areas of land are cleared for crops or when the land is overgrazed by too many animals. The removal of the natural plant cover and the corresponding loss of root systems open the way for climatic conditions to "attack" the soil, which gradually deteriorates from the onslaught of hot summer sun, occasional violent rainstorms, and wind. If a prolonged drought occurs under such conditions, disaster can strike.

The American Dust Bowl Occurred During the 1930s

The inhabitants of a wide region of the central United States vividly experienced the effects of wind on soil erosion during the 1930s. Throughout the late 19th and early 20th centuries, farmers removed much of the native grasses to plant wheat. Then, between 1930 and 1937, the semiarid lands stretching from Oklahoma and Texas into Canada received 65 percent less annual precipitation than was normal. The rugged prairie grasses that had been replaced with crops could have survived these conditions, but there was not enough water to support wheat. The prolonged drought caused crop failures, which left fields barren and vulnerable to wind erosion.

Winds from the west swept across the barren, exposed soil, causing dust storms of incredible magnitude. Colorado and Oklahoma topsoil was blown eastward for hundreds of kilometers. Women hanging out clean laundry in Georgia went outside later to find it dust-covered. Bakers in New York City had to keep freshly baked bread away from open windows and the flying dust, which even discolored the ocean several hundred kilometers off the Atlantic coast.

The Dust Bowl occurred during the Great Depression, and many farmers went bankrupt. Farmers abandoned their dust-choked land and dead livestock and migrated west to the promise of California. The novel *The Grapes of Wrath* by John Steinbeck movingly portrays their plight.

Although the United States no longer has a dust bowl, soil erosion is still a major problem there and elsewhere. The first global assessment of soil conditions, released in 1992, was based on a 3-year study sponsored by the United Nations Environment Program. It reported that 1.96 billion hectares (4.84 billion acres)—an area equal to 17 percent of Earth's total vegetated land—had been degraded since World War II. Soil erosion is responsible for much of this degradation.

Soil erosion caused by water. Gullies, which form as a result of erosion damage from water, expand rapidly because the erosion accelerates as they serve as channels for the runoff of rainwater. (*USDA/Soil Conservation Service*)

Roots Possess a Root Cap and Root Hairs

Each root tip is covered by a **root cap**, a protective thimble-like layer many cells thick that covers the delicate root apical meristem (Figure 6-3a; also see Figure 5-9). As the root grows, pushing its way through the soil, parenchyma cells of the root cap slough off and are replaced by new cells, formed by the root apical meristem toward its outer side.

The root cap also appears to be involved in orienting the root so that it grows downward. When a root cap is removed experimentally, the root apical meristem grows a new cap. However, until the root cap has regenerated, the root grows randomly rather than in the direction of gravity.

Root hairs are short-lived, single-celled extensions of epidermal cells near the growing root tip. Root hairs form continually in the area of cell maturation closest to the root tip, to replace those that are dying off at the more mature end of the root hair zone. Each root hair is short (typically less than 1 centimeter, or 0.4 inch, in length), but root hairs are quite numerous. They greatly raise the absorptive capacity of the root by increasing the surface area of the root that is in contact with the moist soil (Figure 6-3b). Soil particles are coated with a microscopically thin layer of water in which minerals are dissolved. The root hairs establish intimate contact with soil particles and absorb much of this water.

Herbaceous Dicot Roots Can Be Distinguished from Monocot Roots by the Arrangement of Vascular Tissues

Although considerable variation exists in roots, they all possess an outer protective covering (the epidermis), a cortex for storage of starch and other organic molecules, and vascular tissues for conduction. Let us consider first the structures of herbaceous dicot roots and monocot roots.

The Central Core of Vascular Tissue in Most Herbaceous Dicot Roots Lacks Pith

The buttercup root is a representative dicot root with primary growth (Figure 6-4). Like other parts of this herbaceous dicot, the roots are covered by a single layer of protective tissue, the **epidermis**. The root hairs are a modification of the root epidermis that enables it to absorb more water from the soil. The root epidermis does not secrete a thick, waxy cuticle—particularly in the region of root hairs, where a cuticle would impede the absorption of water from the soil. Both the lack of a cuticle and the presence of root hairs increase absorption. However, most of the water that enters the root moves along the cell walls rather than entering the cells. One of the major components of cell walls is cellulose, which absorbs

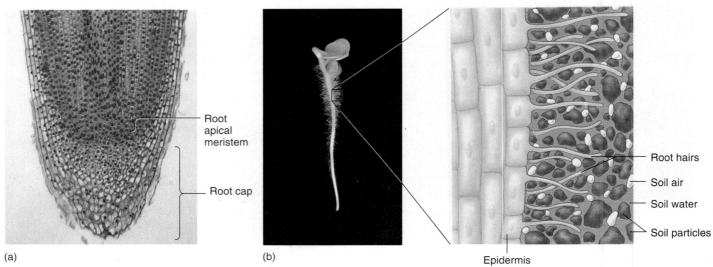

(a)   (b)   Root apical meristem

Root cap

Root hairs

Soil air

Soil water

Soil particles

Epidermis

Figure 6-3   Structures unique to roots. (a) The root cap of an onion (*Allium cepa*) root. The root cap protects the root's apical meristem. **(b)** Root hairs on a radish (*Raphanus sativus*) seedling. Each delicate hair is a single-celled extension of the root epidermis. Root hairs increase the surface area in contact with the soil. (a, *Gail and Jim Nachel/Dembinsky Photo Associates;* b, *Dennis Drenner*)

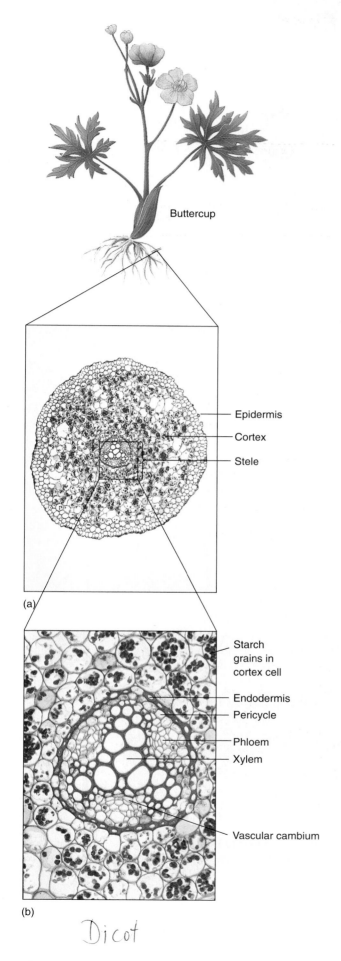

Buttercup

Epidermis

Cortex

Stele

(a)

Starch grains in cortex cell

Endodermis

Pericycle

Phloem

Xylem

Vascular cambium

(b)

Dicot

Figure 6-4   Structure of a herbaceous dicot root.
(a) Cross section of a buttercup (*Ranunculus* sp) root. Note that the bulk of the root is the cortex. **(b)** A close-up of the stele of the buttercup root. Note the solid core of vascular tissues. (a, b, *James Mauseth, University of Texas*)

water like a sponge. Cotton balls, which are almost pure cellulose, illustrate the absorptive property of this material. (Chapter 10 will explain *why* water moves from the soil into the root.)

The **cortex** of a herbaceous dicot root, which is composed primarily of loosely arranged parenchyma cells with large intercellular (*between*-cells) spaces, makes up the bulk of the root. The root cortex usually lacks supporting collenchyma cells, although it may develop some supporting sclerenchyma cells as it ages. The primary function of the root cortex is storage. A microscopic examination of the parenchyma cells that form the cortex often reveals numerous starch grains. Starch, an insoluble carbohydrate composed of glucose units, is the most common form of stored food in plants.

The large intercellular air spaces, another feature of root cortex, provide a pathway for water uptake and allow for aeration of the root. The oxygen that root cells need for aerobic respiration diffuses from air spaces in the soil to the intercellular spaces of the cortex, and from there to the cells of the root.

The inner layer of the cortex, the **endodermis,** controls the amounts and kinds of minerals that enter the xylem in the root's interior. Harmful minerals may be blocked, whereas beneficial minerals are absorbed into the xylem and transported to the stem, leaves, and other parts of the plant.

Structurally, the endodermis differs from the rest of the cortex. Endodermal cells fit snugly against each other, and each cell has a special bandlike region on its radial and transverse walls, called a **Casparian strip** (Figure 6-5). If you think of the endodermis as a hollow cylinder constructed of bricks, the endodermal cells correspond to the bricks, and the Casparian strips correspond to the mortar between them. Casparian strips contain *suberin,* a fatty material that is waterproof.

When dissolved minerals enter the root cortex from the epidermis, they move in solution along two pathways, the apoplast and symplast, until they reach the endodermis (Figure 6-6). The **apoplast** is a pathway that consists of the interconnected porous cell walls along which water flows freely. The **symplast** is a continuum of interconnected cell *interiors*—that is, a pathway from the cytoplasm of one cell to that of an adjacent cell by way of cytoplasmic connections.

Up to this point, most of the dissolved minerals may have traveled along the apoplast instead of passing through a plasma membrane or entering the cytoplasm of a root cell. However, the waterproof Casparian strip on the radial and transverse walls of endodermal cells prevents water and min-

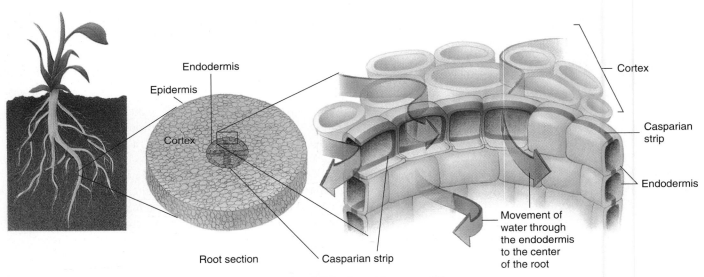

Endodermis

Epidermis

Cortex

Cortex

Casparian
strip

Endodermis

Movement of
water through
the endodermis
to the center
of the root

Root section

Casparian strip

Figure 6-5   A few cells of the endodermis. Note the Casparian strip around the radial and transverse walls. The endodermis controls uptake of minerals by the root.

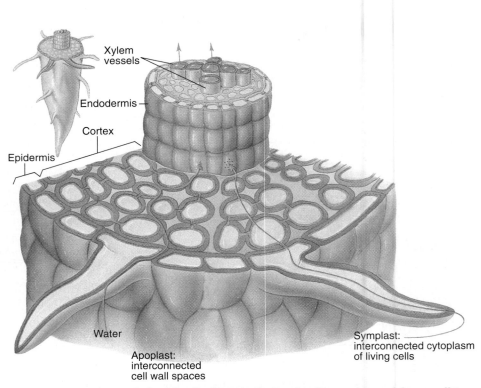

Xylem
vessels

Endodermis

Cortex

Epidermis

Water

Apoplast:
interconnected
cell wall spaces

Symplast:
interconnected cytoplasm
of living cells

Figure 6-6   Water and dissolved minerals that enter the root travel from cell to cell along the interconnected porous cell walls (the apoplast) or from one cell's cytoplasm to another through cytoplasmic connections (the symplast). On reaching the endodermis, water and minerals can continue to move into the root's center only if they pass through the plasma membrane and enter an endodermal cell. The Casparian strip blocks the passage of water and minerals along the endodermal cell wall.

People

Important Root Crops

Many roots are storage organs, which store the products of photosynthesis and are important sources of food for human consumption. Root crops are predominantly taproots, including carrots, beets, sugar beets, parsnips, turnips, rutabagas, and radishes. These taproot crops are *biennials,* plants that accumulate starch and sugar in their roots during the first year's growth and then use it to reproduce during the second year. However, the roots are harvested as annuals—that is, after the first year's growth. Only a few root crops—sweet potatoes, yams, and cassava—are fibrous roots.

Worldwide, more cassava and sweet potatoes are grown than all other roots combined (see figure). Cassava (*Manihot esculenta*), also called manioc, is a tropical American plant that is grown in many tropical countries worldwide for its edible starchy roots, which resemble very large sweet potato roots. It is a mainstay in the diets of millions of people. Tapioca is a granular starch squeezed out of cassava roots and used to make puddings and to thicken soups, primarily in North America and Europe. Sweet potatoes (*Ipomoea batatas*), also American in origin, are widely planted not only in the Americas but also in West Africa, China, India, and the Pacific Islands. Sweet potatoes are often confused

with yams, a root crop belonging to a different genus (*Dioscorea*).

Because roots are generally rich in carbohydrates and certain vitamins but poor in proteins, people must consume them with other foods in order to achieve a balanced diet. The most nutritious roots are sweet potatoes and yams, both of which contain about 5 percent protein and are rich in vitamins A and D in addition to iron, calcium, and other minerals.

The sugar beet (*Beta vulgaris*) is an important agricultural product that provides 35 percent of the world's sugar. Sugar beets were developed during the 18th century from a white variety of the common beet. Using **artificial selection**, the process by which humans deliberately enhance

desirable features over time by selecting which individuals to cross, scientists increased the sugar content of these beet roots from about 2 percent to around 20 percent. Interestingly, Napoleon supported the development of sugar beets because he wanted to end England's monopoly on cane sugar in France. The sugar produced by sugar beets is sucrose, the same as that in sugarcane.

In addition to their use as important food crops, some roots are used as flavorings. For example, root beer gets its characteristic flavor from dried greenbrier (*Smilax* spp) roots, as does sarsaparilla, which was once widely used as a medicinal tonic.

Cassava (*Manihot esculenta*) and sweet potato (*Ipomoea batatas*) are very important tropical root crops.

erals from continuing along the cell walls. Water enters endodermal cells by osmosis, whereas minerals enter the endodermal cells by passing through carrier proteins in their plasma membranes. Thus, the endodermis controls the movement of minerals into the root, even though it is an internal tissue and minerals pass through other tissues to reach it.

Just inside the endodermis is a single layer of cells called the **pericycle**, which gives rise to lateral roots (Figure 6-7).

The pericycle is composed of parenchyma cells that remain meristematic. Lateral roots originate when a portion of the pericycle starts dividing. As it grows, the lateral root breaks through several layers of root tissue (endodermis, cortex, and epidermis) before entering the soil. Each lateral root has all the structures and features of the larger root from which it branches: root cap, root hairs, epidermis, cortex, endodermis, pericycle, xylem, and phloem. In addition to producing lateral

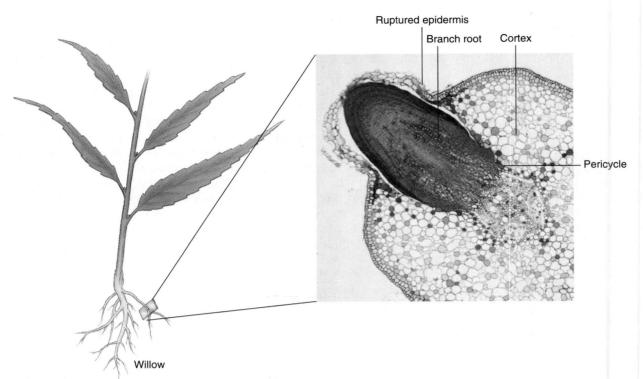

Figure 6-7   A multicellular lateral root emerges from a larger root. Lateral roots originate at the pericycle. (*James Mauseth, University of Texas*)

roots, the pericycle is involved in forming the lateral meristems that produce secondary growth in woody roots (to be discussed shortly).

At the center of a primary dicot root is a cylinder of vascular tissues known as a **stele** (Figure 6-4b). **Xylem**, the centermost tissue, often has two, three, four, or more extensions, or "xylem arms." **Phloem** is located in patches between the xylem arms. The xylem and phloem of the root have the same functions as in the rest of the plant: xylem conducts water and dissolved minerals, and phloem conducts carbohydrates (sucrose). After passing through the endodermal cells, water enters the root xylem, often at one of the xylem arms. Up to this point the pathway of water has been horizontal, from the soil into the center of the root:

Root hair → epidermis → cortex (apoplast or symplast pathway) → endodermis → pericycle → xylem of root

Once water enters the xylem, it is transported upward through root xylem into stem xylem and throughout the rest of the plant.

Phloem conducts dissolved sugar (sucrose) from the leaves, where it is made by photosynthesis, to the root, where it is stored, or from the root to other parts of the plant, where it is used for growth and maintenance of tissues. The **vascular cambium**, which gives rise to secondary tissues, is sandwiched between the xylem and phloem. The primary dicot root lacks a **pith**, a center core of parenchyma tissue within the vascular cylinder.

Xylem Does Not Form the Central Tissue in Some Monocot Roots

Although monocot roots exhibit considerable variation in structure when compared to dicot roots, they possess the same basic tissues. Starting at the outside of a greenbrier root (a representative monocot root) and moving inward, we find epidermis, then cortex, endodermis, and pericycle (Figure 6-8). Unlike herbaceous dicot roots, a monocot root does not have a solid cylinder of vascular tissues in its center. Instead, the phloem and xylem are in separate alternating strands that in cross section are arranged in a circle around the centrally located pith.

Because the vast majority of monocots do not have secondary growth, there is no vascular cambium in monocot roots. Despite their lack of secondary growth, monocots that are long-lived, such as palms, often have thickened roots. The cortex expands in such plants.

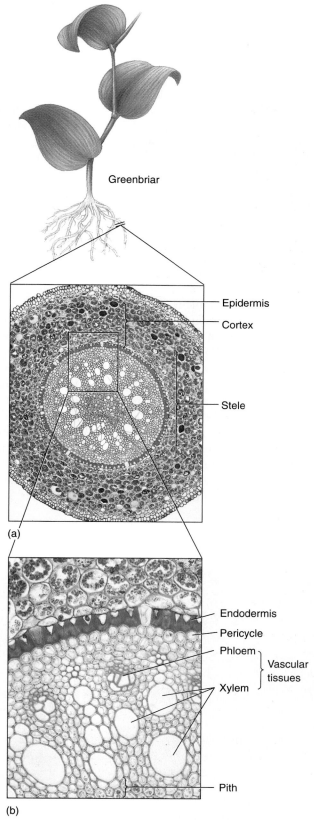

Greenbriar

Epidermis

Cortex

Stele

(a)

Endodermis

Pericycle

Phloem ⎫
⎬ Vascular
Xylem ⎭ tissues

Pith

(b)

Figure 6-8   Structure of a monocot root. (a) Cross section of a greenbriar (*Smilax* sp) root. Note the extensive cortex. **(b)** Close-up of a portion of the center of the root, showing the vascular tissues and pith. (a,b, *James Mauseth, University of Texas*)

WOODY PLANTS HAVE ROOTS WITH SECONDARY GROWTH

Plants that produce stems with secondary growth—that is, wood and bark—also produce roots with secondary growth. Gymnosperms and woody dicots have primary growth at apical meristems, while secondary growth occurs at lateral meristems. The production of secondary tissues occurs some distance back from the root tips and is the result of the activity of two lateral meristems, the vascular cambium and the cork cambium. Major roots of trees are often massive and possess both wood and bark. In temperate climates, the wood of both roots and stems exhibits annual rings in cross section. Chapter 7 discusses secondary growth in greater detail.

SOME ROOTS ARE SPECIALIZED FOR UNUSUAL FUNCTIONS

Adventitious roots often arise from stem nodes (regions of the stem where leaves are attached). Many adventitious roots that are aerial are adapted for functions other than anchorage, absorption, conduction, or storage. **Prop roots** are adventitious roots that develop from branches or from a vertical stem and grow downward into the soil to help support the plant in an upright position (Figure 6-9). Prop roots are more common in monocots than in dicots. Corn and sorghum, both monocots, are herbaceous plants that produce prop roots. Some tropical dicot trees, such as red mangrove and banyan trees, also produce prop roots.

Other aerial roots have additional functions. In swampy or tidal environments where the soil is flooded or waterlogged, for example, roots often grow upward until they are above the high-tide level. Even though roots live in the soil, they still require oxygen for aerobic respiration. A flooded soil is depleted of oxygen, so these aerial "breathing" roots, known as **pneumatophores**, may assist in getting oxygen to the submerged roots. Pneumatophores have a well-developed system of internal air spaces that is continuous with the submerged parts of the root, presumably allowing gas exchange. Black mangrove, white mangrove, and bald cypress are examples of plants with pneumatophores (Figure 6-10).

Epiphytes (plants that grow attached to other plants) and climbing plants have aerial roots that anchor the plant to the bark, branch, or other surface on which it grows. The aerial roots of some epiphytes are specialized for functions other than anchorage. Certain epiphytic orchids, for example, have photosynthetic roots that form the bulk of the plant. Epiphytic roots may absorb moisture, as well. Some parasitic epiphytes, such as mistletoes, have roots that penetrate the

— Branch

— Prop roots

(a)

— Stem

— Prop roots

(b)

Figure 6-9   Prop roots are adventitious roots that provide additional support. (a) The banyan tree (*Ficus benghalensis*) has aerial prop roots that eventually form trunks that support the branches. A single banyan tree can cover several acres of land. **(b)** Prop roots in corn (*Zea mays*) arise near the base of the stem. (a, *James Mauseth, University of Texas;* b, *Dennis Drenner*)

the strangler fig grew is often killed as the strangler fig grows around it, competing with it for light and other resources. Eventually the strangler fig becomes a self-supporting tree.

Plants that produce corms or bulbs (underground stems or buds specialized for asexual reproduction) often have wiry **contractile roots** in addition to their "normal" roots. The contractile roots grow into the soil and then contract (the root cells shorten or totally collapse on themselves), thus pulling the corm or bulb deeper into the soil (Figure 6-12). In bulbs and corms, each succeeding year's growth is *on top of* the preceding year's growth. As a result, bulbs and corms tend to move upward in the soil over time. Without contractile roots, they would eventually be exposed at the soil's surface. Contractile roots are common in monocots, but certain dicots and ferns also possess them.

Some roots reproduce asexually by producing **suckers,** which are aboveground stems that develop from adventitious buds on the roots. Each sucker grows additional roots and becomes an independent plant when the parent plant dies. Examples of plants that form suckers include black locust, pear, apple, cherry, red raspberry, and blackberry. Some weeds—field bindweed, for example—can produce many suckers. These plants are difficult to control, because pulling the plant out of the soil seldom removes all of the roots. The roots of field bindweed can grow as deep as 3 meters (10 feet) in the soil. In fact, in response to a wound the roots produce additional suckers, which can be a considerable nuisance.

Pneumatophores

Figure 6-10   Many plants living in swampy areas have pneumatophores, aerial roots thought to be adapted for aeration. Black mangrove (*Avicennia germinans*) produces pneumatophores that may provide oxygen for roots buried in anaerobic (oxygen-deficient) mud. (*Visuals Unlimited/Nada Pecnik*)

host-plant tissues and absorb nutrients (Figure 6-11). One plant that starts its life as an epiphyte is the strangler fig, which produces long aerial roots that eventually reach the ground and anchor the plant in the soil. The tree on which

(a)

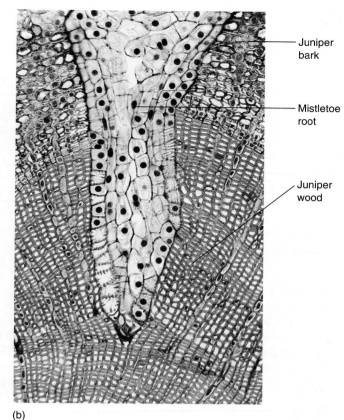

Juniper bark

Mistletoe root

Juniper wood

(b)

Figure 6-11   Specialized roots of parasitic epiphytes.
(a) Clumps of mistletoe (*Phoradendron serotinum*) have invaded a host oak (*Quercus* sp) tree in Florida. Mistletoe is a
parasite that obtains nutrients from the host on which it lives.
This photograph was taken in winter after the oak shed its
leaves, so that the extent of mistletoe invasion would be
obvious. **(b)** Micrograph of a parasitized juniper (*Juniperus*
sp) branch, showing a mistletoe root. (a, *James L. Castner;*
b, *C. Calvin, Portland State University*)

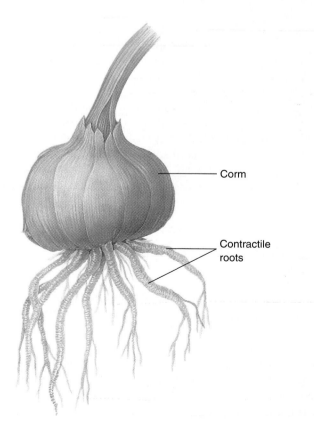

Corm

Contractile roots

**Figure 6-12   Plants that produce corms or bulbs often
have <u>contractile roots</u>.** During succeeding seasons, contractile roots pull the corm (*shown*) or bulb deeper and
deeper until it reaches some optimum depth.

PLANTS AND THE Environment

Desertification

Rangelands are grasslands, in both temperate and tropical climates, that serve as important areas of food production for humans by providing fodder for domestic animals such as cattle, sheep, and goats. Grasses, the predominant vegetation of rangelands, have a fibrous root system, in which many diffuse roots anchor the plant in the soil. Such plants hold the soil in place quite well, thereby reducing soil erosion. If animals eat only the upper portion of the grass, the bottom part can continue to develop, enabling the plant to recover and grow to its original size again.

The **carrying capacity** of a rangeland is the maximum number of animals that the rangeland plants can support in a sustainable fashion. When the carrying capacity of a rangeland is exceeded, grasses and other plants are **overgrazed**; that is, the grazing animals consume so much of the plant that it cannot recover and therefore dies. Overgrazing results in barren, exposed soil that is susceptible to erosion.

Most of the world's rangelands occur in semiarid areas that have extended natural periods of drought. Under normal conditions, native grasses can survive a drought; if the drought becomes severe, the top portion of the plant dies back while the underground part of the plant remains alive. Although the grasses of a drought-stricken rangeland appear to be dead, their extensive, living root systems hold the soil in place. When the rains return, the underground parts send forth new stems and leaves.

When overgrazing occurs in combination with a period of extended drought, once-fertile rangeland can be converted to desert. The lack of plant cover due to overgrazing allows winds to erode the soil. Even when the rains return, the land is so degraded that it cannot recover. Water erosion removes the little remaining topsoil, and the land turns into sand dunes. This conversion of rangeland to desert is called **desertification**. Desertification ruins economically productive land and destroys wildlife habitat.

Desertification is related to overpopulation. From 1980 to 1986, a disasterous drought struck the arid lands of East Africa, particularly in Ethiopia, Sudan, and Mozambique. Approximately a decade earlier, a devastating drought had occurred in the African Sahel, which extends south of the Sahara from Senegal to Somalia. In both cases, the people living in these lands suffered greatly. Many children and adults starved to death after their crops failed. These arid lands have always had periodic droughts, but what made the droughts during this period so devastating was that too many people (and livestock) were attempting to live on such ecologically fragile land. The people overwhelmed the land and degraded it; they chopped down most of the trees for firewood, and their livestock severely overgrazed the rangeland. Had many fewer people been farming and ranching in such a marginally productive area, they might have been spared the horrors of starvation.

ROOTS FORM RELATIONSHIPS WITH OTHER SPECIES

As roots of certain trees grow through the soil, they sometimes encounter roots of other trees of the same or different species. When this occurs, the two types of roots may grow together to form a natural **graft**—as, for example, between a birch and a maple. (Figure 6-13). Because the two trees' vascular tissues are connected in the graft, dissolved sugars and other materials such as hormones pass between them. Root grafts have been observed in more than 160 species of trees.

The roots of most plant species form a mutually beneficial relationship with certain soil fungi. These associations, known as **mycorrhizae**, enable plants to absorb adequate amounts of certain essential minerals (for example, phosphorus) from the soil. The threadlike body of the fungal partner extends into the soil, extracting minerals well beyond where the roots grow. Minerals absorbed from the soil by the fungus travel to the roots, and carbohydrate molecules produced by photosynthesis in the plant travel to the fungus. Mycorrhizae often enhance plant growth, and when mycorrhizae are not present, neither the fungus nor the plant grows as well (Figure 6-14).

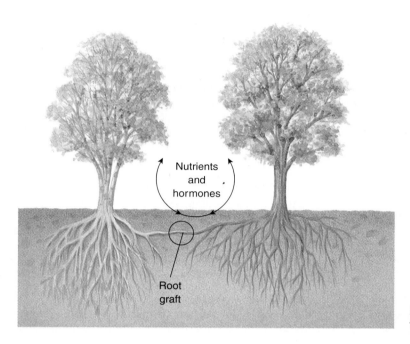

Figure 6-13   **Root grafts in trees often occur between two species, such as birch (*left*) and maple (*right*).**

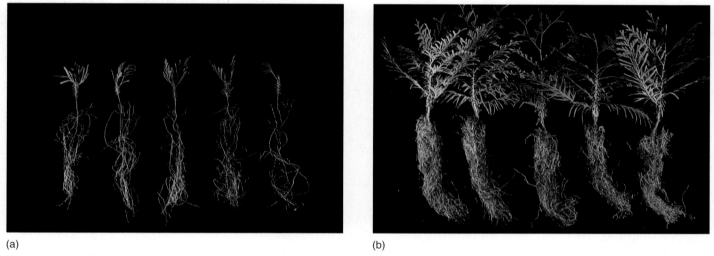

(a)   (b)

Figure 6-14   Western red cedar (*Thuja plicata*) seedlings respond to mycorrhizae.
(a) Control plants grown in low phosphorus in the absence of the fungus. **(b)** These seedlings were grown under conditions identical to the control, except that their roots formed mycorrhizal associations. (a, b, *Courtesy of Randy Molina, U.S. Forest Service*)

The roots of some plants form an association with nitrogen-fixing bacteria. Swellings called **nodules** develop on the roots and house millions of the bacteria. Like mycorrhizae, the association between nitrogen-fixing bacteria and the roots of plants is mutually beneficial. The bacteria receive the products of photosynthesis from the plants while helping the plant to meet its nitrogen requirement. Because nitrogen is an essential part of biologically important molecules such as proteins and nucleic acids, organisms—plants included—must have nitrogen in order to survive.

Casuarinas

STUDY OUTLINE

I. Two types of root systems exist in plants.
   A. A taproot system has one main root (the taproot) with many smaller lateral roots coming out of it.
   B. A fibrous root system has several to many roots of the same size that develop from the end of the stem. Because fibrous roots do not come from preexisting roots, they are said to be adventitious.
II. The main functions of roots are anchorage, absorption, conduction, and storage.
III. Roots have several unique structures.
   A. Each root tip has a root cap, a protective thimble-like layer many cells thick that covers the delicate root apical meristem. The root cap also appears to be involved in orienting the root so that it grows downward.
   B. Root hairs are short-lived, single-celled extensions of the epidermal cells near the growing root tip. Root hairs increase the surface area of the root that is in contact with the moist soil.
IV. Primary roots possess an outer protective covering (epidermis), ground tissues (cortex and, in certain roots, pith), and vascular tissues (xylem and phloem).
   A. The epidermis protects the root, and its root hairs aid in absorption of water and dissolved minerals.
   B. The cortex consists of parenchyma cells that often store starch.
   C. The endodermis, the innermost layer of the cortex, controls uptake of minerals into the xylem of the root.
      1. The cells of the endodermis possess a Casparian strip around their radial and transverse walls that is impermeable to water and dissolved minerals.
      2. Minerals therefore pass through carrier proteins in the plasma membranes of the endodermal cells.
   D. The pericycle is the origin of lateral roots.
   E. The xylem conducts water and dissolved minerals, and the phloem conducts dissolved sugar.

V. There are some structural differences between monocot and herbaceous dicot roots.
   A. The xylem and phloem of herbaceous dicot roots form a solid mass in the center of the root. In contrast, monocot roots often have a pith in the center of the root.
   B. Monocot roots lack a vascular cambium and therefore do not have secondary growth.
VI. Roots of gymnosperms and woody dicots develop secondary tissues (wood and bark).
VII. Roots may be modified for specialized functions.
   A. Prop roots are adventitious roots that develop from branches or from a vertical stem and grow downward into the soil to help support the plant in an upright position.
   B. Pneumatophores are aerial "breathing" roots that may assist in getting oxygen to the submerged roots.
   C. Certain epiphytes have roots that are modified to photosynthesize, absorb moisture, or—if the plant is parasitic—penetrate a host's tissues and absorb nutrients.
   D. Corms and bulbs often have contractile roots that grow into the soil and then contract, thereby pulling the corm or bulb deeper into the soil.
   E. Some roots reproduce asexually by producing suckers, aboveground stems that develop from adventitious buds on the roots.
VIII. Roots of one species of plant often form associations with other species.
   A. A root graft is a natural union between the roots of trees of the same or different species.
   B. Mycorrhizae are mutually beneficial associations between roots and soil fungi.
   C. Root nodules are swellings that develop on certain roots and house millions of nitrogen-fixing bacteria.

SELECTED KEY TERMS

adventitious, p. 93
Casparian strip, p. 97
contractile roots, p. 102
cortex, p. 97
desertification, p. 104
endodermis, p. 97

epidermis p. 96
fibrous root system, p. 93
mycorrhizae, p. 104
nodules, p. 105
pericycle, p. 99
phloem, p. 100

pith, p. 100
pneumatophore, p. 101
prop root, p. 101
root cap, p. 96
root graft, p. 104
root hair, p. 106

stele, p. 100
sucker, p. 102
taproot system, p. 93
vascular cambium, p. 100
xylem, p. 100

REVIEW QUESTIONS

1. What are the two types of root systems?
2. List several functions of roots, and describe the tissue(s) responsible for each function.
3. How would you distinguish between a root hair and a small lateral root?
4. If you were examining a cross section of a primary root of a flowering plant, how would you determine whether the plant is a dicot or a monocot?
5. Trace the pathway of water from the soil through the various root tissues in a herbaceous dicot root.
6. Distinguish among the following specialized roots: storage root, prop root, aerial "breathing" root, photosynthetic root, and contractile root.

7. Label the tissues of this herbaceous dicot root. Give at least one function for each tissue.

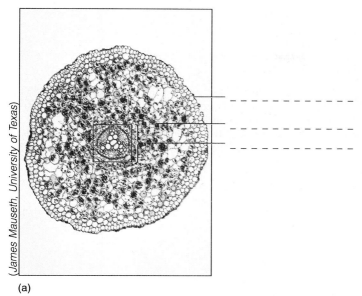

(a)

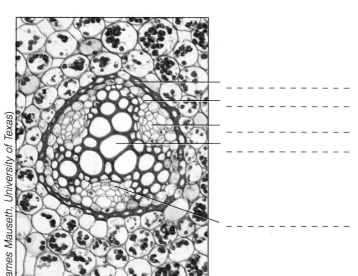

(b)

8. Label the tissues of this monocot root. Give at least one function for each tissue.

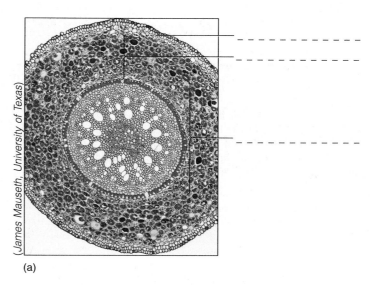

(a)

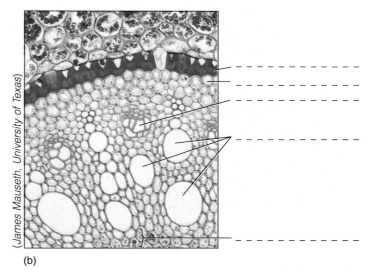

(b)

p. 97

THOUGHT QUESTIONS

1. A mesquite root is found penetrating a mine shaft about 150 feet below the surface of the soil. How might you determine *when* the root first grew into the shaft? (*Hint:* Mesquite is a woody plant.)

2. A barrel cactus that is 2 feet tall and 1 foot in diameter has roots more than 10 feet long. However, all of the plant's roots are found in the soil at a depth of 2 to 6 inches. What possible adaptive value does such a shallow root system confer on a desert plant?

3. Someone gives you a plant structure that was found growing in the soil and asks you to determine whether it is root or an underground stem. How would you identify the plant part without a microscope? With a microscope?

4. The root cap is thought to orient the root so that it grows downward, in the direction of gravity. Why might it be advantageous for roots to grow downward?

5. Some biologists have suggested that the roots of all the plants living in a forest are a single functional unit. Explain.

6. The American Dust Bowl is sometimes portrayed as a natural disaster brought on by drought and high winds. Present a case for the point of view that this disaster was caused less by nature than by humans.

Suggested Readings

Feldman, L.J., "The Habits of Roots," *Bioscience* Vol. 38, No. 9, October 1988. How roots interact with their soil environment.

Gillis, A.M., "Should Cows Chew Cheatgrass on Commonlands?" *Bioscience* Vol. 41, No. 10, November 1991. A new range war pits cattle growers against conservationists over degraded rangelands.

Moore, P., "Upwardly Mobile Roots," *Nature* Vol. 341, September 21, 1989. An essay on unusual root behaviors, including roots that grow out of the soil and up the trunk of a tree to obtain nutrients leaching from the forest canopy.

Sachs, A., "Dust to Dust," *World Watch* Vol. 7, No. 1, January/February 1994. Could another dust bowl occur in the plains states? The author builds a case for such an event.

PLANT ORGANS: STEMS

During colonial times, one of of the most highly prized trees in the forests of eastern North America was the eastern white pine (*Pinus strobus*). Because its trunk grows so straight and tall, the eastern white pine made an ideal mast for sailing ships, and representatives of the English crown reserved the very best trees for the Royal Navy. The wood of eastern white pine is lightweight and straight-grained. It contains less resin (a clear to yellowish viscous liquid that exudes from trees, possibly to deter plant-eating insects) than the wood of other pine species. These characteristics make the wood suitable for house construction (frames, interior woodwork, flooring, and paneling), as well as for fashioning crates, boxes, inexpensive furniture, railroad ties, and many other items.

Because of the species' suitability for masts and home construction, Europeans in North America took their first lumber from the vast forests of eastern white pine. Although eastern white pines in virgin (uncut) forests often grew 61 meters (200 feet) or more in height, they were logged to such an extent that few of these giant trees remain today. By the end of the 19th century, overlogging with no attention to conservation had taken its toll, and almost all of

A plantation of eastern white pine (*Pinus strobus*). Most trees are 40 to 50 years old. (*U.S. Forest Service photo by B. Peacock*)

the virgin tracts of eastern white pine were gone. Since then, certain areas have been reforested extensively with eastern white pine, but none of these trees can yet approach the majestic size of those in the original forests. Today eastern white pines rarely exceed heights of 37 meters (120 feet), and most are less than 200 years old.

Pines are evergreen conifers (cone-bearing trees and shrubs). About 64 species of pines are native to North America, and many of them, like the eastern white pine, are valuable sources of lumber for everything from matchsticks to telephone poles. Today other pines (for example, southern pines such as loblolly pine and western pines such as Ponderosa pine) are much more extensively logged than eastern white pine. Pines are also an important source of the pulp used in the manufacture of paper. The sticky resin in pine wood provides turpentine (a solvent for oil-based paints) and rosin (a material used to make varnishes, inks, sealants, and soaps). Many pines are also grown for ornamental purposes, as well as to provide shade or windbreaks. In the natural environment, pines provide food and shelter for wildlife. (See Plants and People: The Importance of Wood, on page 121, for additional uses of trees.)

After reading this chapter, you should be able to:

1. *Describe three functions of stems.*
2. *Label a cross section of a herbaceous dicot stem and describe the functions of each tissue.*
3. *Label a cross section of a monocot stem and describe the functions of each tissue.*
4. *Distinguish between the structures of stems and roots.*
5. *Outline the transition from primary growth to secondary growth in a woody stem. List the two lateral meristems, and describe the tissues that arise from each.*
6. *Discuss the economic importance of wood and the ecological importance of forests.*

THE MAIN FUNCTIONS OF STEMS ARE SUPPORT, CONDUCTION, AND PRODUCTION OF NEW STEM TISSUES

Stems perform three main functions for plants. First, they support leaves and reproductive structures. The upright position of most stems and the arrangement of leaves on them enable each leaf to absorb maximum light for use in photosynthesis. Reproductive structures (flowers and fruits) are positioned on stems in areas accessible for insects, birds, or wind, which transfer pollen from flower to flower and help disperse seeds and fruits.

Second, stems provide internal transport. Stems conduct water and dissolved minerals from the roots, where they were absorbed from the soil, to the leaves and other plant parts. Stems also conduct the carbohydrates that were produced in the leaves by photosynthesis to the roots and other parts of the plant. Remember, however, that the stem is not the only plant organ that conducts materials. The vascular system is continuous throughout all parts of the plant, and conduction occurs in roots, stems, leaves, and reproductive structures.

Third, stems produce new living tissue. Stems continue to grow throughout a plant's life, forming buds that produce new leaves and reproductive structures. In addition to performing the main functions of support, conduction, and production of new tissue, some stems are modified for asexual reproduction (discussed later in the chapter). Also, certain stems are specialized to store food or, if green, to manufacture carbohydrates by photosynthesis.

Figure 7-1   Variation in stems.
(a) Morning glory (*Ipomoea pur-purea*), with its trailing, twining stem, is often grown to cover a fence or trellis. **(b)** The stem of a Ponderosa pine (*Pinus ponderosa*) grows toward the sunlight at Bryce Canyon, Utah. (a, *Carlyn Iverson;* b, *Doug Locke/Dembinsky Photo Associates*)   (a)   (b)

A Woody Twig Exemplifies the External Structure of All Stems

Stems exhibit varied forms, ranging from ropelike vines to massive tree trunks (Figure 7-1). They link a plant's roots to its leaves and are usually aboveground (although many plants have underground stems). Stems can be either herbaceous (consisting of soft, nonwoody tissues) or woody (developing extensive hard tissues of wood and bark). Most stems are circular in cross section, although a few, such as mint stems, are square.

Although stems exhibit great variation in structure and growth, they all have **buds**, which are undeveloped shoots that contain embryonic meristems. At the tip of a stem is a **terminal bud**. When a terminal bud is dormant (that is, unopened and not actively growing), it is covered and protected by an outer protective layer of **bud scales**, which are modified leaves. **Axillary buds**, also called *lateral buds,* are found in the **axils**— the upper angles between leaves and the stem to which they are attached. When terminal and axillary buds grow, they form stems that bear leaves and/or flowers. The area on a stem where each leaf is attached is called a **node**, and the region of a stem between two successive nodes is an **internode**.

A woody twig of a deciduous tree (a tree that sheds its leaves annually) can be used to model stem structures (Figure 7-2). The terminal bud is covered by bud scales that protect its delicate tip during dormancy. When the bud resumes growth, the bud scales fall off, leaving **bud scale scars** on the stem where they were attached. Temperate plants form terminal buds once a year, at the end of the growing season. Therefore, the number of groups of terminal bud scale scars on a twig reveals its age.

A **leaf scar** shows where a leaf was attached on the stem, and the vascular (conducting) tissue that extends from the stem out into the leaf forms **bundle scars** within the leaf scar. Axillary buds develop above the leaf scars. Also, the bark of a woody twig has **lenticels**, sites of loosely arranged cells that allow gas exchange to occur. Lenticels look like tiny marks, or specks, on the bark of a twig and are often used as an aid in identifying the plant.

Stems Originate and Develop at Meristems

Recall from Chapter 5 that plants undergo two different types of growth. Primary growth is an increase in the length of a plant and occurs at **apical meristems** at the tips of stems and roots. Secondary growth is an increase in the girth (circumference) of a plant and is due to the activity of **lateral meristems** along the sides of stems and roots. The new tissues formed by the lateral meristems are called *secondary tissues* to distinguish them from the *primary tissues* produced by apical meristems.

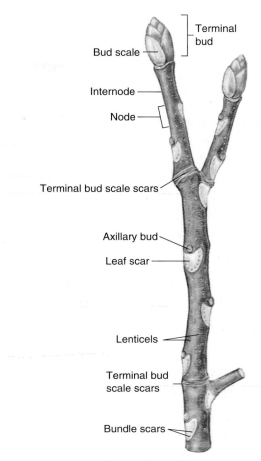

Figure 7-2 labels: Terminal bud · Bud scale · Internode · Node · Terminal bud scale scars · Axillary bud · Leaf scar · Lenticels · Terminal bud scale scars · Bundle scars

Figure 7-2   The external structure of a woody twig. The age of a woody twig can be determined by counting the number of groups of terminal bud scale scars (don't count those on side branches). How old is this twig?

All plants have primary growth; some plants have both primary and secondary growth. Recall that stems with only primary growth are herbaceous, while those with both primary and secondary growth are woody. (Some herbaceous stems—for example, geranium—also have limited amounts of secondary growth.) A woody plant increases in length by primary growth at the tips of its stems and roots, while the older stems and roots farther back from the tips increase in girth by secondary growth. In other words, at the same time that secondary growth is adding wood and bark (thereby causing the stem to thicken), primary growth (which increases the length of the plant) continues.

Herbaceous Dicot Stems and Monocot Stems Can Be Distinguished by the Arrangement of Vascular Bundles

Although stems vary considerably, they all possess an outer protective covering, one or more types of ground tissue, and vascular tissues (xylem and phloem). Let us consider first the structures of herbaceous dicot stems and monocot stems.

The Vascular Bundles of Herbaceous Dicot Stems Are Arranged in a Circle in Cross Section

The young sunflower stem is a representative herbaceous dicot stem exhibiting primary growth (Figure 7-3). The **epidermis**, an outer covering, provides protection to herbaceous stems as it does to leaves and herbaceous roots. The epidermis is usually covered by a cuticle, a waxy layer that reduces water loss from the surface.

Inside the epidermis is the **cortex**, a layer several cells thick that is part of a plant's ground tissue system. The cortex is a complex tissue that may contain parenchyma, collenchyma, and sclerenchyma cells (see Chapter 5). As might be expected from the various types of cells that it contains, the cortex in a herbaceous dicot stem has three functions: photosynthesis, storage, and support. If the stem is green, photosynthesis occurs within chloroplast-containing parenchyma cells in the cortex. Parenchyma cells in the cortex also store starch grains and crystals. Collenchyma and sclerenchyma in the cortex provide strength and structural support to the stem.

The vascular tissues provide conduction and support. When a herbaceous dicot stem is viewed in cross section, the vascular tissues appear as **vascular bundles** arranged in a circle. These vascular bundles extend as long strands throughout the length of the stem and are continuous with the vascular tissues of both roots and leaves. The central cylinder of a stem, which includes xylem, phloem, and often pith, is called a **stele**.

Each vascular bundle contains both **xylem** tissue, which transports water and dissolved minerals from the roots to the leaves, and **phloem** tissue, which transports dissolved carbohydrates (sucrose), often from the leaves to the roots. The xylem is usually on the inner side of the vascular bundle, and the phloem is on the outside. Sandwiched between the xylem and the phloem is a single layer of cells, the **vascular cambium**, a lateral meristem that is responsible for secondary growth.

Because stems grow through the air, they have to be much stronger than roots in order to support the plant body. Fibers occur in both xylem and phloem, although they are usually more extensive in phloem. These fibers add considerable strength to the herbaceous stem. In sunflowers and certain other herbaceous dicot stems, the phloem contains a cluster of fibers called a **phloem fiber cap**, which helps strengthen the stem. The phloem fiber cap is not found in all herbaceous dicot stems.

At the center of the herbaceous dicot stem is **pith**, a tissue composed of large, thin-walled parenchyma cells that function primarily for storage. Due to the arrangement of the vascular tissues in bundles, there is no distinct separation of cortex and pith between the vascular bundles. Table 7-1 summarizes the major differences between roots and stems in herbaceous dicots.

Vascular Bundles Are Scattered Throughout Monocot Stems

A monocot stem, as exemplified by the herbaceous stem of corn, is covered by an epidermis with its waxy cuticle. As in herbaceous dicot stems, the vascular tissues run in strands throughout the length of the stem (Figure 7-4). In cross sec-

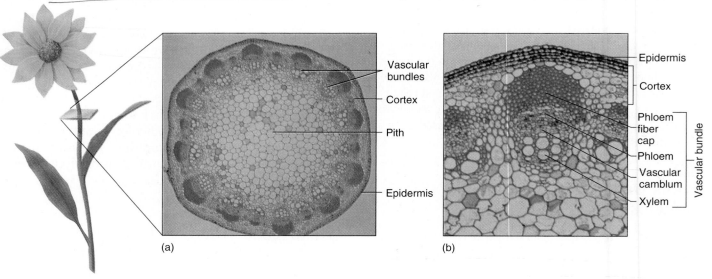

Sunflower

Figure 7-3   Primary growth in a dicot stem. (a) Cross section of a sunflower (*Helianthus annuus*) stem, showing the organization of tissues. The vascular bundles are arranged in a circle. **(b)** Close-up of a vascular bundle. The xylem is toward the stem's interior, and the phloem toward the outside. Each vascular bundle is "capped" by a batch of fibers for additional support. (a,b, *Dennis Drenner*)

TABLE 7-1   General Differences Between Herbaceous Dicot Roots and Stems*

Roots	Stems
No nodes or internodes	Nodes and internodes
No leaves or buds	Leaves and buds
Nonphotosynthetic	Photosynthetic
No pith	Pith
No cuticle	Cuticle
Root cap	No cap
Root hairs	Trichomes
Pericycle	No pericycle
Endodermis	Endodermis rare
Branches form internally from the pericycle	Branches form externally from lateral buds

*Some exceptions to these general differences exist.

tion, however, the vascular bundles, which contain xylem toward the inside and phloem toward the outside, are not arranged in a circle, as they are in herbaceous dicot stems, but are scattered throughout the stem. Each vascular bundle is usually enclosed in a bundle sheath of sclerenchyma cells for support. The tissue in which the vascular tissues are embedded, called **ground tissue**, performs the same functions as cortex and pith in herbaceous dicot stems. The monocot stem does not have distinct areas of cortex and pith.

The lateral meristems (vascular cambium and cork cambium), which give rise to secondary growth, do not occur in monocot stems. Monocots have only primary growth and do not produce the secondary tissues wood and bark. Although some monocots, such as palm trees, attain considerable size, they do so by a modified form of primary growth rather than by secondary growth. The stems of monocots such as bamboo and palms contain a great deal of sclerenchyma tissue, however, and may be extremely hard.

WOODY PLANTS HAVE STEMS WITH SECONDARY GROWTH

Woody plants undergo secondary growth, an increase in the girths of stems and roots that results from the activity of two lateral meristems, the vascular cambium and the cork cambium. Among flowering plants, only woody dicots (such as apple, hickory, and maple) have secondary growth. Gymnosperms (such as the conifers pine, juniper, and spruce) also have secondary growth. Table 7-2 summarizes the relationships of meristems and tissues in a woody dicot stem.

Cells in the **vascular cambium** divide and produce two complex tissues: (1) secondary xylem (wood), to replace pri-

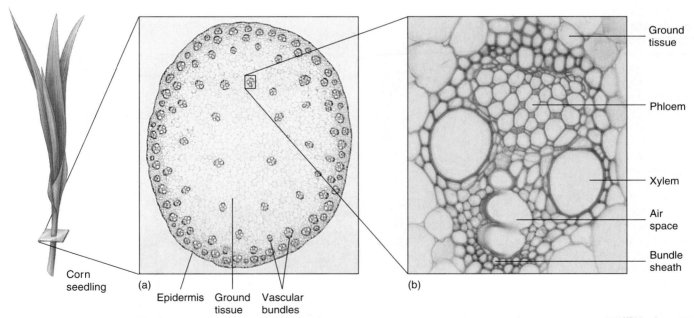

(a)   Corn seedling

Epidermis   Ground tissue   Vascular bundles

(b)

Ground tissue
Phloem
Xylem
Air space
Bundle sheath

Figure 7-4   Arrangement of stem tissues in corn (*Zea mays*), a monocot. (a) Cross section of stem, showing the scattered vascular bundles. **(b)** Close-up of a vascular bundle. The air space is where the first xylem elements formed. The entire bundle is enclosed in a bundle sheath of sclerenchyma for additional support. (a,b, *Dennis Drenner*)

TABLE 7-2   Development in a Woody Dicot Stem

Apical Meristem	Primary Tissues	Lateral Meristems	Secondary Tissues
Meristematic cells	Primary xylem	Vascular cambium	Secondary xylem (wood)
	Primary phloem		Secondary phloem (inner bark)
	Cortex	Cork cambium*	Cork parenchyma (phelloderm)*
	Pith		Cork cells*
	Epidermis		

*Collectively known as periderm (outer bark).

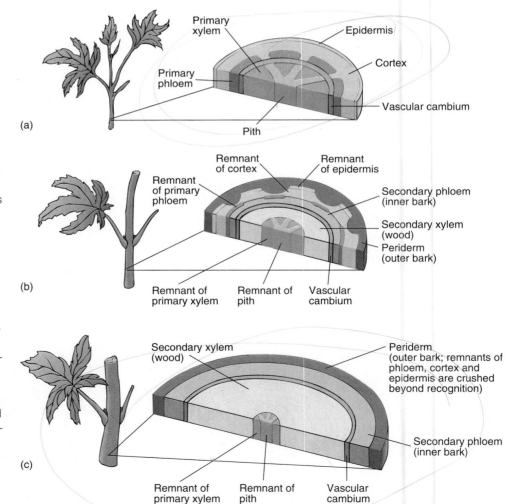

Figure 7-5 Development of secondary growth in the dicot stem is revealed in cross section. Vascular cambium and the tissues it produces are shown; cork cambium is not depicted. **(a)** A herbaceous dicot stem. Vascular cambium is sandwiched between primary xylem and primary phloem in each vascular bundle. At the onset of secondary growth, vascular cambium arises in the parenchyma between the vascular bundles, forming a cylinder of meristematic tissue that appears as a circle in cross section (*shown*). **(b)** Vascular cambium begins to divide, forming secondary xylem on the inside and secondary phloem on the outside. Note that primary xylem and primary phloem in the original vascular bundles are separated in this process. **(c)** A young woody stem. Vascular cambium produces more secondary xylem than secondary phloem.

(a)

Primary xylem
Epidermis
Primary phloem
Cortex
Vascular cambium
Pith

(b)

Remnant of cortex
Remnant of epidermis
Remnant of primary phloem
Secondary phloem (inner bark)
Secondary xylem (wood)
Periderm (outer bark)
Remnant of primary xylem
Remnant of pith
Vascular cambium

(c)

Secondary xylem (wood)
Periderm (outer bark; remnants of phloem, cortex and epidermis are crushed beyond recognition)
Secondary phloem (inner bark)
Remnant of primary xylem
Remnant of pith
Vascular cambium

mary xylem, and (2) secondary phloem (inner bark), to replace primary phloem. Primary xylem and primary phloem cannot transport materials indefinitely and so must be replaced if the plant is to survive long-term.

The second lateral meristem, the **cork cambium**, divides to produce cork cells and cork parenchyma (*phelloderm*). The cork cambium and the tissues it produces are collectively called **periderm** (outer bark). Periderm functions as a replacement for the epidermis, which splits apart as the stem increases in girth.

Vascular Cambium Gives Rise to Secondary Xylem and Secondary Phloem

Primary tissues in very young stems of woody plants are organized similarly to those in herbaceous dicot stems, with the vascular cambium being a thin layer of cells sandwiched between the xylem and the phloem in the vascular bundles. Once secondary growth commences, however, the internal structure of the stem changes considerably (Figure 7-5). Although the vascular cambium is not initially a solid cylinder of cells, it becomes continuous when production of secondary

tissues begins. This continuity develops because certain parenchyma cells between the vascular bundles retain the ability to divide. These cells, along with the vascular cambium cells in each vascular bundle, form a ring of vascular cambium.

When a cell in the vascular cambium divides, one of the daughter cells remains meristematic—that is, it remains as a part of the vascular cambium. The other cell may divide again several times, but it eventually stops dividing and develops into mature secondary tissue.

Cells in the vascular cambium divide to produce tissues in two directions. The cells formed from the dividing vascular cambium form secondary xylem, or wood, toward the *inside* and secondary phloem, or inner bark, toward the *outside* (Figure 7-6). The vascular cambium is thus a thin layer of cells sandwiched between the wood and inner bark, the two tissues it produces.

What happens to the original primary tissues of the stem once secondary growth occurs? As the stem increases in thickness, the relative orientations of the original primary tissues change. For example, secondary xylem and secondary phloem are laid down between the primary xylem and primary phloem within each vascular bundle. Therefore, as the vascular cam-

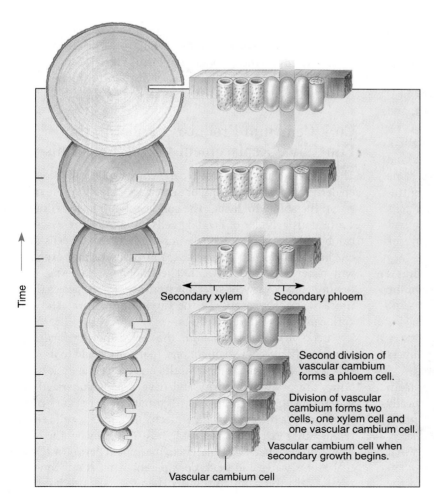

Secondary xylem ← → Secondary phloem

Second division of vascular cambium forms a phloem cell.

Division of vascular cambium forms two cells, one xylem cell and one vascular cambium cell.

Vascular cambium cell when secondary growth begins.

Vascular cambium cell

Time

Figure 7-6   Radial view of a dividing vascular cambium cell. Note that the vascular cambium divides in two directions, forming secondary xylem to the inside and secondary phloem to the outside. These cells differentiate to form the mature cell types associated with xylem and phloem. As secondary xylem accumulates, the vascular cambium moves outward, and the woody stem increases in diameter. (To study the figure, start at the bottom and move up.)

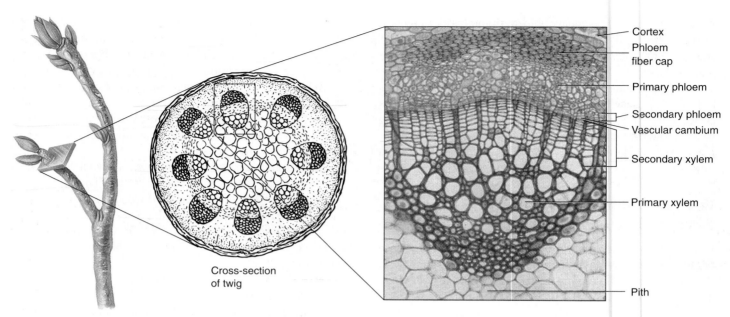

Figure 7-7   Cross section through part of a magnolia (*Magnolia* sp) stem, showing a vascular bundle that has been split apart by secondary growth. Compare this vascular bundle with the one in Figure 7-3b, which has primary growth only. (*Dennis Drenner*)

bium continues to form secondary tissues, the primary xylem and primary phloem in each vascular bundle are separated from each other (Figure 7-7). The primary tissues outside the cylinder of secondary growth (that is, the primary phloem, cortex, and epidermis) are subjected to the pressures produced by secondary growth and are crushed, split apart, and gradually sloughed off.

Secondary tissues replace the primary tissues in function. In the woody plant, secondary xylem conducts water and dissolved minerals from roots to leaves and contains the same types of cells found in primary xylem: water-conducting tracheids and vessel elements in addition to parenchyma cells and fibers. The arrangement of the different cell types in secondary xylem results in distinctive wood characteristics for each species.

Secondary phloem conducts dissolved carbohydrates (sucrose)—for example, from the place of manufacture in the leaves to a place of storage in the stem or roots. The same types of cells found in primary phloem are also found in secondary phloem: sieve-tube members, companion cells, parenchyma cells, and fibers. There are usually more fibers in secondary phloem than in primary phloem, however.

Whereas the secondary xylem and secondary phloem transport water, minerals, and carbohydrates vertically throughout the plant, materials must also move horizontally (that is, sideways, or laterally). These materials move laterally through **rays**, which are chains of parenchyma cells that radiate out from the center of the stem. The vascular cambium forms the rays, which are often continuous from the

secondary xylem to the secondary phloem. Water and dissolved minerals travel laterally through rays from the secondary xylem to the secondary phloem. Likewise, rays form pathways for the lateral transport of dissolved carbohydrates from the secondary phloem to the secondary xylem.

Cork Cambium Produces Periderm, the Functional Replacement for the Epidermis

The cork cambium arises from parenchyma cells in the cortex, epidermis, primary phloem, or secondary phloem; these cells retain the ability to divide. The cork cambium forms either a continuous cylinder of dividing cells (similar to the vascular cambium) or a series of small arcs of meristematic cells that cut into successively deeper layers of tissue (which explains why outer bark is furrowed). The thickness, patterns, and texture of bark vary considerably from species to species (Figure 7-8), largely because of the varying growth rates of the cork cambia of different species.

Like the vascular cambium, the cork cambium divides to form new tissues in two directions, to its inside and to its outside. Cork cells, which form to the outside of the cork cambium, are dead at maturity and have heavily suberized, or waterproofed, walls. The cork cells produced by the cork cambium protect the stem against mechanical injury, mild fires, temperature extremes, and water loss. To its inside, the cork cambium produces cork parenchyma (phelloderm), a tissue that stores water and food. Cork parenchyma is only one to several cells thick, much thinner than the cork cell layer.

(a)   (b)   (c)   (d)

Figure 7-8   Bark varies considerably from one species to another. (a) Bur oak
(*Quercus macrocarpa*) bark is deeply fissured. **(b)** Shagbark hickory (*Carya ovata*) has
a rough, "shaggy" bark. **(c)** Bark from Norway pine (*Pinus resinosa*) is scaly. **(d)** Paper
birch (*Betula papyrifera*) has a peeling bark. (a,c,d, *Carlyn Iverson;* b, *R.C. Bonner*)

Cork cells are impermeable to water and gases, yet the internal tissues of the woody stem must be able to exchange gases with the surrounding atmosphere. As a stem thickens from secondary growth, the epidermis, including the stomata that allowed gas exchange for the herbaceous stem, dies. Stomata are replaced by lenticels, areas of the cork in which the cork cells are loosely arranged, permitting gas exchange (Figure 7-9).

The bark of cork oak (*Quercus suber*), native to the western Mediterranean region, is commercially important. Harvesters carefully strip away the outer bark so as not to damage the cork cambium, which regenerates new cork cells in sufficient amounts to be harvested every 10 years. The main use of cork today is to stopper bottles; cork is also found in gaskets, life preservers, floats for fishing lines, flooring, and insulation.

**Figure 7-9   Cross section through the periderm of a
calico flower (*Aristolochia elegans*) stem, showing a
lenticel.** (*James Mauseth, University of Texas*)

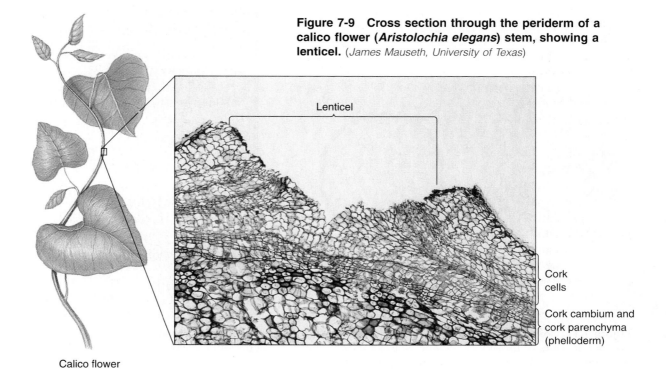

Lenticel

Cork
cells

Cork cambium and
cork parenchyma
(phelloderm)

Calico flower

Common Terms Associated with Wood Can Be Explained on the Basis of Plant Structure

If you've ever examined different types of lumber, you may have noticed that some trees have wood with two different colors (Figure 7-10). **Heartwood,** the older wood in the center of the trunk, is typically a brownish-red color. A microscopic examination of heartwood reveals that its vessels and tracheids are plugged with pigments, tannins, gums, resins, and other materials. Therefore, heartwood no longer functions in conduction. The functional secondary xylem—that is, the part that conducts water and dissolved minerals—is the **sapwood,** the younger, lighter-colored wood closest to the bark. Heartwood is denser than sapwood and provides structural support for trees. There is some evidence that heartwood is also more resistant to decay.

Botanically speaking, **hardwood** is the wood of flowering plants—that is, woody dicots—and **softwood** is the wood of conifers. Pine and other conifers typically have wood that lacks fibers and vessel elements. The only kind of conducting cell in conifer wood is the tracheid. These structural features generally make conifer woods softer than the woods of flowering plants, although a lot of variation occurs from one species to another. The balsa (*Ochroma pyramidale*) tree, for example, is a "hardwood" whose extremely light, soft wood is used to fashion model airplanes.

Plants that grow in temperate climates where both a growing period and a period of dormancy (during winter)

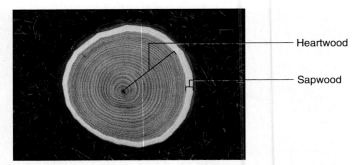

Figure 7-10   Cross section through a butternut (*Juglans cinerea*) trunk, revealing the darker heartwood in the center of the tree and the lighter sapwood. The sapwood is the functioning xylem that conducts water and dissolved minerals. (*Carlyn Iverson*)

occur each year, exhibit **annual rings,** or **growth increments,** which are evident in wood cross sections as concentric circles (Figure 7-11). To determine the age of a woody stem in a temperate zone, simply count the annual rings. In the tropics, environmental conditions determine the presence or absence of rings: trees growing in the moist humid tropics do not produce rings, whereas those growing in areas with pronounced wet and dry seasons do. Thus, in tropical trees, rings are not reliable for determining age.

Examination of annual rings with a magnifying lens reveals that there is no line separating one year's growth from the

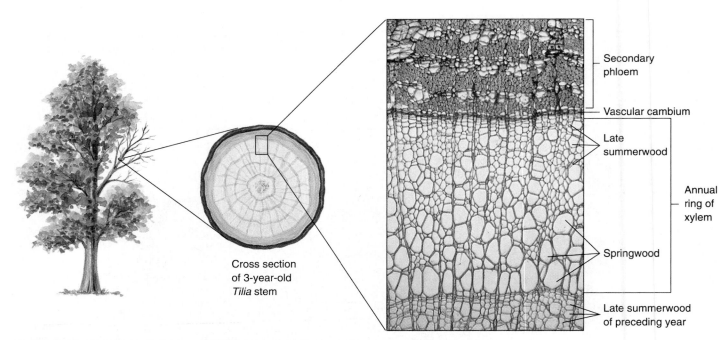

Figure 7-11   Portion of basswood (*Tilia americana*) stem, in cross section, showing one annual ring, or growth increment. Note the differences in size between the vessel elements of springwood and late summerwood. (*Dennis Drenner*)

Focus On
Tree Ring Analysis

In temperate climates, the age of a tree can be established by counting the number of annual rings. Analysis of tree rings reveals other useful information as well. For example, the size of each ring varies according to local weather conditions, including precipitation and temperature. Sometimes the variation in tree rings can be attributed to a single environmental factor, and similar patterns appear in the rings of different tree species over a large geographical area. For example, trees in the Southwest of the United States have similar ring patterns due to variations in the amount of annual precipitation. Years in which adequate precipitation occurs produce larger rings of growth, whereas years of drought produce much smaller rings.

It is possible to study the sequence of rings back in time for several thousand years. First, a **master chronology,** a complete sample of rings dating back as far as possible, is developed (see figure). A small core of wood is bored out of the trunk of an old, living tree to obtain a sample of rings. The oldest rings of this sample, toward the center of the tree, are matched with some of the youngest rings (toward the outside) of an older tree or even an old piece of lumber from a house. Scientists can develop a master chronology of the area by examining older and older sections of wood, even those obtained from prehistoric dwellings, and overlapping their matching ring sequences. The longest master chronology is of bristlecone pines in the western United States; it goes back almost 9000 years.

Dendrochronology, the study of both visible and microscopic details of tree rings, has been used extensively in several related fields. Astronomers have attempted to correlate annual ring patterns with cycles of sunspots, which are giant magnetic fields on the surface of the sun that vary in number on an 11-year cycle. An American astronomer, Andrew Douglass, pioneered this work during the early 20th century.

Tree ring analysis has been extremely useful in investigations of prehistoric sites of Native Americans in the Southwest. For example, tree ring analysis at the Cliff Palace in Mesa Verde National Park, which dates back to the year AD 1073, indicates that an extended drought forced the original inhabitants to abandon their homes.

Climatologists use tree ring data to study past climatic patterns, for example, to reconstruct annual temperatures in the Sierra Nevada of California for the past 2000 years. Researchers are also analyzing tree rings for other disciplines, including ecology (to study changes in a forest community over time), environmental science (to study the effects of air pollution), and geology (to locate the epicenters and estimate the magnitudes of old earthquakes).

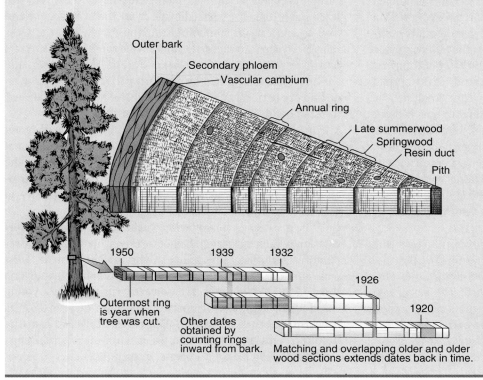

Tree ring dating. Scientists develop a master chronology by comparing progressively older pieces of wood from the same geographical area. They can then accurately determine the age of a sample by matching its rings to the master chronology. Ring matching, once a painstakingly tedious job, is usually done by computer today.

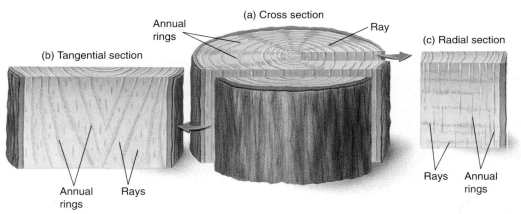

Figure 7-12   A block of wood showing (a) cross, (b) tangential, and (c) radial sections. Both tangential and radial sections are longitudinal sections. A tangential section is cut in a plane that does not pass through the center of the stem but instead passes at a right angle to a radius. A radial section is cut in a plane that passes through the center along a radius of the stem. The appearances of rays and annual rings are distinctive for each section.

next. The appearance of a ring in cross section is due to differences in cell size and cell wall thickness between secondary xylem formed at the end of one year's growth and that formed at the beginning of the next year's growth. In the spring, when water is plentiful, wood formed by the vascular cambium has thin-walled, large-diameter conducting cells and few fibers and is appropriately called *springwood*. As summer progresses and water becomes less plentiful, the wood that forms, known as *late summerwood,* has thicker-walled, narrower conducting cells and many fibers. The differences between the late summerwood of one year and the springwood of the following year give the appearance of rings. We can obtain a lot of information about climates in past times from the study of the annual rings of ancient trees (see Focus On: Tree Ring Analysis).

The appearance of wood differs according to how it is cut (Figure 7-12). In *cross section,* annual rings appear as concentric rings, and rays appear as straight lines radiating from the center of the stem. In *tangential section,* annual rings appear as vertical lines that often come together in a V-shape, and rays appear as specks or very short vertical lines. In *radial section,* annual rings appear as lines running the length of the wood, and rays appear as horizontal strips.

As a woody stem increases in girth over the years, the branches that it bears grow along with it as long as they are alive. After a branch dies and falls off, the stem grows outward and surrounds the base of the branch. The base of an old branch embedded in younger stem xylem is called a **knot.** It is possible for a knot to contain bark as well as wood. The presence of knots in wood lessens its commercial value, except when the knots are desired for ornamental purposes. Some plants, such as knotty pine, are valued for their high production of knots.

Forests Are Ecologically Important

Approximately one third of the Earth's total land area is covered by forests. Forests play an essential role in the global carbon cycle. For example, photosynthesis by trees removes large quantities of carbon dioxide from the atmosphere and fixes it into carbon compounds. At the same time, the trees release oxygen into the atmosphere. Tree roots hold vast tracts of soil in place, reducing erosion. Forests absorb, hold, and slowly release water, thereby regulating the flow of water, even during dry periods, and helping to control floods. In addition, forests provide essential wildlife habitat.

In the United States, not including Alaska, there are approximately 2 million hectares (5 million acres) of national forests, which are owned by the public and managed by the federal government. National forests have multiple uses, including recreation, livestock forage, water resources, fish and wildlife habitat, and timber harvest.

The greatest problem facing world forests today is **deforestation**, which is the removal of forest without sufficient replanting. Deforestation results in the replacement of forest with agricultural and other nonforest land. Only about one third of harvested trees are replaced, usually with young seedlings of fast-growing species. Although deforestation is occurring around the world, it is particularly acute in the humid tropics (see Plants and the Environment: The Deforestation of Tropical Rain Forests on page 122).

People

The Importance of Wood

Not counting firewood, the most common wood product is lumber. Lumber from particular tree species has unique qualities that make it suitable for certain types of products. For example, oak, which is heavy and strong, is fashioned into furniture, doors, coffins, and flooring. Ash is used in baseball bats, oars, skis, tool handles, and bentwood furniture. Maple is valuable for bowling pins and bowling alleys, dance floors, and toys. Birch is used to make clothespins, toothpicks, and spools. Wood paneling, hockey sticks, church pews, and gymnasium equipment are often made from elm. Violins and pianos contain red spruce, which has exceptional resonance. Fences, shingles, clothes closets, and pencils are made from cedar, while major uses of teak include furniture, chests, and shipbuilding. Shingles, siding, and garden furniture such as picnic tables continue to be made from redwood, even though these

important trees are endangered by overharvesting.

The second most common use of wood after lumber is for paper pulp. Almost 90 percent of the world's pulpwood (wood used to make paper) comes from conifers. Sawmills convert wood to pulp by mechanically grinding it up and adding strong chemicals to free the "fibers" (actually tracheids). The slushy pulp is spread on a moving wire screen to dry and is then pressed, trimmed, and rolled into large rolls. Some of the many types of paper are newsprint; book paper (which you are reading from right now); absorbent paper for blotters and filters; tissue paper for toilet tissue, paper towels, and paper napkins; and paperboard for boxes and containers. In addition to paper, pulpwood is used to make products such as rayon, cellophane, and fiberboard.

Plywood is an important wood product, often used in place of lum-

ber, that consists of two or more veneers (thin layers of wood sliced off a log in sheet form) that are glued together. Plywood is strong, lightweight, and less likely to split than solid wood. Most furniture made today contains some plywood covered by a veneer of a more valuable wood.

Sawmills used to discard crooked trees, branches, wood chips, and shavings. Today such materials are pressed together to make sturdy boards such as fiberboard (made from pulp) and particle board (made from wood chips). Fiberboard is used for containers and insulation, whereas particle board has applications similar to those of plywood.

A number of useful chemicals are derived from wood. Wood cellulose is used to make industrial alcohols, certain plastics, and synthetic fibers. Other chemical products produced from wood are acetic acid, wood tar, and adhesives.

VINES HAVE WEAK, ELONGATED STEMS

Vines are weak-stemmed plants that depend on other plants for support. Because vines do not have to expend resources to produce structurally strong stems, they are able to grow extremely rapidly, ascending through the shaded understory to the sunlit canopy of the forest, where they produce luxuriant foliage.

Vines have a variety of adaptations that fit their climbing lifestyle. As newly germinated plants, many vine seedlings grow *away* from sunlight rather than toward it. In growing toward the darkest part of its environment, a vine seedling usually encounters a large tree, which it then ascends. Along their trunks, woody vines (known as *lianas*) often produce

special roots with adhesive pads that stick to the bark of the host tree. Herbaceous vines frequently have tendrils, modified leaves or stems that wrap around supports, while other vines are *twiners,* with stems that grow spirally around their host as they ascend it.

Vines are most numerous and diverse in tropical forests, particularly tropical rain forests, where both herbaceous vines and lianas flourish. Lianas grow from the upper branches of one forest tree to those of another, connecting the tops of the trees and providing walkways for many of the canopy's animal residents. They and other vines provide nectar and fruit to many tree-dwelling animals.

Temperate forests, boreal (far-northern) forests, and island forests have far fewer vines than do tropical rain forests. Some biologists have suggested that vines are less common in tem-

(Text continues on page 124)

Environment

The Deforestation of Tropical Rain Forests

Tropical rain forest prevails in tropical places where the climate is very moist throughout the year—on the order of 200 to 450 centimeters (79 to 177 inches) of precipitation annually. Tropical rain forests are found in Central and South America, Africa, and Southeast Asia, but almost half of them are in just three countries: Brazil in South America, Zaire in Africa, and Indonesia in Southeast Asia (see figure).

All over the world, tropical rain forests are being cleared—cut for timber or firewood or burned to make pasture or agricultural land. More than half of all the world's tropical rain forests had been destroyed by the early 1990s. What is left, some 6 million square kilometers (about 2.3 million square miles), is less than the area of the United States, and each year an area larger than the size of Washington state is cleared. In the mid-1990s, humans are destroying tropical forests at the rate of 0.54 hectares (1.3 acres) per second. Currently, the highest *rate* of deforestation is in Southeast Asia, followed by Central and South America, including the Caribbean, and then Africa.

What Happens When Tropical Rain Forests Disappear?
When tropical rain forests are harvested or destroyed, they no longer make valuable contributions to the environment or to the people who depend on them. Destruction of tropical rain forest particularly threatens native people, whose cultures and ways of life depend on the forests.

Deforestation results in decreased soil fertility and increased soil erosion. When forest is removed, the total amount of surface water that flows into rivers and streams increases. However, because this water flow is no longer regulated by the forest, the affected region experiences alternating periods of flood and drought.

Deforestation can also cause the extinction of many plant and animal species. Tropical species often live

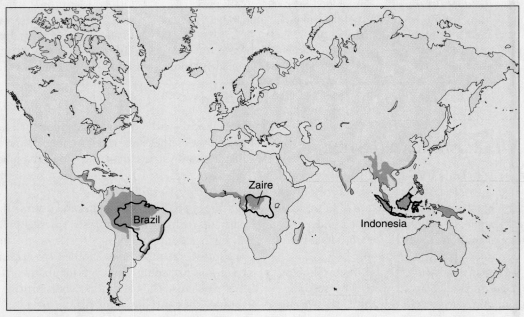

The distribution of tropical rain forests (*green areas*). Rain forests exist in portions of South and Central America, Africa, and Asia.

within a very limited area in a forest, so they are particularly vulnerable to habitat destruction or modification. Wildlife in temperate areas, including migratory birds and butterflies, also suffers from tropical deforestation, because the animals spend part of each year in these tropical areas.

Deforestation induces regional and global climatic changes. Trees release substantial amounts of moisture into the air; about 97 percent of the water that a plant's roots absorb from the soil is evaporated directly into the atmosphere and falls back to the Earth as precipitation. Consequently, when forest is removed, rainfall declines and droughts occur more frequently in that region. Tropical deforestation also contributes to an increase in global temperature.

Why Are Tropical Rain Forests Disappearing?

The three main causes of tropical deforestation are subsistence agriculture, commercial logging, and cattle ranching. Subsistence agriculture, in which each family produces enough food to feed itself, accounts for more than half of tropical rainforest loss. Subsistence farmers often practice **slash-and-burn agriculture** in which they first cut down the forest and allow it to dry, then burn the area and immediately plant crops. The yield from the first crop is often quite high, because the minerals that were in the trees are available in the soil after the trees are burned. However, soil productivity declines rapidly so that subsequent crops are poor. In a very short time, the people farming

the land must move to new forest and repeat the process.

Slash-and-burn agriculture on a small scale, with periods of 20 to 100 years between cycles, is actually *sustainable*—that is, it can be carried out indefinitely without overusing the soil and other natural resources. But when *millions* of people simultaneously try to obtain a living in this way, the land is not allowed to lie uncultivated for adequate recovery periods, and so the forest does not recover.

More than 20 percent of tropical deforestation is the result of commercial logging. Vast tracts of tropical trees are being removed for export abroad, including to the United States. Most tropical countries are allowing commercial logging to proceed at a much higher rate than is sustainable.

More than 10 percent of destruction of tropical rain forests occurs to provide open rangeland for cattle. Much of the beef raised on ranches cleared from tropical forests is exported to fast-food chains. After the forests are cleared, cattle can be grazed on the land for 6 to 10 years, after which time shrubby plants, known as scrub savanna, take over the range.

What Is Being Done to Curb Deforestation?

Several environmental programs attempt to preserve tropical forests or at least slow their loss by promoting economic uses of forests other than the harvesting of timber. For example, the Suriname Biodiversity Prospecting Initiative, announced in 1993 by Conservation International, pays money to indigenous people in

Suriname, a small country in South America, for new medicines delivered from the plants the natives identify in rain forests. The boycotting of beef from cattle raised in pastures that were once rain forests is an example of a consumer action campaign designed to slow tropical deforestation.

Existing international forestry agreements have been ineffective at curbing forest loss. The statement of principles on deforestation formulated at the 1992 United Nations Conference on Environment and Development, for example, was originally to have been a legally binding treaty that would stop developing nations from burning tropical forests. Developing countries objected that the treaty unfairly focused on tropical forests and did not address the logging of old-growth forests in the United States, Canada, and Europe. When a compromise could not be reached, the treaty was scrapped in favor of a weaker statement that is not legally binding.

Unfortunately, all of these efforts have had only a slight impact on the rate of tropical deforestation, and strong international measures are needed to help save the remaining forests.

perate and boreal forests because they are less resistant to droughts and fires than are the trees that grow there. Vines in tropical rain forests generally do not have to adapt to long droughts or forest fires. Vines are thought to be relatively uncommon on islands because the seeds of most vines, which are dispersed by wind, cannot reach islands to colonize them.

SOME STEMS FUNCTION IN ASEXUAL REPRODUCTION

In **asexual reproduction**, a single individual may split, bud, or fragment, giving rise to offspring that are genetically similar to the parent. Asexual reproduction in flowering plants does not usually involve the formation of flowers, fruits, and seeds. In plants that reproduce asexually, the stems, leaves, and roots can form offspring, usually when part of an existing plant becomes separated from the rest of the plant. This part subsequently grows to form a complete, independent plant. Asexual reproduction always involves only one parent, and there

Figure 7-14   Potato (*Solanum tuberosum*) plants form rhizomes, which enlarge at the ends into tubers. Growers propagate potatoes by planting whole tubers or pieces of tubers; each piece must contain at least one "eye," or axillary bud. (*Carlyn Iverson*)

is no fusion of reproductive cells (gametes); the offspring are virtually identical to one another and to the parent plant.

Flowering plants have evolved many methods of asexual reproduction, a number of which are modified stems (rhizomes, tubers, bulbs, corms, and stolons). A **rhizome** is a horizontal, underground stem that may or may not be fleshy. Fleshiness indicates that the stem is used for storing food materials such as starch (Figure 7-13). Although rhizomes resemble roots, they are really stems, as indicated by the presence of scale-like leaves, buds, nodes, and internodes. Rhizomes frequently branch in different directions, and over time, the old portion of the rhizome dies, and the two branches eventually separate to become two distinct plants. Irises, bamboos, ginger, and many grasses are examples of plants that reproduce asexually by forming rhizomes.

Some rhizomes produce greatly thickened ends called **tubers**, which are fleshy underground stems enlarged for food storage. When the connection between a tuber and its parent plant breaks, often as a result of the death of the parent plant, the tuber grows into a separate plant. White potatoes and elephant's ear (*Caladium* sp) are examples of plants that produce tubers (Figure 7-14). The "eyes" of a potato are actually axillary buds, evidence that the tuber is an underground stem rather than a storage root like sweet potatoes or carrots.

A **bulb** is a shortened underground bud in which fleshy storage leaves are attached to a short stem (Figure 7-15a; also see Figure 8–16d). A bulb is rounded and is covered by paper-like bulb scales, which are modified leaves. It frequently forms small daughter bulbs (bulblets) that are initially attached to the parent bulb; when the parent bulb dies and rots away, each daughter bulb can become established as a separate plant. Lilies, tulips, onions, and daffodils are some of the plants that form bulbs.

Figure 7-13   The iris, a popular garden plant, has a horizontal underground stem called a rhizome. New shoots arise from axillary buds that develop along the rhizome.

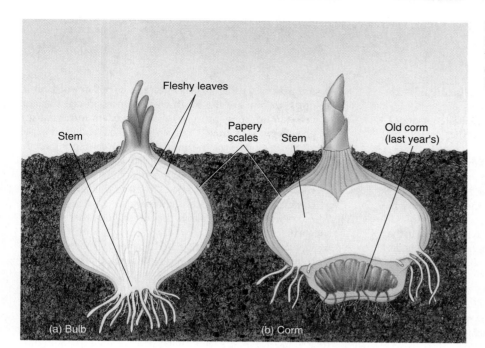

Figure 7-15 Bulbs and corms. (a) A bulb is an underground stem to which are attached overlapping, fleshy leaves. **(b)** The entire corm is stem tissue, in contrast to the bulb, which is mostly leaf tissue.

Fleshy leaves

Stem

Papery scales

Stem

Old corm (last year's)

(a) Bulb

(b) Corm

A **corm** is a very short, erect underground stem that superficially resembles a bulb (Figure 7-15b). Unlike the bulb whose food is stored in underground leaves, the corm's storage organ is a thickened underground stem covered by papery scales (modified leaves). Axillary buds that will give rise to new corms frequently develop on a corm; the death of the parent corm separates these daughter corms, which then become established as separate plants. Familiar garden plants that produce corms include crocus, gladiolus, and cyclamen.

Stolons, or runners, are horizontal aboveground stems that grow along the ground's surface and are characterized by long internodes (Figure 7-16). Buds develop along the stolon, and each bud gives rise to a new plant that roots in the ground. When the stolon dies, the daughter plants live separately. The strawberry plant produces stolons.

New plant

Stolon (runner)

Adventitious root

Figure 7-16 Strawberries reproduce asexually by forming stolons, or runners.

STUDY OUTLINE

I. The main functions of stems are support, conduction, and production of new stem tissues.

II. Woody twigs exemplify the external structure of all stems.

   A. Buds are undeveloped shoots that contain embryonic meristems. A terminal bud develops at the tip of a stem. Axillary (lateral) buds develop in the axils of leaves.

   B. A dormant bud is covered and protected by bud scales. When the terminal bud resumes growth, the bud scales covering the terminal bud fall off, leaving terminal bud scale scars.

   C. The area on a stem where each leaf is attached is called a node, and the region of a stem between two successive nodes is an internode.

   D. A leaf scar shows where each leaf was attached on the stem.

   E. Lenticels are sites of loosely arranged cells that allow for gas exchange.

III. A herbaceous stem consists of an epidermis, vascular tissue, and cortex and pith, or ground tissue.

   A. The epidermis is a protective outer layer. It is covered by a water-conserving cuticle. Stomata permit gas exchange.

   B. Xylem conducts water and dissolved minerals, and phloem conducts dissolved carbohydrates (sucrose).

   C. The cortex, pith, and ground tissue function primarily for storage.

IV. Although herbaceous stems all have the same basic tissues, the arrangement of the tissues varies considerably.
  A. Herbaceous dicot stems have the vascular bundles arranged in a circle in cross section and have a distinct cortex and pith.
  B. Monocot stems have scattered vascular bundles and ground tissue instead of distinct cortex and pith.
V. Secondary growth occurs in woody dicots and conifers.
  A. The vascular cambium occurs between the primary xylem and primary phloem.
    1. Vascular cambium produces secondary xylem (wood) to the inside and secondary phloem (inner bark) to the outside.
    2. Secondary xylem conducts water and dissolved minerals in the woody stem.
    3. Secondary phloem conducts dissolved carbohydrates in the woody stem.
  B. The cork cambium arises near the stem's surface.
    1. The cork cambium produces cork parenchyma (phelloderm) to the inside and cork cells to the outside.
    2. Cork parenchyma (phelloderm) functions primarily for storage in the woody stem.
    3. Cork cells are the functional replacement for epidermis in the woody stem.
  C. Rays are chains of parenchyma cells that radiate out from the center of the stem and form pathways for the lateral transport of materials between secondary xylem and secondary phloem.

D. Annual rings, which make concentric circles in wood cut in cross section, are the result of differences in cell size and cell wall thickness between secondary xylem formed at the end of one year's growth (late summerwood) and that formed at the beginning of the next year's growth (springwood).
E. Heartwood, the older, darker-colored wood in the center of a tree, no longer functions in conduction. Sapwood is the younger, lighter-colored wood closest to the bark. Sapwood conducts water and dissolved minerals.
VI. Vines are weak-stemmed plants that depend on other plants for support.
  A. Vines, which may be either herbaceous or woody, grow extremely rapidly and have a variety of adaptations that fit their climbing lifestyle.
  B. Vines are most numerous and diverse in tropical forests, particularly tropical rain forests.
VII. Many stems are specialized for asexual reproduction.
  A. A rhizome (example: iris) is a horizontal underground stem.
  B. A tuber (example: white potato) is an underground stem greatly enlarged for storing food.
  C. A bulb (example: onion) is an underground bud with a shortened stem and fleshy storage leaves.
  D. A corm (example: crocus) is an underground stem that superficially resembles a bulb.
  E. Stolons (example: strawberry) are aboveground horizontal stems with long internodes.

SELECTED KEY TERMS

annual rings, p. 118
apical meristem, p. 111
bud, p. 111
bulb, p. 124
cork cambium, p. 115
corm, p. 125

cortex, p. 112
deforestation, p. 120
epidermis, p. 112
ground tissue, p. 113
heartwood, p. 118
lenticel, p. 111

periderm, p. 115
phloem, p. 112
pith, p. 112
ray, p. 116
rhizome, p. 124

sapwood, p. 118
stolon, p. 125
tuber, p. 124
vascular cambium, p. 112
xylem, p. 112

REVIEW QUESTIONS

1. List several functions of stems, and describe the tissue(s) responsible for each function.
2. If you were examining a cross section of a herbaceous stem of a flowering plant, how would you determine whether it is a dicot or a monocot?
3. What happens to the primary tissues of a stem when secondary growth occurs?
4. When a strip of bark is peeled off a tree branch, what tissues are removed?
5. Distinguish between the vascular cambium and the cork cambium, and describe the tissues that arise from each.
6. What is an annual ring, and how is it formed?
7. Distinguish among these specialized stems: rhizome, tuber, bulb, corm, and stolon.
8. Label the tissues of this herbaceous dicot stem. Give at least one function for each tissue.

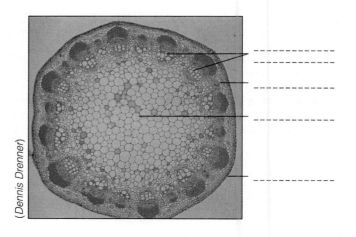
(Dennis Drenner)

9. Label the tissues of this monocot stem. Give at least one function for each tissue.

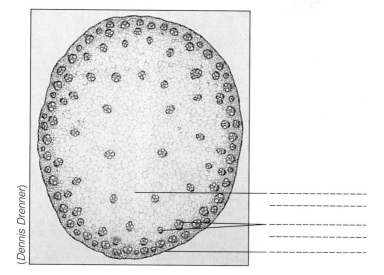

(Dennis Drenner)

- - - - - - - - - - - - -

- - - - - - - - - - - - -

- - - - - - - - - - - - -

- - - - - - - - - - - - -

- - - - - - - - - - - - -

THOUGHT QUESTIONS

1. When secondary growth commences, certain cells become meristematic and begin to divide. Could a tracheid ever do this? A sieve tube member? Why or why not?
2. Why does the wood of many tropical trees lack annual rings?
3. Why is hardwood more desirable than softwood for making furniture? Explain on the basis of the structural differences between hardwood and softwood.

4. Someone gives you a plant structure that was found growing in the soil and asks you to determine whether it is a root or an underground stem. How would you do this without a microscope? With a microscope?

SUGGESTED READINGS

Angier, N., "Warming? Tree Rings Say Not Yet," *The New York Times,* December 1, 1992. How the Laboratory of Tree Ring Research at the University of Arizona deciphers the history of Earth's climate by analyzing tree rings.

Houghton, R.A., "Deforestation," in *Encyclopedia of Environmental Biology,* Vol. 1, New York, Academic Press, 1995, pp. 449–461. This general article focuses on local, regional, and global effects of deforestation.

Marchand, P.J., "Waves in the Forest," *Natural History,* February 1995. In Japan and North America, crescent-shaped bands of dead and dying trees spread through high-elevation fir forests. These dieback zones, sometimes called fir waves, appear to be a natural phenomenon.

Moffett, M., "These Plants Claw and Strangle Their Way to the Top,"

Smithsonian, September 1993. Spectacular photographs accompany this article on vines that grow in tropical rain forests.

Repetto, R., "Deforestation in the Tropics," *Scientific American,* April 1990. Examines government policies that encourage the destruction of tropical forests.

Scuderi, L.A., "A 2000-Year Tree Ring Record of Annual Temperatures in the Sierra Nevada Mountains," *Science* Vol. 259, March 5, 1993. A somewhat technical description of how tree rings were used to reconstruct the mean temperatures for the past 2000 years in the Sierra Nevada of California.

Thybony, S., "Dead Trees Tell Tales," *National Wildlife* Vol. 25, No. 2, August–September 1987. An enjoyable account of the potential uses of tree ring analysis.

CHAPTER 8

PLANT ORGANS: LEAVES

What leaf has had the greatest impact on human history? On human health? Many scholars would agree that the answer to both questions is tobacco (*Nicotiana tabacum*), a leaf crop that humans have cultivated for centuries.

Christopher Columbus and other early European explorers discovered the natives of North and South America growing tobacco, drying and curing the leaves, and then smoking, chewing, or snuffing them. Various tribes used tobacco for medicinal purposes as well as for religious rituals. Shamans (medicine men) of Amazonian tribes, for example, fasted and smoked large quantities of tobacco to produce hallucinations, which they interpreted as visits from spirits.

When Columbus introduced tobacco to Europe, it did not gain immediate acceptance. Within a century, however, many Europeans had adopted the habit of smoking, its popularity enhanced by tobacco's supposed medicinal values. It was touted as a cure for everything from headaches to snakebites and was also used as an aphrodisiac. Although it has no medicinal uses today, tobacco's popularity has continued to the present. Currently millions of people in virtually every country in the world smoke tobacco.

In the 1600s farmers in Virginia and other English colonies began raising tobacco to meet the demand in Europe, and many of them grew very wealthy. As a "cash crop," tobacco provided the grower with more profit per acre than any food crop. Today tobacco leaf still commands a high price per

Tobacco (*Nicotiana tabacum*) growing in a field in eastern Kentucky. (*Adam Jones/Dembinsky Photo Associates*)

pound, and in the United States alone tens of thousands of farmers raise tobacco on more than 800,000 hectares (2 million acres) of land.

The physiologically active ingredients of tobacco are nicotine and related alkaloids, which are colorless, bitter chemicals that affect the central nervous system. When inhaled, nicotine, which is habit-forming, is absorbed through the lung membranes and produces a stimulating sensation. Pure nicotine, however, is poisonous; absorbed in large amounts, it produces nausea, dizziness, and hallucinations. Nicotine is the active ingredient in certain insecticides.

Medical evidence is overwhelming that tobacco smoking causes serious health problems, including heart disease, emphysema, chronic bronchitis, and cancer. The life of a 30-year-old who smokes 15 cigarettes each day, for example, is shortened by an average of more than 5 years. The risk of developing atherosclerosis (hardening of the arteries) is two to six times greater in smokers than in nonsmokers and is directly related to the number of cigarettes smoked daily. A woman who smokes when pregnant doubles the risk of miscarriage and infant death, and babies born to smoking mothers are often underweight. According to the American Cancer Society, more than 75 percent of all deaths from lung cancer are caused by cigarette smoking. If everyone in the United States stopped smoking, more than 300,000 lives would be saved each year.

LEAVES VARY GREATLY IN FORM

Leaves are the most variable plant organ—so variable that plant biologists had to develop an entire set of terminology to describe their shapes, margins (edges), vein patterns, and ways of attaching to stems. Because each leaf is characteristic of the plant on which it grows, many plants can be partially or completely identified on the basis of their leaves alone. Leaves may be round, needle-like, scalelike, cylindrical, heart-shaped, fan-shaped, or thin and narrow. They vary in size from the leaves of the raffia palm (*Raphia ruffia*), which often grow to more than 20 meters (65 feet) in length, to those of the smallest duckweed (*Wolffia microscopica*), which are so small that 32 of them laid end-to-end have a length of 2.5 centimeters (1 inch).

Most leaves are composed of two parts, a blade and a petiole (Figure 8-1). The broad, flat portion of a leaf is the **blade**, and the stalk that attaches the blade to the stem is the **petiole**. Some leaves lack petioles. Many species also have one or two **stipules**, leaflike outgrowths at the base of the petiole.

Leaves may be **simple** (having a single blade) or **compound** (having a blade with two or more leaflets) (Figure 8-2). Compound leaves may be **pinnately compound** (the leaflets are borne on an axis that is a continuation of the petiole) or **palmately compound** (the leaflets arise from a common point at the end of the petiole).

Sometimes it is difficult to tell whether a plant has formed one compound leaf or a small stem bearing several simple leaves. One easy way to determine whether a plant has simple or compound leaves is to look for axillary buds, which develop in the axils of leaves (Figure 8-1). An axillary bud forms at the base of a leaf, whether it is simple or compound. However, axillary buds never develop at the bases of leaflets. Also, the leaflets of a compound leaf lie in a single plane (you can lay a compound leaf flat on a table), whereas the simple leaves on a stem are rarely arranged in one plane.

Leaves are arranged on a stem in one of three possible ways (Figure 8-3). Plants such as walnut have an **alternate** leaf arrangement, with one leaf at each node. *Nodes* are the areas on a stem where leaves are attached. In an **opposite** leaf arrangement, as occurs in lilac and maple, two leaves grow at each node. In a **whorled** leaf arrangement, as occurs in catalpa, three or more leaves grow at each node.

Leaf blades may possess **parallel** venation, in which the primary veins run approximately parallel to one another, or **netted** veins, which are branched in such a way that they resemble a net. Parallel veins are generally found in monocots, whereas netted veins are generally characteristic of dicots (Figure 8-4). Netted veins can be **palmately netted**, so that several major veins radiate out from one point, or **pinnately netted**, so that major veins branch off along the entire length of the main vein.

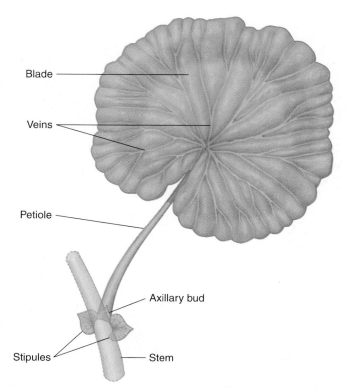

Figure 8-1   The geranium leaf, a "typical" leaf, consists of a blade and petiole. Two stipules emerge at the base of the geranium leaf.

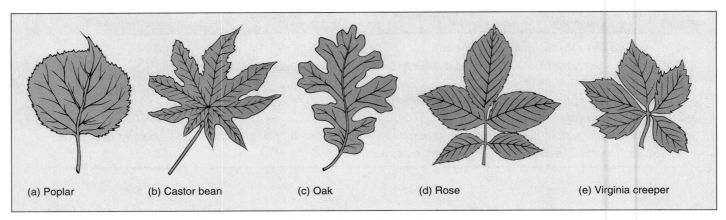

(a) Poplar   (b) Castor bean   (c) Oak   (d) Rose   (e) Virginia creeper

Figure 8-2   Leaves may be simple or compound. Simple leaves **(a,b,c)** have a variety of sizes and shapes. Compound leaves may be pinnately compound **(d)** or palmately compound **(e)**.

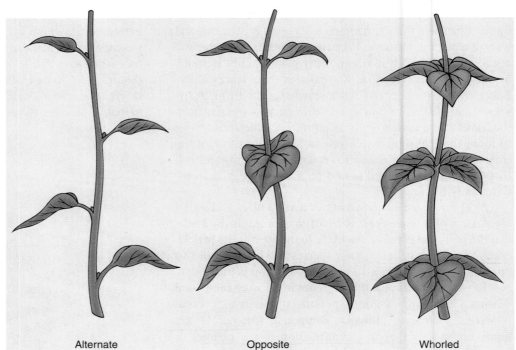

Figure 8-3   Leaf arrangement may be alternate, opposite, or whorled, depending on the number of leaves at each node.

Alternate   Opposite   Whorled

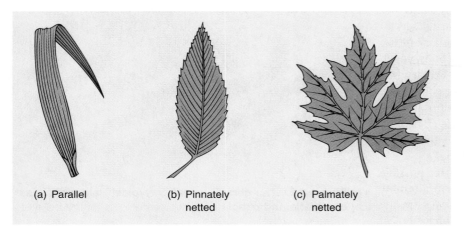

(a) Parallel   (b) Pinnately netted   (c) Palmately netted

Figure 8-4   Venation patterns in leaves. (a) Kentucky bluegrass has parallel veins, which are characteristic of monocot leaves. Siberian elm **(b)** and silver maple **(c)** have netted veins. Siberian elm is pinnately netted, and silver maple is palmately netted.

Epidermis, Mesophyll, Xylem, and Phloem Are the Major Tissues of the Leaf

The leaf is a complex organ composed of several different tissues that are organized to optimize photosynthesis (Figure 8-5). Because the leaf blade has upper and lower sides, it is covered by two epidermal layers, an **upper epidermis** and a **lower epidermis**. Most cells in these layers are living parenchyma cells that lack chloroplasts and are relatively transparent. One interesting feature of leaf epidermal cells is that the cell wall facing outward from the leaf is somewhat

thicker than the cell wall facing the inside. This extra thickness may give the plant additional protection against injury or water loss.

A leaf has a large surface area exposed to the atmosphere; as a result, water loss by evaporation from the leaf's surface is unavoidable. However, epidermal cells secrete a waxy layer, the **cuticle**, that reduces water loss. The cuticle varies in thickness in different plants, in part due to environmental conditions. As one might expect, the leaves of plants adapted to hot, dry climates have very thick cuticles. Furthermore, the exposed (and warmer) upper epidermis generally has a thicker cuticle than the shaded (and cooler) lower epidermis on the same leaf.

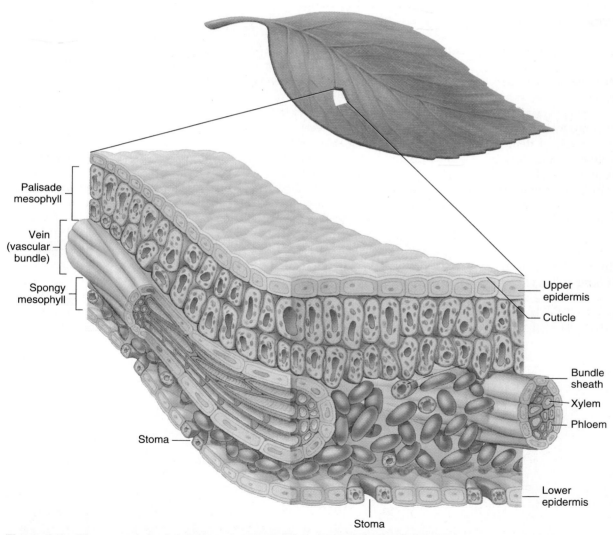

Figure 8-5   Diagram of the internal arrangement of tissues in a typical leaf blade.
The blade is covered by an upper and lower epidermis. The photosynthetic tissue, called mesophyll, is often arranged into palisade and spongy layers. Veins branch throughout the mesophyll.

PLANTS AND People

Important Leaf Crops

Numerous vegetable crops are grown for their edible leaves. Leaves, especially dark green leaves, are particularly good sources of vitamins A and C and the minerals iron and calcium. Cabbage, lettuce, spinach, celery, and rhubarb are representative leaf crops. Also, the leaves of many herbs contain pungent chemicals that make these leaves useful for flavoring food.

Cabbage (*Brassica oleracea*) is a biennial that is harvested as an annual. Thought to be native to the eastern Mediterranean region, it has been grown as a vegetable crop for more than 4000 years. A head of cabbage consists of a shortened stem with numerous short-petioled leaves overlapping a small terminal bud. Some cabbage varieties are green, and others are purple. Cabbage is eaten raw, cooked, or fermented (as sauerkraut).

Lettuce (*Latuca sativa*) is an annual, native to the Mediterranean region, that has been cultivated for more than 2500 years. There are three types of lettuce: head lettuce (for example, iceberg lettuce), which forms a tight, rounded head; cos or romaine type (for example, romaine lettuce), which forms a loose cylindrical head; and leaf type (for example, red lettuce and bibb lettuce), which does not form a head. Lettuce is eaten fresh as a salad vegetable.

Spinach (*Spinacia oleracea*) is an annual native to the eastern Mediterranean region. It has been cultivated for at least 2000 years for its dark green leaves. The two main types of spinach are smooth-leaf and wrinkled-leaf. Spinach is eaten fresh as a salad vegetable or cooked.

Celery (*Apium graveolens*) is a biennial that is harvested as an annual. Native to Europe, celery was first used for medicinal purposes; folklore indicates that it may be useful in treating diabetes. Only later was it accepted as a food. Celery has been a common leaf crop since the early 1600s. The edible part is the thick, fleshy petiole, which is eaten raw or cooked.

Rhubarb (*Rheum rhaponticum*), native to Siberia, is a herbaceous perennial. Rhubarb used to be taken as a folk medicine to expel parasitic worms from the intestinal tract. Today it is cultivated for its thick, fleshy petioles, which are stewed and eaten in sauces and pies. The leaf blades, however, contain toxic levels of oxalic acid and should never be consumed.

Many herbs that people use to flavor food come from leaves—for example, basil, bay leaves, marjoram, and tarragon. Basil (*Ocimum basilicum*) is native to India. When dried, basil leaves have a distinctive aroma that complements many foods, particularly those containing tomatoes. Bay leaves (*Laurus nobilis*) come from an evergreen tree native to the Mediterranean. Bay leaves add a pungent flavor to soups, chowders, and fish. Marjoram (*Majorana hortensis*), which is native to the Mediterranean region, has been used medicinally for centuries. Today the dried leaves are added to almost any food except sweets. Tarragon (*Artemisia dracunculus*) comes from a perennial shrub that originated in Asia. Tarragon is responsible for the distinctive taste of Béarnaise sauce.

| Spinach | Celery | Iceberg lettuce | Rhubarb | Cabbage |

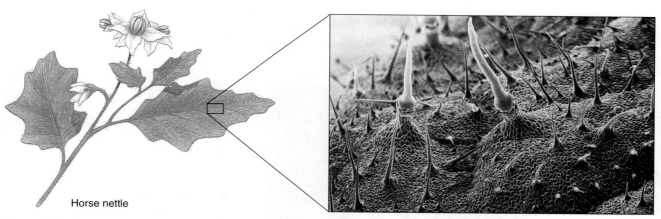

Horse nettle

Figure 8-6   Scanning electron micrograph of a horse nettle (*Solanum carolinense*) leaf. The epidermis is often covered with trichomes that may limit the transpiration of water, discourage herbivores, sting, or perform other functions. When an animal brushes against the nettle or attempts to eat it, the trichomes on the leaves and stems readily break off inside the animal's skin. The trichomes inject irritating substances into the skin that produce a stinging sensation. (*Biophoto Associates*)

The epidermises of many leaves are covered with various hairlike structures called **trichomes** (Figure 8-6). Some leaves, such as those of the popular cultivated plant called lamb's ears (*Stachys byzantina*), have so many trichomes that they actually feel fuzzy. Trichomes help reduce water loss from the leaf surface by retaining a layer of moist air next to the leaf. Trichomes also reflect excessive sunlight, thereby protecting the plant from overheating. Some trichomes secrete stinging irritants (that deter plant eaters) and other substances or excrete excess salts absorbed from a salty soil.

The epidermis is typically covered with minute pores, or **stomata** (sing. *stoma*), for gas exchange. Each stoma is flanked by two specialized epidermal cells called **guard cells** (see Figure 5-7). Changes in the shape of each pair of guard cells open and close the stoma. Guard cells are usually the only epidermal cells that have chloroplasts, but the functional significance of these chloroplasts is not clear. Stomata are especially numerous on the lower epidermis—about 100 stomata per square millimeter although the actual number varies greatly—and in many cases occur *only* on the lower surface. This adaptation reduces water loss because stomata on the lower epidermis are shielded from direct sunlight.

The photosynthetic tissue of the leaf, called the **mesophyll**, is sandwiched between the upper epidermis and the lower epidermis. The word "mesophyll" comes from Greek and means "the middle of the leaf." Mesophyll cells, which are parenchyma cells packed with chloroplasts, are very loosely arranged with lots of air spaces between them that facilitate gas exchange. In many plants, the mesophyll is divided into two sublayers. Toward the upper epidermis the cells are stacked more closely together in a **palisade** layer. In the lower portion, the cells are more loosely and more irregularly arranged in a **spongy** layer. The two layers have different primary functions. Palisade mesophyll is the primary site of photosynthesis in the leaf. Although the spongy mesophyll is also a site for photosynthesis, its primary function is diffusion of gases (particularly CO_2) within the leaf.

Palisade mesophyll may be further organized into one, two, three, or even more rows of cells. The presence of additional layers of palisade mesophyll is an adaptation due in part to environmental conditions. Leaves exposed to direct sunlight are thicker because they contain more rows of palisade mesophyll than do shaded leaves on the same plant. In direct sunlight, the light is strong enough to effectively penetrate multiple layers of palisade mesophyll, enabling all layers to photosynthesize efficiently.

The **veins** (vascular bundles) of a leaf extend through the mesophyll. Branching is extensive, and no mesophyll cell is very far from a vein. Each vein contains two types of vascular tissue, xylem and phloem. **Xylem**, which conducts water and dissolved minerals, is usually located on the upper side of a vein, toward the upper epidermis, whereas **phloem**, which conducts dissolved sugars, is usually confined to the lower side of a vein.

A vein is usually surrounded by one or more layers of nonvascular cells, called a **bundle sheath**. Bundle sheaths are composed of parenchyma or sclerenchyma cells (see Chapter 5). Frequently the bundle sheath has support columns, called **bundle sheath extensions**, that extend through the mesophyll from the upper epidermis to the lower epidermis (Figure 8-7). Bundle sheath extensions may be composed of parenchyma, collenchyma, or sclerenchyma cells.

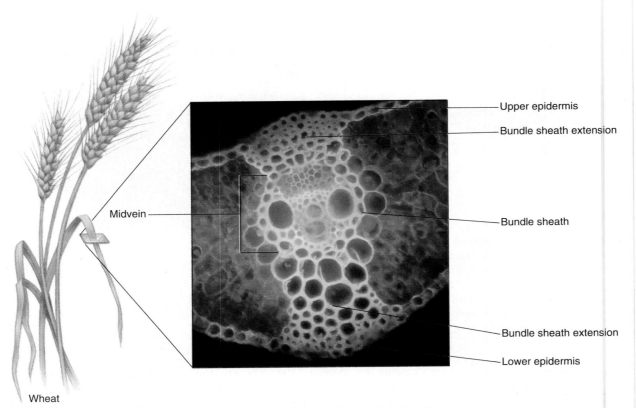

Figure 8-7   Cross section of a wheat (*Triticum aestivum*) midvein, showing a bundle sheath extension. (*Phil Gates, University of Durham/Biological Photo Service*)

Leaf Structure Differs in Dicots and Monocots

The leaf of a **dicot** is usually composed of a broad, flattened blade and a petiole and exhibits netted venation. In contrast, rather than a petiole, **monocots** often have narrow leaves that wrap around the stem in a sheath. Parallel venation is characteristic of monocot leaves.

The internal leaf anatomies of dicots and certain monocots also differ (Figure 8-8). The mesophyll of dicots and many monocots typically contains both palisade and spongy layers. In contrast, mesophyll in certain monocot leaves (corn and other grasses) is not differentiated into palisade and spongy tissue. Because dicots have netted veins, a cross section of a dicot blade often shows veins in cross section as well as lengthwise. In cross section, the parallel venation pattern of monocot leaves produces evenly spaced veins.

The upper epidermises of the leaves of certain grasses have large, thin-walled cells called **bulliform cells** on both sides of the midvein (Figure 8-9). These cells may be involved in the rolling or folding inward of the leaf during drought. When water is plentiful, bulliform cells are turgid (swollen with water) and the leaf is open. When bulliform cells lose water (as they may during a drought), the leaf folds inward, decreasing its surface area exposed to the air, an action that reduces water loss by evaporation.

Differences between the guard cells in dicot leaves and certain monocot leaves also occur (Figure 8-10). The guard cells of dicots and many monocots are shaped like tiny kidney beans. Other monocot leaves (those of grasses, reeds, and sedges) have guard cells shaped like dumbbells. The difference in structure between the guard cells in dicots and those in monocots affects how the cells swell or shrink to open or close the pore. Guard cells of either monocots or dicots may be associated with special epidermal cells called **subsidiary cells** that are structurally different from all other epidermal cells.

STRUCTURE IS RELATED TO FUNCTION IN LEAVES

The primary function of leaves is to collect radiant energy and convert it to the chemical energy stored in the bonds of organic molecules such as glucose. During this process, called **photosynthesis**, plants take relatively simple inorganic molecules (carbon dioxide and water) and convert them to sugar. Oxygen is given off as a by-product.

How is leaf structure related to the leaf's primary function, photosynthesis? Most leaves are thin and flat, a shape

(*Text continues on page 137.*)

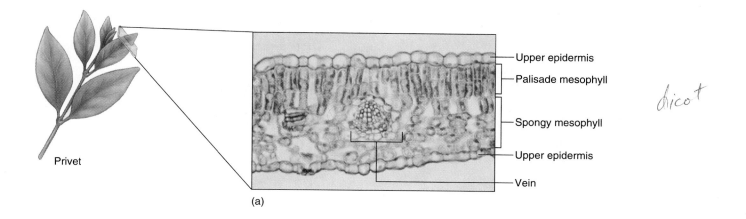

Upper epidermis
Palisade mesophyll
Spongy mesophyll
Upper epidermis
Vein

Privet

dicot

(a)

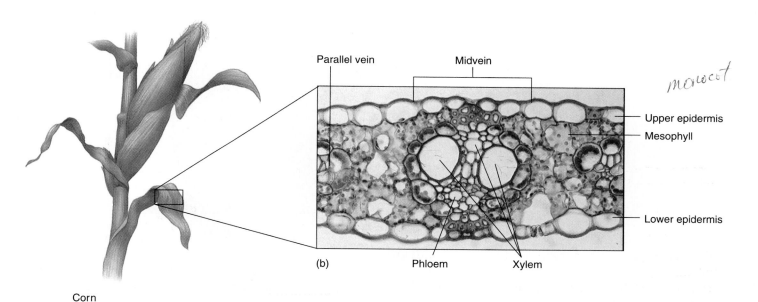

Parallel vein   Midvein

monocot

Upper epidermis
Mesophyll
Lower epidermis

(b)   Phloem   Xylem

Corn

Figure 8-8   Leaf cross sections. (a) Privet (*Ligustrum vulgare*), a dicot, has a mesophyll with distinct palisade and spongy sections. **(b)** Corn (*Zea mays*), a monocot. Note
the absence of distinct regions of palisade and spongy mesophyll. Also evident is the
evenly spaced parallel venation characteristic of monocots. (a, *James Mauseth, University
of Texas;* b, *Dwight R. Kuhn*)

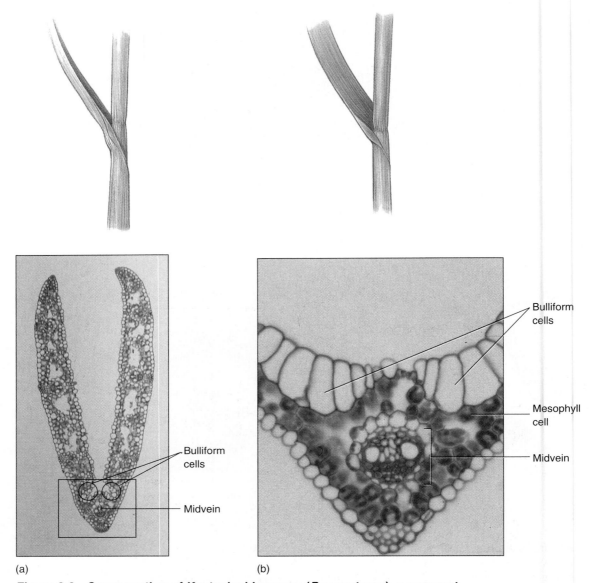

(a)   (b)

Figure 8-9   Cross section of Kentucky bluegrass (*Poa pratense*), a monocot.
(a) A folded leaf blade. The inconspicuous bulliform cells occur in the upper epidermis on both sides of the midvein. **(b)** An expanded leaf blade. A higher magnification of the midvein region shows the enlarged, turgid bulliform cells. (a, b, *Dennis Drenner*)

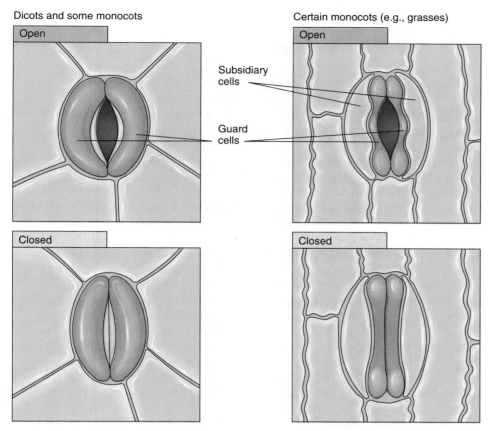

Dicots and some monocots

Open

Closed

Certain monocots (e.g., grasses)

Open

Closed

Subsidiary cells

Guard cells

Figure 8-10   Variation in guard cells. Guard cells of dicots and many monocots are bean-shaped. Other monocot guard cells (those of grasses, reeds, and sedges) are narrow in the center and thicker at the ends. Each narrow guard cell is associated with a special epidermal cell called a subsidiary cell.

that allows maximum absorption of light energy and efficient internal diffusion of gases. As a result of their ordered arrangement on the stem, leaves efficiently catch the sun's rays. The leaves of plants form an intricate green mosaic, bathed in sunlight and atmospheric gases.

The epidermis of a leaf is relatively transparent and allows light to penetrate to the interior of the leaf, where the photosynthetic tissue, the mesophyll, is located. Stomata, which dot the surfaces of the leaf, permit the exchange of gases between the atmosphere and the leaf's internal tissues. Carbon dioxide, a raw material of photosynthesis, diffuses into the leaf through stomata, and the oxygen produced during photosynthesis rapidly diffuses out of the leaf through stomata. Stomata, adapted to optimize gas exchange, also permit other gases, including air pollutants, to enter the leaf (see Plants and the Environment: The Effects of Air Pollution on Leaves).

Water required for photosynthesis is obtained from the soil and transported in the xylem to the leaf, where it moistens the surfaces of mesophyll cells. The loose arrangement of the mesophyll tissue, with its air spaces between cells, allows for rapid diffusion of carbon dioxide to the surfaces of the mesophyll cells. The carbon dioxide then dissolves in a film of water at the cell surfaces before diffusing into the cells.

The veins not only supply water to the photosynthetic tissue (by way of the xylem) but carry (in the phloem) the dissolved sugar produced during photosynthesis to other plant parts. Bundle sheaths and bundle sheath extensions associated with the veins provide additional support; they prevent the leaf, which is structurally weak because of the large amount of air space in the mesophyll, from collapsing under its own mass.

Environment

The Effects of Air Pollution on Leaves

The air we breathe is often dirty and contaminated with many pollutants, particularly in urban areas. **Air pollution** consists of gases, liquids, or solids present in the atmosphere at high enough levels to harm humans, other animals, plants, or materials. Although air pollutants can come from natural sources—as, for example, when lightning causes a forest fire or when certain trees emit volatile organic compounds that react with other air pollutants to form low-level ozone—human activities make a major contribution to global air pollution. Motor vehicles and industry are the two main human sources of air pollution.

All plant parts can be damaged by air pollution, but leaves are particularly susceptible because of their structure and function (see figure). The thin blade provides a large surface area that comes into contact with the surrounding air. The thousands of tiny stomatal pores that dot the epidermis and control gas exchange with the atmosphere also permit pollutants to diffuse into the leaf. Just as lungs—the organs of gas exchange in humans—often show the effects of air pollution, the leaves are the part of the plant affected most.

Trees dramatically demonstrate the effect of air pollution on biological longevity. According to the Urban Forest Forum, which is sponsored by the U.S. Forest Service, the average life span of a tree in a rural setting, where air pollution is generally low, is 150 years. In the most ideal urban setting, a tree typically lives only 60 years, and in a normal city setting, the average life span is 32 years. Trees in a downtown setting, where air quality is lowest, live only 7 years on average. Note, however, that air pollution is not the only cause of short life span

in urban trees. Other factors include inadequate room for growth, polluted runoff from streets and sidewalks, and vandalism.

Many studies have shown that high levels of most forms of air pollution reduce the overall productivity of crop plants. The most significant pollutant in terms of yield loss is low-level ozone, a toxic gas produced when sunlight catalyzes a reaction between pollutants emitted by motor vehicles and industries. In plants, ozone inhibits photosynthesis because it damages the photosynthetic cells sandwiched in the middle of the leaf, probably by altering the permeability of their cell membranes. Exposure to low levels of air pollution often causes a decline in photosynthesis without any other symptoms of injury. Damage to leaves and other obvious symptoms appear at much higher levels of air pollution.

Plants can decline and die from the combination of air pollution, including acid rain (see Plants and the Environment: Acid Rain, in

Chapter 2), with other environmental stresses such as low winter temperatures, prolonged droughts, insects, and bacterial, fungal, and viral diseases. More than half of the red spruce trees in the mountains of the northeastern United States, for example, have died since the mid-1970s. Other tree species (sugar maples, for example) are also dying. Many trees that are not dead are exhibiting symptoms of **forest decline**, characterized by gradual deterioration and often eventual death. The general symptoms of forest decline are reduced vigor and reduced growth, but some plants exhibit specific symptoms, such as yellowing of needles in conifers.

Air pollutants may or may not be the *primary* stress that results in forest decline, but the presence of air pollution certainly reduces plant resistance to other stress factors. When one or more stresses weaken a tree, an additional stress may be decisive in causing death.

Plants exposed to air pollution exhibit a variety of symptoms, including reduced growth. Compare the size and overall vigor of the radish (*Raphanus sativus*) grown in clear air (*left*) and the radish damaged by air pollution (*right*). Note the extensive leaf damage. (*Visuals Unlimited/Science VU*)

Each Type of Plant Has Leaves Adapted to Its Environment

The environment to which a particular plant is adapted affects its leaf structure. Although aquatic plants and plants adapted to dry conditions both perform photosynthesis and have the same basic leaf anatomy, their leaves are modified to enable them to survive their distinct environmental conditions. The leaves of water lilies, for example, have stomata on the upper epidermis rather than the lower epidermis. (How does this feature adapt water lilies to an aquatic environment?) The petioles of water lilies are long enough to allow the blade to float on the water's surface (Figure 8-11). These petioles and other submerged parts have an internal system of air ducts through which oxygen moves from the floating leaves to the underwater roots and stems, which live in a poorly aerated environment.

The leaves of **conifers**—an important group of trees and shrubs that includes pine, spruce, fir, redwood, and cedar—are waxy needles. Most conifers are evergreen, which means that they produce and lose leaves throughout the year rather than during certain seasons. Conifers dominate a large portion of Earth's land area, particularly in northern forests and in mountains. Their needles have structural adaptations that help them survive winter, which is the driest part of the year. Winter is dry even in areas of heavy snows, because roots cannot absorb water from the soil when the soil temperature is low.

Figure 8-12 is a cross section of a pine needle. Note that the needle is somewhat thickened rather than flat and blade-like. This shape, which results in less surface area exposed to the air, presumably reduces water loss. Other features that

Blade   Petiole

Figure 8-11   An unusual view of water lily (*Nymphaea* sp) leaves shows their petioles as well as their blades, which float on the water's surface. Water lilies have special adaptations, such as stomata on the upper epidermis of the blade rather than on the lower epidermis, that enable them to thrive in a watery environment. (*Frans Lanting/Minden Pictures*)

help conserve water include the thick waxy cuticle and sunken stomata (which permit gas exchange while minimizing water loss). Thus, needles help conifers tolerate the dry conditions that exist for part of the year. With the warming and snowmelt of spring, water becomes available again, and the needles quickly resume photosynthesis. In contrast, plants that are **deciduous**—that is, shed all of their leaves during a particular season—have a lag period before resuming growth, because they first have to grow new leaves.

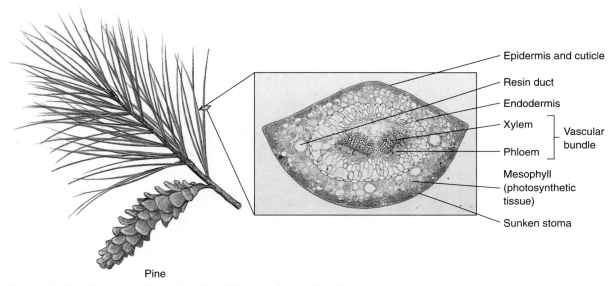

Epidermis and cuticle
Resin duct
Endodermis
Xylem
Phloem
} Vascular bundle
Mesophyll (photosynthetic tissue)
Sunken stoma

Pine

Figure 8-12   Cross section of a pine (*Pinus* sp) needle. (*Dennis Drenner*)

OPENING AND CLOSING OF STOMATA ARE DUE TO CHANGES IN TURGIDITY OF THE GUARD CELLS

When water moves into guard cells from surrounding cells, they become turgid (swollen), and the inner cell walls bend outward at the center, producing a pore. When water leaves the guard cells, they become flaccid (limp) and collapse against one another, closing the pore. Chapter 10 discusses the actual mechanism of stomatal opening and closing.

Stomata are usually open during the day, when carbon dioxide is required for photosynthesis, and closed at night, when photosynthesis is shut down. Although daylight and darkness trigger the opening and closing of guard cells, other environmental factors, including carbon dioxide concentration, are involved. A low concentration of CO_2 in the leaf induces stomata to open even in the dark. The effects of light and CO_2 concentration on stomatal opening are interrelated. Photosynthesis, which occurs in the presence of light, reduces the internal concentration of CO_2 in the leaf, triggering stomatal opening. Another environmental factor that affects stomatal opening and closing is severe water stress. During a prolonged drought, stomata remain closed even during the day. Stomatal opening and closing are also under hormonal control.

The opening and closing of stomata appear to be regulated by an internal biological clock that somehow measures time. For example, plants placed in continual darkness persist in opening and closing their stomata at more or less the same times each day. Biological rhythms that follow an approximate 24-hour cycle are known as **circadian rhythms**.

LEAVES LOSE WATER BY TRANSPIRATION AND GUTTATION

Despite leaf adaptations such as the cuticle, approximately 99 percent of the water that a plant absorbs from the soil is lost by evaporation from the leaves and, to a lesser extent, the stems. Loss of water vapor from aerial plant parts by evaporation is called **transpiration**.

The cuticle is extremely effective in reducing water loss from transpiration. It is estimated that only 1 to 3 percent of the water lost from a plant passes directly through the cuticle. Most transpiration occurs through the stomata. The numerous stomatal pores that are so effective in gas exchange for photosynthesis also provide openings through which water vapor escapes. In addition, the loose arrangement of the mesophyll cells provides a large surface area within the leaf from which water can evaporate.

Environmental factors influence the rate of transpiration. At higher temperatures, plant surfaces lose more water. Light increases the transpiration rate, in part because it triggers the opening of stomata and in part because it increases the temperature of the leaf. Wind and dry air increase transpiration, but humid air *decreases* transpiration because the air is already saturated, or nearly so, with water vapor.

Although transpiration may seem wasteful, particularly to farmers in arid lands, it is an essential process. Transpiration is responsible for water movement in plants, and without it, water from the soil would not reach the leaves (see Chapter 10). There may be some additional benefits from the large amount of water plants lose by transpiration. First, tran-

Figure 8-13 Temporary wilting in squash (*Cucurbita pepo*) leaves. (a) In the late afternoon of a hot day and **(b)** recovery the following morning. Note that wilting helps reduce the surface area from which transpiration occurs. The plants recovered by absorbing water from the soil during the night, when transpiration was negligible. (a,b, *Carlyn Iverson*)   (a)   (b)

spiration, like sweating in humans, cools the leaves and stems. When water changes from a liquid into a vapor, it absorbs a great deal of heat. As the water molecules leave the plant as water vapor, they carry this heat with them. Thus, the cooling effect of transpiration may prevent the plant from overheating, particularly in direct sunlight. On a hot summer day, for example, the internal temperature of leaves is measurably less than that of the surrounding air. (See Plants and the Environment: Trees, Transpiration, and Climate.)

Another possible benefit of transpiration is that it ensures that the plant will have sufficient essential minerals. The water a plant transpires is initially absorbed from the soil, where it is not pure but is a dilute solution of dissolved mineral salts. Water leaves the plant during transpiration, but minerals do not. Many of these minerals are required for the plant's growth. Some biologists have suggested that transpiration enables a plant to take in enough water to provide itself with adequate essential minerals and that plants cannot satisfy their mineral requirements if the transpiration rate is not high enough.

There is no doubt, however, that under certain circumstances transpiration can harm a plant. On hot summer days, plants frequently lose more water by transpiration than they can take in from the soil. Their cells lose turgor, and the plant wilts (Figure 8-13). If a plant can recover overnight, because of the combination of negligible transpiration (recall that stomata are closed) and absorption of water from the soil, it is said to have experienced *temporary wilting*. Most plants recover from temporary wilting with no ill effects. In cases of prolonged drought, however, the soil may not contain sufficient moisture to permit recovery from wilting. A plant that cannot recover overnight is said to be *permanently wilted* and will die unless water is supplied immediately.

Some Plants Exude Water as a Liquid

Many leaves have **hydathodes**, special openings at the tips of leaf veins through which liquid water is literally forced out. This loss of liquid water, known as **guttation**, occurs when transpiration is negligible and available soil moisture is high. Guttation frequently occurs at night, because the stomata are closed but water continues to move into the roots by osmosis. People sometimes think erroneously that the early-morning water droplets on leaves are dew or water condensation from the air, rather than guttation (Figure 8-14).

LEAF ABSCISSION ALLOWS PLANTS IN TEMPERATE CLIMATES TO SURVIVE WINTER

All trees shed leaves. Many conifers, for example, shed their needles year-round. The leaves of deciduous plants, however,

Figure 8-14   Guttation in lady's mantle (*Alchemilla vulgaris*). Many people mistake guttation for early-morning dew. (*Dennis Drenner*)

change color and **abscise**, or fall off, at the same time each year—as winter approaches in temperate climates, or before the dry period in tropical climates with pronounced wet and dry seasons. In temperate forests, for example, most woody plants with broad leaves shed their leaves in the fall; this helps them survive the low temperatures of winter. During winter, the plant's metabolism, including its rate of photosynthesis, slows down a great deal. As a result, plants have little need for leaves at that time.

Another reason for abscission is related to a plant's water requirements, which become critical during the winter. When the ground chills, absorption of water by the roots is inhibited. When the ground freezes, *no* absorption occurs. If the broad leaves stayed on the plant during the winter, the plant would continue to lose water by transpiration but would be unable to replace it with water absorbed from the soil.

Leaf abscission is a complex process that involves many physiological changes, all of which are initiated and carried out by changes in the levels of plant hormones. As autumn approaches, the plant reabsorbs sugar (much of it obtained from the breakdown of starch), and many essential minerals—such as nitrogen, phosphorus, and possibly potassium—are transported from the leaves into the woody tissues. Chlorophyll is broken down, and some of the accessory pigments in the chloroplasts of leaf cells (the orange carotenes and yellow xanthophylls) become evident. These pigments are always present in the leaf but are masked by the green chlorophyll. In addition, red water-soluble pigments may be synthesized and stored in the vacuoles of leaf cells in some species; their function is unknown. The various combinations of these pigments are responsible for the brilliant colors of autumn landscapes in temperate climates.

Environment

Trees, Transpiration, and Climate

An area's **climate** is the average weather conditions that occur in that area over a period of years. Factors that determine an area's climate include temperature and precipitation—both of which are influenced by transpiration, the loss of water vapor from the aerial portions of plants.

Transpiration by trees, for example, influences the local temperature of a forest. If you have ever walked into a forest on a hot summer day, you have probably noticed that the air there is cooler and moister than the air outside the forest. This differ-ence occurs because transpiration, like human sweating, is a biological cooling process. Water must absorb a lot of heat before it changes from a liquid to a vapor. When water evapo-rates through the stomata, the water vapor carries this heat with it. Thus, water transpired from the surfaces of leaves and stems has a cooling effect, not only for the plant but for the local area.

Transpiration is an important part of the **hydrologic cycle**, in which water cycles from the oceans and land to the atmosphere and back to the oceans and land. Water that evaporates from the surfaces of leaves and stems forms clouds in the atmosphere. Thus, transpiration eventually results in precipitation. As you might expect, forest trees release substantial amounts of moisture into the air by transpiration (see figure). Conse-quently, when a large forest is burned or cut down, rainfall declines and droughts become more common in that region. Several studies have documented the alteration of precip-itation patterns following deforesta-tion in the tropics.

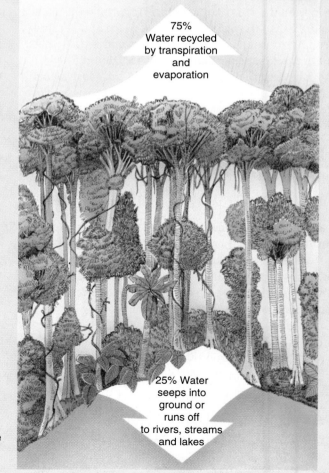

75%
Water recycled
by transpiration
and
evaporation

25% Water
seeps into
ground or
runs off
to rivers, streams
and lakes

Forests play an important role in the hydrologic cycle by returning much precipitation water to the atmos-phere by transpiration.

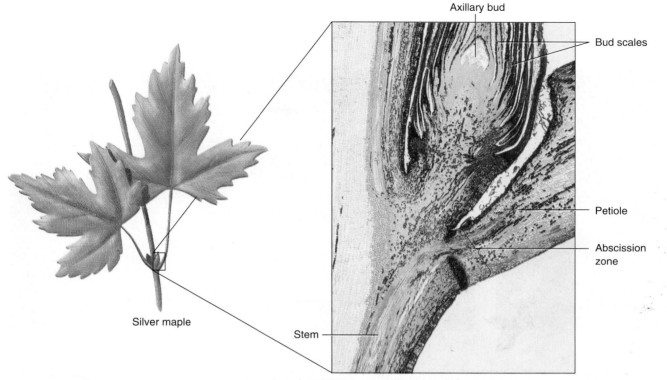

Figure 8-15 Longitudinal section through a silver maple (*Acer saccharinum*) branch, showing the base of the petiole. Note the abscission zone, where the leaf will abscise from the stem. An axillary bud with its protective bud scales is evident above the petiole. (*James Mauseth, University of Texas*)

In Many Leaves, Abscission Occurs at an Abscission Zone Near the Base of the Petiole

The area where a leaf petiole detaches from the stem, called the **abscission zone**, is structurally different from surrounding tissues. It is composed primarily of thin-walled parenchyma cells and, because it contains few fibers, is anatomically weak (Figure 8-15). As autumn approaches, a protective layer of cork cells develops on the stem side of the abscission zone. These cells have suberin, a waxy waterproof material, impregnated in their walls. Enzymes then dissolve the **middle lamella** (the "cement" that holds the primary cell walls of adjacent cells together) in the abscission zone.

Once this process is completed, there is nothing holding the leaf to the stem but a few cells of the xylem tissue. A sudden breeze is enough to cause the final break, and the leaf detaches. The protective layer of cork remains, sealing off the area and forming a leaf scar.

LEAVES WITH FUNCTIONS OTHER THAN PHOTOSYNTHESIS EXHIBIT MODIFICATIONS IN STRUCTURE

Although photosynthesis is the main function of leaves, certain leaves possess special modifications for functions other than photosynthesis. Some plants have leaves specialized for pro-

tection. **Spines**, modified leaves that are hard and pointed, may be found on many desert plants, such as cacti (Figure 8-16a). In the cactus, the main organ of photosynthesis is the stem rather than the leaf. Spines discourage animals from eating the succulent (water-filled) stem tissue of the cactus.

Many vines have **tendrils**, which are usually specialized leaves that grasp and hold onto other structures (Figure 8-16b). A vine is a climbing plant whose stem cannot support its own weight, and so it often possesses tendrils that help keep it attached to the structure on which it is growing. Some tendrils, such as those on a grapevine, are specialized stems.

The winter buds of a dormant woody plant are covered by protective **bud scales**, modified leaves that protect the delicate meristematic tissue of the bud from being injured and drying out (Figure 8-16c; also see Figure 7–2).

Leaves may also be modified for storage of water or food. For example, a **bulb** is a short underground stem to which large, fleshy leaves are attached (Figure 8-16d). Onions and tulips form bulbs. Many plants adapted to arid conditions—for example, string-of-beads (*Senecio rowleyanus*), jade plant (*Crassula arborescens*), and medicinal aloe (*Aloe barbadensis*)—have fleshy, succulent leaves for water storage (Figure 8-16e). These leaves are usually green and photosynthesize as well.

Plants living in unusual environments often have specialized foliar adaptations. The flowerpot plant (*Dischidia rafflesiana*) is a small tropical plant that is an **epiphyte**—a plant that grows attached to another, larger plant, which it uses for support. The flowerpot plant lives high in the canopy of the

(a)   (b)   (c)

(d)   (e)

Figure 8-16 Leaf modifications. (a) The leaves of cacti, such as this barrel cactus (*Ferocactus* sp), are modified as spines for protection. **(b)** Tendrils of a bur cucumber (*Echinocystis lobata*) vine grasp grass stems and aid in climbing. Tendrils may be modified leaves or stems. **(c)** Overlapping bud scales protect buds. Shown here are a terminal bud and two axillary buds of a maple (*Acer* sp) twig. **(d)** The leaves of bulbs such as the onion (*Allium cepa*) are fleshy for storage of food materials and water. **(e)** Some plants, such as cobweb houseleek (*Sempervivum arachnoideum*), have thick succulent leaves modified for water storage as well as photosynthesis. (a,e, *James Mauseth, University of Texas;* b, *Stephen P. Parker/Photo Researchers, Inc.;* c,d, *Dennis Drenner*)

tropical rain forest, attached to the bark of a tree trunk. It has two kinds of leaves; some are normal, and others are modified to form hollow containers, or "flowerpots," about 10 centimeters deep (Figure 8-17). Water collects in the pot from condensation of water vapor that enters the leaf through stomata. Ant colonies that live in the pots collect debris that releases minerals as it decomposes. The plant grows adventitious roots into its own pots from nearby stems or petioles and in this way is able to absorb water and dissolved minerals high above the ground.

Stems

Pots
(modified
leaves)

(a)

Root

(b)

Figure 8-17   The flowerpot plant *(Dischidia rafflesiana)*.
(a) The leaves of the flowerpot plant are modified to hold
water and organic material carried in by ants. **(b)** A cutaway
view of a pot removed from a plant reveals the special root
that absorbs water and dissolved minerals inside the pot.
(a,b, *David A. Steingraeber*)

Some of the Most Remarkable Examples of Modified Leaves Are Those of Insectivorous Plants

Insectivorous plants are plants that capture insects. Most insectivorous plants grow in soil that is deficient in certain essential minerals. These plants meet some of their mineral requirements by digesting insects and other small animals. The leaves of insectivorous plants are adapted to attract, capture, and digest animal prey, usually insects.

Some insectivorous plants have passive traps. The leaves of a pitcher plant, for example, are shaped so that rainwater collects in them, forming a reservoir that also receives digestive enzymes secreted by the plant (Figure 8-18). Some pitchers are quite large; in the tropics, for example, a pitcher plant may be large enough to hold 1 liter (approximately 1 quart) or more of liquid.

An insect attracted by the odor or nectar of the pitcher may lean over the edge and fall in. Although it may make repeated attempts to escape, the insect is prevented from crawling out by slippery walls and a row of stiff spines that point downward around the lip of the pitcher. The insect eventually drowns, and part of its body is digested and absorbed.

The Venus flytrap is an insectivorous plant with active traps. Its leaves resemble tiny bear traps (see Figure 1-6). Each side of the leaf contains three small hairs. If an insect alights and brushes against two of the hairs, the trap springs shut with amazing rapidity. After the insect has died and been digested, the trap reopens, and the indigestible remains fall out.

Figure 8-18   A common pitcher plant (*Sarracenia purpurea*) growing in its natural environment. The insectivorous pitcher plant has leaves modified to form a pitcher that collects water, in which the plant's prey drown. Notice the dead beetle in the "pitcher." (*Bill Lea/Dembinsky Photo Associates*)

STUDY OUTLINE

I. Leaves exhibit variation.
  A. Leaves typically consist of a broad, flat blade and a stalklike petiole. Many leaves have small, leaflike outgrowths from the base, called stipules.
  B. Leaves may be simple or compound; may have alternate, opposite, or whorled leaf attachment; and may have parallel or netted (either pinnately netted or palmately netted) venation.
II. Leaf structure is adapted for its primary function, photosynthesis.
  A. Most leaves have a broad, flattened blade that is very efficient in collecting the sun's radiant energy.
  B. The transparent epidermis allows light to penetrate into the mesophyll, where photosynthesis occurs.
    1. Stomata are small pores in the epidermis that permit both gas exchange, needed for photosynthesis, and transpiration.
    2. The waxy cuticle coats the epidermis, enabling the plant to survive in the dry conditions of a terrestrial environment.
  C. The mesophyll tissue contains air spaces that permit rapid diffusion of carbon dioxide and water into, and oxygen out of, mesophyll cells.
  D. Leaf veins have xylem to conduct water and essential minerals to the leaf, and phloem to conduct sugar produced by photosynthesis to the rest of the plant.
III. Monocot and dicot leaves can be distinguished on the basis of external structure and internal anatomy.
  A. Monocot leaves have parallel venation.
  B. Dicot leaves have netted venation.
IV. Stomata open during the day and close at night. Multiple factors, such as light or darkness, CO_2 concentration, and a circadian rhythm within the plant, affect stomatal opening and closing.
V. Transpiration is the loss of water vapor from plants.
  A. Transpiration occurs primarily through the stomata.
  B. Environmental factors such as temperature, wind, and relative humidity affect the rate of transpiration.
  C. Transpiration may be both beneficial and harmful to the plant.
VI. Guttation, the release of liquid water from leaves, occurs through special structures called hydathodes when transpiration is negligible and available soil moisture is high.
VII. Leaf abscission (when leaves fall off a plant) is a complex process involving physiological and anatomical changes.
VIII. Leaves may be modified for functions other than photosynthesis.
  A. Spines are leaves adapted to provide protection.
  B. Tendrils are leaves modified for grasping and holding onto other structures (to support weak stems).
  C. Bud scales are leaves modified to protect the delicate meristematic tissue of buds.
  D. Bulbs are short underground stems with fleshy leaves specialized for storage.
  E. Some leaves are modified to trap insects.

SELECTED KEY TERMS

abscission, p. 141
air pollution, p. 138
blade, p. 129
bud scales, p. 143
bulb, p. 143
bundle sheath, p. 133
compound leaf, p. 129
cuticle, p. 131

deciduous, p. 139
dicot, p. 134
epidermis, upper and lower, p. 131
forest decline, p. 138
guard cells, p. 133
guttation, p. 141
hydathode, p. 141

mesophyll, p. 133
monocot, p. 134
palisade mesophyll, p. 133
petiole, p. 129
photosynthesis, p. 134
simple leaf, p. 129
spine, p. 143

spongy mesophyll, p. 133
stipule, p. 129
stoma, p. 133
tendril, p. 143
transpiration, p. 140
trichome, p. 133
vein, p. 133

REVIEW QUESTIONS

1. Draw diagrams to demonstrate simple versus compound leaves; alternate, opposite, and whorled leaf arrangement; and parallel, palmately netted, and pinnately netted venation.
2. Give the general equation for photosynthesis (see Chapter 4), and discuss how the leaf is organized to deliver the raw materials and products of photosynthesis.
3. What leaf structure is related to both photosynthesis and transpiration? How is it tied to each physiological process?
4. Relate how water is involved in stomatal opening and closing.
5. Discuss at least two ways in which certain leaves adapted to conserve water during the course of evolution.
6. How do environmental factors (sunlight, temperature, humidity, and wind) influence the rate of transpiration? How does the environment influence stomatal opening and closing?
7. Why do many woody plants lose their leaves in autumn?
8. Discuss the specialized features of the leaves of insectivorous plants.
9. Discuss some the possible causes of tree decline.

10. Label the following diagram.

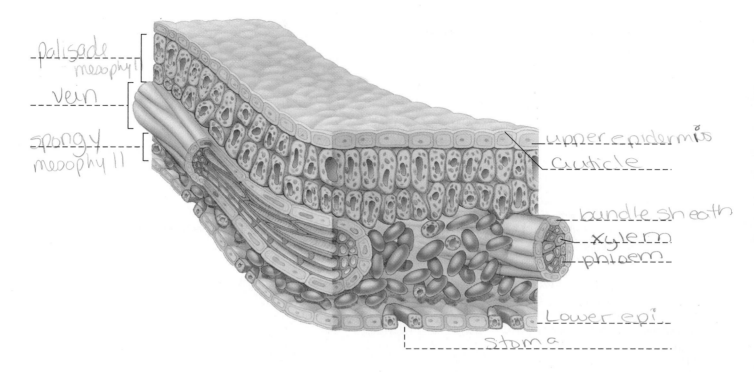

palisade mesophyll
vein
spongy mesophyll
upper epidermis
cuticle
bundle sheath
xylem
phloem
Lower epi
stoma

THOUGHT QUESTIONS

1. Stomata open in response to light. Could the pigment involved in stomatal opening be chlorophyll? Why or why not?
2. Given that xylem is situated toward the upper epidermis in leaf veins and phloem toward the lower epidermis, and the vascular tissue of a leaf is continuous with that of the stem, suggest one possible arrangement of vascular tissues in the stem that accounts for the arrangement of vascular tissue in the leaf.
3. Suppose you observe a micrograph of a leaf cross section and are asked to determine which side is covered by the upper epidermis and which side by the lower epidermis. What characteristics would you look for in order to make this decision?
4. What might be some of the advantages of a plant that has a few very large leaves? Disadvantages? What might be some advantages of a plant that has many very small leaves? Disadvantages? In what kinds of habitats might you expect to find such plants?

SUGGESTED READINGS

Dale, J.E., "How Do Leaves Grow?" *BioScience* Vol. 42, No. 6, June 1992. Leaf research is branching out in new directions—away from describing leaf variations and toward understanding changes that occur during leaf development.

Lipske, M., "Forget Hollywood: These Bloodthirsty Beauties Are for Real," *Smithsonian,* December 1992. An examination of some of the relationships between insectivorous plants and animals.

Raloff, J., "When Nitrate Reigns," *Science News* Vol. 147, No. 6, February 11, 1995. This article dispels the myth that nitrates and other forms of nitrogenous air pollution are good for trees because they provide fertilizer. In actuality, an excess of such nitrogen compounds promotes forest decline.

Smith, W.H., "Air Pollution and Forests," *Encyclopedia of Environmental Biology* Vol. 1, pp. 37–51, New York, Academic Press, 1995. Heavy metals, acid deposition, and oxidants such as nitric oxide are examined in the context of forest health.

CHAPTER 9

FLOWERS, FRUITS, AND SEEDS

The common ragweed (*Ambrosia artemisiifolia*) is one of the most frequently encountered weeds in the north-eastern and central parts of North America. Found along roadsides, in fields, in woods, and in backyards, this plant grows to about 1.5 meters (5 feet) in height and has deeply dissected leaves and clusters of minute greenish flowers borne on a slender, hairy stem. During late summer, ragweed makes many people miserable with allergies, because it releases large quantities of pollen grains into the air as it reproduces.

If you suffer from allergic rhinitis, commonly called hay fever, you are not alone. About 40 million people in the United States endure the sneezing and the itchy, watery eyes associated with this condition. Because one of the primary causes of hay fever is pollen grains suspended in the air, many sufferers blame their discomforts on any plant that happens to be in bloom when they exhibit symptoms. Consequently, roses, sunflowers, and goldenrod are often accused of causing allergies. Contrary to popular opinion, however, the pollens of these showy flowering plants generally do not cause allergic reactions. Plants that produce large, colorful petals are pollinated by insects or other animals; their large pollen grains do not get into the air in appreciable quantities, and so they do not cause hay fever.

Hay fever is caused by inhalation of the pollen of certain wind-pollinated plants, that is, plants that rely on wind to transfer their pollen from one flower to another. When inhaled, the

Spikes of the inconspicuous, pollen-producing male flowers of ragweed (*Ambrosia artemisiifolia*). (*Marion Lobstein*)

pollen stimulates the body to release histamine and other substances, which cause inflammation and other symptoms of allergy. Wind-pollinated plants must produce vast amounts of tiny pollen grains to ensure that at least some of them will land on flowers of the same species and result in successful reproduction. Not all pollens of wind-pollinated plants cause allergic reactions, however. Pines, for example, are wind-pollinated, yet allergies to pine pollen are rare.

People with allergies suffer at different times during the growing season, depending on which plants are releasing pollen and on whether they are sensitive to the pollens of those plants. In early spring, many trees release pollen before their leaves are fully developed. Trees that cause allergic reactions include oaks, ashes, walnuts, maples, alders, and elms. If you suffer from pollen allergies in late spring and early summer, you are probably allergic to pollen from grasses such as bluegrass, timothy, and redtop. Interestingly, most of our major grass crops (corn, rice, and wheat) do not cause allergies in humans. In late summer and early fall, ragweed is the primary culprit in the East, whereas saltbush and Russian thistle (tumbleweed) are problems in dry areas of the West.

Most people are born with some resistance to pollen allergies, but many become sensitized by repeated contact. For that reason, a move to a different geographic location often temporarily halts the suffering from one pollen allergy, although it frequently gives rise to another!

148

ONE REASON FOR THE SUCCESS OF FLOWERING PLANTS IS THEIR REPRODUCTIVE FLEXIBILITY

Many flowering plants can reproduce both sexually and asexually. Sexual reproduction entails the fusion of reproductive cells (eggs and sperm cells called **gametes**). The union of gametes, which is called **fertilization**, occurs within the ovary of the flower. After fertilization, flowering plants produce seeds and fruits.

The offspring of plants that reproduce sexually exhibit a great deal of genetic variation. They may resemble one of the parent plants, both of the parents, or neither of the parents. Sexual reproduction offers the advantage of new combinations of genes (the units of heredity) not found in either parent. These new gene combinations might make an individual plant better suited to its environment. (Chapters 12 and 14 give the details of how this variation occurs.)

In flowering plants, asexual reproduction does not usually involve the formation of flowers, seeds, and fruits but instead involves the formation of offspring from the plant's vegetative structures—stems, leaves, and roots. In asexual reproduction, part of an existing plant separates from the rest of the plant (usually by the death of tissues) and grows to form a complete, independent plant. Because asexual reproduction involves only one parent and no fusion of gametes occurs, the offspring of asexual reproduction are virtually identical to each other and to the parent plant from which they came.

FLOWERS ARE INVOLVED IN SEXUAL REPRODUCTION

A flower is a reproductive shoot usually composed of four parts—sepals, petals, stamens, and carpels—arranged in whorls (that is, in circles) on the end of a flower stalk, or **peduncle** (Figure 9-1). The peduncle may terminate in a single flower

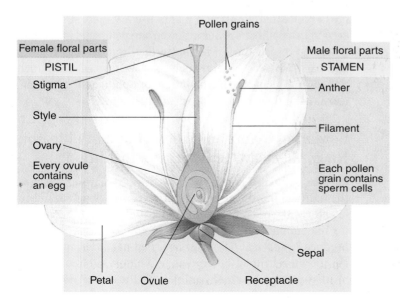

Figure 9-1   The flower bears the organs for sexual reproduction in flowering plants. Cutaway view of a "typical" flower.

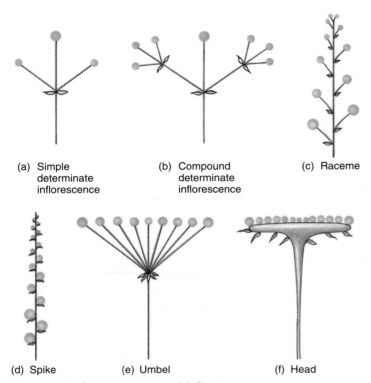

(a) Simple determinate inflorescence

(b) Compound determinate inflorescence

(c) Raceme

(d) Spike

(e) Umbel

(f) Head

Figure 9-2   Selected types of inflorescences.

or a cluster of flowers known as an **inflorescence** (Figure 9-2).

The tip of the peduncle enlarges to form a **receptacle** that bears some or all of the flower parts. Sepals and petals consist of modified leaves that are sterile, and stamens and carpels consist of fertile modified leaves. All four floral parts are important in the reproductive process, but only the stamens (the "male" organs) and carpels (the "female" organs) are fertile and produce gametes.

A flower that has all four parts—sepals, petals, stamens, and carpels—is said to be a **complete** flower; an **incomplete** flower lacks one or more of these four parts. A flower that possesses both stamens *and* carpels is described as **perfect**; an **imperfect** flower has stamens *or* carpels, but not both.

Sepals, which constitute the lowest and outermost whorl on a floral shoot, are leaflike in shape and form and are often green (Figure 9-3a). Sepals cover and protect the flower parts when the flower is a bud. As the blossom opens from a bud, the sepals fold back to reveal the more conspicuous petals. The collective term for all the sepals of a flower is **calyx**.

The whorl just above the sepals consists of **petals**, which are broad, flat, and thin (like sepals and leaves) but vary tremendously in shape and are frequently brightly colored. Petals play an important role in ensuring that sexual reproduction occurs. Sometimes petals are fused to form a tube (for example, trumpet honeysuckle flowers, Figure 9-3b) or other floral shape (for example, snapdragons). The collective term for all the petals of a flower is **corolla**.

Just inside the petals are the **stamens**, the "male" reproductive organs (Figure 9-3c). Each stamen is composed of a thin stalk called a **filament**, at the top of which is an **anther**, a saclike structure in which pollen grains form. Each pollen grain produces two cells surrounded by a tough outer wall. One cell generates two male gametes, known as **sperm cells**, and the other produces a **pollen tube** through which the sperm cells travel to reach the ovule.

In the center of most flowers are one or more **carpels**, the "female" reproductive organs that bear **ovules**, which are structures that have the potential to develop into seeds. For sexual reproduction to occur, pollen must be transferred from an anther of one plant to a carpel, usually of another flower of the same species. The carpels of a flower may be separate or fused into a single structure. The female part of the flower is also referred to as a **pistil**. A pistil may consist of a single carpel (making it a *simple* pistil) or a group of fused carpels (making it a *compound* pistil). Each pistil has three sections: a **stigma**, on which the pollen lands; a **style**, a necklike structure through which the pollen tube grows; and an **ovary**, a juglike structure that contains one or more ovules (Figure 9-3c). Each ovule contains cells that form one female gamete (an **egg**) and two **polar nuclei**. The egg and both polar nuclei participate directly in fertilization.

THE FIRST STEP TOWARD FERTILIZATION: POLLEN IS TRANSFERRED FROM ANTHER TO STIGMA

Before fertilization can occur, pollen grains must travel from the anther (where they form) to the stigma. The transfer of pollen from anther to stigma is known as **pollination**. Plants are **self-pollinated** if pollination occurs within the same flower or within a different flower on the same plant. When pollen is transferred to a flower on another plant, the flower has been **cross-pollinated**. Flowering plants use a variety of agents to accomplish pollination. Some flowers are pollinated by animals, including insects (bees, flies, butterflies, moths, wasps), birds, and bats, whereas others, such as the ragweed plant mentioned in the chapter introduction, are pollinated by wind. A few aquatic flowers are pollinated via water.

Flowering Plants and Their Animal Pollinators Have Affected One Another's Evolution

Flowers pollinated by animals have various features to attract them, including the visual attraction of showy petals and the olfactory attraction of scent (Figure 9-4). One of the rewards for the animal pollinator is food. Nectar is a sugary solution produced in special floral glands called **nectaries** and used as an energy-rich food by animal pollinators. Pollen is also a protein-rich food for many animals. As the animal moves from

Petals   Sepals

(a)

(b)

(c)

Figure 9-3   The four parts of a flower. (a) The leaflike sepals of a rose flower cover and protect the inner flower parts. **(b)** The five petals of each trumpet honeysuckle (*Lonicera sempervirens*) flower are fused to form a tubular corolla. **(c)** A twinleaf (*Jeffersonia diphylla*) flower has eight stamens. Note the rounded green ovary in the center of the flower. (a, *James Mauseth, University of Texas;* b, *James L. Castner;* c, *Marion Lobstein*)

Figure 9-4   Devil's tongue (*Amorphophallus* sp) produces a floral odor, disagreeable to humans, that causes a visible reaction. The chemicals that give rise to the odor react with hydrochloric acid on the filter paper to produce a precipitate that is visible. Devil's tongue is pollinated by beetles and flies. (*J.M. Patt and B.J.D. Meeuse*)

flower to flower searching for food, it inadvertently carries pollen on its body parts, thus facilitating sexual reproduction in plants.

Plants pollinated by insects often have blue or yellow petals. The insect eye does not perceive color in the same manner as does the human eye. Insects see very well in the violet, blue, and yellow ranges of visible light but do not perceive red as a distinct color. Consequently, flowers pollinated by insects are not usually red. Insects, unlike humans, can see ultraviolet radiation as a color called *bee's purple.* Many flowers have dramatic ultraviolet markings called **nectar guides** that are invisible to humans but direct insects to the center of the flower, where the pollen and nectar are located (Figure 9-5).

(a)

(b)

Figure 9-5   Many insect-pollinated flowers have nectar guides that are invisible to humans but very conspicuous to insects. (a) An evening primrose (*Oenothera parviflora*) flower as seen by the human eye is solid yellow. **(b)** The same flower viewed under ultraviolet radiation provides clues about how the insect eye perceives it. The outer, light-appearing portions of the petals appear purple to a bee's eyes, whereas the inner parts appear yellow. These differences in coloration draw attention to the center of the flower, where the pollen and nectar are located. (a, b, *Thomas Eisner*)

Insects have a well-developed sense of smell, and many insect-pollinated flowers have strong scents that may be either pleasant or foul to humans. The carrion plant, for example, is pollinated by flies and smells like the rotting flesh in which flies like to lay their eggs. As flies move from one smelly flower to another looking for a place to lay their eggs, they transfer pollen.

Birds such as hummingbirds are important pollinators (Figure 9-6a). Flowers pollinated by birds are usually red, orange, or yellow since birds see well in this region of visible light. Because birds do not have strong sense of smell, bird-pollinated flowers usually lack a scent.

Bats, which feed at night and do not see well, are important pollinators, particularly in the tropics, where they are most

(a)

(b)

Figure 9-6   Animal pollinators. (a) A ruby-throated hummingbird pollinating a trumpet vine (*Campis radicans*) flower. **(b)** Bats are important pollinators in the tropics. The pollen grains on the fur of this lesser long-nosed bat will be carried to the next plant, where pollination will occur. (a, *Dan Dempster/Dembinsky Photo Associates*; b, *Merlin D. Tuttle, Bat Conservation International*)

abundant (Figure 9-6b). Bat-pollinated flowers are night-blooming and have dull white petals and strong scents, usually of fermented fruit. Nectar-feeding bats are attracted to the flowers by the scent and lap up the nectar with their long, extensible tongues. As they move from flower to flower, they transfer pollen. Other animals, including snails and small rodents, sometimes pollinate plants.

Animal pollinators and the plants they pollinate have had such a close, interdependent relationship over time that they have affected the evolution of certain physical and behavioral features in each other. The term **coevolution** describes such mutual adaptation, in which two different organisms interact so closely that they become increasingly adapted to one another.

During the time plants were coevolving specialized features such as petals, scent, and nectar to attract pollinators, animal pollinators coevolved specialized body parts and behaviors that enabled them to aid pollination and obtain nectar and pollen as a reward. Coevolution is responsible, for example, for the hairy bodies of bumblebees, which catch and hold the sticky pollen for transport from one flower to another. Coevolution is also responsible for the long, curved beaks of certain honeycreepers, Hawaiian birds that insert their beaks into tubular flowers to obtain nectar (Figure 9-7). The long, tubular corollas of the flowers that honeycreepers visit also came about through coevolution.

Animal behavior has also coevolved. The flowers of certain orchids (*Ophrys* sp), for example, resemble female wasps

Figure 9-8   Wind is an agent of pollination for a number of flowering plants. A cluster of grass flowers with their dangling stamens, which catch air currents for efficient pollen dissemination. (*James Mauseth, University of Texas*)

in coloring and shape (see Figure 16–8a). The resemblance between the flowers of one *Ophrys* species and female wasps is so strong that male wasps mount the flowers and attempt to copulate with them. During this misdirected activity, a pollen sac usually becomes attached to the back of the wasp. When the frustrated wasp departs and attempts to copulate with another orchid flower, the pollen is transferred to that flower.

Some Flowering Plants Depend on Wind To Disperse Their Pollen

Flowering plants such as grasses, ragweed, and oaks are pollinated by wind and produce many, often inconspicuous, flowers (Figure 9-8). Wind-pollinated plants do not produce large, colorful petals, scent, or nectar. Some have stigmas that are large and feathery, presumably to make it easier to trap windborne pollen. Because wind pollination is a "hit-or-miss" affair, the likelihood of pollen landing on a stigma of the same species of flower is slim. Wind-pollinated plants therefore produce large quantities of pollen.

FERTILIZATION OCCURS, FOLLOWED BY SEED AND FRUIT DEVELOPMENT

Once pollen has been transferred from anther to stigma, one of the two cells in the pollen grain grows a thin pollen tube down through the style and into the ovule in the ovary. The second cell within the pollen grain divides to form two male gametes, the sperm cells, which move down the pollen tube and enter the ovule (Figure 9-9). After fertilization has occurred, the ovule develops into a seed, and the ovary surrounding it develops into a fruit. Chapter 25 considers fertilization and the flowering plant life cycle in greater detail.

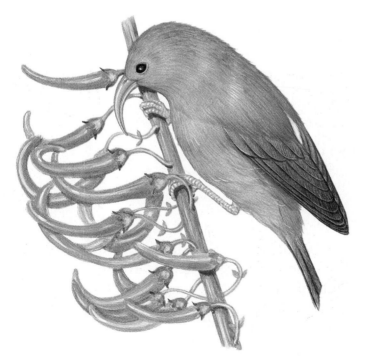

Figure 9-7   The gracefully curved bill of the 'i'iwi, one of the Hawaiian honeycreepers, enables it to sip nectar from flowers of the lobelia (*Lobelia* sp). The 'i'iwi bill fits perfectly into the long, tubular lobelia flowers.

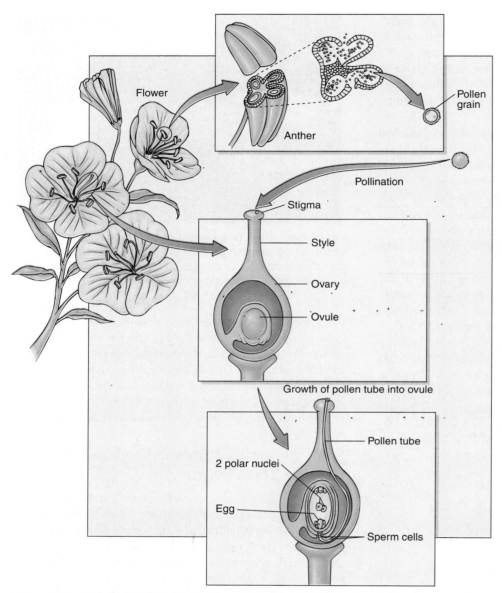

Figure 9-9   After pollination, a pollen tube grows through the style to the ovule in the ovary. Two sperm cells move down the pollen tube and enter the ovule, where they participate in fertilization.

Embryonic Development in Seeds Follows an Orderly and Predictable Path

Flowering plants produce a young plant embryo complete with stored nutrients in a compact package, the **seed**, that develops after fertilization occurs. The nutrients in seeds not only are used by germinating plant embryos but provide food for animals, including humans (see Plants and People: Seed Banks). Development of the embryo following fertilization is possible because of the constant flow of nutrients from the parent plant into the developing seed.

The Mature Seed Contains an Embryonic Plant and Storage Materials To Be Used During Germination

A mature seed consists of an embryonic plant and nutrients stored in either the endosperm or the cotyledons. **Endosperm** is the nutritive tissue that surrounds the embryonic plant in a seed. The seed, in turn, is surrounded by a tough, protective **seed coat** and enclosed within a fruit.

The mature embryo within the seed consists of a short embryonic root (the **radicle**), an embryonic shoot, and one or

PLANTS AND People

Seed Banks

Deep in the frozen soil of an arctic island that is part of Norway lies an international treasury. This bank does not, however, contain any money, gold, or jewels. Instead, numerous seeds of thousands of different kinds of plants collected from all over the world are stored here, frozen in a giant deep freeze as a safeguard against future catastrophic plant extinctions (see figure).

Most seed collections, called seed banks, help to preserve the genetic variation in crops. Farmers typically discontinue planting local varieties when newer, improved varieties become available. The newer varieties have desirable genetic characteristics (greater yield, for example), but the discontinued local varieties also contain valuable genes that people may need in the future.

More than 100 seed banks exist around the world, and collectively they hold more than 3 million samples. The Svalbard International Seed Bank in Norway, like other seed banks, offers the advantage of storing a large amount of plant genetic material in a very small space. Seeds stored in seed banks are safe from habitat destruction, climate change, and general neglect. There have even been some instances of seeds from seed banks being used to reintroduce extinct plant species to the wild.

Seed banks have some disadvantages, as well. Seeds do not remain alive indefinitely and must be germinated periodically so that new seeds can be collected. In addition, accidents such as fires or power failures can result in the permanent loss of the genetic diversity represented by the seeds in a seed bank. Also, the seeds

of many types of plants—avocados and coconuts, for example—cannot be stored because they do not tolerate being dried out, which is a necessary step before they are frozen. If 20 percent or more moisture remains in a seed when it is frozen, it will probably die. Some seeds cannot be stored successfully because they remain viable for only a short period of time, such as a few months or even just a few days.

Perhaps the most important disadvantage of seed banks is that plants stored in this manner remain stagnant in an evolutionary sense. They do not adapt in response to

changes in their natural environments. As a result, when they are reintroduced into the wild, they may be less fit for survival than they were in their original habitat. Despite their shortcomings, seed banks are increasingly viewed as an important method of safeguarding seeds for future generations.

Other international efforts to preserve plant diversity are also being planned and implemented. For example, some farmers may soon be paid to set aside some of their land for cultivating local varieties of crops, thereby preserving the diversity in agriculturally important plants.

Small vials of seeds from the seed bank in Svalbard, Norway. (*Courtesy of Evan Bratberg, Agricultural University of Norway*)

two seed leaves, or **cotyledons** (monocots have a single cotyledon, and dicots have two cotyledons) (Figure 9-10). The short portion of the embryonic shoot connecting the radicle to one or two cotyledons is known as the **hypocotyl**. The shoot apex, or terminal bud, positioned above the point of attachment of the cotyledon(s) is the **plumule** (also known as the *epicotyl*).

Because the embryonic plant is nonphotosynthetic, it must have its own nutrients during germination (early growth) until it becomes self-sufficient. The cotyledons of many plants function as storage organs and become large and thick as they absorb the food reserves (starches, oils, and proteins) initially produced as endosperm. These seeds have little or no endosperm at maturity. Examples of seeds with thick and fleshy cotyledons are peas, beans, squash seeds, sunflower seeds, and peanuts. Other plants—wheat and corn, for example—have thin cotyledons that function primarily to help the young plant digest and absorb food stored in the endosperm.

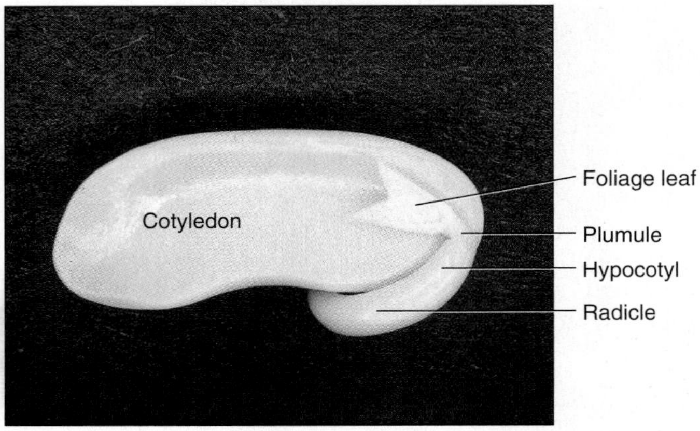

Seed size varies considerably in flowering plants, from the microscopic, dustlike seeds of orchids to the giant seeds of the double coconut (*Lodoicea seychellarum*), which weigh as much as 27 kilograms (almost 60 pounds). Despite this variation among species, seed size is a remarkably constant trait within a species (see Focus On: Relating Seed Size to a Plant's Habitat).

Figure 9-10 A dissected bean (*Phaseolus vulgaris*) seed showing the radicle, hypocotyl, cotyledon, plumule, and leaf produced by the terminal bud. (The seed coat and one of its two cotyledons were removed). This particular seed has begun germinating, so its radicle is larger than that of a nongerminating seed. (*James Mauseth, University of Texas*)

Labels in figure: Cotyledon, Foliage leaf, Plumule, Hypocotyl, Radicle

(a)

Fruits Are Mature, Ripened Ovaries

After fertilization takes place within the ovule, the ovule develops into a seed, and the ovary surrounding it develops into a **fruit**. For example, a pea pod is a fruit, and the peas within it are seeds. A fruit may contain one or more seeds; some orchid fruits contain several thousand to a few million seeds!

There are several types of fruits, which differ as a result of variations in the structure or arrangement of the flowers from which they formed. The four basic types of fruits are simple fruits, aggregate fruits, multiple fruits, and accessory fruits (Table 9-1).

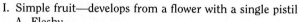

TABLE 9-1   Selected Types of Fruits

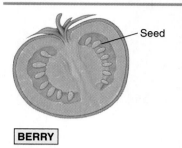

BERRY

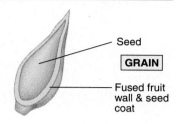

Seed

GRAIN

Fused fruit wall & seed coat

I. Simple fruit—develops from a flower with a single pistil
   A. Fleshy
     1. Berry—soft and fleshy throughout; usually has many seeds (tomato, grape)
     2. Drupe—has a hard, stony pit; has a single seed (peach, cherry)
   B. Dry, splits open to release seeds
     1. Follicle—splits open along one side (milkweed, columbine)
     2. Legume—splits along two sides (bean, pea)
     3. Capsule—splits along many sides or pores (poppy, iris)
   C. Dry, does not split open
     1. Grain—single-seeded; seed fully fused to fruit wall (wheat, corn)
     2. Achene—single-seeded; seed attached to fruit wall at base only (sunflower)
     3. Nut—single-seeded; has a hard, thick fruit wall (acorn, chestnut)

II. Aggregate fruit—develops from a flower with many separate ovaries (raspberry, blackberry)

III. Multiple fruit—develops from an inflorescence—that is, from many flowers borne on a common floral stalk; the ovaries of these flowers fuse together to form a single fruit (pineapple, mulberry)

IV. Accessory fruit—develops from a flower in which the receptacle or floral tube enlarges and becomes part of the mature fruit (apple, strawberry)

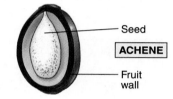

Seed

ACHENE

Fruit wall

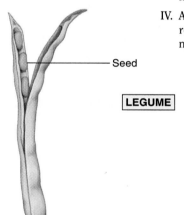

Seed inside pit

Pit

DRUPE

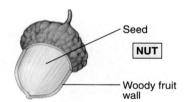

Seed

NUT

Woody fruit wall

Seed

LEGUME

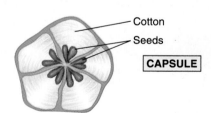

Cotton

Seeds

CAPSULE

Seed

AGGREGATE FRUIT

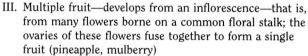

Seeds

FOLLICLE

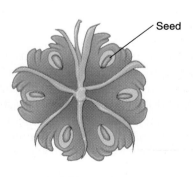

Seed

MULTIPLE FRUIT

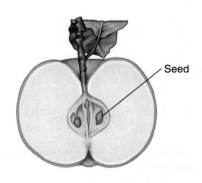

Seed

ACCESSORY FRUIT

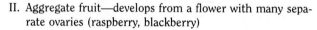

(a)

(b)

(c)

(d)

Figure 9-11   Simple fruits that are fleshy at maturity. (a,b) The tomato (*Lycopersicon lycopersicum*) and the blueberry (*Vaccinium* sp), representative berries, are composed of soft tissues throughout. **(c,d)** The peach (*Prunus persica*) and avocado (*Persea americana*) are drupes. In drupes, the hard, stony pit that surrounds the single seed is part of the fruit. (a, *Dennis Drenner;* b,c, *Marion Lobstein;* d, *Carlyn Iverson.*)

Most fruits are simple fruits. A **simple fruit** develops from a single pistil of a single flower. At maturity, simple fruits may be fleshy or dry. Two examples of simple fleshy fruits are berries and drupes (Figure 9-11). A **berry** is a fleshy fruit that has soft tissues throughout and contains a few to many seeds; a tomato is a berry, as are grapes, blueberries, cranberries, and bananas. Many so-called berries do not fit the botanical definition. Strawberries, raspberries, and mulberries, for example, are not berries.

A **drupe** is a simple fleshy (or fibrous) fruit that contains a hard, stony pit surrounding a single seed. Examples of drupes include peaches, plums, olives, avocados, and almonds. The shell of the almond is actually the stony pit, which is exposed when the rest of the almond fruit is removed.

Many simple fruits are dry at maturity. Some mature dry fruits split open, usually along seams called sutures, to release their seeds (Figure 9-12). A milkweed pod is an example of a **follicle**, a simple dry fruit that splits open along *one* suture to release its seeds. A **legume** is a simple dry fruit that splits open along *two* sutures (the top and bottom). Pea pods are legumes, as are green beans, although both are generally harvested before the fruit has dried out and split open. In peas

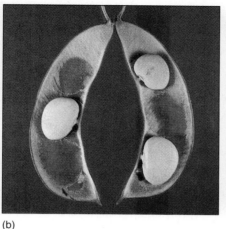

(a)

(b)

(c)

Figure 9-12 Simple dry fruits that split open to release their seeds. (a) The milkweed (*Asclepias syriaca*) follicle, which is dry at maturity, splits open along one suture to release its seeds. **(b)** A lima bean (*Phaseolus limensis*) fruit is a legume that splits open along two sutures at maturity. **(c)** These red buckeye (*Aesculus pavia*) capsules split along several sutures. (a, *Carlyn Iverson;* b, *R. Calentine/CBR Images;* c, *Patti Murray*)

the seeds are usually removed from the fruit and consumed, whereas in green beans the entire fruit is eaten. A **capsule** is a simple dry fruit that splits open along *multiple* sutures or pores. Poppy and cotton fruits are capsules.

Other simple dry fruits—**grains**, for example—do not split open at maturity (Figure 9-13a). Each grain contains a single seed. Because the seed coat is fused to the fruit, a grain appears to be a seed rather than a fruit. Kernels of corn and wheat are fruits of this type, called **caryopses**.

An **achene** is a similar fruit in that it is simple and dry, does not split open at maturity, and contains a single seed (Figure 9-13b). However, in an achene the seed coat is not fused to the fruit wall. Instead, the single seed is attached to the fruit wall at one point only. Therefore, an achene can be separated from its seed. The sunflower fruit is an example of an achene. One can peel off the fruit wall (the "shell") to reveal the sunflower seed within.

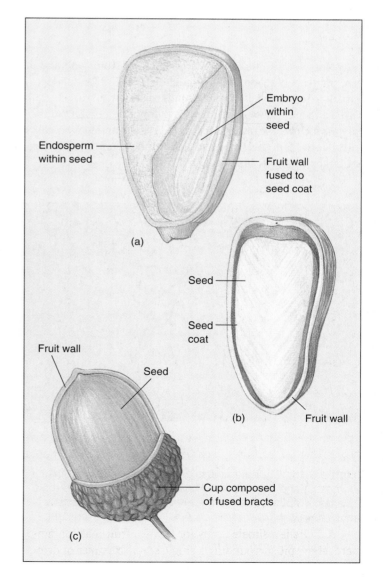

Figure 9-13 Simple dry fruits that do not split open. (a) The corn (*Zea mays*) fruit is a grain, or caryopsis. In grains, the fruit wall is fused to the seed coat. **(b)** A sunflower (*Helianthus annuus*) fruit is an achene. Its seed coat is attached to the fruit wall at one spot only, and it is possible to peel off the fruit wall, separating it from the seed. **(c)** An oak (*Quercus* sp) acorn is a nut. A nut has a hard fruit wall that surrounds a single seed.

Figure 9-14 Blackberries (*Rubus* sp) are aggregate fruits. (a) Cutaway view of a blackberry flower, showing the many separate carpels in the center of the flower. **(b)** A developing blackberry fruit is an aggregate of tiny drupes. The little "hairs" on the blackberry are remnants of stigmas and styles. **(c)** Developing fruits at various stages of maturity. (*Dennis Drenner*)

Nuts are simple dry fruits that have a stony wall and do not split open at maturity (Figure 9-13c). Unlike achenes—small fruits derived from a simple pistil—nuts are usually large and are often derived from a compound pistil. Examples of nuts include chestnuts, acorns, and hazelnuts. Many so-called nuts do not fit the botanical definition. Peanuts and Brazil nuts, for example, are seeds, not nuts.

Aggregate fruits are a second main type of fruit. An **aggregate fruit** forms from a single flower that contains several to many separate (free) carpels (Figure 9-14). After fertilization, each ovary from each individual carpel enlarges. As the ovaries enlarge, they may fuse together to form a single fruit. Raspberries, blackberries, and magnolia fruits are examples of aggregate fruits.

(a)

(b)

Figure 9-15 Multiple fruits. Both **(a)** pineapple (*Ananas comosus*) and **(b)** mulberry (*Morus* sp) are formed from the ovaries of many separate flowers that fused to become a multiple fruit. Mulberry flowers are imperfect and contain either stamens or pistils. The inflorescence of female flowers from which the mulberry fruit develops is also shown. (a, *Courtesy of Dole Packaged Foods Co.*)

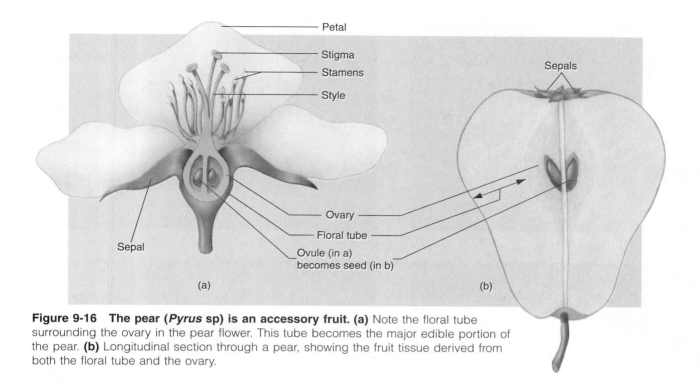

Figure 9-16   The pear (*Pyrus* sp) is an accessory fruit. (a) Note the floral tube surrounding the ovary in the pear flower. This tube becomes the major edible portion of the pear. **(b)** Longitudinal section through a pear, showing the fruit tissue derived from both the floral tube and the ovary.

A third type of fruit is the **multiple fruit**, which forms from the ovaries of many flowers that grow in proximity on a common floral stalk. The ovary from each flower fuses with nearby ovaries as it develops and enlarges after fertilization. Pineapples, figs, and mulberries are multiple fruits (Figure 9-15).

Accessory fruits are the fourth type of fruit. They differ from other fruits in that other plant tissues in addition to ovary tissue make up the fruit. For example, the edible portion of a strawberry is the red, fleshy receptacle. Apples and pears are accessory fruits called **pomes**; the outer part of each of these fruits is an enlarged **floral tube**, consisting of receptacle tissue, along with portions of the calyx, that surrounds the ovary (Figure 9-16).

Seed Dispersal Is Highly Varied In Flowering Plants

The seeds and fruits of flowering plants are dispersed by wind, animals, water, and explosive dehiscence. Effective methods of seed dispersal have made it possible for certain plants to expand their geographical ranges. In some cases the actual agent of dispersal is the seed, whereas in others it is the fruit. In tumbleweeds, such as Russian thistle, the *entire plant* is the agent of dispersal because it detaches and blows across the ground, scattering seeds as it bumps along. Tumbleweeds are lightweight but have a lot of wind resistance and are sometimes carried for miles by the wind.

Wind disperses the seeds of many plants (Figure 9-17a,b). Plants such as maple have winged fruits adapted for wind dispersal. Light, feathery plumes are other structures that allow seeds or fruits to be transported by wind, often for considerable distances. Both dandelion fruits and milkweed seeds have this type of adaptation.

Some plants have special structures that aid in the dispersal of their seeds and fruits by animals (Figure 9-17c,d). The spines and barbs of cockleburs and similar fruits often get caught in animal fur and are dispersed as the animal moves about. Fleshy edible fruits are also adapted for animal dispersal. As animals eat these fruits, the seeds are either discarded or swallowed. Seeds that are swallowed have thick seed coats and are not digested; instead, they pass through the digestive tract and are deposited with the animal's feces some distance away from the parent plant. In fact, some seeds will not germinate unless they have passed through an animal's digestive tract.

Animals such as squirrels and many bird species also help to disperse acorns and other fruits and seeds by burying them for winter use. Many buried seeds are never used by the animal and germinate the following spring. (Also see Focus On: Seed Dispersal by Ants.)

The coconut is an example of a fruit adapted for dispersal by water. It has air spaces and corky floats that make it buoyant and capable of being carried by ocean currents for thousands of kilometers. When it washes ashore, the seed within the fruit germinates and grows into a coconut palm tree.

(a)

(b)

(c)

(d)

Figure 9-17   Methods of seed and fruit dispersal. (a) The fruits of sugar maple (*Acer saccharum*) have wings for wind dispersal. **(b)** The feathery plumes of a milk-weed (*Asclepias* sp) seed make it buoyant for dispersal by wind. **(c)** Burdock (*Arctium minus*) burs, which are hooked fruits, are carried away from the parent plant after they become matted in animal fur or clothing. **(d)** Fleshy fruits such as strawberries (*Fragaria* sp) are eaten by animals such as this meadow vole. The seeds are frequently swallowed whole and pass unharmed through the animal's digestive tract. (a, *James Mauseth, University of Texas;* b, *David Cavagnaro;* c, *DPA/Dembinsky Photo Associates;* d, *Dwight R. Kuhn*)

Focus On

Seed Dispersal by Ants

For a plant species to survive, it must be able to disperse its seeds to places where they can successfully germinate and grow. Plants lend themselves to a variety of dispersal methods (such as wind, animals, and water) that increase the chances of their seeds landing in suitable locations. Regardless of how seeds are dispersed from the parent plant, most seeds either land in places that are unsuitable for growth or are eaten by animals, such as mice and squirrels, shortly after being dispersed.

The survivability of the seeds of some plants is enhanced by a dispersal method in which the seeds are buried underground, where they are less likely to be eaten by animals. For many plants, the role of burying seeds is performed by ants, which collect the seeds and take them underground to their nests. Ants disperse and bury seeds for hundreds of different plant species in almost every terrestrial environment, from northern coniferous forests to tropical rain forests to deserts.

Both ants and flowering plants benefit from this association. The ants ensure the reproductive success of the plants whose seeds they bury, and the plants supply food to the ants. Many of the seeds that ants collect

and take underground have a special protruding structure called an **elaiosome**, or *oil body* (see figure). Ants carry the seeds underground before removing the elaiosomes, which are a nutritious food for them. Once they

remove an elaiosome from a seed, the ants discard the undamaged seed in an underground refuse pile, which happens to be rich in organic material such as ant droppings and dead ants and thus contains the minerals required by young seedlings. Hence, ants not only bury seeds away from animals that might eat them but also place them in rich soil that is ideal for seed germination and seedling growth.

Bloodroot (*Sanguinaria canadensis*) seeds have elaiosomes, or oil bodies. The golden-brown part is the seed proper, and the white part is the elaiosome. (*Marion Lobstein*)

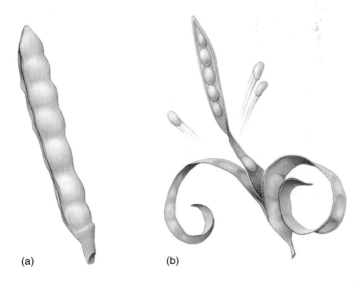

(a)   (b)

Some seeds are dispersed not by wind, animals, or water but by *explosive dehiscence*, in which the fruit bursts open suddenly, and quite often violently, to forcibly discharge its seeds. Pressures due to differences in turgor or to drying out cause these fruits to burst open suddenly. The fruits of plants such as touch-me-not (*Impatiens noli-tangere*) and bitter cress (*Cardamine pratensis*) split open so explosively that their seeds are scattered a meter or more (Figure 9-18).

Figure 9-18 Explosive dehiscence in bitter cress (*Cardamine pratensis*). (a) An intact fruit before it has split open. **(b)** The fruit splits open with explosive force, propelling the seeds some distance from the plant.

163

STUDY OUTLINE

I. Asexual reproduction involves the formation of offspring without fusion of gametes. The offspring produced by asexual reproduction are virtually identical to the single parent plant.
II. The offspring produced by sexual reproduction are genetically variable.
   A. The flower is the structure in which sexual reproduction occurs. A flower may contain sepals, petals, stamens, and carpels (pistils).
      1. Sepals cover and protect the parts of the flower when it is a bud.
      2. Petals play an important role in attracting animal pollinators to the flower.
      3. Stamens produce pollen grains.
      4. Each carpel bears one or more ovules.
   B. Each pollen grain contains two cells. One generates two male gametes (sperm cells), and the other produces a pollen tube through which the sperm cells will reach the ovule.
   C. An egg and two polar nuclei form in the ovule.
III. Pollination is the transfer of pollen from anther to stigma.
   A. Some plants rely on animals to transfer pollen.
      1. Coevolution occurs when two different organisms (such as flowering plants and their animal pollinators) form such an interdependent relationship that they affect the course of each other's evolution.
      2. Flowers pollinated by insects are often yellow or blue and possess a scent.
      3. Bird-pollinated flowers are often yellow, orange, or red and do not have a strong scent.
      4. Bat-pollinated flowers often have dusky white petals and possess a scent.
   B. Plants that are wind-pollinated often have smaller petals or lack petals altogether, and they do not produce a scent or nectar. Wind-pollinated flowers make copious amounts of pollen.
IV. After pollination, fertilization (fusion of gametes) occurs.
V. The seed and fruit develop as a result of successful fertilization.
   A. A mature seed contains a young plant embryo and nutritive tissue (stored in the endosperm or cotyledons) for use during germination.
      1. The seed is covered by a protective seed coat.
      2. The mature flowering plant embryo consists of a radicle, a hypocotyl, one or two cotyledons, and a plumule (epicotyl).
   B. Seeds of flowering plants are enclosed within fruits, which are mature, ripened ovaries.
      1. Simple fruits develop from a single flower with a single pistil. Some simple fruits are fleshy at maturity (for example, grapes and peaches), whereas others are dry (for example, milkweed pods and wheat grains).
         a. Berries and drupes are simple fleshy fruits.
         b. Follicles, legumes, and capsules are simple dry fruits that split open at maturity.
         c. Grains, achenes, and nuts are simple dry fruits that do not split open at maturity.
      2. An aggregate fruit, such as a raspberry, develops from a single flower with many separate ovaries.
      3. A multiple fruit, such as a pineapple, develops from many ovaries of many flowers growing in proximity on a common axis.
      4. In accessory fruits, such as strawberries and apples, the major part of the fruit consists of tissue other than ovary tissue.
VI. Seeds and fruits of flowering plants are adapted for various means of dispersal, including wind, animals, water, and explosive dehiscence.

SELECTED KEY TERMS

accessory fruit, p. 161
achene, p. 158
aggregate fruit, p. 160
berry, p. 158
capsule, p. 158
carpel, p. 150
coevolution, p. 153
cotyledon, p. 156

drupe, p. 158
endosperm, p. 154
fertilization, p. 149
follicle, p. 158
fruit, p. 156
grain, p. 158
inflorescence, p. 150
legume, p. 158

multiple fruit, p. 161
nut, p. 160
ovary, p. 150
ovule, p. 150
peduncle, p. 149
petal, p. 150
pistil, p. 150
pollination, p. 150

receptacle, p. 150
seed, p. 154
sepal, p. 150
simple fruit, p. 158
stamen, p. 150
stigma, p. 150
style, p. 150

REVIEW QUESTIONS

1. Distinguish between sexual and asexual reproduction.
2. Is an imperfect flower also incomplete? Is an incomplete flower also imperfect? Explain your answers.
3. What is the difference between pollination and fertilization? Which process occurs first?
4. Distinguish between seeds and fruits. From what part of the flower does each develop?

5. Distinguish among simple, aggregate, multiple, and accessory fruits, and give examples of each.
6. Explain some of the features possessed by seeds and fruits that are dispersed by animals.

7. Label the following diagram, and give the function of each part.

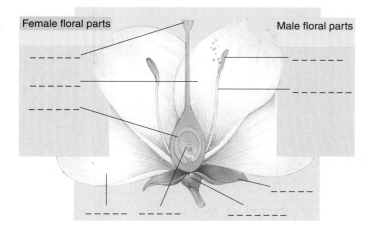

Female floral parts

Male floral parts

THOUGHT QUESTIONS

1. Might sexual or asexual reproduction be more beneficial in each of the following sets of circumstances, and why?
   a. A perennial (a plant that lives more than 2 years) in a stable environment
   b. An annual (a plant that lives 1 year) in a rapidly changing environment
   c. A plant adapted to an extremely narrow climate range

2. Draw pictures to show the kinds of flowers that might form simple, aggregate, multiple, and accessory fruits.
3. How is sexual reproduction in humans similar to sexual reproduction in flowering plants? How is it different?
4. Could seed dispersal by ants be considered an example of coevolution? Why or why not?

SUGGESTED READINGS

Cox, P.A., "Water-Pollinated Plants," *Scientific American,* October 1993. Some aquatic plants rely on water currents to transfer pollen to female flowers.

Handel, S.N., and A.J. Beattie, "Seed Dispersal by Ants," *Scientific American,* August 1990. How plants induce ants to disperse seeds without harming them.

Kearns, C.A., and D.W. Inouye, "Pistil-Packing Flies," *Natural History,* April 1993. Flies are effective pollinators of many kinds of flowering plants.

Shell, E., "Seeds in the Bank Could Stave Off Disaster on the Farm," *Smithsonian,* January 1990. How seed banks are used to improve important crop plants.

Stein, B.A., "Sicklebill Hummingbirds, Ants, and Flowers," *Bio-Science* Vol. 42, No. 1, January 1992. Examines two different plant-animal interactions (between ants and flowers and between hummingbirds and flowers) to obtain clues about the structural features of plants.

Stolzenburg, W., "The Lonesome Flower," *Nature Conservancy,* March/April 1992. The only natural pollinator for a certain flowering plant growing on the Hawaiian Islands has disappeared, presumably due to extinction. As a result, the plant is also doomed to extinction unless conservation efforts are successful.

Young, J.A., "Tumbleweed," *Scientific American,* March 1991. The tumbleweed was introduced to North America, where it has spread throughout much of the West.

MINERAL NUTRITION AND TRANSPORT IN PLANTS

The coastal redwood (*Sequoia sempervirens*), the state tree of California, is an immense evergreen with deeply grooved, spongy, reddish-brown bark that may be 30 centimeters (12 inches) thick on a mature tree. It produces flat bluish-grey needles and bears its seeds in small, oval cones that mature in one season; needles and cones are about 2.5 centimeters (1 inch) in length. Coastal redwoods require the high humidity and cool temperatures provided by ocean fogs and do not grow farther inland than the fogs will penetrate. Consequently, their growth is restricted to a narrow ribbon of land about 800 kilometers (500 miles) long in the foothills along southwestern Oregon and the northern California coast and no more than 30 kilometers (19 miles) from the Pacific Ocean.

Coastal redwoods are among the world's tallest trees. One tree was reported to have measured 117 meters (385 feet) in height and 8.5 meters (28 feet) in diameter. Most mature redwood trees are 61 to 83 meters (200 to 275 feet) tall, with trunks measuring 3 meters (10 feet) in diameter. As a coastal redwood ages, the lower branches die and fall off,

The majestic redwoods are some of the world's tallest plants. (*Gary R. Bonner*)

leaving a tapering, columnar trunk that extends about 30 meters (100 feet) above the ground. The trunk spreads out at its base to provide support.

In addition to being noted for their gigantic size, coastal redwoods are known for their extreme longevity. They mature in 400 to 500 years, and individuals 1500 or more years old are not uncommon. The oldest known coastal redwood is 2200 years old.

Redwoods are named for their reddish heartwood, which is straight-grained, attractive, and resistant to decay. Redwood lumber is highly valued for such uses as shingles, furniture, interior paneling, doors, windowsills, and fence posts. These majestic trees have been overexploited for their valuable timber, and most of the original forests have been cut down. As a result, only a few virgin groves of coastal redwoods remain, and conservationists strongly support their protection. About 23,470 hectares (58,000 acres) of redwood forest have been set aside in the Redwood National Park. Muir Woods National Monument in California contains one of the most famous groves of stately coastal redwoods.

The coastal redwood is an appropriate plant to feature in this chapter because of its immense size. How do water and dissolved minerals move from the roots to the needles at the top of the coastal redwood? How does sugar produced in photosynthesis move from the needles to the living cells in the roots? We now consider these questions and examine the soil from which plants obtain water and minerals.

SOIL ANCHORS PLANTS AND PROVIDES WATER AND DISSOLVED MINERALS

Soil is the ground underfoot, a relatively thin layer of the Earth's crust that has been modified by the natural actions of weather, wind, water, and organisms, among other things. It is easy to take soil for granted. We walk on and over it throughout our lives, but we rarely stop to think about how important it is to our survival. Vast numbers and kinds of organisms colonize soil and depend on it for shelter and food. Plants anchor themselves in soil and draw water and essential minerals from it. Thirteen of the 16 different elements essential for plant growth are obtained directly from the soil. Plants could not survive without soil, and because we depend on plants for our food, humans could not exist without soil, either.

Soils form from rock (called parent material) in the Earth's crust that is gradually broken down into smaller and smaller particles by biological, chemical, and physical **weathering processes** (Figure 10-1). Two important factors that work together in the weathering of rock are climate and organisms. Lichens, which are "dual organisms" composed of a fungus and an alga or cyanobacterium, produce acids that etch tiny cracks, or fissures, in the rock surface. Water seeps into these cracks. If the parent material is in a temperate climate, the alternate freezing and thawing during winter enlarge the cracks by breaking off particles of the rock. Small plants can then become established and send their roots into the larger cracks, fracturing the rock further.

Topography—which is a region's surface features, such as the presence or absence of mountains and valleys—is also involved in soil formation. Steep slopes often have little or no soil on them, because gravity continually transports soil and rock down the slopes; runoff from precipitation tends to amplify erosion on steep slopes. Moderate slopes, on the other hand, may restrict downhill transport, thereby encouraging the formation of deep soils.

The disintegration of solid rock into finer and finer mineral particles and the accumulation of organic material in the soil take a very long time, usually thousands of years. Soil forms continually through the weathering of parent material beneath soil that has already formed. The thickness of soil varies from a thin film on very young lands, near the poles and on the tops of mountains, to more than 3 meters (10 feet) on very old lands, such as tropical rain forests.

Soil Is Composed of Inorganic Minerals, Organic Matter, Soil Organisms, Soil Atmosphere, and Soil Water

Soil has five distinct components: inorganic mineral particles, organic matter, soil organisms, water, and air. The inorganic mineral particles, which come from weathered rock, constitute most of what we call soil. Because different rocks are composed of different minerals, soils vary in mineral composition. Also, soils formed from the same kind of parent material may not develop in the same way, because other factors such as weather, topography, and kinds of organisms differ.

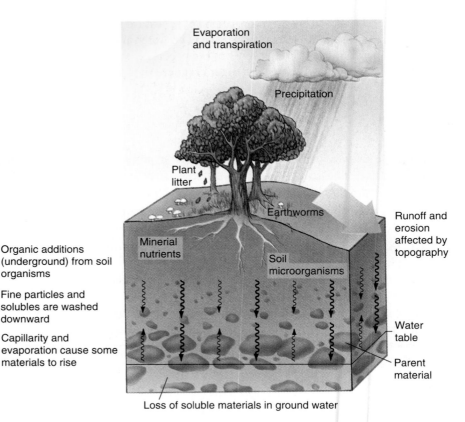

Evaporation
and transpiration

Precipitation

Plant
litter

Earthworms

Runoff and
erosion
affected by
topography

Organic additions
(underground) from soil
organisms

Minerial
nutrients

Soil
microorganisms

Fine particles and
solubles are washed
downward

Capillarity and
evaporation cause some
materials to rise

Water
table

Parent
material

Figure 10-1   Weather, climate, topography, and organisms interact with the Earth's crust to form soil, the complex material that supports life on land.

Loss of soluble materials in ground water

The texture of a soil is determined by the amounts (measured in percentages by weight) of different-sized inorganic mineral particles—sand, silt, and clay. The size assignments for sand, silt, and clay are arbitrary; they give soil scientists a way to classify soil texture. Any particles larger than 2 millimeters in diameter, called stones, are not considered soil particles because they have no direct value for plants. The largest soil particles are called **sand** (0.02 to 2 millimeters in diameter), the medium-sized particles are called **silt** (0.002 to 0.02 millimeter in diameter), and the smallest particles are

called **clay** (less than 0.002 millimeter in diameter) (Figure 10-2). Sand particles are large enough to be seen easily with the eye, silt particles (about the size of flour particles) are barely visible to the eye, and individual clay particles are too small to be seen with an ordinary light microscope; they can be seen only under an electron microscope.

The clay component of a soil is particularly important in determining many of its characteristics, in part because clay particles have the greatest surface area for chemical reactions. If the surface areas of about 450 grams (1 pound) of clay par-

Silt   Sand

Clay

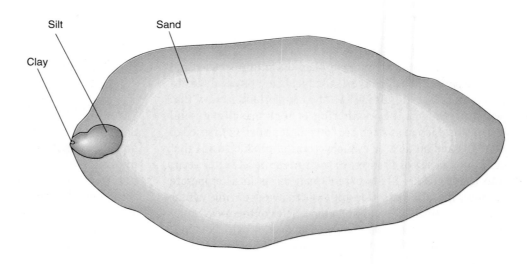

Figure 10-2   Relative sizes of soil particles.

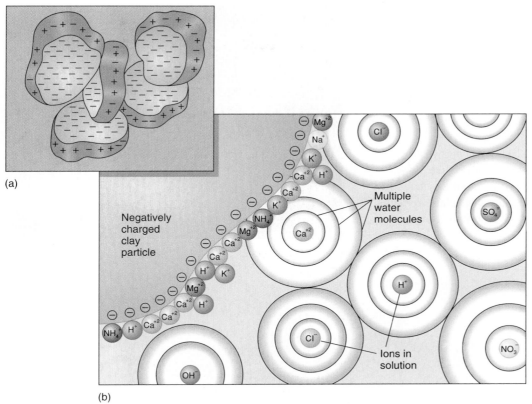

(a)

(b)

Figure 10-3   Clay particles help hold positively charged mineral ions in the soil, where they can be absorbed by plant roots. (a) Positively charged ions are attracted to the broad, negatively charged surfaces of clay particles. **(b)** A close-up of part of a clay particle and the thin film of water around it. The numerous water molecules are represented by circles surrounding mineral ions that are dissolved in the soil water. Note the large number of positively charged ions attracted to the surface of the clay particle.

ticles were laid out side by side, they would occupy 1 hectare (2.5 acres). Each clay particle has negative electrical charges on its outer surface that attract and bind positively charged mineral ions (Figure 10-3). Many of these mineral ions, such as potassium (K^+) and magnesium (Mg^{2+}), are essential for plant growth. Their interactions with clay particles "hold" them in the soil, where they are available for use by plants. The plant's roots absorb them along with water, which forms a film around the soil particles.

Soil always contains a mixture of different-sized particles, but the proportions vary from one soil to another. A **loam**, which is an ideal agricultural soil, has an optimal combination of different soil particle sizes: approximately 40 percent each of sand and silt and about 20 percent of clay. Generally speaking, the larger particles provide aeration and the smaller particles bind together into clumps and hold minerals and water. Soils with larger proportions of sand are not as desirable for most plants, because they do not hold water and mineral ions well; plants grown in such soils are more susceptible

to drought. Soils with larger proportions of clay are also not desirable for most plants, because they have poor drainage. Clay soils tend to get compacted and eliminate spaces that could have been filled by water and air.

A soil's organic matter consists of the wastes and remains of soil organisms

The organic matter in soil is composed of litter (dead leaves and branches on the soil's surface), droppings (animal dung), and the remains of dead plants, animals, and microorganisms that are in various stages of decomposition. Microorganisms, particularly bacteria and fungi, that inhabit the soil decompose organic matter—that is, break it down into simpler materials. During decomposition, essential minerals are released into the soil, from which they may be absorbed by new plants. Organic matter increases the soil's water-holding capacity by acting much like a sponge. For this reason, gardeners often add organic matter to soils—especially sandy soils, which are naturally low in organic matter (see Plants and People: Managing

Figure 10-4   Humus is partially decomposed organic material (primarily from plant and animal remains). Soil that is rich in humus has a loose, somewhat spongy structure with several properties (such as increased water-holding capacity) that are beneficial for plants and other organisms living in it. (*USDA/Soil Conservation Service*)

Your Lawn and Garden at Home). The partly decayed organic portion of the soil is referred to as **humus** (Figure 10-4).

Some soils are dominated by organic matter. Peat soils, which consist almost entirely of undecayed plant remains, may form where poor drainage limits decomposition by soil-dwelling bacteria.

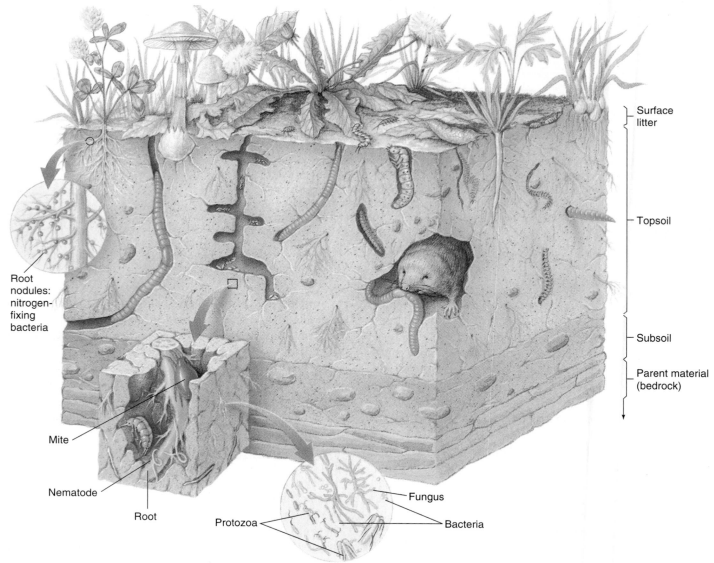

Figure 10-5   The diversity of life in fertile soil is remarkable. The life forms include plants, algae, fungi, earthworms, flatworms, roundworms, insects, spiders and mites, bacteria, and burrowing animals such as moles and groundhogs.

The organisms living in the soil form a complex community

A single teaspoon of fertile agricultural soil may contain millions of microorganisms such as bacteria, fungi, algae, protozoa, and microscopic worms. Many other organisms also colonize soil, including earthworms, insects, plant roots, and larger animals such as moles, gophers, snakes, and groundhogs (Figure 10-5). Most numerous in soil are bacteria, which number in the hundreds of millions per gram of soil. Bacteria and fungi are particularly essential in decomposing dead organic material in the soil. They are also important in nutrient cycles, which are the processes by which matter cycles from the living world to the nonliving physical environment and back again. Nutrient cycles break down or build up chemicals needed by plants (Figure 10-6). For example, microorganisms are involved in most steps in the nitrogen cycle, which supplies nitrogen, an essential part of proteins and nucleic acids, to organisms. Chapter 26 covers nutrient cycles such as the nitrogen cycle in more detail.

Worms are some of the most important organisms living in soil. The earthworm, probably one of the most familiar soil inhabitants, eats soil and obtains energy and raw materials by

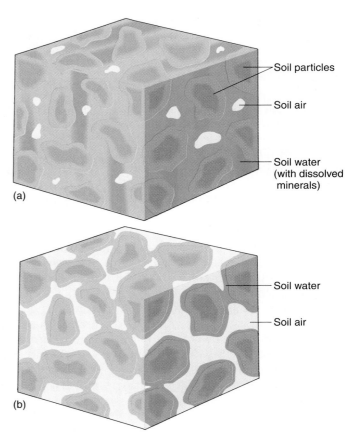

(a)

(b)

Figure 10-7   Pore space is occupied by varying amounts of soil air and soil water. (a) In a wet soil, most of the pore space is filled with water. **(b)** In a dry soil, a thin film of water is tightly bound to soil particles, and most of the pore space is occupied by soil air.

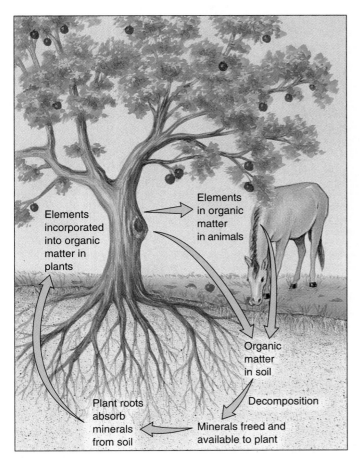

Figure 10-6   Mineral nutrients cycle from the soil to organisms and then back to the soil.

digesting humus. **Castings**, bits of soil that have passed through the gut of an earthworm, are deposited on the soil surface, bringing minerals from deeper layers to upper layers. Earthworm tunnels serve to aerate the soil, and the worms' waste products and corpses add organic material to deeper layers of the soil.

Ants live in the soil in enormous numbers, constructing tunnels and chambers that help to aerate it. Members of soil-dwelling ant colonies forage on the surface for bits of food, which they carry back to their nests. Not all of this food is eaten, however, and its eventual decomposition helps increase the organic matter in the soil.

Approximately 30 to 60 percent of the volume of soil is occupied by pore space between soil particles

A healthy soil has numerous pore spaces of different sizes around and among the soil particles. Pore spaces are filled with varying proportions of soil air and soil water, both of which are necessary to produce a moist but well-drained soil that sustains plants and other soil-dwelling organisms (Figure 10-7). Generally speaking, water is held in the smaller

Managing Your Lawn and Garden at Home

You can put to use some of what you have learned in this chapter when managing your lawn and garden at home. By using compost and mulch, you can maintain and improve your soil. Gardeners often dispose of grass clippings, leaves, and other plant refuse by either bagging it for garbage collection or burning it. Neither action is desirable, however; treating yard wastes as garbage contributes more material to our already over-burdened landfills, and the burning of yard wastes contributes to air pollution.

These materials do not have to be treated as wastes. They can be a valuable resource for making **compost**, a natural soil-and-humus mixture that improves not only soil fertility but soil structure. Grass clippings, leaves, weeds, sawdust, coffee grounds, animal manure, ashes from the fireplace or grill, shredded newspapers, fruit peelings and other meat- and cheese-free leftovers, and eggshells are just some of the materials that can be composted, that is, transformed into compost by microbial action.

To make a compost heap, spread a 15- to 30-centimeter (6- to 12-inch) layer of grass clippings, leaves, or other plant material in a shady area, sprinkle it with garden fertilizer or a thin layer of animal manure, and cover it with several centimeters of soil. Add layers as you collect more

organic debris. Water the material thoroughly and turn it over with a pitchfork each month to aerate it. Although it is possible to make compost by just heaping it in layers, it is more efficient to construct a compost enclosure. An enclosed compost heap is also less likely to attract animals. When the compost is uniformly dark in color, is crumbly, and has a pleasant "woodsy" odor, it is ready to be used. The time required for decomposition varies from 1 to 6 months depending on the temperature, the materials used, and how often the heap is turned and watered.

Whereas compost is mixed into soil to improve the soil's fertility, **mulch** is placed on the surface of soil, around the bases of plants (see figure). Mulch helps control weeds and increases the amount of water in the upper levels of the soil by reducing evaporation. It lowers the soil

temperature in the summer and extends the growing season slightly by providing protection against cold in the fall. It also decreases soil erosion by lessening precipitation runoff.

Although mulches can consist of inorganic materials such as plastic sheets or gravel, natural mulches of grass clippings, straw, chopped corn-cobs, or shredded bark have the added benefit of increasing the organic content of the soil. Grass clippings are a very effective mulch when placed around the bases of garden plants because they mat together, making it difficult for weeds to become established. You must replace grass mulches often, however, because they decay rapidly. Some gardeners prefer mulches of more expensive materials such as shredded bark, because they take longer to decompose and are more attractive.

Mulches discourage the growth of weeds and help keep the soil damp. Organic mulches have the added benefit of gradually decaying, thereby increasing soil fertility. (*USDA/Soil Conservation Service*)

pores and air in the larger pores. After a prolonged rain, almost all of the pore spaces may be filled with water, but water drains rapidly from the larger pore spaces, drawing air from the atmosphere into those spaces.

Soil air contains the same gases as atmospheric air, although usually in different proportions. Generally, as a result of cellular respiration by roots and other soil organisms, less

oxygen and more carbon dioxide are present in soil air than in atmospheric air. Among the important gases in soil air are (1) oxygen (O_2), required by roots and soil organisms for respiration; (2) nitrogen (N_2), used by nitrogen-fixing soil organisms; and (3) carbon dioxide (CO_2), involved in soil weathering. Carbon dioxide dissolves in water to form carbonic acid, H_2CO_3, a weak acid that helps to weather rock.

Soil water contains low concentrations of dissolved mineral salts that enter the roots of plants when they absorb water. Soil water not absorbed by roots percolates (drains downward) through soil, carrying dissolved minerals with it. The removal of dissolved materials from soil as water percolates is called **leaching**. Some substances are completely leached out of the soil because they are so soluble that they migrate all the way down to the groundwater. Recall that not *all* soluble materials leach out of the soil, because soil particles, particularly clay particles, bind them. It is also possible for water to move *upward* through the soil and carry dissolved materials with it.

Soil pH Affects Soil Characteristics and Plant Growth

As discussed in Chapter 2, soil acidity is measured with the pH scale, which runs from 0 (extremely acidic) through 7 (neutral) to 14 (extremely alkaline). The pH of most soils ranges from 4 to 8, but some soils are outside this range. The soil of the Pygmy Forest in Mendocino County, California, is extremely acidic, with a pH of 2.8 to 3.9. At the other extreme, certain saline soils in Death Valley, California, have a pH of 8.5.

Soil pH affects plants and in turn is influenced by those organisms. Plants are affected by soil pH partly because the solubilities of certain minerals vary with differences in pH, and plants can absorb soluble mineral elements but not insoluble forms. At a low pH, for example, the aluminum and manganese in soil water are more soluble, sometimes becoming toxic to plants because they are absorbed by the roots in

Figure 10-9 A hydroponics experiment at the University of California at Davis. The plants are grown in a liquid solution of mineral nutrients through which air is bubbled to allow the roots to respire. (*L. Migdale, Science Source/Photo Researchers, Inc.*)

greater concentrations than are good for the plant. (See Plants and the Environment: Soil Salinization for an examination of soils with too many minerals.) Other mineral salts that are essential for plant growth, such as calcium phosphate, become less soluble at higher pH levels.

Soil pH greatly affects the availability of nutrients (Figure 10-8). An acidic soil has less ability than an alkaline soil to bind positively charged ions to itself. As a result, certain minerals that are essential for plant growth, such as potassium (K^+), leach more readily from acidic soil. The optimum soil pH for most plant growth is 6.5 to 7.5, because most essential elements needed by plants are available in that range.

Soil Provides Most of the Minerals Found in Plants

More than 90 naturally occurring elements exist on Earth, and more than 60 of these, including elements as common as carbon and as rare as gold, have been found in plant tissues. Not all of these elements are considered essential for plant growth, however.

How do biologists determine whether an element is essential? One of the most useful methods of testing whether or not an element is essential is **hydroponics**, which is the cultivation of plants in aerated water to which mineral salts have been added (Figure 10-9). It is impossible to conduct mineral

Figure 10-8 An azalea leaf exhibits the effects of iron deficiency as a result of growing in an alkaline soil. Azaleas that grow in acidic soils seldom exhibit iron deficiency. (*James Mauseth, University of Texas*)

PLANTS AND THE Environment

Soil Salinization

Soils found in arid and semiarid areas often contain high concentrations of inorganic mineral salts. In these areas, the amount of water that drains into lower soil layers is minimal because the little precipitation that falls quickly evaporates. In contrast, humid climates have enough precipitation to leach salts out of the soils and into waterways and groundwater. Irrigation of agricultural fields often results in their becoming increasingly salty. Also, when irrigated soil becomes waterlogged, salts may move upward from groundwater to the soil surface, where they are deposited as a crust.

Most plants cannot obtain all the water they need from salty soil, because such soil produces a water balance problem. Under normal con-ditions, the dissolved materials in the watery cytoplasm of plant cells give them a concentration of water lower than that of soil. As a result, water moves by osmosis from the soil into plant roots. When soil water contains a large quantity of dissolved salts, however, its concentration of water may be lower than that in plant cells; consequently, water moves *out* of plant roots and into the salty soil (see figure). Not surprisingly, most plants cannot survive under these conditions. Plant species that thrive in saline soils have special adaptations that enable them to tolerate the high amount of salt. Most crops, unless they have been genetically selected to tolerate high salt, are not produc-tive in saline soil.

In principle, the way to remove excess salt from saline soils is to add enough water to leach it away. Although this sounds straightforward, it is extremely difficult and in many cases impossible. For one thing, saline soils usually occur in arid or semiarid lands where water is in short supply. Also, many soils don't have good drainage properties, so adding lots of water simply causes them to become waterlogged. Another factor that should not be overlooked is that, even if the salt is flushed out of the soil, it has to go somewhere. The excess salt is usually carried to groundwater or to rivers and streams, where it becomes a water pollutant.

The effect of saline soil on water absorption by the roots of plants. (a) Normally there is a net movement of water (*black arrows*) into root cells from the soil. (b) When soil contains a high amount of salt, water moves *out* of the roots into the soil, even when the soil is wet.

nutrition experiments by growing plants in soil, because soil is too complex and contains too many elements. Hydroponics has other applications in addition to its scientific use (see Focus On: Commercial Hydroponics).

If biologists suspect that a particular element is essential for plant growth, they grow plants in a nutrient solution that contains all known essential elements *except* the one in question. If plants grown in the absence of that element are unable to develop normally or to complete their life cycle, the element may be essential (Figure 10-10). Biologists use additional criteria to confirm whether an element is essential. For example, the element must be shown to have a direct effect on the plant's metabolism, and it must be demonstrated that the element is essential for a wide variety of plants.

Sixteen Elements Are Essential for Plant Growth

Sixteen elements have been demonstrated to be essential for plant growth (Table 10-1). Nine of these elements are required in fairly large quantities (greater than 0.05 percent dry weight) and are therefore known as **macronutrients**. These include carbon, hydrogen, oxygen, nitrogen, phosphorus,

Figure 10-10   A hydroponically grown tomato plant exhibits the effects of a nitrogen deficiency (*right*). All other known essential elements were supplied in solution to this plant. A control tomato grown in a complete mineral solution (*left*) was healthy and symptom-free. (*Edgar E. Webber*)

TABLE 10-1   Essential Elements for Plant Growth

Element (Symbol)	Source	Significant function in plants
Macronutrients*		
Carbon (C)	Air (as CO_2)	Part of important biological molecules
Hydrogen (H)	Water	Part of important biological molecules
Oxygen (O)	Water, air (as O_2)	Part of important biological molecules
Nitrogen (N)	Soil	Part of important biological molecules
Phosphorus (P)	Soil	In genetic molecules and energy molecules
Calcium (Ca)	Soil	Part of cell walls
Magnesium (Mg)	Soil	In chlorophyll molecules
Sulfur (S)	Soil	In proteins and vitamins
Potassium (K)	Soil	Involved in ionic balance of cells
Micronutrients†		
Chlorine (Cl)	Soil	Involved in ionic balance of cells
Iron (Fe)	Soil	Involved in photosynthesis and respiration
Manganese (Mn)	Soil	Involved in respiration and nitrogen metabolism
Copper (Cu)	Soil	Involved in photosynthesis
Zinc (Zn)	Soil	Involved in respiration and nitrogen metabolism
Molybdenum (Mb)	Soil	Involved in nitrogen metabolism
Boron (B)	Soil	Exact role unclear

*Required in quantities greater than 0.05 percent dry weight.
†Required in trace amounts.

Focus On

Commercial Hydroponics

A greenhouse with lettuce growing hydroponically. (*Hank Morgan/Photo Researchers, Inc.*)

Scientists have used hydroponics, the practice of growing plants in an aerated solution of mineral salts, to determine which elements are essential for plant growth. Initially, entrepreneurs hailed hydroponics as the scientific way to grow plants in places where soil was poor or unavailable. The expenses involved in commercial cultivation of produce for human consumption, however, prevented hydroponics from becoming more than a curiosity. Recent technical improvements have revived interest in commercial hydroponics (see figure).

Hydroponics has great potential in several places. It is being tried experimentally in desert countries in the Middle East, where the soil is too arid to support cultivation and water is unavailable for irrigation. When plants are grown hydroponically in greenhouses, little water is used in comparison with traditional irrigation methods. Hydroponics is also being tried in temperate climates, particularly to produce crops in winter.

Hydroponics has other advantages. First, it is possible to grow crops hydroponically under conditions in which disease-causing microorganisms and insect pests are completely absent. This means that the crops are not exposed to chemical pesticides. Hydroponics can also be used to grow crops near their area of use, thus reducing transportation costs. On average, most foods consumed by Americans travel an average of 2100 kilometers (1300 miles).

The main disadvantage of hydroponics is the expense. Plants grown hydroponically must be supplied with a nutrient solution that is continually monitored and adjusted. Heating and lighting costs are high. Aeration of the roots was a major expense until developments such as the nutrient-film technique helped cut costs. In the nutrient-film technique, plants are grown in trenches produced by two sheets of plastic through which a nutrient solution trickles. In this way, the roots get adequate aeration. The nutrient solution is saved and reused, which cuts down on water and mineral costs.

Although hydroponics will probably never replace traditional agriculture, it has proven to be a viable alternative in certain situations. As new techniques are developed, hydroponics may become even more common.

---

potassium, sulfur, calcium, and magnesium. The remaining seven **micronutrients** are needed in very small (trace) amounts for normal plant growth and development. These include iron, boron, manganese, copper, molybdenum, chlorine, and zinc.

Four of the 16 elements—carbon, oxygen, hydrogen, and nitrogen—come from water or gases in the atmosphere. Carbon is obtained from carbon dioxide (CO_2) in the atmosphere during photosynthesis. Oxygen is obtained from atmospheric oxygen (O_2) and water. Water also supplies hydrogen to the plant. Plants absorb their nitrogen from the soil in the form of ions of nitrogen salts—nitrate (NO_3^-) and ammonium (NH_4^+). The nitrogen in the nitrogen salts comes from atmospheric nitrogen (N_2) and is converted to nitrate and ammonium by various microorganisms in either the soil or the root nodules of certain plants. The remaining 12 essential elements are obtained from the soil as dissolved mineral ions. Their ultimate source is the rock from which the soil formed.

Carbon, hydrogen, and oxygen are found in the structures of all biologically important molecules, including lipids, carbohydrates, nucleic acids, and proteins. Nitrogen is part of proteins, nucleic acids, and chlorophyll. Phosphorus is critical for plants because it is found in nucleic acids, phospholipids (an essential part of cell membranes), and energy transfer molecules such as ATP. Calcium plays a key structural role as a component of the middle lamella (the cementing layer between the cell walls of adjacent plant cells). Calcium has also been implicated in a number of physiological roles in plants, such as alteration of membrane permeability. Magnesium is part of the chlorophyll molecule. Sulfur is essential because it is found in certain amino acids and vitamins.

Potassium, which plants use in fairly substantial amounts, is not found in a specific compound or group of compounds in plant cells. Instead, it remains as free K^+ ions and plays a key physiological role in maintaining the turgidity of cells because it is osmotically active. The presence of K^+ in cytoplasm causes the cell to have a greater solute concentration than surrounding cells. As a result, water passes through the plasma membrane into the cell by osmosis. Potassium is also involved in the opening and closing of stomata (discussed later in the chapter).

Another element that has a role in the turgor balance of cells is chlorine. In addition to its osmotic role, the chloride (Cl^-) ion, which is present in very minute amounts in plants, is essential for photosynthesis.

Five of the micronutrients (iron, manganese, copper, zinc, and molybdenum) are associated with various plant enzymes (often as enzymatic activators) and are involved in certain enzymatic reactions. Although the role of boron in plants is unclear, some data suggest that it is involved in the transport of carbohydrates across cell membranes and affects calcium utilization.

In addition to the 16 essential elements, several other elements have been demonstrated to be essential for specific plants. Nickel is involved in enzymatic reactions in legumes such as peas and beans; sodium is probably essential for some plants adapted to a salty soil, some desert plants, sugar beets, bluegrass, and plants with C_4 photosynthesis; silicon enhances the growth of various grasses. After further evaluation, one or more of these elements may be added to the list of essential elements.

Fertilizers Can Replace Essential Elements Missing from the Soil

In a natural ecosystem, essential minerals removed from the soil by plants are returned when the plants or the animals that have eaten the plants die and decompose. An agricultural system disrupts this pattern (Figure 10-11). Crops, which contain minerals, are removed from the cycle when they are harvested, so that they fail to decay and release their nutrients back into the soil. Thus, soil that is farmed eventually loses its fertility; that is, it loses its ability to produce abundant crops.

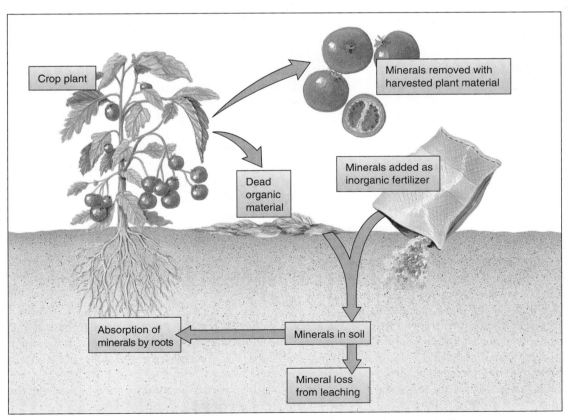

Figure 10-11   As plant and animal remains decompose in natural environments, nutrients are cycled back to the soil for reuse. In agriculture, much of the plant material is harvested. Because the mineral nutrients in the harvested portions are unavailable to the soil, the nutrient cycle is broken. For this reason, fertilizer must be added to the soil periodically.

In a similar manner, homeowners often mow their lawns and remove the clippings, preventing decomposition and cycling of minerals that were in the grass blades (see Plants and People: Managing Your Lawn and Garden at Home).

Plant growth is usually limited by the essential material (water, sunlight, or some essential element) in shortest supply. This phenomenon is sometimes called the concept of **limiting factors**. The three elements that are most often limiting factors for plants are nitrogen, phosphorus, and potassium. To sustain the productivity of agricultural soils, fertilizers are periodically added to replace the minerals that limit plant growth.

The two main types of fertilizers are organic and inorganic. *Organic fertilizers* include such natural materials as cow manure, crop residues, bone meal, sewage sludge, and compost. Green manure, a special type of organic fertilizer, is actually a crop such as alfalfa or sweet clover that is grown in the soil and then deliberately plowed under to decompose, rather than harvested. Frequently, the plant grown as green manure has nitrogen-fixing bacteria living in root nodules and thereby increases the amount of nitrogen in the soil. Organic fertilizers are complex, and their exact compositions vary, but the mineral nutrients in them become available to plants only as the organic material decomposes. Consequently, organic fertilizers are slow-acting and long-lasting.

Inorganic fertilizers are manufactured from chemical compounds, and their exact compositions are known. Because they are soluble, they are available to plants immediately. Inorganic fertilizers are available in the soil for only a short period of time, however, because they quickly leach away. Most inorganic fertilizers contain the three elements that are usually the limiting factors in plant growth—nitrogen, phosphorus, and potassium. The numbers on fertilizer bags (for example, 10,20,20) tell the relative concentrations of each of the three elements (N, P, and K). One advantage of inorganic fertilizers over organic fertilizers is that one knows precisely how much of which elements are being applied to the soil. Varying the relative concentrations of nitrogen, phosphorus, and potassium causes different growth responses in plants. When growing a lettuce crop, for example, it is best to use a fertilizer with a high nitrogen content, because that stimulates vigorous vegetative growth (growth of leaves, stems, and roots) rather than reproduction (growth of flowers, fruits, and seeds). The application of a fertilizer with a high nitrogen content around tomato plants, however, causes a low production of tomatoes. Although the roots, stems, and leaves of the tomato plants grow vigorously with a high-nitrogen fertilizer, reproduction is *not* stimulated; the lush plants form few flowers and therefore few fruits.

It is environmentally sound to avoid or limit the use of manufactured fertilizers, for several reasons. First, because of their high solubility, inorganic fertilizers are very mobile and may leach into groundwater or waterways, polluting the water. Second, manufactured fertilizers do not improve the water-holding capacity of the soil as organic fertilizers do. Third, the manufacture of inorganic fertilizers requires a great deal of fossil fuels and is therefore not energy-conserving. Another advantage of organic fertilizers is that, in ways that are not completely understood, they change the types of organisms living in the soil, sometimes suppressing microorganisms that cause certain plant diseases.

TRANSPORT IN PLANTS OCCURS IN XYLEM AND PHLOEM

We have seen that roots obtain water and dissolved minerals from the soil. Once in the roots, these materials are transported up the roots into the stems, leaves, and other structures. Furthermore, the sugar molecules manufactured in the leaves by photosynthesis are transported in solution (dissolved in water) throughout the plant, including down to the roots.

Water and dissolved minerals are transported from roots to other parts of the plant in xylem, whereas dissolved sugar is **translocated** in phloem (Figure 10-12). Xylem transport and phloem translocation do not resemble the transport of materials in animals because in plants nothing *circulates* in a system of vessels. Water and minerals transported in xylem travel only upward, whereas translocation of dissolved sugar may occur upward or downward in separate phloem cells. Also, xylem transport and phloem translocation differ from internal transport in animals in that plants lack a heart. In plants, movement in both xylem and phloem is driven largely by natural physical processes rather than by a pumping organ.

How, exactly, do materials travel in the continuous system of the plant's vascular tissues? First we examine water and its movement through the plant. Later we will discuss the transport of dissolved sugar.

WATER AND MINERALS ARE TRANSPORTED IN XYLEM

Water and dissolved minerals are transported within tracheids and vessel elements, which are hollow dead cells in the xylem tissue (see Figure 5-5). The movement of water in the xylem is the most rapid of any in plants (Table 10-2).

Water initially moves horizontally into the roots from the soil, passing through several tissues until it reaches the xylem. Once the water is in the tracheids and vessel elements of the root xylem, it travels upward through a continuous network of xylem cells from root to stem to leaf. The dissolved min-

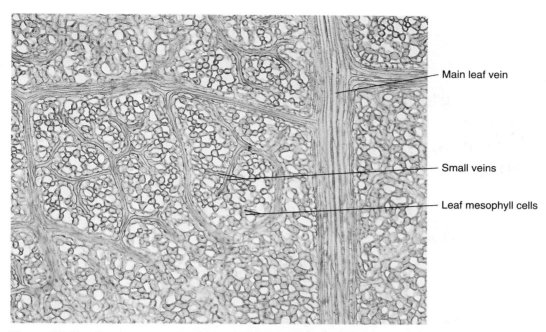

Figure 10-12   A network of veins in leaf tissue. Transport of materials throughout the plant body involves xylem and phloem, which are found in leaves as veins that branch extensively. No leaf cell is far from a vein. (*James Mauseth, University of Texas*)

erals are carried along passively in the water, although their initial uptake from the soil may be active (to be discussed shortly). The plant does not expend any energy of its own to transport water, which moves as a result of natural physical processes.

How does water move to the tops of plants? It might be either pushed up from the bottom of the plant or pulled up to the top of the plant. Actually, although both mechanisms exist, current evidence indicates that most water is transported through the xylem by being *pulled* to the top of the plant.

According to the Tension-Cohesion Theory, Transpiration in the Leaves Pulls Water Up the Stem

According to the **tension-cohesion theory,** water is pulled up the plant as a result of a *tension* produced at the top of the plant (Figure 10-13). This tension, which resembles that produced by sucking a liquid up a straw, is caused by the evaporative pull of transpiration. Transpiration, the evaporation of water vapor from the aerial parts of plants, is an essential process that is responsible for water movement in plants. Most water loss from transpiration takes place through the numerous stomata present on leaf and stem surfaces. The tension extends from the leaves, where most transpiration occurs, down the stems and into the roots. It pulls water from root cells into the root xylem and draws water up the xylem to leaf cells that have lost water as a result of transpiration. As water is pulled upward from the roots, additional water from the soil is drawn into the roots.

This upward pulling of water is possible only as long as there is an unbroken column of water in the xylem throughout the plant. Water forms an unbroken column in a plant because of the **cohesiveness of water molecules.** In addition, the adhesion of water to the walls of the xylem cells is an important factor in maintaining an unbroken column of water. Thus, the cohesive and adhesive properties of water enable it to form an unbroken column, which can be pulled.

TABLE 10-2   Xylem and Phloem Transport Rates in Selected Plants*

Plant	Maximum rate in xylem (cm/hr)	Maximum rate in phloem (cm/hr)
Conifer	120	48
Woody dicot	4,400	120
Herbaceous dicot/monocot	6,000	168–660
Herbaceous vine	15,000	72

*Xylem and phloem rates are from different plants within each general group and should be used for comparative purposes only. Adapted from Mauseth, J.D., *Botany: An Introduction to Plant Biology,* 2nd ed., Philadelphia, Saunders College Publishing, 1995.

☆ transpiration pull
cohesive & adhesive (capillary)

Figure 10-13   The tension-cohesion theory.
The evaporative pull of transpiration draws water up the plant from the soil to the atmosphere. (*Top*) Water vapor diffuses from the surfaces of leaf mesophyll cells to the drier atmosphere through the stomata. This produces a tension that pulls water out of the leaf xylem toward the mesophyll cells. (*Middle*) The cohesion of water molecules, caused by hydrogen bonding, allows water to be pulled up the narrow vessels and tracheids of the xylem in the stem. (*Bottom*) This movement, in turn, pulls water up the root xylem, which forms a continuous column of water from root xylem to stem xylem to leaf xylem. As water moves upward in the root, it produces a pull that causes soil water to diffuse into the root.

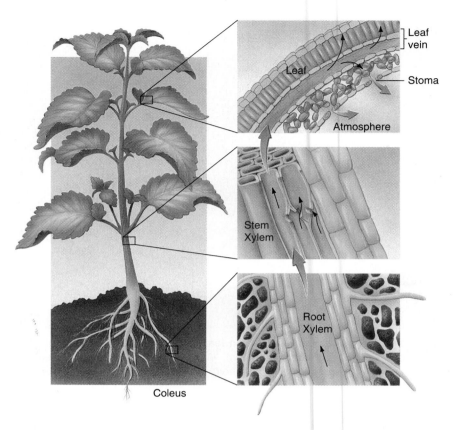

Coleus

Is tension-cohesion powerful enough to explain the rise of water in the tallest plants? Plant biologists have calculated that the tension produced by transpiration is strong enough to pull water up 150 meters (500 feet) in tubes of the same diameter as xylem vessels. Because the tallest trees are no more than 117 meters (375 feet) high, the tension-cohesion theory easily accounts for their water translocation. Tension-cohesion is considered the dominant mechanism of xylem transport in most plants.

The potassium ion mechanism explains stomatal opening and closing

Both the gas exchange needed for photosynthesis and the loss of water by transpiration occur through open stomata. These processes occur during the day, when stomata are open. At night, stomata typically close.

Data from numerous experiments and observations suggest that stomata open and close by the **potassium ion (K^+) mechanism**. Light somehow triggers an influx of potassium ions (K^+) into the guard cells from the surrounding cells of the epidermis (Figure 10-14). This movement of potassium ions, which occurs through special channels, or gates, in the guard cells' plasma membranes, has been experimentally measured and verified.

The increased concentration of K^+ in the guard cells lowers the relative concentration of water in those cells. As a result, water passes into the guard cells from surrounding epidermal cells by osmosis. The increased turgidity of the guard cells changes their shape, and the pore opens.

In the late afternoon or early evening, stomata close by a reversal of the process.[1] The potassium ions are pumped out of the guard cells into the surrounding epidermal cells. Water leaves the guard cells by osmosis; the cells lose their turgidity; and, as they collapse, the pore closes.

Root Pressure Pushes Water in a Root Up Through the Stem

In the less important mechanism for water transport, known as **root pressure**, water that moves into a plant's roots from the soil is *pushed* up through the xylem toward the top of the plant. Water moves into the roots by osmosis. The accumulation of water in root tissues produces a pressure that forces the water up the xylem.

Root pressure is a real phenomenon in plants. Guttation, the forceful release of liquid water through special openings at the tips of leaf veins, is a manifestation of root pressure. Plant biologists have measured root pressure, however, and

---

[1]Stomatal closure may not occur by an *exact* reversal of stomatal opening. There is evidence that Ca^{2+} triggers stomatal closure but inhibits stomatal opening. The actual mechanism whereby Ca^{2+} exerts this effect is under investigation.

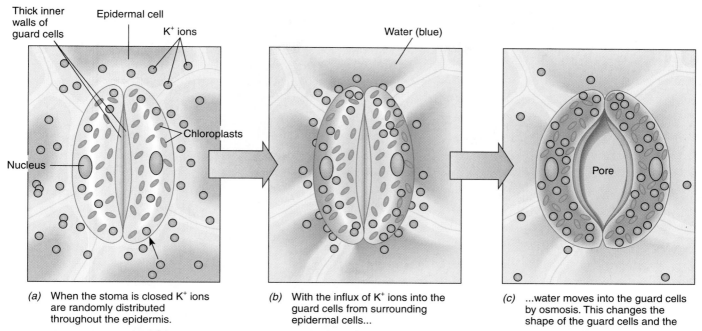

(a) When the stoma is closed K⁺ ions are randomly distributed throughout the epidermis.

(b) With the influx of K⁺ ions into the guard cells from surrounding epidermal cells...

(c) ...water moves into the guard cells by osmosis. This changes the shape of the guard cells and the

Figure 10-14   The mechanism of stomatal opening and closing. Movement of K⁺ ions into and out of the guard cells affects stomatal opening and closing.

have found that it is not a strong enough force to explain the rise of water to the tops of coastal redwoods and other tall trees. Root pressure exerts an influence in smaller plants, particularly in the spring when the soil is very wet, but it clearly does not cause water to rise 100 meters or more in the tallest plants. Furthermore, root pressure does not occur to any appreciable extent in summer, when there may be little water in the soil; yet the upward movement of water in trees is greatest during hot summer days.

Roots Selectively Absorb Minerals

Minerals are available to plants as ions dissolved in soil water. Because the concentrations of various mineral ions in xylem sap (fluid in the xylem) are different from those in soil water, plants appear to selectively accumulate the mineral ions they require. Most minerals are thought to enter the root by passing from one cell's cytoplasm to the next (the *symplast*) or along the interconnected cell walls (the *apoplast*) until they reach the Casparian strip of the endodermis (see Chapter 6). At this point minerals enter the endodermal cells by passing through their plasma membranes.

Dissolved mineral ions pass through plasma membranes by active transport. In active transport, the mineral ions move *against* the concentration gradient (that is, from an area of *low* concentration to an area of *high* concentration of that mineral) through special channels in the membrane. One of many reasons root cells require sugar and oxygen for cellu-

lar respiration is that active transport of mineral ions across biological membranes requires the expenditure of energy, usually in the form of ATP. From the endodermis, mineral ions enter the xylem of the root (precisely *how* this happens is not known) and are conducted to the rest of the plant.

 ## SUGAR IN SOLUTION IS TRANSLOCATED IN PHLOEM

The sugar produced during photosynthesis is converted to the disaccharide sucrose (common table sugar) before being loaded into the phloem and translocated to the rest of the plant. Sucrose is the predominant photosynthetic product carried in phloem. Translocation in phloem is not as swift as xylem transport (Table 10-2).

Fluid within phloem tissue moves both upward and downward. Sucrose is translocated in the phloem from the *source,* an area of excess sugar supply (usually a leaf), to the *sink,* an area of storage or metabolism (such as a root, an apical meristem, a fruit, or a seed).

The Pressure-Flow Hypothesis Explains Phloem Translocation

Phloem translocation is explained by the **pressure-flow hypothesis**, in which dissolved sugar moves in phloem by means of a pressure gradient (that is, a difference in pressure). The

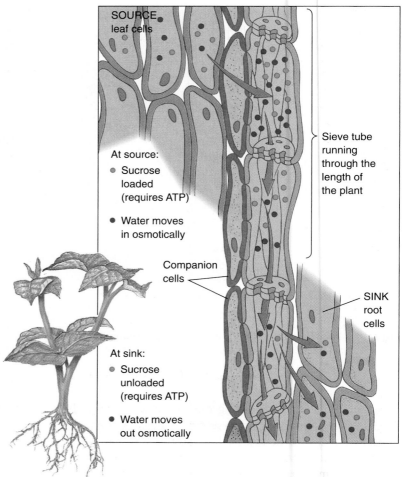

SOURCE
leaf cells

At source:
- Sucrose loaded (requires ATP)
- Water moves in osmotically

Companion cells

At sink:
- Sucrose unloaded (requires ATP)
- Water moves out osmotically

Sieve tube running through the length of the plant

SINK root cells

Figure 10-15   The pressure-flow hypothesis for phloem translocation. Sugar is actively loaded into the sieve-tube member at the source. As a result, water moves osmotically into the sieve-tube member. At the sink, the sugar is actively unloaded, and water leaves the sieve-tube member by osmosis. The gradient of sugar from source to sink causes transport from the area of higher pressure (the source) to the area of lower pressure (the sink).

pressure gradient exists between the source, where the sugar is loaded into the phloem, and the sink, where the sugar is removed from the phloem.

At the source, the dissolved sucrose moves from a leaf's mesophyll cells, where it was manufactured, into the companion cells of the phloem by active transport, which requires ATP (Figure 10-15). Once the sugar is in the companion cell, it readily moves into a sieve-tube member (see Figure 5-6) through the many cytoplasmic connections between the two cells. As a result of the increase in dissolved sugars in the sieve tubes, water moves by osmosis into the sieve tubes, increasing the pressure inside them. This pressure pushes the sugar solution through the phloem much as water is forced through a hose.

At its destination at the sink, sugar is actively unloaded from the sieve-tube members, with ATP again being required. With the loss in sugar, water moves out of the sieve tubes by osmosis and into surrounding cells. This decreases the pressure inside the sieve tubes at the sink.

Thus, the pressure-flow hypothesis explains the movement of dissolved sugar in phloem by means of a pressure gradient. It is the difference in sugar concentrations between the source (where a high sugar concentration causes a high pressure) and the sink (where a low sugar concentration causes a low pressure) that causes translocation in phloem, as water and dissolved sugar flow down the pressure gradient.

The actual translocation of dissolved sugar in the phloem does not require metabolic energy. However, both the loading of sugar at the source and the unloading of sugar at the sink require energy derived from ATP. The ATP energy is used to transport the sugar across cell membranes.

Although the pressure-flow hypothesis adequately explains current data on phloem translocation, much remains to be learned about this complex process. Phloem translocation is difficult to study in plants. Because phloem cells are under pressure, cutting into the phloem to observe it relieves the pressure and causes the contents of the sieve-tube members to be sucked against one end wall. Aphids, small wingless

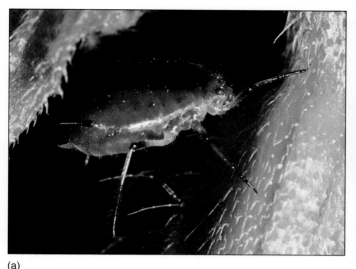
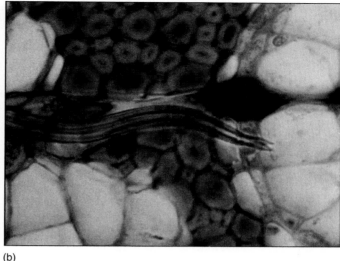

(a)   (b)

Figure 10-16 Aphids, tiny insects about 6 millimeters long, have been used to help study translocation in phloem. (a) A mature aphid feeding on a stem. **(b)** A microscopic view of phloem, showing the aphid mouthpart that has penetrated a sieve-tube member. (a, *Dwight R. Kuhn;* b, *M.H. Zimmerman,* Science *133:73–79 [fig. 4], January 13, 1961,* © *1988 by the American Association for the Advancement of Science*)

insects that insert their mouthparts into phloem sieve tubes to feed, have been a useful tool in phloem research (Figure 10-16). The pressure in the punctured phloem drives the sugar solution through the aphid's mouthpart and into its digestive system. When the aphid's mouthpart is severed from its body by a laser beam, the sugar solution continues to flow through the mouthpart at a rate proportional to the pressure in the phloem; this rate can be measured.

Researchers have also obtained much useful information about phloem translocation by using radioactive tracers. When a leaf is exposed to $^{14}CO_2$, the radioactive carbon is fixed into the sugars produced by photosynthesis. The scientists can follow the translocation of the sugar in phloem by employing special techniques. After exposure to the $^{14}CO_2$, the stem tissue is freeze-dried, sliced into thin sections, and placed on photographic film. The part of the film in contact with radioactive substances is exposed. After development, the exact locations of the radioactive sugars in the sieve tubes can be determined.

STUDY OUTLINE

I. Soil is the complex material in which plants root.
   A. Factors that influence soil formation include parent material, climate, organisms, the passage of time, and topography.
   B. Soil is composed of inorganic minerals, organic matter, organisms, soil air, and soil water.
      1. Inorganic mineral particles (sand, silt, and clay) come from weathered parent material and constitute most of what we call soil.
      2. Organic matter consists of litter, droppings, and the remains of dead plants, animals, and microorganisms that are in various stages of decomposition.
      3. Soil-dwelling organisms include bacteria, fungi, algae, protozoa, microscopic worms, earthworms, insects, plant roots, and larger animals such as moles, gophers, snakes, and groundhogs.
      4. Pore spaces of different sizes around and among the soil particles are filled with varying proportions of soil air and soil water. Plants and other soil-dwelling organisms need both soil air and soil water.

II. Plants require 16 essential elements for normal growth.
   A. Nine of the essential elements are macronutrients: carbon, oxygen, hydrogen, nitrogen, potassium, phosphorus, sulfur, magnesium, and calcium.
   B. Seven of the essential elements are micronutrients: iron, boron, manganese, copper, zinc, molybdenum, and chlorine.
   C. Essential elements are part of the structure of biological molecules, are important in the ionic balance of cells, and are involved in enzymatic reactions.
   D. Some essential elements are added to the soil as organic or inorganic fertilizer.
III. Water and dissolved minerals move upward in xylem from the root to stems and leaves.
   A. The tension-cohesion theory explains the rise of water in even the largest plants.
      1. The evaporative pull of transpiration causes a tension at the top of the plant.

2. As a result of the tension caused by transpiration, water moves from the soil to root xylem, to stem xylem, to leaf xylem, and finally to the atmosphere.

3. The column of water that is pulled up through the plant is unbroken as a result of the cohesive and adhesive properties of water.

B. The potassium ion mechanism explains the opening and closing of stomata.

1. Light triggers an influx of potassium ions (K^+) into the guard cells. The resulting osmotic movement of water into the guard cells causes their shape to change, thus forming a pore.

2. Stomata close when potassium ions leave the guard cells and water flows osmotically out of these cells. The guard cells become less turgid and collapse, closing the pore.

C. Root pressure caused by the movement of water into roots from the soil helps explain the rise of water in small plants, particularly during the spring, when the soil is wet.

IV. Dissolved sugar is translocated upward or downward in the phloem.

A. Sucrose is the predominant sugar translocated in the phloem.

B. The pressure-flow hypothesis explains the movement of materials in phloem.

1. Sugar is actively loaded (requiring ATP) into the sieve tubes at the source. As a result, water moves into these sieve tubes by osmosis, increasing the pressure inside the sieve tubes at the source.

2. Sugar is actively unloaded (requiring ATP) from the sieve tubes at the sink. As a result, water exits these sieve tubes by osmosis, decreasing the pressure inside the sieve tubes at the sink.

3. The flow of materials between the source and the sink is driven by the pressure gradient produced by water entering the phloem at the source and water leaving the phloem at the sink.

SELECTED KEY TERMS

clay, p. 168
compost, p. 172
humus, p. 170
hydroponics, p. 173
leaching, p. 173

limiting factor, p. 178
loam, p. 169
macronutrients, p. 175
micronutrients, p. 176
mulch, p. 172

potassium ion mechanism, p. 180
pressure-flow hypothesis, p. 181
root pressure, p. 180

sand, p. 168
silt, p. 168
tension-cohesion theory, p. 179
translocation, p. 178
weathering processes, p. 167

REVIEW QUESTIONS

1. List the five components of soil, and tell how each is important to plants.

2. Explain how weathering processes convert rock to soil.

3. What criteria have biologists used to determine which elements are essential for plant growth?

4. Distinguish between macronutrients and micronutrients. know them

5. Compare organic and inorganic fertilizers.

6. Name and briefly describe the theory used to explain the rise of water in the tallest trees.

7. Describe root pressure. What are its limitations?

8. Describe the pressure-flow hypothesis of sugar movement in phloem, including the activities at source and sink.

9. Explain why saline soils are physiologically dry for plants even when they are physically wet. essay

THOUGHT QUESTIONS

1. How would you design an experiment to determine whether gold is essential for plant growth? What would you use for an experimental control?

2. Why can overwatering the soil damage a plant?

3. Why must hydroponic solutions be aerated?

4. The San Joaquin Valley in California, with its mild climate and rich soil, is one of the nation's most fertile agricultural areas. It must be irrigated, however, because not much rain falls there. In recent years the productivity of the San Joaquin Valley has declined. Explain the probable cause.

SUGGESTED READINGS

Brown, J.C., and V.D. Jolley, "Plant Metabolic Responses to Iron-Deficiency Stress," *BioScience* Vol. 39, No. 8, September 1989. An in-depth discussion of the factors that make iron in the soil available to roots.

Feldman, L.J., "The Habits of Roots," *BioScience* Vol. 38, No. 9, October 1988. How roots interact with their soil environment.

CHAPTER 11

GROWTH RESPONSES AND REGULATION OF GROWTH

Grapes are one of the world's most important fruit crops. They are eaten fresh as table grapes, squeezed for juices and wines, dried to make raisins, and processed into jams, jellies, and canned fruit. The European grape (*Vitis vinifera*), which has been cultivated for thousands of years, probably originated in the Caucasus region of Europe (between the Black and Caspian Seas), but humans have introduced it to temperate areas worldwide. More recently, grapes have spread into subtropical areas. Today there are more than 5000 different grape varieties that are descended from the European grape. These different varieties may have black, purple, blue, red, pink, golden, pale green, or white fruits.

Although more than 90 percent of cultivated grapes are varieties of the European grape, other important species of the genus *Vitis* originated in North America. The North American species are often characterized by more robust odor and taste, by the fact that their skins separate from the fruit pulp easily, and by the resistance of their roots to low temperatures and insect and fungal pests. Frequently, disease-susceptible European vines are grafted to disease-resistant American rootstock to produce high-quality European grapes.

The grape plant is a climbing deciduous woody vine with coiled tendrils that attach the plant to its supports. Grape flowers are small, green, fragrant, and borne in clusters. After

A new variety of red grape (*Vitis vinifera*) under development in California. (*Richard Shiell/Dembinsky Photo Associates*)

self-pollination and fertilization, clusters of fruits develop that vary in size, form, and color depending on the variety. Botanically, the fruits are berries and usually contain seeds within a soft, juicy pulp. Thompson seedless grapes, first developed in the United States, and other seedless varieties are widely consumed because they are easier to eat.

Growers increase the size of grapes by spraying them with a natural plant hormone called *gibberellin* during their development. Almost all Thompson seedless grapes grown for direct human consumption are treated with gibberellin, which lengthens the branches that bear flowers and thereby gives the fruit clusters additional space in which to grow. Another advantage of gibberellin treatment of grapes is that the longer stems allow air to circulate around the individual berries, keeping them dry and preventing destructive fungi from attacking them.

In this chapter we consider the role of chemicals called hormones in regulating all aspects of plant growth (increase in size) and development (progressive changes during an organism's life). These processes include not only seed germination and the growth of seedlings into mature plants but also a plant's responses to changes in various aspects of its environment, including temperature, light, gravity, and touch.

EXTERNAL AND INTERNAL FACTORS AFFECT SEED GERMINATION AND EARLY GROWTH

In Chapter 9 we learned how pollination and fertilization are followed by seed and fruit development. Each seed develops from an ovule and contains an embryonic plant and food to provide nourishment for the embryo during germination, when the seed sprouts. A mature seed—that is, a seed in which the embryo is fully developed—is often dormant (not actively growing) and may not germinate immediately, even if growing conditions are ideal.

Numerous factors influence whether or not a seed germinates. Many of these are environmental cues, including the presence of water and oxygen, proper temperature, and sometimes the presence of light penetrating the soil surface. No seed germinates, for example, unless it has absorbed water. The embryo in a mature seed is dehydrated, and a watery environment in cells is necessary for active metabolism. When a seed germinates, its metabolic machinery is turned on, and numerous materials are synthesized and degraded. Therefore, water is an absolute requirement for germination.

The absorption of water by a dry seed that precedes germination is known as **imbibition**. As a seed imbibes water, it often swells to several times its original, dry size (Figure 11-1). Cells imbibe water by osmosis, which is the movement of water across a membrane from an area of high concentration to an area of low concentration, and by adsorption of water onto and into materials such as cellulose, pectin, and starches within the seed. Water is attracted and bound to these materials by adhesion, the attraction between unlike materials.

Seed germination and subsequent growth also require a great deal of energy. Because plants obtain this energy by converting the energy of fuel molecules stored in the seed's

endosperm or cotyledons to ATP by aerobic respiration, oxygen is usually needed during germination. Some plants, such as rice, can respire without oxygen during the early stages of germination and seedling growth. This enables rice plants to grow and become established in flooded soil, an environment that would suffocate most young plants.

Temperature is another environmental factor that affects germination. In a group of seeds of the same species, some germinate at each temperature in a broad range of temperatures. Each plant species, however, has an optimal, or ideal, temperature at which the largest number of seeds germinates. For most plants, the optimal germination temperature falls between 25° and 30°C (77° and 86°F). Some seeds, such as those of apples, require prolonged exposure to cold before their seeds germinate at any temperature.

Some of the environmental factors that are needed for seed germination help ensure the survival of the young plant.

Figure 11-1   Pinto bean (*Phaseolus vulgaris*) seeds (*left*) before and (*right*) after imbibition. (*Marion Lobstein*)

The requirement of a prolonged cold period ensures that seeds germinate in the spring rather than in the winter. Some plants—especially those with tiny seeds, such as lettuce—require light for germination. A light requirement ensures that a tiny seed germinates only if it is close to the surface of the soil. If such a seed germinates several inches below the soil surface, it may not have enough food reserves to grow to the surface. On the other hand, if this light-dependent seed remains dormant until the soil is disturbed and it is brought to the surface, it has a much greater likelihood of survival.

In certain seeds, internal factors, which are under genetic control, prevent germination even when all external conditions are favorable. Many seeds are dormant either because the embryo is immature and must develop further or because certain chemicals are present. The presence of such chemical inhibitors helps ensure the survival of the plant. The seeds of many desert plants, for example, often contain high levels of abscisic acid (discussed later in this chapter). Abscisic acid is washed out only when rainfall is sufficient to support the plant's growth after the seed germinates. Some seeds, such as those of legumes, have extremely hard, thick seed coats that prevent water and oxygen from entering, thereby inducing dormancy. *Scarification,* the process of scratching or nicking the seed coat (physically with a knife or chemically with an acid) before sowing it, induces germination in these plants.

Dicots and Monocots Exhibit Characteristic Patterns of Early Growth

Once conditions are right for seed germination, the first part of the plant to emerge from the seed is the radicle, or embryonic root. As the root grows and forces its way through the soil, it encounters considerable friction from soil particles. The delicate apical meristem of the root tip is protected by a root cap.

The plant shoot is next to emerge from the seed. Stem tips are not protected by anything like a root cap, but plants have ways to protect the delicate stem tip as it grows up through the soil to the surface. The stem of a bean seedling (a dicot), for instance, curves over to form a hook so that the stem tip and cotyledons are actually *pulled* up through the soil (Figure 11-2). Corn and other grasses (monocots) have a special sheath of cells called a **coleoptile** that surrounds the young shoot (Figure 11-3). First the coleoptile pushes up through the soil, and then the leaves and stem grow up through the middle of the coleoptile.

Certain parts of a plant grow throughout its life. This **indeterminate growth**—the ability to grow indefinitely—is characteristic of stems and roots, both of which arise from apical meristems. Theoretically, stems and roots could continue to grow forever. Other parts of a plant, such as leaves

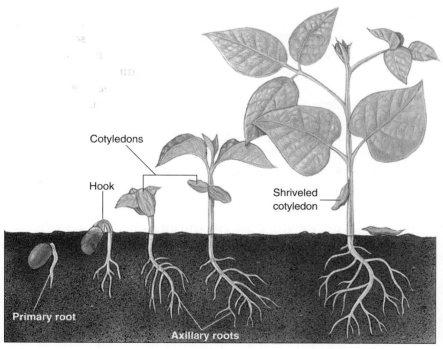

Cotyledons

Hook

Shriveled cotyledon

Primary root

Axillary roots

Figure 11-2   Seed germination and growth of a young soybean (*Glycine max*), a dicot. Note the hook in the stem of the young seedling, which protects the delicate stem tip as it moves up through the soil. Once the stem has emerged from the soil, the hook straightens.

Figure 11-3   Seed germination and growth of a young corn (*Zea mays*) plant. Note the coleoptile, a sheath of cells that emerges first from the soil. The stem and leaves grow up through the middle of the coleoptile.

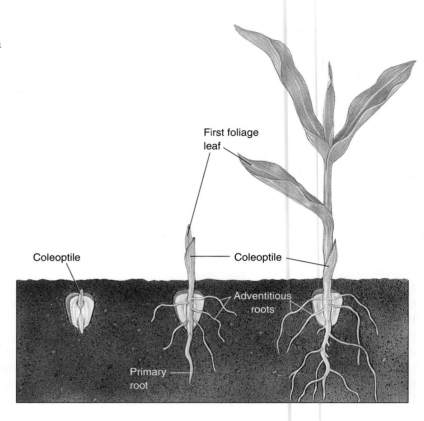

and flowers, have **determinate growth**; that is, they stop growing after reaching a certain size. The size of each of these structures varies from species to species and from individual to individual depending on the plant's genetic programming and on environmental conditions, such as availability of sunlight, water, and essential minerals.

PLANT GROWTH AND DEVELOPMENT ARE UNDER GENETIC AND ENVIRONMENTAL CONTROLS

The ultimate control of plant growth and development is genetic. If the genes required for development of a particular trait—for example, the shape of the leaf, the color of the flower, or the type of root system—are not present, that characteristic does not develop. When a particular gene is present, its *expression,* that is, how it exhibits itself as an observable feature in an organism, is determined by a variety of factors. The *location* of a cell in the young plant body, for example, has a profound effect on gene expression during development, causing some genes in that cell to be turned off and others turned on. Thus, cell location helps to determine differentiation, or what each cell ultimately becomes.

Environmental cues, such as changing daylength, variation in precipitation, and temperature, also exert an important

influence on gene expression, as they do on all aspects of plant growth and development. The initiation of sexual reproduction is often under environmental control, particularly in temperate latitudes. Such control is important for the plant's survival, because the timing of sexual reproduction is critical for reproductive success: all flowering plants in temperate climates must flower and form seeds before the onset of winter induces dormancy. A number of plants detect changes in the relative amounts of daylight and darkness that accompany the changing seasons and flower in response to those changes. Other plants have temperature requirements that induce sexual reproduction.

Flowering May Be Initiated by Changes in Light and Dark Periods

Photoperiodism is any response of a plant to the relative lengths of daylight and darkness. Flowering is one of several physiological activities that are photoperiodic in many plants. Plants are classified into four main groups on the basis of how photoperiodism affects their flowering: short-day, long-day, intermediate-day, and day-neutral plants.

Short-day plants (also called **long-night plants**) flower when the night length is equal to or greater than some critical length (Figure 11-4). This critical length varies considerably from one plant species to another but falls between 12 and 14 hours for many. The initiation of flowering in short-day plants is due not to the short period of daylight but to the long, uninterrupted period of darkness. Examples of short-day plants are chrysanthemum, aster, ragweed, and poinsettia, which typically flower in late summer or fall. Short-day plants can detect the lengthening nights of late summer or fall, and they flower at that time.

Long-day plants (also called **short-night plants**) flower when the night length is equal to or less than some critical length. Plants such as clover, spinach, black-eyed Susan, and lettuce flower in late spring or summer and are long-day plants. These plants can detect the shortening nights of spring and early summer, and they flower at that time.

Intermediate-day plants flower when they are exposed to days and nights of intermediate length. Sugarcane and several other grasses are intermediate-day plants. These plants do not flower when daylength is either too long or too short.

Some plants, called **day-neutral plants**, initiate flowering not in response to seasonal changes in the amounts of daylight and darkness but in response to some other type of stimulus, external or internal. Tomato, dandelion, string bean, and pansy are examples of day-neutral plants. Many of these plants originated in the tropics, where daylength does not vary appreciably during the year.

Phytochrome Detects Varying Periods of Daylength

For a plant or any organism to have a biological response to light, it must have a light-sensitive substance called a **photoreceptor** to absorb the light. The photoreceptor for photoperiodism and a number of other light-initiated responses of plants

Figure 11-4 The chrysanthemum (*Chrysanthemum* sp) is a short-day plant and therefore flowers only when exposed to long nights. Two identical cuttings from the same plant were planted in separate pots. The plant on the left, which flowered, received short days (8 hours of daylight) and long nights (16 hours of darkness) for several weeks. The plant on the right, which remained vegetative, received long days (16 hours of daylight) and short nights (8 hours of darkness) during the same time period. (*Dennis Drenner*)

is a blue-green pigment called **phytochrome**. Phytochrome is present in cells of all the vascular plants examined so far. It exists in two forms and readily converts from one form to the other upon absorption of light of specific wavelengths. One form, designated P_r (for *red*-absorbing *p*hytochrome), strongly absorbs red light with a relatively short wavelength (660 nm). In the process, the shape of the molecule changes to the second form of phytochrome, P_{fr} (Figure 11-5)—so

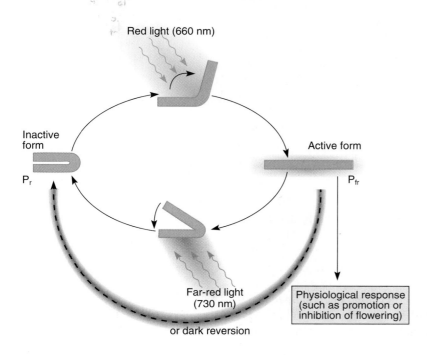

Red light (660 nm)

Inactive form

Active form

P_r

P_{fr}

Far-red light (730 nm)

Physiological response (such as promotion or inhibition of flowering)

or dark reversion

Figure 11-5   Phytochrome, a pigment that occurs in two forms designated P_r and P_{fr}, readily converts from one form to the other.

designated because it absorbs *far-red* light, which is red light with a relatively long wavelength (730 nm). When P_{fr} absorbs far-red light, it reverts back to the original form, P_r. The P_{fr} form of phytochrome is less stable than the P_r form, so it also reverts slowly to P_r in the dark. P_{fr} is the form of phytochrome that triggers physiological responses such as flowering.

What does a pigment that absorbs red light and far-red light have to do with daylight and darkness? Sunlight is composed of the entire spectrum of visible light in addition to ultraviolet and infrared radiation. Because sunlight contains more red light than far-red light, however, when a plant is exposed to sunlight, its level of P_{fr} increases. During the night, its level of P_{fr} slowly decreases as it reverts back to P_r.

In short-day plants, the P_{fr} form of phytochrome *inhibits* flowering, so these plants need long nights in order to flower. The long period of darkness allows the P_{fr} to revert back to P_r, so the plant has some minimum time during the 24-hour period with *no* P_{fr} present. This absence of P_{fr} initiates flowering in short-day plants. Biologists have experimented with short-day plants by growing them under a short-day/long-night regimen and interrupting the night with a short burst of red light (Figure 11-6). Exposure to red light for a few minutes in the middle of the night prevents flowering in short-day plants. This effect occurs because the brief exposure to red light converts some of the phytochrome from the P_r form to the P_{fr} form. Therefore, the plant does not have a sufficient period of nighttime without any P_{fr}.

In long-day plants, the active form of phytochrome, P_{fr}, *induces* flowering. The long days cause these plants to produce predominantly P_{fr}. During the short nights, some P_{fr} changes to P_r, but because the night is short, the plant still has a sufficient level of P_{fr} present to flower.

Day lengths

	Short nights	Long nights	Long nights with interruption
Short-day plant	Vegetative growth only	Flowers	Vegetative growth only
Long-day plant	Flowers	Vegetative growth only	Flowers

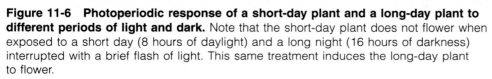

Figure 11-6   Photoperiodic response of a short-day plant and a long-day plant to different periods of light and dark. Note that the short-day plant does not flower when exposed to a short day (8 hours of daylight) and a long night (16 hours of darkness) interrupted with a brief flash of light. This same treatment induces the long-day plant to flower.

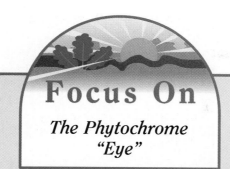

Focus On

The Phytochrome "Eye"

Can plants sense the presence of nearby plants? The answer is yes: they not only detect the presence of their neighbors, but they react to nearby plants by changing the way they grow and develop.

Although a plant does not possess a real eye, it can use its phytochrome "eye" to perceive changes in the ratio of red to far-red light that results from the presence of nearby plants. The leaves of neighboring plants absorb much more red light than they do far-red. (Recall from Chapter 4 that the green pigment chlorophyll, used in photosynthesis, strongly absorbs red light.) Therefore, in a densely plant-populated area, the ratio of red light to far-red light (*R/FR*) decreases, and far-red light is reflected from leaves of nearby plants. As a result, the phytochrome "eye" absorbs more far-red light, which in turn triggers a series of responses that cause the plant to grow taller.

When a plant is using much of its resources for stem elongation, it has fewer resources to allocate for new leaves and branches. However, for a plant that is shaded by its neighbors, a rapid increase in stem length is advantageous because, once this plant is taller than its neighbors, it obtains a larger share of unfiltered sunlight.

Plant biologists are puzzled by the observation that P_{fr} inhibits flowering in short-day plants but induces flowering in long-day plants. Why different plants respond so differently to P_{fr} is unknown at this time. Nevertheless, the importance of phytochrome to plants cannot be overemphasized. Timing of daylength and darkness is the most reliable way for plants to measure the change from one season to the next. This measurement is crucial for their survival, particularly in environments where the climate goes through a regular, annual pattern of favorable and unfavorable seasons.

Phytochrome Is Involved in Many Other Responses to Light

Phytochrome is involved in the light requirement that some seeds have for germination. Seeds with a light requirement must be exposed to red light. Exposure to red light converts P_r to P_{fr}, and germination occurs. Other physiological functions under the influence of phytochrome include "sleep" movements in leaves (discussed shortly), shoot dormancy, leaf abscission, and pigment formation in flowers, fruits, and leaves. (See Focus On: The Phytochrome "Eye.")

Temperature May Affect Reproduction

Certain plants have a temperature requirement that must be met if they are to flower. The promotion of flowering by exposure to low temperature for a period of time is known as **vernalization**. The part of the plant that must be exposed to low temperature varies. The moist seeds of some plants must be exposed to low temperature for a period of several weeks in order for flowering to be induced after the seeds germinate and grow. For other plants, recently germinated seedlings have a "cold" requirement.

In some plants, the requirement of a low-temperature period is absolute, meaning that they will not flower unless they have been vernalized. Other plants flower sooner if exposed to low temperatures but will still flower at a later date if they are not exposed to low temperatures.

Examples of plants with a low-temperature requirement include annuals such as winter wheat, which grow, reproduce, and die in 1 year, and biennials such as carrots, which take 2 years to complete their life cycles. Winter wheat is planted in the fall and germinates at that time. The young seedlings are exposed to cold during the winter and subsequently flower after resuming growth the following spring.

Carrots and other biennials grow vegetatively (that is, their leaves, stems, and roots grow) the first year and store surplus food in their roots (Figure 11-7). If the roots are not harvested, the plants flower and reproduce sexually during the second year, after exposure to the low temperatures of winter. Carrots growing in a warm environment and not exposed to low temperatures continue vegetative growth indefinitely and do not initiate sexual reproduction.

An external stimulus to which a plant responds, such as low temperature, may be moderated and influenced by internal conditions, such as hormone levels in the plant. It is possible, for example, to eliminate the low-temperature requirement for flowering in biennials by treating the plants with gibberellin, a plant hormone that is discussed later in this chapter.

(a)   (b)

Figure 11-7   Temperature requirements for flowering. (a) The carrot (*Daucus carota*),
a biennial plant, grows vegetatively the first year, storing surplus food in its storage root.
(b) After the plant remains dormant through the winter, the energy stored in the root is
used during the plant's second year to flower and reproduce. (a, *Carlyn Iverson;* b, *Jack
Bostrack/CBR Images*)

A BIOLOGICAL CLOCK INFLUENCES MANY PLANT RESPONSES

Plants, animals, and microorganisms appear to have an internal timer, or biological clock, that approximates a 24-hour cycle. Such internal cycles, known as **circadian rhythms** (from the Latin *circum,* "around," and *diurn,* "daily"), help the organism respond to the time of day. In contrast, photoperiodism enables an organism to respond to the time of year.

In the absence of external cues, circadian rhythms repeat every 20 to 30 hours. In nature, the rising and setting of the sun reset the biological clock so that the cycle repeats every 24 hours. For many plants, phytochrome has been implicated as the photoreceptor involved in resetting the biological clock.

One example of a circadian rhythm in plants is the opening and closing of stomata, independent of light and darkness: plants placed in continual darkness for extended periods continue to open and close their stomata on an approximate 24-hour cycle.

"Sleep" movements observed in the common bean and other plants are another example of a circadian rhythm in plants (Figure 11-8). During the day, bean leaves are horizontal for optimal light absorption, but at night the leaves fold down or up, a movement that orients them perpendicular to their daytime position. The biological significance of sleep movements is unknown at this time. These movements occur independently of Earth's 24-hour cycle. If bean plants are placed in continual darkness or continual light, sleep movements continue on an approximate 24-hour cycle.

Why do plants and other organisms exhibit circadian rhythms? Predictable environmental changes—sunrise and sunset, for example—occur during the course of each 24-hour period. These predictable changes may be important to an individual organism, causing it to change its behavior (in the case of animals) or its physiological activities. It is thought that circadian rhythms help an organism to synchronize repeatable daily activities so that it carries them out at an appropriate time each day.

CHANGES IN TURGOR CAN INDUCE TEMPORARY PLANT MOVEMENTS

The sensitive plant (*Mimosa pudica*) dramatically folds its leaves and droops in response to touch (or to an electrical, chemical, or thermal stimulus) (Figure 11-9). The response,

(a)

(b)

Figure 11-8   Sleep movements in the common bean (*Phaseolus vulgaris*). (a) Leaf position at noon. **(b)** Leaf position at midnight. It is not known why some plants exhibit sleep movements, but they occur on an approximate 24-hour cycle, regardless of the amount of light or darkness to which the plants are exposed. (a,b, *Dennis Drenner*)

which typically occurs in a few seconds, spreads throughout the plant even if only one leaflet is initially aroused. When a sensitive plant leaf is touched, an electrical impulse moves down the leaf to special cells housed in an organ at the base of the petiole, called the **pulvinus** (plural, *pulvini*). The pulvinus is a somewhat swollen joint that acts as a hinge. When the electrical signal reaches cells in the pulvinus, it induces a chemical signal that increases membrane permeability to certain ions. A loss of *turgor* (rigidity or distension caused by absorption of water) occurs in certain pulvinus cells as potassium ions exit through the now-permeable plasma membrane, causing water to leave the cells by osmosis. The sudden change in turgor causes the leaf movement. Such **turgor movements** are temporary and reversible. The movement of potassium ions and water back into the pulvinus cells causes the plant part to return to its original position, although recovery takes several to many minutes longer than the original movement.

The mechanism by which the Venus flytrap leaf closes is similar to the mechanism of the sensitive plant. An electrical signal, which moves much more rapidly than in the sensitive plant, induces a chemical signal that causes a movement of potassium ions out of certain cells, followed by the exit of water. The loss of turgor causes the leaf to snap shut.

(a)

(b)

Figure 11-9   Turgor movements in the sensitive plant (*Mimosa pudica*). The sensitive plant **(a)** before and **(b)** several seconds after being touched. Note how the leaves have folded and drooped. **(c)** How the drooping occurs. Pulvini occur in three areas: the base of each leaflet, the base of each cluster of leaflets, and the base of each leaf. Only changes in the pulvini at the bases of leaflets are shown. (*Top right*) A section through two leaflets, showing their pulvini when the leaf is undisturbed. (*Bottom right*) A section through the two leaflets, showing how a loss of turgor produces the folding of the leaves. (a,b, *Dennis Drenner*)

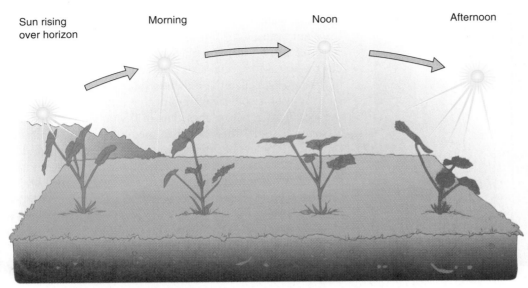

Sun rising over horizon   Morning   Noon   Afternoon

Figure 11-10   Solar tracking. Note that all the leaves are oriented so that they are perpendicular to the sun's rays throughout the day. The orientation of the leaves toward light in solar tracking is due to changes in turgor.

Changes in turgor are also responsible for **solar tracking**, the ability of the leaves or flowers of certain plants, such as sunflower, soybean, and cotton, to follow the sun's movement across the sky (Figure 11-10). Frequently, the leaves of such plants arrange themselves so that they are perpendicular to the sun's rays, regardless of the time of day or the sun's position in the sky. This allows for maximal light absorption. Many solar trackers have pulvini at the bases of their petioles. Changes in turgor in the cells of the pulvinus help position the leaf in its proper orientation relative to the sun.

A TROPISM IS GROWTH IN RESPONSE TO AN EXTERNAL STIMULUS FROM A SPECIFIC DIRECTION

A plant may respond to an external stimulus, such as light, gravity, or touch, by directional growth. Such a directional growth response, called a **tropism**, is a *permanent* change in the position of a plant part. Tropisms may be positive or negative, depending on whether the plant grows toward (positive) or away from (negative) the stimulus.

Phototropism is the growth of a plant caused by the direction of light. Most growing shoot tips exhibit positive phototropism by bending (growing) toward light, something you may observe if you place houseplants near a sunny window (Figure 11-11).

Growth in response to the direction of gravity is called **gravitropism**. Stems (that is, shoot tips) generally exhibit negative gravitropism by growing away from the center of the

Earth, whereas root tips exhibit positive gravitropism (Figure 11-12). The root cap is apparently the site of gravity perception in roots.

Special cells in the root cap contain large starch grains known as **statoliths** that collect toward the bottoms of the cells in response to gravity, and for many years it was thought that these statoliths initiated the gravitropic response. If the root is moved—as, for example, when a potted plant is laid

Figure 11-11   Stems exhibit positive phototropism when they grow in the direction of light. This coleus (*Coleus* sp) plant was exposed to light from the direction of the window for several weeks. (The rest of the room was darkened.) The bending toward light was caused by greater elongation on the shaded side of each stem than on the light side. (*Dennis Drenner*)

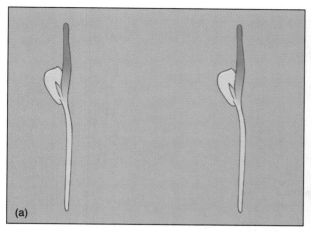

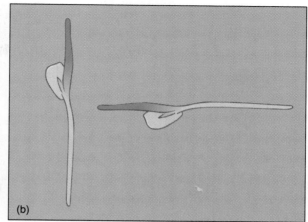

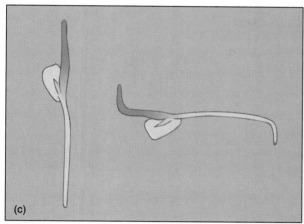

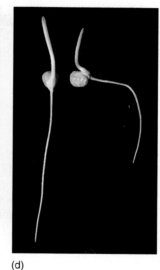

Figure 11-12   Stems exhibit negative gravitropism, and roots exhibit positive gravitropism. (a) Two corn (*Zea mays*) seeds were germinated at the same time. The plant on the left served as a control. **(b)** The plant on the right was turned on its side on day 3, orienting the shoot and root in a horizontal instead of vertical direction. **(c)** In 1 hour, the root and shoot tips had curved. **(d)** In 24 hours, the new root growth was downward and the new shoot growth was upward. Bending of both root and stem was caused by greater elongation on one side of the stem or root than on the other. (*Dennis Drenner*)

(d)

on its side—the statoliths tumble to a new position, always settling in the direction of gravity. The side of the root opposite where the statoliths have collected begins elongating shortly thereafter.

Despite the movement of statoliths in response to gravity, their role in gravitropism has recently been questioned. Timothy Caspar and his colleagues at Michigan State University developed a mutant plant species that lacks statoliths in its root cap cells. When placed on their sides, roots of these mutant plants still responded gravitropically, indicating that roots do not need statoliths to respond to gravity.

Thigmotropism is growth in response to a mechanical stimulus, such as contact with a solid object. The twining or curling growth of tendrils, which helps attach a plant such as a vine to some type of support, is an example of thigmotropism (see Figure 8-16b). Tropisms in plants may also be caused by other stimuli in the environment, such as water, temperature, chemicals, and oxygen. Because tropisms are growth responses, they cause permanent changes in the positions of plant parts.

HORMONES ARE CHEMICAL MESSENGERS THAT REGULATE PLANT GROWTH AND DEVELOPMENT

Plant **hormones** are organic chemical compounds produced in one part of a plant and transported to another part, where they elicit some kind of physiological response. The study of plant hormones and their effects is very challenging and difficult because hormones are effective in extremely small amounts and each hormone elicits *many* different responses. These features contrast with the actions of animal hormones, which are usually very specific.

In addition, the effects of different plant hormones overlap so that it is difficult to determine which hormone, if any, is the primary cause of a particular response. Moreover, plant hormones may stimulate a certain response at one concentration and inhibit that same response at a different concentration. Biologists have thus far identified five classes of plant

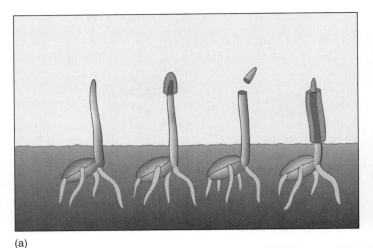

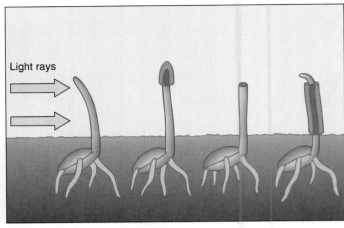

(a)

(b)

Figure 11-13   The Darwins' experiments with coleoptiles of canary grass (*Phalaris canariensis*) seedlings. (a) Some plants were uncovered, some were covered only at the tip, some had the tip removed, and some were covered everywhere but at the tip. **(b)** After exposure to light coming from one direction, the uncovered plants and the plants with uncovered tips (*far right*) grew toward the light. The plants with covered tips (*center left*) or tips removed (*center right*) did not bend toward light.

hormones: auxin, gibberellins, cytokinins, ethylene, and abscisic acid. Together these hormones control the growth and development of a plant.

Auxin Promotes Cell Elongation

Charles Darwin, best known for developing the theory of natural selection to explain evolution, also provided the first evidence for the existence of auxin. The experiments that Darwin and his son Francis performed in the 1880s involved positive phototropism, the directional growth of plants toward light. The plants they used were newly germinated canary grass seedlings (Figure 11-13). As in all grasses, the first part of a canary grass seedling to emerge from the soil is the coleoptile, a protective sheath that encircles the stem. When coleoptiles are exposed to light from only one direction, they bend toward the light. The bending occurs close to the tip of the coleoptile.

The Darwins tried to influence this bending in several ways. For example, they covered the tip of the coleoptile as soon as it emerged from the soil. Even though they covered that part of the coleoptile *above* where the bend would occur, the plants treated in this manner did not bend. On other plants, they removed the coleoptile tip and found that bending did not occur. When the bottom of the coleoptile was shielded from the light, however, the coleoptile still bent toward light. From these experiments, the Darwins concluded that "some influence is transmitted from the upper to the lower part, causing it to bend."

It took many years before this influence, a substance named **auxin** (from the Greek word for "enlarge" or "increase"), was successfully extracted and identified. Auxin was originally defined as any of a group of compounds that stimulate phototropic curvature in coleoptiles and stems. A broader definition for auxin is that it is a group of natural and artificial hormones that regulate growth, often by promoting cell elongation. The main auxin found in plants is **indoleacetic acid (IAA)**.

The movement of auxin in the plant is said to be *polar*—that is, it moves downward from its site of production, usually the shoot apical meristem. Young leaves and seeds are also sites of auxin production.

Auxin promotes cell growth by triggering cell elongation, an effect it apparently exerts by changing the cell walls so that they can expand. Auxin's effect on cell elongation also provides an explanation for phototropism. When a plant is exposed to a light from only one direction, the auxin migrates to the shaded side of the stem before moving down the stem by polar transport. As a result, the cells on the shaded side of the stem elongate more than the cells on the light side, and the stem bends toward the light (Figure 11-14). Auxin is thought to be involved in gravitropism and thigmotropism as well.

Auxin exerts other effects on plants. For example, some plants tend to branch out very little when they grow. Growth in these plants occurs almost exclusively from the apical meristem rather than from axillary buds, which do not develop as long as the terminal bud is present. Such plants are said to exhibit **apical dominance**—the inhibition of axillary buds

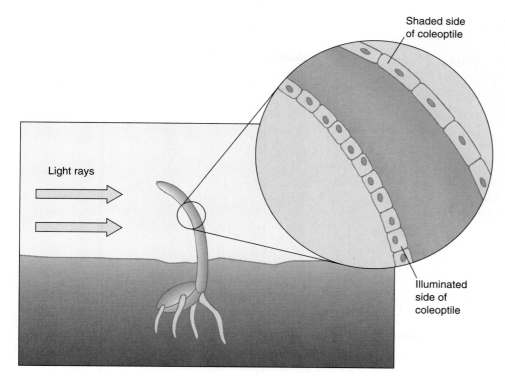

Shaded side
of coleoptile

Light rays

Illuminated
side of
coleoptile

Figure 11-14   Phototropism is caused by the unequal distribution of auxin. Auxin travels down the side of the stem or coleoptile *away* from the light, causing cells on the shaded side to elongate. Therefore, the stem or coleoptile bends toward light.

by the apical meristem (Figure 11-15). In plants with strong apical dominance, it appears that auxin produced in the apical meristem inhibits the development of axillary buds into actively growing shoots. When the apical meristem is pinched off, the auxin source is removed, and axillary buds grow into branches.

Auxin produced by developing seeds stimulates the development of the fruit. When auxin is applied to certain flowers in which fertilization has not occurred (and, therefore, in which seeds are *not* developing), the ovary enlarges and develops into a seedless fruit. Seedless tomatoes have been produced in this manner.[1] Auxin is not the only hormone involved in fruit development, however.

Some manufactured, or synthetic, auxins have been made that have structures similar to indoleacetic acid. Synthetic auxin is used to stimulate root development on stem cuttings for asexual propagation, particularly of woody plants with horticultural importance (Figure 11-16). Other synthetic auxins, 2,4-D and 2,4,5-T, are used as selective herbicides (weed killers). The compounds 2,4-D and 2,4,5-T kill plants with broad leaves but, for reasons not completely understood at this time, do not kill grasses. Both herbicides are similar in structure to IAA and disrupt the plants' normal growth

processes. Both cause exaggerated growth in some plant parts and growth inhibition in others. Because many of the world's most important crops are grasses (for example, wheat, corn, and rice), both 2,4-D and 2,4,5-T can be used to kill broadleaf weeds that compete with these crops. However, because of its association with dioxin, the use of 2,4,5-T is no longer allowed in the United States. (See Plants and the Environment: Herbicide Applications and the Vietnam War for a discussion of the environmental impact and health effects of herbicide spraying during the Vietnam War.)

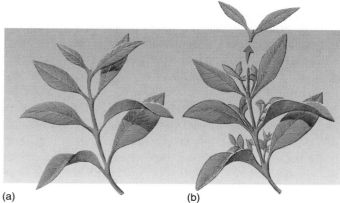

(a)   (b)

Figure 11-15   Auxin inhibits the development of axillary buds. (a) When the tip of the plant (the source of auxin) is intact, the axillary buds do not develop. **(b)** The tip of the plant has been removed. Because there is no auxin moving down from the stem tip, axillary buds develop into branches.

---

[1]Not all seedless fruits are produced by treatment with auxin. For example, in Thompson seedless grapes (discussed in the chapter introduction) fertilization occurs, but the embryos abort, and therefore the seeds fail to develop.

Figure 11-16 Honeysuckle (*Lonicera fragrantissima*) cuttings treated with a synthetic auxin. Many adventitious roots formed on honeysuckle cuttings placed in a higher auxin concentration (*left*), whereas fewer roots formed in a lower auxin concentration (*middle*). Cuttings placed in water (*right*) served as a control and did not form roots in the same time period. (*Visuals Unlimited/Joe Eakes, Color Advantage*)

Gibberellins Promote Stem Elongation

In the 1920s a Japanese biologist was studying a disease of rice in which the young rice seedlings grow extremely tall and spindly, fall over, and die. The cause of the disease, dubbed the "foolish seedling" disease, was a fungus that produces a chemical substance called **gibberellin**. Not until after World War II did scientists in Europe and North America learn of the exciting work done by the Japanese. During the 1960s gibberellins were isolated from healthy plants and determined to be hormones that are involved in many normal plant functions. The symptoms of the "foolish seedling" disease were caused by an abnormally high gibberellin concentration in the plant tissue.

As in the "foolish seedling" disease, gibberellins promote stem elongation in many plants. When gibberellin is applied to a plant, particularly a plant that normally has a very short stem, this elongation may be spectacular. Some corn and pea plants that are dwarf as a result of one or more mutations (changes in their genetic material) grow to a normal height when treated with gibberellin (Figure 11-17). Gibberellins are also involved in **bolting**, the rapid elongation of a floral stalk that occurs naturally in many plants when they initiate flowering (Figure 11-18). In all these cases, gibberellins cause stem elongation by stimulating cells to divide as well as elongate. The actual mechanism of cell elongation appears to be different from that caused by auxin, however.

Gibberellins are involved in several reproductive processes in plants. They stimulate flowering, particularly in long-day plants. In addition, they can substitute for the low-temperature requirement that biennials have before the initiation of flowering. If gibberellins are applied to biennials during their first

Figure 11-17 Effect of the continued application of gibberellin on normal and dwarf corn (*Zea mays*) plants. *From left to right:* dwarf, untreated; dwarf, treated with gibberellin; normal, treated with gibberellin; normal, untreated. Note that the dwarf plant responded to gibberellin much more dramatically than the normal plant. In fact, dwarf plants treated with gibberellin resemble normal plants in their growth rate. This dwarf variety is a mutant with a single recessive gene that impairs gibberellin metabolism. (*Courtesy of B.O. Phinney, University of California, Los Angeles*)

Figure 11-18 Bolting in Indian blanket (*Gaillardia pulchella*). Many biennials grow as a rosette (a circular cluster of leaves close to the ground) during their first year (*left*) and then bolt when they initiate flowering in their second year (*right*). The rapid stem elongation is triggered by gibberellin. (*Robert E. Lyons*)

Environment

Herbicide Applications and the Vietnam War

One of the many controversial aspects of the Vietnam War was the defoliation program carried out by the United States in South Vietnam. From 1961 to 1971, the United States sprayed herbicides over large areas of South Vietnam to expose suspected hiding areas and destroy crops planted by the Vietcong and North Vietnamese troops (see figure). The three mixtures of herbicides used were designated Agent White, Agent Blue, and Agent Orange.

The negative impact of these herbicides on the environment is still being felt today. It is estimated that between 20 and 50 percent of the ecologically important mangrove forests of South Vietnam were destroyed. Shrubs have replaced these forests, which may take decades to return. Moreover, approximately 30 percent of the nation's commercially valuable hardwood forests were killed and have been replaced by bamboo and weedy grasses. It has been estimated that the crops that were destroyed during the 10-year spraying period could have fed 600,000 people per year.

In addition to the ecological damage, the herbicide sprays caused health problems in the native people and members of the U.S. military who were exposed to them in the Vietnamese jungles. Agent Orange,

The spraying of herbicides over a forested area beside a South Vietnamese highway during the Vietnam War. This photo was taken in 1966. (*Archive Photos*)

a mixture of two herbicides (2,4-D and 2,4,5-T), also contained minute traces of dioxin, an extremely dangerous poison formed during the manufacture of 2,4,5-T. High doses of dioxin have been shown to cause birth defects in animals. Reportedly, the number of birth defects and stillbirths in Vietnam increased during the period of herbicide spraying. It also appears that American veterans who were exposed to high levels of Agent Orange have more health problems than do other veterans.

In 1993 the National Academy of Science's Institute of Medicine released the results of a comprehensive review of more than 6000 scientific articles on the effects of exposure to herbicides. It concluded that Vietnam veterans with any of three types of cancer (soft-tissue carcinoma, non-Hodgkin's lymphoma, and Hodgkin's disease) and two skin diseases (chloracne and porphyria cutanea tarda) should receive compensation for their exposure to Agent Orange.

year of growth, flowering occurs without the period of low temperature. Gibberellins, like auxin, affect the development of fruits. As mentioned in the chapter introduction, gibberellins are applied to several varieties of grapes to produce larger fruits.

Gibberellins are also involved in the germination of seeds in many plants. In plants with light or low-temperature requirements for seed germination, such as lettuce, oats, and tobacco, the application of gibberellins can substitute for the specific environmental requirement.

Cytokinins Promote Cell Division

During the 1940s and 1950s, a number of researchers using **tissue culture**, a technique for isolating cells from plants and growing them in a nutrient medium, were trying to find substances that might induce plant cells to divide. It was discovered that cells would not divide without some substance found in coconut milk. Because coconut milk has a complex chemical composition, the division-inducing substance was not chemically identified for some time. Finally, the active substance was isolated and called **cytokinin** because it induces cell division, or cytokinesis. In 1963 the first naturally occurring cytokinin was identified from corn, and since that time several similar molecules have been identified from other plants.

Cytokinins promote cell division and differentiation from young, relatively unspecialized cells to mature, more specialized cells in intact plants. They are also a required ingredient in any plant tissue culture medium and must be present in order for cells to divide. In tissue culture, cytokinins interact with auxin during the formation of plant organs such as roots and stems (Figure 11-19). For example, in tobacco tissue culture, a high ratio of cytokinin to auxin induces shoots to form, whereas a low ratio of cytokinin to auxin induces roots to form.

Cytokinins and auxin also interact in the control of apical dominance. Here their relationship is antagonistic: auxin inhibits the growth of axillary buds, and cytokinin promotes their growth.

One very interesting effect of cytokinins on plant cells is to delay **senescence**, or aging. Plant cells, like all living cells, go through a natural aging process. This process is accelerated in cells of plant parts that are cut, such as cut stems. It is thought that plants must have a continual supply of cytokinins from the roots. Cut stems, of course, lose their source of cytokinins and therefore age and die rapidly. When cytokinins are sprayed on leaves of a cut stem, they remain green, while any leaves that are not sprayed turn yellow and die (Figure 11-20). Despite their involvement in many aspects of plant growth and development, cytokinins have few commercial applications.

Ethylene Stimulates Abscission and Fruit Ripening

Ethylene, a gaseous hormone produced by plants, was first discovered in 1934. Many diverse plant processes are influenced by ethylene. Ethylene inhibits cell elongation, promotes

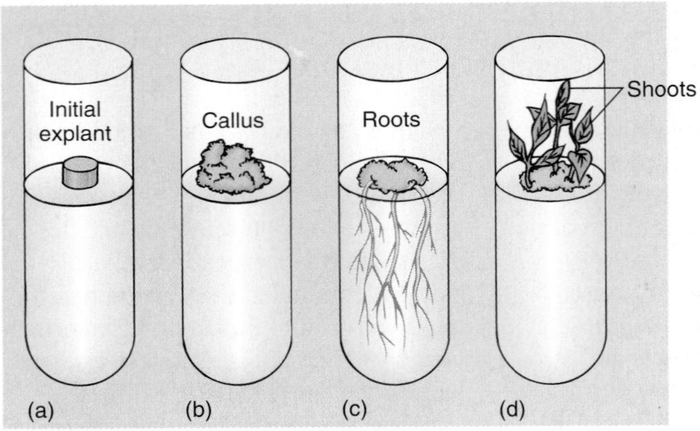

Figure 11-19   Propagating tobacco by tissue culture.
(a) A fragment of undifferentiated tissue from the center of a tobacco stem is placed in a culture medium. A complete plant can be formed from the tissue fragment because each cell of the fragment contains all the genetic information for the entire organism. Varying amounts of auxin and cytokinin in the culture media produce different growth responses.
(b) Nutrient agar containing an equivalent amount of both auxin and cytokinin caused cells to divide and form a callus (a clump of undifferentiated tissue). **(c)** When callus is transplanted to a medium with a higher relative amount of auxin, roots differentiate. **(d)** Shoot growth is stimulated by a medium containing a higher relative amount of cytokinin. Plants grown by tissue culture techniques can be transferred to soil and grown normally.

Figure 11-20   Senescence was delayed in the green philodendron (*Philodendron scandens*) leaf by repeated applications of cytokinin. Compare this leaf with the other leaf, which is aging because it was not treated with cytokinin. (*Courtesy of A.C. Leopold, Cornell University; from A.C. Leopold and M. Kawase*, Amer. Journ. Bot. *51: 294–298, 1964*)

the germination of seeds, and is involved in plant responses to wounding or invasion by disease-causing microorganisms.

Ethylene also has a major role in many aspects of senescence, including the ripening process in fruits. As a fruit ripens, it produces ethylene, which triggers an acceleration of the ripening process. This induces the fruit to produce more ethylene, which further accelerates ripening. The expression "One rotten apple spoils the lot" is true. A rotten apple is one that is overripe and produces large amounts of ethylene, which initiates a "domino effect" as it diffuses and triggers the ripening process in nearby apples. Ethylene is used commercially to promote the uniform ripening of bananas. Bananas are picked while green and shipped to their destination, where they are exposed to ethylene before they are delivered to grocery stores.

Ethylene has been implicated as the hormone that induces leaf abscission. However, abscission is actually influenced by two plant hormones that are antagonistic toward each other, ethylene and auxin. As a leaf ages (when autumn approaches, for deciduous trees in temperate climates), the level of auxin in the leaf decreases. Concurrently, cells in the abscission zone begin producing ethylene. To further complicate the process, it is possible that cytokinins may be involved in abscission. Cytokinins, like auxin, decrease in concentration as leaf tissue ages.

Abscisic Acid Promotes Bud and Seed Dormancy

Abscisic acid was discovered simultaneously in 1963 by two independent research teams. Its name is an unfortunate choice because abscisic acid is primarily involved in dormancy and does not induce abscission in most plants. Abscisic acid is sometimes referred to as the plant "stress hormone" since it promotes change in plant tissues that are stressed, or exposed to unfavorable conditions such as freezing, high salt levels, and droughts. (Ethylene also affects plant responses to certain stresses.) The effect of abscisic acid on plants suffering from water stress is understood best. The level of abscisic acid increases dramatically in the leaves of plants that are exposed to severe drought conditions. The high level of abscisic acid in the leaves triggers the closing of stomata, which saves the water that would normally be transpired through the stomata, thereby increasing the plant's likelihood of survival.

The onset of winter could also be considered a type of stress on plants. As winter approaches, woody plants cease growing, and protective coverings of bud scales form over terminal buds. Abscisic acid promotes these adaptations. Another winter adaptation that involves abscisic acid is dormancy in seeds. Many seeds have high levels of abscisic acid in their tissues and are therefore unable to germinate until the abscisic acid washes out. In a corn mutant that is unable to synthesize abscisic acid, the seeds are not dormant. Instead, they

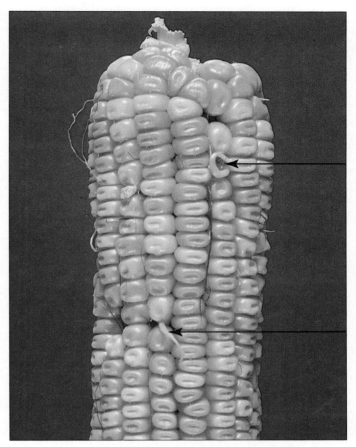

Figure 11-21 Inability to produce abscisic acid can prevent seed dormancy in corn (*Zea mays*). Some of the white kernels have germinated prematurely, while still on the ear, to produce roots (*see arrows*). (*Courtesy of M.G. Neuffer*)

germinate as soon as the embryos are mature, even while attached to the ear (Figure 11-21).

The evidence that abscisic acid is the only hormone involved in both bud and seed dormancy is not conclusive, particularly because the addition of gibberellin reverses the effects of dormancy. In seeds, the level of abscisic acid decreases during the winter, and the level of gibberellin increases. Cytokinins have also been implicated in breaking dormancy. Once again we see that a single physiological activity such as dormancy may be controlled in plants by the interaction of several hormones. The plant's actual response may be the result of changing ratios of hormones rather than the effect of each individual hormone. Table 11-1 summarizes the interactions among plant hormones during various aspects of growth and development.

Florigen Is a Hypothetical Hormone

Experiments in which different tobacco species are grafted together indicate that both flower-promoting and flower-

TABLE 11-1   Some of the Interactions Among Plant Hormones During Various Aspects of Plant Growth

Physiological Activity	Auxin	Gibberellin	Cytokinin	Ethylene	Abscisic Acid	Other Factors for Some Plants
Seed germination		Promotes	Possible role		Inhibits	Cold requirement, light requirement
Growth of seedling into mature plant	Cell elongation, formation of plant organs[1]	Cell division and elongation	Cell division and differentiation, formation of plant organs[1]			
Apical dominance	Inhibits lateral bud development		Promotes lateral bud development	Possible role		
Initiation of reproduction (flowering)		Stimulates flowering in some plants[2]	Possible role			Cold requirement, photoperiod requirement
Fruit development and ripening	Development	Development		Promotes ripening		Light requirement (for pigment formation)
Leaf abscission	Inhibits		Inhibits	Promotes		Light requirement
Winter dormancy of plant		Breaks	Possible role		Promotes	Light requirement
Seed dormancy		Breaks	Possible role		Promotes	

[1]In plant tissue culture.

[2]Gibberellin cannot be considered *the* flowering hormone. There is evidence for a flowering hormone that has not yet been isolated and characterized.

inhibiting substances may exist. *Nicotiana silvestris* is a long-day plant, and a variety of *Nicotiana tabacum* is a day-neutral plant. When a long-day tobacco is grafted to a day-neutral tobacco and exposed to short nights, both plants flower (Figure 11-22). The day-neutral tobacco flowers sooner than it normally would. It has been suggested that a flower-promoting substance may be induced in the long-day tobacco and transported to the day-neutral tobacco through the graft union, causing the day-neutral tobacco to flower sooner than expected. The hypothetical flower-promoting substance is known as **florigen**.

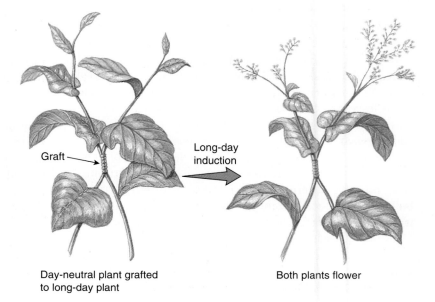

Figure 11-22 Evidence of a flower-promoting substance. When a long-day tobacco (*Nicotiana silvestris*) is grafted to a day-neutral tobacco (*N. tabacum*) and both plants are exposed to a long-day, short-night regimen, they both flower. The day-neutral plant flowers sooner than it normally would, presumably because a flower-promoting substance passes from the long-day plant to the day-neutral one through the graft.

Graft

Long-day induction

Day-neutral plant grafted to long-day plant

Both plants flower

When a long-day tobacco is grafted to a day-neutral tobacco and exposed to long nights, neither plant flowers. As long as these conditions continue, the day-neutral plants do not flower even when they would normally do so. In this case, it has been suggested that the long-day tobacco may produce a flower-inhibiting substance that is transported to the day-neutral tobacco through the graft union, preventing the day-neutral tobacco from flowering. The many attempts that have been made to isolate and chemically characterize substances that are clearly identifiable as flower promoters (florigen) or flower inhibitors have been unsuccessful.

STUDY OUTLINE

I. Both internal and external factors affect seed germination.
   A. External environmental factors that may affect seed germination include requirements for oxygen, water, temperature, and light.
   B. Internal factors affecting whether a seed germinates include the maturity of the embryo and the presence or absence of chemical inhibitors.
II. Plant growth and development are controlled not only by internal genetic factors but also by external environmental factors.
   A. The location of a cell in the young plant body affects gene expression during development by causing some genes in that cell to be turned off and others turned on.
   B. Many factors in the physical environment (such as changing daylength, variation in precipitation, and temperature) determine gene expression and affect plant growth and development.
III. Many plants flower in response to specific cues from the environment.
   A. Photoperiodism is the response of plants to the duration and timing of light and dark.
      1. Flowering is a photoperiodic response, with some plants being short-day plants, some being long-day plants, and others being intermediate-day plants. In day-neutral plants, flowering is not affected by photoperiod.
      2. The photoreceptor in photoperiodism is phytochrome, a blue-green pigment with two forms, P_r and P_{fr}.
   B. Vernalization is the promotion of flowering by exposure of seeds or young seedlings to low temperatures.
IV. Circadian rhythms are regular rhythms in growth or activities of a plant that approximate the 24-hour day and are reset by the rising and setting of the sun.
V. Turgor movements and tropisms are the two kinds of plant movements in response to external stimuli.
   A. Temporary changes in turgor in special cells produce turgor movements. One phenomenon caused by turgor movements is solar tracking, the ability of leaves or flowers to follow the sun.
   B. Tropisms are directional growth responses and are permanent.
      1. Phototropism is the growth of a plant in response to the direction of light.
      2. Gravitropism is the growth of a plant in response to the influence of gravity.
      3. Thigmotropism is the growth of a plant in response to contact with a solid object.
VI. Plants respond to hormones, chemical messengers that regulate plant growth and development.
   A. Hormones are effective in extremely small amounts.
   B. The functions of plant hormones overlap.
   C. Many physiological activities of plants may be due to the interactions of several hormones rather than the effect of a single hormone.
   D. There are five classes of plant hormones.
      1. Auxin is involved in cell elongation, tropisms, apical dominance, and fruit development.
      2. Gibberellins are involved in stem elongation, flowering, and seed germination.
      3. Cytokinins promote cell division and differentiation, delay senescence, and interact with auxin in apical dominance.
      4. Ethylene has roles in the ripening of fruits, leaf abscission, and senescence.
      5. Abscisic acid is the stress hormone. It is involved in stomatal closure due to water stress and in bud and seed dormancy, among other plant processes.
   E. Grafting experiments indicate the possible existence of both flower-promoting (florigen) and flower-inhibiting substances, neither of which have been successfully isolated.

SELECTED KEY TERMS

REVIEW QUESTIONS

1. What factors influence the germination of seeds? Explain the role that each of these factors plays in germination, and discuss why the plant responds the way it does.
2. Why are plant growth and development so sensitive to environmental cues?
3. What is phytochrome? Describe its role in flowering.
4. Define vernalization.
5. Distinguish between photoperiodism and circadian rhythms.
6. Distinguish between turgor movements and tropisms.
7. What is a hormone?
8. How is auxin involved in phototropism?
9. Discuss the various hormones that are involved in each of the following physiological processes: (a) germination of seeds, (b) stem elongation, (c) ripening of fruits, (d) abscission of leaves, (e) dormancy of seeds.
10. Summarize the roles of auxin, gibberellins, cytokinins, ethylene, and abscisic acid.

THOUGHT QUESTIONS

1. If apple seeds are planted in a tropical area where there is no winter, can they germinate successfully? Why or why not? NO
2. Predict whether flowering will occur in each of the following situations. Explain each answer.
   a. A short-day plant is exposed to 15 hours of daylight and 9 hours of darkness.
   b. A short-day plant is exposed to 9 hours of daylight and 15 hours of darkness. Yes
   c. A short-day plant is exposed to 9 hours of daylight and 15 hours of darkness, with a 10-minute flash of red light in the middle of the night. NO enough to change r to Fr
3. How might solar tracking benefit a plant? Under what conditions could solar tracking *harm* a plant? turger, dry leaves out.
4. Why might some plants have sleep movements? Are sleep movements an example of turgor movements or a tropism? Explain. unknown

Pr – Pfr
Phyton-chrome

5. What are the benefits to a plant when its stems exhibit positive phototropism and its roots exhibit positive gravitropism? photosynthesis roots grow down
6. As a result of apical dominance, axillary buds closest to the shoot apex tend to be inhibited the most and therefore grow the least, whereas those farthest from the shoot apex grow more. Would a "Christmas tree" such as spruce be considered to have this type of apical dominance? How about a "bushy" tree such as maple? Explain your answers. Yes strong auxin.
7. The nursery industry uses a plant hormone to induce dormancy in late summer so that plant material can be dug at that time rather than in autumn. (It is less traumatic for a plant to be moved when dormant than when it is actively growing.) Based on what you have learned in this chapter, which hormone do you think is used? Explain your answer. abscissic acid to promote dormancy

SUGGESTED READINGS

Chapin, F.S. III, "Integrated Responses of Plants to Stress," *BioScience,* Vol. 41, January 1991. Plants respond to many environmental stresses with altered growth.

Evans, M.L., R. Moore, and K.H. Hasenstein, "How Roots Respond to Gravity," *Scientific American,* December 1986. A closer look at gravitropism.

"Leaves with Clocks," in "Breakthroughs" section of *Discover,* September 1993. Sleep movements are another example of circadian rhythms in plants.

Mores, P.B., and N.H. Chua, "Light Switches and Plant Genes," *Scientific American,* April 1988. How an environmental cue is related to gene expression.

Rensberger, B., "Getting to the Root of Plant Growth," *Washington Post,* July 13, 1992. Why roots grow downward and shoots grow upward.

PART 3

THE CONTINUITY OF PLANT LIFE

CHAPTER *12*

THE MOLECULAR BASIS OF INHERITANCE

You may have seen the science fiction movie "Jurassic Park," in which the leading characters create living dinosaurs from ancient dinosaur DNA (the genetic material of organisms) that has been preserved in amber for millions of years. Although amber deposits have provided interesting information about extinct animals and plants, it is not possible to grow dinosaurs from dinosaur DNA except in movies. Currently, humans cannot develop any animal, living or extinct, from a single cell or its DNA.

On the other hand, a complete plant *can* be developed from a single plant cell through the use of **tissue culture techniques**, methods of growing large numbers of cells in a sterile, synthetic culture medium. These techniques make it possible to grow an entire plant that looks and functions like any other individual of that species from a single cell—a root cell, for example. This ability of a single cell to develop into an entire organism is called **totipotency** (from the Latin *toti* and *poten,* meaning "all powerful").

Although totipotency in plants was suggested in the early 1900s, it was not actually demonstrated until the late 1950s, when a plant biologist, F.C. Steward, placed some cells from

A clump of undifferentiated carrot cells, when transferred to a different culture medium, grows tiny roots and shoots. (*Visuals Unlimited/E. Webber*)

a carrot root into a culture medium containing sugar, minerals, vitamins, and coconut milk, which was known to contain certain growth factors for plants. In this culture medium, the carrot cells grew and divided but did not differentiate; that is, they did not develop into the many different types of cells found in carrot plants. When the carrot root cells were transferred to a slightly different culture medium, however, differentiation occurred, and some of the cells formed tiny stems with leaves. Roots later formed on those shoots, and the carrot plants were transplanted to soil.

Thus, Steward demonstrated that each carrot cell contained the necessary genetic information to grow and develop into an entire carrot plant. Since Steward's pioneering work with carrots, plant tissue culture techniques have been successfully used to generate many different kinds of plants, from African violets to California redwoods. Tissue culture techniques have the potential to aid in the improvement of crop plants through genetic engineering and to preserve biological diversity by enabling scientists to grow numerous individuals of an endangered species.

The instructions that tell a carrot—or any plant, for that matter—what to develop into, how large to grow, and what pigments to produce are contained in the nucleus of every cell of the plant. The cell's nucleus contains DNA, which stores genetic information. In this chapter we describe the molecular basis of inheritance, that is, the structure of the DNA molecule and how the genetic information in DNA is expressed.

In the next three chapters we continue our study of genetics. Chapter 13 follows the cell's chromosomes (and DNA) when new cells are produced by cell division involving mitosis or meiosis; Chapter 14 presents patterns of inheritance that apply to organisms from carrots to humans; and Chapter 15 discusses the cutting edge of genetic research—the deliberate manipulation of DNA to accomplish human goals.

LEARNING OBJECTIVES

After reading this chapter, you should be able to:

1. *Label a diagram of a DNA molecule, name the units of which the DNA molecule is composed, and identify the three parts of each unit.*
2. *Describe base pairing in DNA molecules. Given the base sequence of one strand of DNA, predict that of a complementary strand of DNA.*
3. *Summarize the process of DNA replication, and explain what is meant by "semiconservative replication."*
4. *Compare the structures of DNA and RNA molecules, and identify the functions of the three types of RNA: transfer RNA, messenger RNA, and ribosomal RNA.*
5. *Outline the flow of genetic information in cells from DNA to protein, and summarize the processes of transcription and translation.*
6. *Explain the universality of the genetic code and its evolutionary significance.*
7. *Explain the effect of mutations on protein synthesis and on evolution.*

CHROMOSOMES CARRY GENETIC INFORMATION IN EUKARYOTES

Each organism contains within it a set of hereditary instructions characteristic of its species—a blueprint for life—which is passed from generation to generation. In order for these instructions to contain all the information necessary for an organism to live and grow and develop, an organism's genetic material must be copied accurately and delivered to each new cell and to each offspring. The instructions embodied in the genetic material express themselves by controlling each cell's structure and functions. Through their control of the cells, the instructions direct the life of the entire organism.

In eukaryotes, these hereditary instructions are stored in the nucleus of every cell. In each nucleus are threadlike structures, called **chromosomes**, that contain **genes**, units of hereditary information. Most genes contain instructions for the manufacture of the proteins that make up the various structural and functional components of the cell.

Each chromosome consists of a single large *deoxyribonucleic acid (DNA)* molecule that contains hundreds or thousands of different genes (Figure 12-1 and Focus On: How Scientists Confirmed That DNA Is the Genetic Material). The DNA molecule is so long that it would become hopelessly

tangled during cell division were it not organized in some fashion. Just as thread is wound around spools to keep it untangled, the DNA molecule is wound around special protein molecules called **histones**, which help keep the DNA organized. The histones and their associated DNA are part of **chromatin**, the DNA and protein complex that makes up the chromosome.

DNA IS COMPOSED OF NUCLEOTIDES

In Chapter 2 we learned that DNA is composed of molecular units called **nucleotides** (see Figure 2-14). Each nucleotide consists of three parts: (1) a five-carbon sugar, **deoxyribose**; (2) a phosphate group, and (3) a nitrogen-containing organic compound called a base. There are four different bases in DNA: **cytosine (C), thymine (T), adenine (A),** and **guanine (G)**. These bases project, like the rungs of a ladder, more or less at right angles from the sugar-phosphate backbone of the DNA molecule.

Much as the sequence of letters determines the information conveyed by words and sentences, the sequence in which the four nitrogen bases appear determines their genetic message. Thus, the sequence -AGGTAC- encodes a different set of information from the sequence -TAGTCC-.

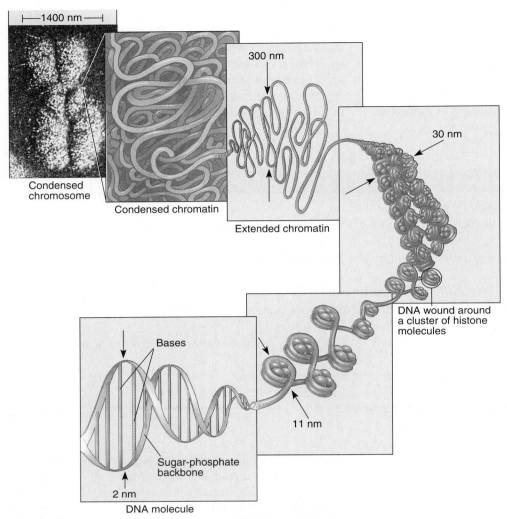

Figure 12-1   Levels of organization in the eukaryotic chromosome. A chromosome consists of chromatin, which is DNA wrapped around protein molecules (histones). (*Visuals Unlimited/K.G. Murti*)

The Two Nucleotide Chains of DNA Form a Double Helix

The structure of DNA was first elucidated by James D. Watson, a cell biologist, and Francis Crick, a biochemist, in 1953. They determined that DNA consists of *two* strands attached to each other by means of their bases as follows: adenine always pairs with thymine, and guanine always pairs with cytosine. The ladder-like double strand of DNA is twisted into a **double helix**, which superficially resembles a spiral staircase (Figure 12-2).

The two strands of a double helix are *complementary* but not identical. This means that if you know the base sequence of one of the two strands, you can predict the base sequence of the other, because adenine on one strand pairs only with thymine on the other strand, and cytosine pairs only with guanine. Under normal circumstances, no other pairing rela-

tionship is possible because of the nature of the hydrogen bonds that form between the bases. For instance, if a portion of one chain reads -AGGCTA-, then the other strand must have the sequence -TCCGAT-.

DNA REPLICATES SEMICONSERVATIVELY

One of DNA's most distinctive (and essential!) properties is that it undergoes precise replication; that is, DNA makes an exact duplicate of itself. Before a cell divides, the two strands of DNA separate, and each one is used as a template, that is, a strand on which a new complementary strand is built. The result of replication is *two* double helix molecules, each identical in base sequence to the original double helix. Each "new" DNA molecule contains one of the original strands plus a

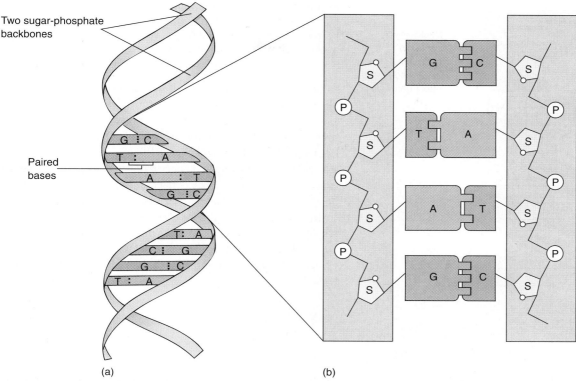

Two sugar-phosphate backbones

Paired bases

(a)   (b)

Figure 12-2   The structure of part of a DNA molecule. (a) DNA is a double helix consisting of two sugar-phosphate backbones joined by their paired bases. **(b)** A small portion of the DNA molecule has been unwound to show complementary base pairing. (P = phosphate group; S = the sugar deoxyribose; G = guanine; C = cytosine; T = thymine; A = adenine.)

newly synthesized strand (Figure 12-3). Because the two original strands are not simply copied and then left behind but are *conserved,* or kept, as part of the two new double helices, this process is known as **semiconservative replication** (*semiconservative* here meaning "keeping a portion the same"). Each new double helix contains one old strand and one new strand.

Replication need not begin at the end of a double helix. Enzymes break the hydrogen bonds linking the two DNA strands at various sites along their length, causing them to separate. As the two strands of DNA separate, they form a Y-shaped region known as the **replication fork**. Many replication forks may occur simultaneously in a DNA molecule (Figure 12-4). Because it occurs in many places at once, the replication of the enormous amount of information contained in DNA can take as little as 20 minutes in very rapidly dividing cells, such as those in a root tip.

After separating to form replication forks, both DNA strands act as templates for the assembly of new complementary strands. This assembly is the task of the enzyme **DNA polymerase**, which adds free nucleotides onto the "unzipped" DNA molecule. As the new complementary strands are pro-

duced, the original double helix separates further, and the newly synthesized strands continue to form hydrogen bonds with the old strands. At the end of the process, there are two identical double helices of DNA.

DNA polymerase is one of the most important enzymes in organisms, because DNA replication could not take place without it. DNA polymerase is a complex of several enzymes that catalyzes the reaction in which nucleotides are joined together in a very precise way. An enzyme speeds up a chemical reaction that would ordinarily occur much more slowly, and enzymes act only on specific substances, or *substrates*. Because DNA polymerase has *four* different nucleotide substrates (A, C, T, G) that it must add to the growing end of a new strand of DNA, it is an especially complex enzyme.

DNA polymerase accomplishes a number of remarkable tasks. DNA polymerase recognizes a specific base on an original strand of a replicating DNA molecule and then identifies the nucleotide base that is its complement (an A for a T, a C for a G, and so on). DNA polymerase brings the complementary nucleotide to the proper end of the new, elongating strand of DNA and catalyzes the chemical reaction that fastens the nucleotide to the new strand. The enzyme then moves along

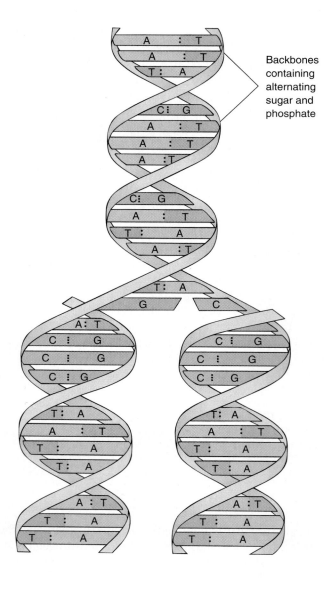

Backbones containing alternating sugar and phosphate

Figure 12-3   The mechanism of DNA replication. The two strands are shown separating, and both are being copied. In the new strand, as in the old, adenine pairs with thymine, and cytosine pairs with guanine.

to the next base on the existing strand and repeats the process all over again. DNA polymerase continues this process at a rate of about 200 bases per second, and it does it for hundreds or thousands of bases every time the cell divides, and occasionally at other times as well!

Errors in replication do occur, but infrequently. Although exact base pairing of DNA is necessary for a regular helix to be produced, improper pairing occasionally produces a distortion in the structure of the helix. However, there is a backup system that catches most of the infrequent errors that DNA polymerase makes. This backup system is a team of enzymes that snips out improperly paired bases and replaces them with the right ones.

MOST GENES CARRY INFORMATION FOR MAKING PROTEINS

The messages in a plant's genes control everything about it by governing its production of protein. Both obvious traits, such as flower color and shape of leaves, and other aspects of a plant's structure and function that are more difficult to observe are the results of protein synthesis.

Proteins are large, complex molecules manufactured from amino acids joined by peptide bonds. A long chain of amino acids is called a **polypeptide**. Some proteins consist of a single polypeptide chain, while others are composed of two or more polypeptide chains.

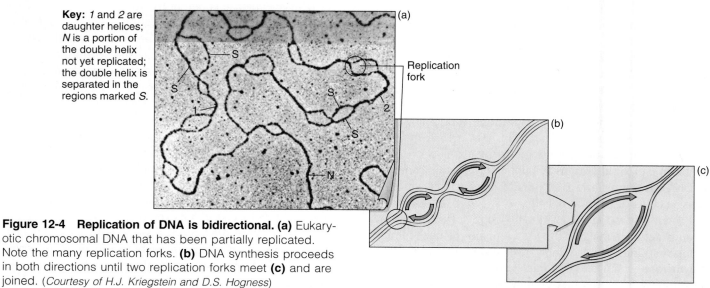

Key: *1* and *2* are daughter helices; *N* is a portion of the double helix not yet replicated; the double helix is separated in the regions marked *S*.

(a)

Replication fork

(b)

(c)

Figure 12-4   Replication of DNA is bidirectional. (a) Eukaryotic chromosomal DNA that has been partially replicated. Note the many replication forks. **(b)** DNA synthesis proceeds in both directions until two replication forks meet **(c)** and are joined. (*Courtesy of H.J. Kriegstein and D.S. Hogness*)

Viruses, a group of disease-producing agents, lack most of the traits of organisms; they are not composed of cells, nor do they metabolize, grow, or move by themselves. A virus can, however, reproduce with the aid of a living cellular host.

All organisms, including microscopic bacteria, seem to be susceptible to viral infection. Viruses that attack bacteria are called **bacteriophages,** or **phages** for short. Like other viruses, a phage consists of a protein coat that surrounds a molecule of nucleic acid. A bacterium's infection begins when a phage first attaches to the bacterial cell wall. Part of the phage then penetrates the bacterial cell's outer coverings and takes over the bacterium's metabolic machinery. The phage causes this machinery to replicate viral genetic material and form new phages like itself. In other words, the phage forces the bacterium host to follow the genetic directions the phage has injected into it. Eventually, what remains of the host cell bursts open, releasing many phage particles. Some of these particles may come in contact with another bacterial cell and repeat the process.

Before 1952 no one knew whether the nucleic acid core or the protein coat (or perhaps both) carried the phage's genetic information into the bacterial cell, but in

that year Alfred Hershey and Martha Chase settled the question. Hershey and Chase worked with *Escherichia coli,* the common intestinal bacterium, and a phage that attacked it. They knew that the phage's deoxyribonucleic acid (DNA) core contained phosphorus but no sulfur. They also knew that the phage's protein coat contained sulfur but no phosphorus. They reasoned that if they could find some sulfur from a phage that had made its way into a bacterial cell, they would know that the phage's protein coat, not its DNA core, had entered the cell. On the other hand, if they found phosphorus inside the bacterium, they would know that the phage's DNA core, not its protein coat, had entered the cell. Whichever substance entered the cell, Hershey and Chase reasoned, was responsible for replicating the virus and was therefore the material that transmitted traits to the next generation.

Hershey and Chase prepared a batch of phage whose protein coats

contained radioactive sulfur (^{35}S). The sulfur would serve as a tracer that would enable them to detect even small quantities of the protein. They also prepared a separate batch of phage whose DNA core contained radioactive phosphorus (^{32}P). When the phage whose protein was labeled with radioactive sulfur was added to a fresh culture of bacteria, the bacteria did not become radioactive. This showed that the protein coat did not enter the bacteria. When the phage whose DNA was labeled with radioactive phosphorus was added to a bacterial culture, however, the bacteria did become radioactive, indicating that viral DNA had entered the bacteria (see figure).

Since phages contain only protein and DNA, Hershey and Chase had investigated all possible genetic material. It was clear that DNA had to be the material that directed the production of new viruses. DNA, not protein, was the genetic material.*

---

*In some viruses RNA, rather than DNA, serves to store genetic information, but RNA is also a nucleic acid.

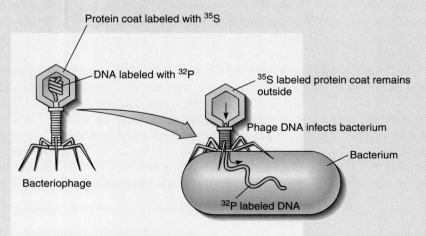

Protein coat labeled with ^{35}S

DNA labeled with ^{32}P

^{35}S labeled protein coat remains outside

Phage DNA infects bacterium

Bacterium

Bacteriophage

^{32}P labeled DNA

The Hershey-Chase experiment. Hershey and Chase grew bacteriophages in the presence of radioactive isotopes, thereby labeling the protein and DNA with different isotopes. They found that only the ^{32}P-labeled DNA entered the bacterium. The DNA provided all the genetic information needed for the synthesis of new phages.

A typical cell produces—continually and simultaneously—hundreds of kinds of proteins. The sequence of bases in a cell's DNA contains the directions, or blueprints, for the production of every single kind of protein molecule found in a plant.

THE GENETIC CODE IS READ AS A SERIES OF TRIPLETS

In Chapter 2 we learned that 20 different amino acids are commonly found in proteins. These amino acids can be thought of as the 20 letters of the protein alphabet. How can the four different nucleotides in the DNA alphabet (A, C, T, G) specify instructions for the 20 amino acids? If single nucleotides were the instructions, DNA could specify the sequence of only four kinds of amino acids in any protein. If pairs of nucleotides were the instructions, there would be only 16 different amino acids, because in a pool of 4 nucleotides there are only 16 unique pairs.

However, trios of nucleotides can and do form the genetic code, because a pool of 4 nucleotides permits 64 unique combinations (AAA, AAC, AAG, AAT, and so on), more than enough to specify the synthesis of the 20 amino acids. Each sequence of three nucleotides—more simply, three bases—is known as a **triplet**, and it is the basic unit of genetic information. With some exceptions, each triplet codes for one amino acid. Some of the 64 triplets are synonymous; that is, they code for the same amino acid. A few triplets do not code for amino acids but have other roles in protein synthesis.

PROTEIN SYNTHESIS IS A TWO-STEP PROCESS

Although the triplet sequence of bases in DNA specifies the order of amino acids in a polypeptide chain, DNA does not produce protein directly. DNA is found in the nucleus, and proteins are made by ribosomes in the cytoplasm, which is outside the nucleus. Instead, a related nucleic acid, *ribonucleic acid (RNA),* translates the information encoded in the DNA into a manufactured protein.

RNA differs from DNA in that each RNA nucleotide contains the sugar **ribose** rather than deoxyribose (Table 12-1 and Figure 12-5). As in DNA, each RNA nucleotide contains one of four bases, although in RNA the base **uracil (U)** replaces the thymine (T) found in DNA. Uracil behaves chemically like thymine and readily forms base pairs with adenine. Another difference between RNA and DNA is that RNA molecules usually do not form double helices but function as single strands, although some RNA molecules are elaborately folded.

In the first stage of protein synthesis, an RNA molecule is made on the basis of the information encoded in DNA. RNA

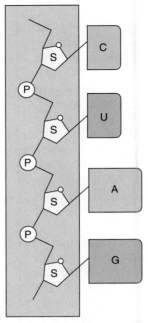

Figure 12-5   The structure of part of an RNA molecule. RNA differs from DNA in that it is composed of a single strand rather than a double helix. The sugar in RNA is ribose, rather than deoxyribose. Three of RNA's four bases—adenine, guanine, and cytosine—are also found in DNA, but the fourth RNA base is uracil rather than thymine. (P = phosphate group; S = the sugar ribose; G = guanine; C = cytosine; U = uracil; A = adenine.)

synthesis is similar to DNA replication in that the DNA molecule unwinds and unzips. One of the DNA strands serves as a template on which RNA nucleotides (A, U, C, G) are assembled in the proper order, based on complementary base pairing. The process of making RNA from a DNA template is called **transcription**, and it takes place in the nucleus in eukaryotic cells.[1]

When the RNA molecule is complete, it detaches from its DNA template, leaves the nucleus, and goes to a ribosome, a cellular structure that is the site of protein synthesis. (The two strands of the DNA molecule rezip and rewind to again form a double helix.) In this way the RNA molecule carries the instructions for the synthesis of a particular polypeptide from DNA in the nucleus to the ribosome in the cytoplasm.

The ribosome reads the message encoded in the RNA molecule and uses it to assemble amino acids in the proper order to produce a specific polypeptide chain. The formation of a polypeptide chain at the ribosome on the basis of the information encoded in RNA is known as **translation**.

---

[1]Some transcription also takes place in chloroplasts and mitochondria, organelles that contain their own DNA and ribosomes.

TABLE 12-1   A Comparison of DNA and RNA

	DNA	RNA
Units	Nucleotides	Nucleotides
Sugar	Deoxyribose	Ribose
Phosphate	Yes	Yes
Bases	Adenine	Adenine
	Guanine	Guanine
	Cytosine	Cytosine
	Thymine	Uracil
Structure	Double helix	Single strand, sometimes folded elaborately
Function	Stores genetic information; used to make RNA	Used to make proteins
Location in cell	Part of chromosomes in nucleus	Made in nucleus; works in cytoplasm (ribosomes)
Structural forms	—	Three forms (messenger RNA, transfer RNA, ribosomal RNA)

To summarize, protein synthesis involves two steps (Figure 12-6). In Step 1, transcription, the genetic message is copied from the DNA to the RNA. In Step 2, translation, the genetic message is translated from the language of nucleic acids into the language of proteins.

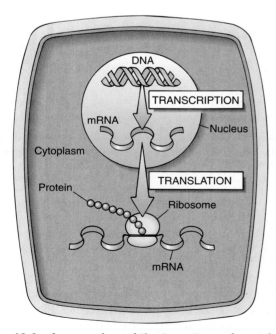

Figure 12-6   An overview of the two steps of protein synthesis, transcription and translation, in a plant cell. The master plans (DNA) are kept in the nucleus. During transcription, copies (mRNA) of the master plans are made and subsequently sent to the protein production area of the cell (the ribosomes). During translation, enzymes use the mRNA to assemble the correct sequence of amino acids into a protein.

Three Kinds of RNA Are Involved in Protein Synthesis

There are three types of RNA molecules: messenger RNA (mRNA), ribosomal RNA (rRNA), and transfer RNA (tRNA). All three types are made in the nucleus from DNA by complementary base pairing, and each has a role in protein synthesis. **Messenger RNA (mRNA)** carries the coded instructions about protein structure from DNA to the ribosome. The ribosome assembles amino acids in the proper sequence into a polypeptide, guided by the information in the mRNA molecule. **Transfer RNA (tRNA)** carries specific amino acids to the ribosome during protein assembly. Ribosomes, the sites of protein synthesis, are composed of various proteins and **ribosomal RNA (rRNA)**. Ribosomal RNA assists in the attachment of the ribosome to the mRNA molecule and in the assembly of amino acids in the proper order to make a polypeptide.

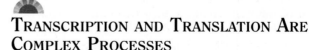

TRANSCRIPTION AND TRANSLATION ARE COMPLEX PROCESSES

The brief overview of transcription and translation in the preceding section provides sufficient information for some introductory botany courses. The following material is more detailed.

Transcription Is the Synthesis of RNA from DNA

In transcription, a sequence of nucleotides on a DNA strand acts as the template for the synthesis of a complementary RNA strand. For example, where a nucleotide containing guanine occurs in the DNA strand, an RNA nucleotide containing cytosine pairs with it and becomes part of the new RNA strand.

An enzyme complex called **RNA polymerase** controls transcription. This remarkable enzyme responds to certain base sequences in the DNA molecule that tell the RNA polymerase which gene to transcribe, which of the two paired DNA strands it should copy, and where it should begin and end transcription.

In preparation for transcription, RNA polymerase unwinds the DNA double helix. Transcription begins when RNA polymerase binds to DNA at a specific base sequence it recognizes as a start signal. RNA polymerase moves along a strand of DNA like a railroad locomotive on a track, assembling a single strand of RNA as it goes. Complementary RNA nucleotides are matched to the bases along the DNA template and added to the growing strand. The RNA strand separates from the DNA template as it forms. Transcription continues until the RNA polymerase comes to a specific nucleotide sequence in the DNA that it recognizes as a stop signal. The RNA molecule then separates completely from the DNA strand.

DNA replication and transcription differ in several important ways. First, in transcription only a single strand of DNA serves as the template, rather than both strands as in DNA replication. Second, RNA contains uracil rather than thymine; thus, during transcription, an RNA nucleotide containing uracil pairs with the DNA nucleotide containing adenine. Third, the RNA molecule detaches from the DNA template rather than forming a double helix with it.

Translation Is the Synthesis of a Polypeptide Specified by mRNA

Recall that the genetic code is based on triplets of nucleotide bases in DNA. This same information is encoded in mRNA as a sequence of three bases known as a **codon**. Each codon in mRNA is complementary to a DNA triplet and codes for a specific amino acid (Table 12-2). During translation, the order of the codons in the mRNA determines the order in which amino acids are added to a growing polypeptide chain.

Translation, which takes place at a ribosome, begins when a strand of mRNA becomes attached to one or more ribosomes. Each ribosome moves along the mRNA strand, and amino acids are added one by one to the growing polypeptide chain. The final step of translation occurs when the completed polypeptide chain is released. The polypeptide chain subsequently performs a specific cellular function; for example, it might function as an enzyme to catalyze a chemical reaction that is needed to produce a certain flower color.

Translation employs all three types of RNA. The mRNA specifies the order of amino acids in the polypeptide, the rRNA and its associated enzymes (as part of the ribosome) play several important roles, and the tRNA identifies and transports amino acids to the ribosome, the site of protein synthesis.

TABLE 12-2   The Genetic Code: Codons of mRNA (as Defined by DNA Triplets) That Specify a Given Amino Acid

Codon	Amino Acid	Codon	Amino Acid
UUU UUC	Phenylalanine	GUU GUC GUA GUG	Valine
UUA UUG	Leucine		
UCU UCC UCA UCG	Serine	GCU GCC GCA GCG	Alanine
UAU UAC	Tyrosine	GAU GAC	Aspartic acid
UAA UAG	STOP† STOP†	GAA GAG	Glutamic acid
UGU UGC	Cysteine	GGU GGC GGA GGG	Glycine
UGA	STOP†		
UGG	Tryptophan		
AUU AUC AUA	Isoleucine	CUU CUC CUA CUG	Leucine
AUG	Methionine (START)*	CCU CCC CCA CCG	Proline
ACU ACC ACA ACG	Threonine		
		CAU CAC	Histidine
AAU AAC	Asparagine	CAA CAG	Glutamine
AAA AAG	Lysine		
AGU AGC	Serine	CGU CGC CGA CGG	Arginine
AGA AGG	Arginine		

*The codon AUG initiates synthesis of a polypeptide and calls for methionine as the first amino acid.

†The three STOP codons signal positions where the ribosome stops reading and terminates the polypeptide chain.

Ribosomal RNA is part of the ribosome

The ribosome is the cell's protein assembly factory. Each ribosome has a small subunit and a large subunit, each of which consists of rRNA and many protein molecules. Ribosomes provide everything needed for translation—mRNA, tRNAs with their associated amino acids, rRNA, and the enzymes necessary to catalyze all the chemical actions required to join the amino acids together with peptide bonds. The ribosome serves, in effect, as an enzymatic "matchmaker" that ensures not only that peptide bonds form but that they form between the correct amino acids in the order specified by mRNA.

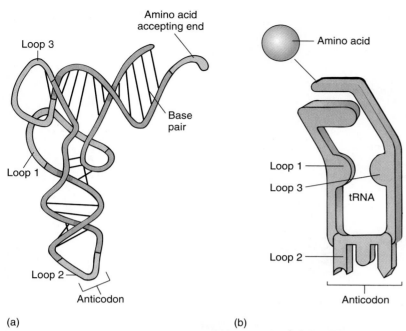

Loop 3

Amino acid
accepting end

Base
pair

Loop 1

Loop 2

Anticodon

(a)

Amino acid

Loop 1

Loop 3

tRNA

Loop 2

Anticodon

(b)

Figure 12-7   Two representations of the structure of tRNA, the molecules that "read" the genetic code. (a) A simplified diagram of the actual shape of a tRNA molecule. Its three-dimensional shape is determined by hydrogen bonds between base-paired regions. One loop contains the anticodon that base-pairs with the mRNA codon. The amino acid is attached to one end of the tRNA molecule. **(b)** A schematic diagram that shows how an amino acid is attached to its tRNA.

Transfer RNA carries amino acids to the ribosome

The individual amino acids that will be assembled into protein are carried to the ribosome by molecules of tRNA. Each tRNA molecule is elaborately folded and can form a temporary chemical bond with a specific amino acid (Figure 12-7). Enzymes, one for each kind of amino acid, ensure that each tRNA molecule links up with the correct amino acid.

Each tRNA contains a trio of bases, called an **anticodon**, that projects from the tRNA molecule. The ribosome recognizes the amino acid because the anticodon of its tRNA carrier matches with a complementary three-base sequence (the codon) on the mRNA strand.

Messenger RNA codons are the instructions the ribosome uses to construct a protein

After the mRNA binds to the ribosome, a special sequence of three bases (the codon AUG) on mRNA signals the start of translation. As the mRNA molecule passes through the ribosome, each codon on mRNA is read in turn. A tRNA molecule whose anticodon pairs with the codon on mRNA by complementary base pairing moves into the ribosome, carrying its specific amino acid. That amino acid is joined by a peptide bond to the amino acid at the end of the growing polypeptide chain (Figure 12-8). The ribosome then moves along the

mRNA strand so that a new codon can be read, and the process repeats.

Termination of protein synthesis is usually accomplished by special stop signals. These codons (UAA, UAG, UGA) do not code for amino acids but instead trigger the separation of the polypeptide chain and mRNA from the ribosome.

Each strand of mRNA may be used to make multiple copies of a particular protein. As many as 10 or 20 ribosomes usually bind to a single strand of mRNA, each one independently producing a separate copy of the same polypeptide (Figure 12-9). After the instructions in mRNA have been used to produce the required number of polypeptide chains, mRNA is degraded to its constituent nucleotides, which may be used to construct new RNA molecules.

THE GENETIC CODE IS SHARED BY ALMOST ALL ORGANISMS

DNA is the molecule of inheritance for all organisms except RNA viruses, and the *order* of bases in a DNA molecule determines the characteristics of the traits for which DNA codes. Thus, different species owe their individual traits to different sequences of bases in their DNA.

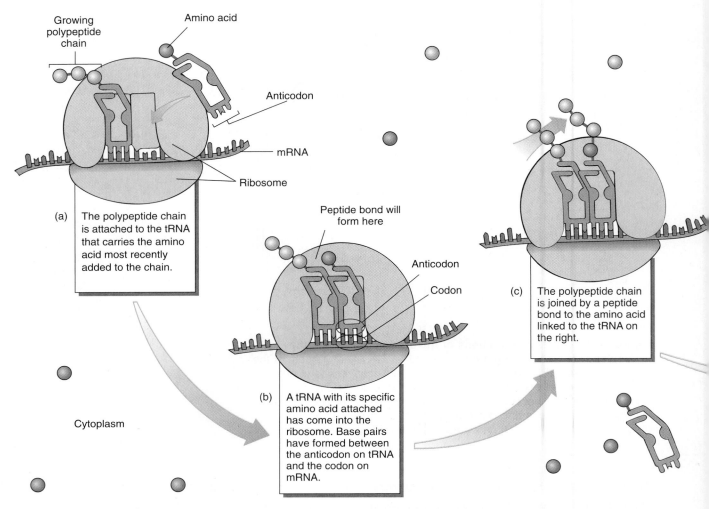

(a) The polypeptide chain is attached to the tRNA that carries the amino acid most recently added to the chain.

(b) A tRNA with its specific amino acid attached has come into the ribosome. Base pairs have formed between the anticodon on tRNA and the codon on mRNA.

(c) The polypeptide chain is joined by a peptide bond to the amino acid linked to the tRNA on the right.

Figure 12-8   During translation, a ribosome progresses along an mRNA strand, adding amino acids to a growing polypeptide chain as specified by the sequence of codons in mRNA.

Figure 12-9   Multiple ribosomes progress along an ▶ mRNA strand, adding amino acids (brought in by tRNA molecules) to growing polypeptide chains as called for by the sequence of codons in the mRNA. (*Courtesy of E. Kiseleva*)

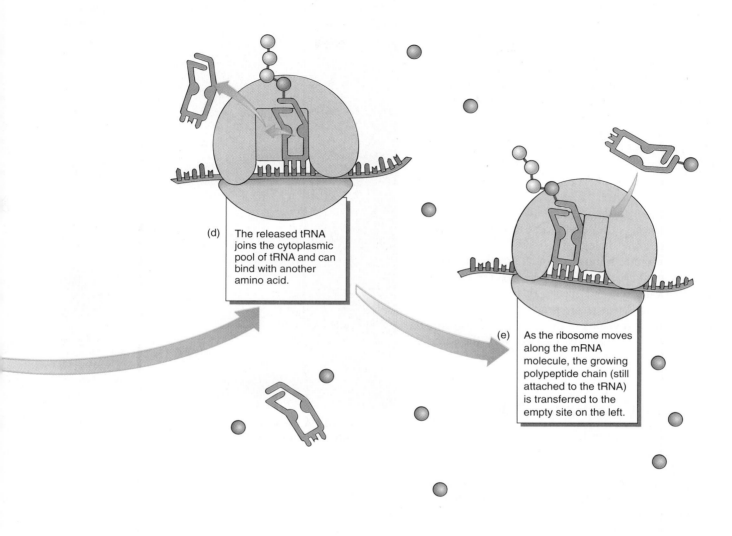

(d) The released tRNA joins the cytoplasmic pool of tRNA and can bind with another amino acid.

(e) As the ribosome moves along the mRNA molecule, the growing polypeptide chain (still attached to the tRNA) is transferred to the empty site on the left.

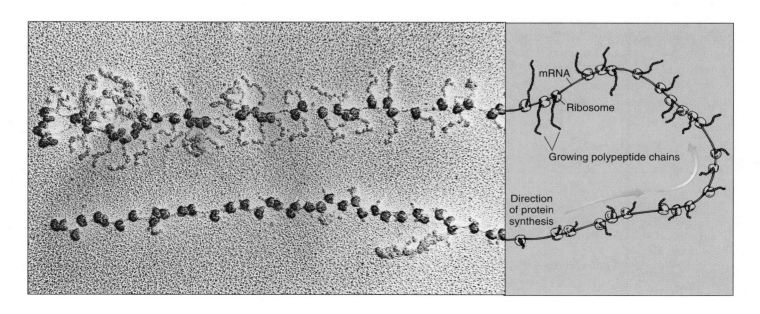

mRNA

Ribosome

Growing polypeptide chains

Direction of protein synthesis

The fact that DNA is the molecule of inheritance in virtually all organisms is evidence that all life is related. Moreover, when two species are closely related, their DNA has a greater proportion of identical sequences of nucleotide bases. Thus, a daisy and a tulip have significant stretches of base sequences that are identical because they are both flowering plants and they shared a common ancestor in the not-so-distant past. On the other hand, a yeast (a single-celled fungus) and a tulip do not have very many identical stretches of bases in their DNA. Yeasts and tulips are very different organisms because they have evolved for hundreds of millions of years along different lines of descent. The fact that they share even a few identical base sequences, however, indicates that they had a common ancestor in the very distant past.

An important aspect of the genetic code is that it is nearly universal throughout the biological world. The same DNA triplets code for the same amino acids in all organisms, with only a very few minor variations (Figure 12-10). The types of protein assembled vary from one species to another, and this variation accounts for the specific traits that make each species separate. The universality of the genetic code is compelling evidence that all organisms evolved from the same early life forms.

MUTATIONS ARE CHANGES IN DNA

What would happen if, during DNA replication, a base not complementary to the template base were placed in the newly forming strand? What if an entire sequence of bases from one DNA molecule moved and became part of another portion of the molecule or of a different DNA molecule? Such a change would be transmitted to all the DNA strands replicated from the strand in which it originally occurred. Thus, all the descendants of that cell would share its genetic difference.

Such accidental changes in genetic information, called **mutations**, do occur. Mutations are not necessarily harmful; some are harmless and some are actually beneficial. Indeed, mutations are the raw material for evolution and sometimes permit new, advantageous traits to arise. These instances are rare, however, because of the already highly adapted nature

Figure 12-10   A gene from a firefly was genetically engineered into this tobacco (*Nicotiana tabacum*) plant. The gene codes for an enzyme that catalyzes the reaction in fireflies that produces light. The tobacco plant was able to express the firefly gene and produce light—a dramatic example of the universality of the genetic code. (*From D.W. Ow, et al.*, Science *234:856–859, 1986. © 1986 by the AAAS*)

of organisms. A random change in the genetic material is unlikely to improve the function of a cell or an organism.

Every organism ever investigated has been shown to be subject to mutation. The wonder is that we do not observe more mutations than we do. Fortunately, cells possess enzymes whose function is to repair or remove damaged sections of DNA. As a result, most mutations are not passed on to other cells.

STUDY OUTLINE

I. The basic plan of every organism and its cells is contained in its DNA, which is located in the nucleus.
  A. DNA is packaged into chromosomes in a highly organized way.
  B. DNA associates with proteins called histones to form chromatin.

II. The structure of DNA has been elucidated.
  A. Watson and Crick worked out the basic structure of DNA in the 1950s.
  B. The DNA molecule is a double helix, with each strand composed of nucleotides. The two strands are joined by hydrogen bonds between complementary bases.

C. Each nucleotide consists of deoxyribose (a sugar), a phosphate group, and an organic base. There are four bases: adenine pairs (that is, forms hydrogen bonds) with thymine, and cytosine pairs with guanine.

III. DNA replicates itself by semiconservative replication, in which each double helix contains an old strand and a newly synthesized strand.
   A. DNA strands unwind during replication.
   B. DNA synthesis takes place at replication forks, Y-shaped regions where the two strands of DNA separate and where DNA synthesis occurs on both strands at once.
   C. DNA polymerase catalyzes the reaction that connects nucleotides in the proper sequence (by complementary base pairing) to form a new strand.

IV. DNA contains instructions for inherited characteristics.
   A. A gene is a sequence of nucleotides in a DNA molecule. Most genes code for specific polypeptides (proteins).
   B. DNA directs protein synthesis through an intermediary, RNA. Thus, the flow of genetic information is from DNA to RNA to proteins.
   C. Transcription and translation are the two processes involved in the use of the information encoded in the nucleotide sequence of DNA to specify the sequence of amino acids in polypeptides.

V. In transcription, one DNA strand serves as a template for the synthesis of a complementary RNA molecule. In eukaryotes, transcription occurs in the nucleus.
   A. RNA polymerase catalyzes the synthesis of RNA on a DNA template.
   B. Three kinds of RNA are transcribed and participate in protein synthesis.

   1. Messenger RNA (mRNA) carries the coded genetic message that specifies the amino acid sequences of a polypeptide from the nucleus to a ribosome.
   2. Transfer RNA (tRNA) molecules carry specific amino acids to the ribosome for attachment to the growing polypeptide chain.
   3. Ribosomal RNA (rRNA) is a structural part of ribosomes.

VI. Genetic information is stored in sequences of three nucleotide bases called triplets in DNA, codons in mRNA, and anticodons in tRNA.
   A. The base sequence in a DNA triplet is complementary to the base sequence in an mRNA codon.
   B. The base sequence in an mRNA codon is complementary to the base sequence in a tRNA anticodon.

VII. In translation, a polypeptide is synthesized according to instructions encoded in mRNA. Translation occurs at the ribosome.
   A. The ribosome attaches to one end of mRNA and begins reading the mRNA where a start codon instructs it to.
   B. Amino acids are carried to the ribosome by tRNA molecules.
   1. Each tRNA molecule attaches to a specific amino acid and carries it to the ribosome.
   2. The ribosome recognizes the anticodon of the tRNA molecule and allows it to base-pair with the mRNA codon.
   3. The amino acid brought in by tRNA forms a peptide bond with the growing polypeptide chain.
   4. The tRNA molecule is then released, and the process is repeated.
   C. A stop codon prompts the mRNA, ribosome, and polypeptide to separate.

SELECTED KEY TERMS

anticodon, p. 215
chromatin, p. 207
chromosome, p. 207
codon, p. 214
DNA polymerase, p. 209
double helix, p. 208

gene, p. 207
genetic code, p. 212
messenger RNA (mRNA), p. 213
mutation, p. 218
nucleotide, p. 207

polypeptide, p. 210
ribosomal RNA (rRNA), p. 213
RNA polymerase, p. 214
semiconservative replication, p. 209

totipotency, p. 206
transcription, p. 212
transfer RNA (tRNA), p. 213
translation, p. 212
triplet, p. 212

REVIEW QUESTIONS

1. Sketch a short section of DNA and label the following parts: double helix, nucleotide, sugar-phosphate backbone, and base-pair rungs. Which base pairs are complementary in DNA?
2. How does the molecular structure of DNA enable it to form exact copies of itself during replication?
3. What is meant by semiconservative replication?
4. Contrast DNA replication with transcription.
5. Compare and contrast the structures of DNA and RNA.
6. Summarize the roles of mRNA, tRNA, and rRNA in protein synthesis.

7. What is the relationship among triplets, codons, and anticodons?
8. What is the logical basis of the fact that the unit of the genetic code is a triplet rather than a single base or pair of bases?
9. Explain the evolutionary significance of the fact that almost all organisms possess the same genetic code.
10. Match these terms with the appropriate letter in the diagram: translation, replication, transcription.

(a) DNA $\xrightarrow{\text{(b)}}$ RNA $\xrightarrow{\text{(c)}}$ protein

---

THOUGHT QUESTIONS

1. What characteristics must a molecule have if it is to serve as genetic material?
2. How is the molecular structure of DNA uniquely adapted to its function as hereditary material?
3. A researcher analyzes the DNA in a dandelion and determines that guanine makes up 20 percent of the bases in the DNA. What is the percentage of cytosine? Of adenine? Of thymine?
4. What would happen to evolution if DNA were always transmitted precisely from generation to generation with no mutations?

5. The following sequence of triplets in DNA codes for which amino acids in a polypeptide chain? (*Hint:* You may use Table 12-2 to help you answer the question, but you must do something *before* using the table.)

   DNA: -TGT-TTT-TCA-GGT-CTA-

SUGGESTED READINGS

Darnell, J.E., Jr., "RNA," *Scientific American,* October 1985. Considers the role of RNA in the translation of nucleic acid information into protein.

Felsenfeld, G., "DNA," *Scientific American,* October 1985. Discusses the structure and organization of DNA.

Radman, M., and R. Wagner, "The High Fidelity of DNA Duplication," *Scientific American,* August 1988. How the cell proofreads for genetic mistakes.

Rennie, J., "DNA's New Twists," *Scientific American,* March 1993. Gene expression is not as simple as it has been presented in this chapter. The more we learn about how DNA functions, the more intriguing the DNA molecule becomes.

Watson, J.D., *The Double Helix,* New York, Atheneum, 1968. Watson's personal account of the experience of making a Nobel Prize–winning discovery.

MITOSIS AND MEIOSIS

The autumn crocus (*Colchicum autumnale*) is a small perennial herb that produces light purple flowers in late summer and autumn. A member of the lily family, the autumn crocus is native to Europe, where it grows in subalpine meadows. It is also widely cultivated in temperate areas as an ornamental and for the chemical colchicine, which is obtained from its seeds and corms.

Colchicine blocks cell division by interfering with the normal function of the spindle, a special structure that aids in the movement of chromosomes during cell division. In cells treated with colchicine, chromosomes duplicate, but they cannot separate and move to the opposite ends of the cell. As a result, cell division stops and the cell ends up with an extra set of chromosomes.

Colchicine is used in plant breeding to induce **polyploidy**, a condition in which cells have more than two sets of chromosomes. Polyploidy is also a naturally occurring phenomenon that is common in many plant species but rare in most animal species. In general, plants are relatively tolerant of extra chromosome sets—30 to 80 percent of all flowering plants are polyploids—whereas extra chromosome sets are often lethal in animals. Polyploid plants often tend to be larger and more vigorous than diploid plants and may have other desirable qualities, such as larger fruits or extra flower petals.

Colchicine can be used to induce polyploidy in cells destined to become gametes—that is, in eggs or sperm cells.

The autumn crocus (*Colchicum autumnale*) produces the powerful drug colchicine, which blocks cell division. (*Patti Murray*)

When gametes with extra chromosome sets unite, they give rise to a new generation of polyploid plants. Thus, plant breeders use colchicine to improve plants by inducing changes in the number of chromosomes.

Colchicine is of great commercial importance, especially in connection with the development of polyploid varieties of ornamental and agricultural plants. For example, colchicine was used to produce the polyploid condition in triticale, a wheat-rye hybrid that has wheat's high protein content and rye's ability to withstand cold and drought. Before treatment with colchicine, triticale plants were sterile; after colchicine treatment, which made them polyploids, they were fully fertile.

In this chapter we examine how deoxyribonucleic acid (DNA) in chromosomes is distributed into daughter cells during cell division. The most complex part of cell division is the division of the nucleus into two nuclei. There are two kinds of nuclear divisions, mitosis and meiosis. *Mitosis,* which results in daughter cells with the same number of chromosomes as the parent cell, is the basis of growth and repair of tissues and of asexual reproduction in plants and other eukaryotes. *Meiosis* is sometimes called reduction division because it reduces the chromosome number by half. Meiosis is required before sexual reproduction can occur and is largely responsible for the variations observed among the offspring of sexual reproduction.

EACH SPECIES HAS A CHARACTERISTIC CHROMOSOME NUMBER

In plants and all other eukaryotes, genes are found on *chromosomes,* threadlike structures in a cell's nucleus that are visible under the microscope only during cell division. You may recall from Chapter 12 that each chromosome consists of proteins and a single large molecule of DNA that contains hundreds or thousands of different genes.

The number of chromosomes within a nucleus varies from one species to another, but every somatic (body) cell of every organism of a given species contains a characteristic number of chromosomes. Human cells have 46 chromosomes, for example, and cabbage cells have 20. A desert-adapted daisy (*Haplopappus gracilis*) has only four chromosomes per cell, and the adder's-tongue fern (*Ophioglossum reticulatum*) has 1262 chromosomes per cell. No simple relationship exists between the number of chromosomes and the size or complexity of an organism.

In somatic cells of plants and other complex organisms, each chromosome occurs as a member of a pair. The 20 chromosomes of a cabbage cell, for example, occur in 10 pairs. The chromosome pairs in a cell are different enough in size and shape that biologists can count and identify them. The two members of a given pair of chromosomes are referred to as **homologous chromosomes**. They carry information governing the same *traits,* although their information is not necessarily the same.

For example, the members of a pair of homologous chromosomes might both carry a gene that specifies flower color, but one chromosome might call for red petals while the other calls for white petals. Homologous chromosomes carry different information because one of each pair originally came from the sperm cell and the other from the egg during sexual reproduction.

A cell in which each chromosome occurs in pairs is called **diploid**, or **2n**. In cabbages, the diploid number is 20 chromosomes, or 10 pairs. Human somatic cells, with 46 chromosomes, are also diploid and have 23 pairs of chromosomes per cell.

The somatic cells of many, but not all, organisms are diploid. A cell that has a single set of unpaired chromosomes is called **haploid**, or **n**. The haploid chromosome number is half of the diploid number. In humans, for example, male and female reproductive cells are haploid ($n = 23$).

THE CELL CYCLE DESCRIBES THE ACTIVITIES OF CELLS THAT ARE GROWING AND DIVIDING

In somatic cells that are capable of dividing, the **cell cycle** is the period from the beginning of one division to the beginning of the next division. The cell cycle may be represented as a circle (like the face of a clock) and includes three main events: interphase, mitosis, and cytokinesis (Figure 13-1). The time period between two successive divisions, represented by a complete revolution of the circle, is the generation time. The generation time varies a great deal from one species to another and from one cell type to another but is usually about 1 day in most plant cells.

One of the main unanswered questions about cell division is how the events of cell division are regulated. This question is of more than passing interest. If biologists can determine the factors that control when cells divide, they may be able to determine how those factors go awry in cancer, a disease in which cell division appears to be uncontrolled.

Cell division involves two main processes, mitosis and cytokinesis. **Mitosis** is the division of the nucleus, and **cytokinesis** is the division of the cytoplasm to form two cells. Before a eukaryotic cell divides by cytokinesis, it nucleus *must* undergo mitosis, a complex division that precisely distributes a complete set of chromosomes to each daughter nucleus. It is by means of mitosis that each new cell contains the identical number and types of chromosomes that were present in the original parent cell.

For the most part, mitosis and cytokinesis take place in localized areas of the plant body called **meristems** (from the Greek *meristo,* meaning "divided"). Meristems occur in the shoot and root tips (the **apical meristems**) and, in some

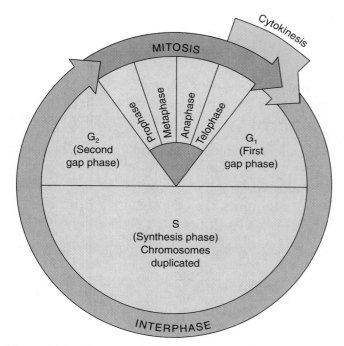

Figure 13-1 The cell cycle. The time required for each stage varies with cell type and species. Most cells spend about 90 percent of their cell cycle in interphase.

plants, in thin cylindrical regions that run the entire lengths of stems and roots except at the tips (the **lateral meristems**) (see Chapter 5). The production and subsequent elongation of new cells by apical meristems causes growing stems and roots to increase in length. Lateral meristems produce additional wood and bark tissues that add girth to—that is, thicken—stems and roots of trees and shrubs. In addition to apical and lateral meristems, plants produce temporary meristems in response to wounding or disease.

Interphase Is the Stage Between Successive Cell Divisions

The cell spends most of its life in **interphase**, a period of active growth and maintenance that precedes mitosis (Figure 13-2a). Because this stage of the cell's life cycle occurs between the phases of successive cell divisions, it is called interphase ("between phases"). During interphase, the cell synthesizes needed materials and grows. Chromosomes undergo duplication during interphase, although it is not readily visible. Then, during mitosis, they condense into visibly separate structures and are distributed to the two daughter nuclei.

MITOSIS

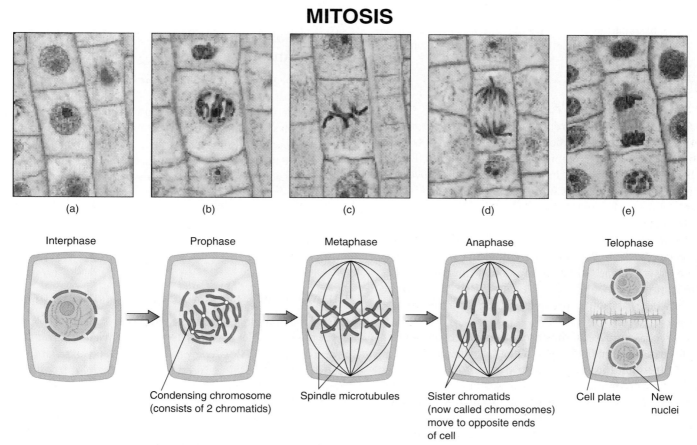

Figure 13-2 Interphase and the stages of mitosis in onion (*Allium cepa*) root tip cells prepared with stains. The diploid number for the cells shown in the diagrams is four. (*Ed Reschke*)

223

Interphase is subdivided into three periods: G_1, S, and G_2 (Figure 13-1). The first period, **G_1**, or the **first gap phase**, is the time between the end of the previous cell division and the beginning of DNA replication. The cell grows during this phase. Late in G_1, the cell synthesizes certain enzymes used in DNA replication. These activities make it possible for the cell to enter the next phase.

The second period of interphase, called the **S phase** (S for "synthesis"), involves the replication of DNA. Other materials, such as proteins that are components of chromosomes, are also synthesized at this time.

Following the completion of the S phase, the cell enters a short **second gap phase**, or **G_2**. At this time, increased protein synthesis occurs as the cell prepares to divide. The beginning of mitosis marks the completion of the G_2 phase.

MITOSIS DISTRIBUTES IDENTICAL CHROMOSOMES INTO TWO NUCLEI

The completion of interphase is signalled by the beginning of mitosis, in which visible changes associated with the division of the nucleus take place. Most other cellular activities, such as protein synthesis, are suspended during mitosis, which is a relatively brief period of the cell's life. Mitosis is divided into four stages: prophase, metaphase, anaphase, and telophase.

Duplicated Chromosomes Condense and Become Visible During Prophase

The first stage of mitosis, **prophase**, begins when chromatin (the threadlike material of which chromosomes are composed) begins to condense and coil into visible chromosomes (Figure 13-2b). In this form, the chromosomes can easily move into position and eventually pass into the daughter cells without becoming hopelessly tangled. Although each chromosome may contain several centimeters of DNA, at mitosis this DNA is condensed into a chromosome that is only 5 to 10 micrometers in length—a 10,000-fold shortening! In early prophase, the nuclear envelope, which is the double membrane that encloses the nucleus, and the nucleolus, the structure in the nucleus where ribosomes are synthesized, break down and disappear.

As prophase proceeds, the chromosomes become shorter and thicker and are individually visible under the light microscope. At this point in mitosis, each chromosome is actually a doubled chromosome (recall that it was duplicated during the preceding S phase) that consists of two identical subunits called **chromatids**. The two chromatids are attached to each other at a constricted region of the chromosome called the **centromere**, whose location is specific for each chromosome and can be at almost any point along the length of the chromosome (Figure 13-3).

During prophase, microtubules (see Chapter 3) organize between the poles (opposite ends of the cell) to form the mitotic **spindle**, a special structure that aids in the movement of chromosomes. Toward the end of prophase, the condensed chromosomes attach to the spindle fibers and move toward the cell's equator, which is midway between the two poles.

During Metaphase, Chromosomes Line Up Along the Cell's Equator

The short period during which the chromosomes are lined up along the equatorial plane of the cell is **metaphase** (Figure 13-2c). The mitotic spindle is now completely visible and can be seen to be composed of numerous fibers that extend from pole to pole.

During metaphase, each chromosome is quite condensed and appears thick and discrete. Because metaphase chromosomes can be seen more clearly than those at any other stage of mitosis, they are typically photographed and studied during this stage. Sometimes scientists compare the physical similarities and differences in chromosomes from cells of two or more species in an effort to understand evolutionary relationships.

Chromosomes Move Toward the Cell's Poles During Anaphase

Anaphase begins with the centromeres of each pair of sister chromatids splitting apart (Figure 13-2d). Each chromatid is now considered an independent chromosome. The separated chromosomes then begin to move toward opposite poles of the cell. (*How* the chromosomes move apart during anaphase is one of the main unanswered questions about cell division.) The chromosomes move rapidly toward the poles, with the centromeres (attached to the spindle fibers) leading the way. The two arms of each chromosome trail behind as the spindle fibers seem to pull on the centromeres. Anaphase ends when complete sets of chromosomes have arrived at opposite ends of the cell.

During Telophase, Two Separate Nuclei Form

During the final stage of mitosis, **telophase**, the chromosomes begin to elongate by uncoiling, becoming invisible chromatin threads (Figure 13-2e). A new nuclear envelope forms around each set of chromosomes, produced at least in part from lipid components of the old nuclear envelope. Nucleoli reappear, and spindle fibers disappear.

Cytokinesis Forms Two Daughter Cells

Cytokinesis, the division of the cytoplasm that usually accompanies mitosis, generally begins at the conclusion of anaphase

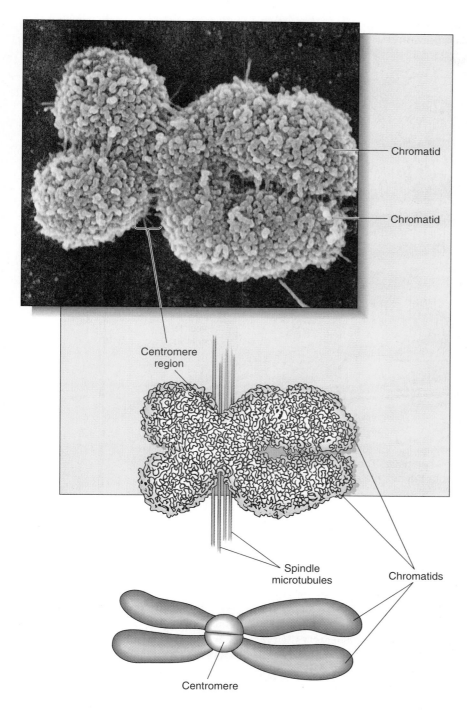

Figure 13-3   A doubled chromosome is composed of two identical sister chromatids attached to one another at the centromere. (*Biophoto Associates/PhotoResearchers, Inc.*)

Chromatid

Chromatid

Centromere region

Spindle microtubules

Chromatids

Centromere

and extends through telophase. In plant cells, cytokinesis occurs through the development of a **cell plate**—a layer that forms between the newly formed nuclei in the equatorial region of the cell. The cell plate grows laterally (sideways) to the edges of the cell, forming two adjacent daughter cells (Figure 13-4). The cell plate is destined to become two new plasma membranes and cell walls that will separate the daughter cells.

Vesicles produced by the Golgi bodies that gather near the equator form the cell plate. The vesicles contain materials to construct both a primary cell wall for each daughter cell and a middle lamella that will cement the primary cell walls together. The vesicle membranes fuse together to become the plasma membrane of each daughter cell.

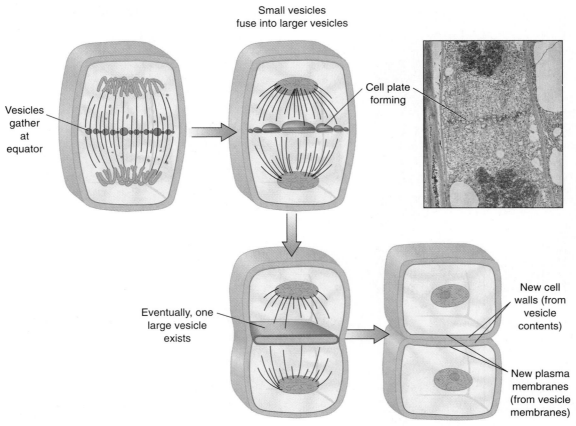

Small vesicles
fuse into larger vesicles

Vesicles
gather
at
equator

Cell plate
forming

Eventually, one
large vesicle
exists

New cell
walls (from
vesicle
contents)

New plasma
membranes
(from vesicle
membranes)

Figure 13-4   Cytokinesis in a plant cell. Stages in the formation of plasma membranes and cell walls from the fusing vesicles of the cell plate. The electron micrograph shows cytokinesis in a maple (*Acer* sp) leaf cell. Note the cell plate, which consists of gathering vesicles. (*T.E. Schroeder/Biological Photo Service*)

SEXUAL REPRODUCTION REQUIRES A DIFFERENT TYPE OF DIVISION: MEIOSIS

In most plants and animals, when certain cells found in the reproductive organs divide, the result is not new somatic cells with a full (diploid) number of chromosomes but cells that become or give rise to reproductive cells, which have *half* the diploid number of chromosomes. These reproductive cells—eggs and sperm cells—are called **gametes**. Instead of becoming part of the body of the organism that produced them, gametes form a complete new organism by uniting with one another. If the female gamete, or egg, unites with the male gamete, or sperm cell, the result is a fertilized egg, or **zygote**, which is the first somatic cell of a completely new organism. This type of reproduction, involving the union of male and female gametes, is called **sexual reproduction**.

Gametes cannot have a full complement of chromosomes, or the zygote resulting from their union would have twice as many chromosomes as it should. A special type of reduction division called **meiosis** reduces the number of chromosomes in reproductive cells by half. As a result of meiosis, each gamete has only one chromosome of each pair, resulting in an egg or sperm cell with a haploid, or *n*, number of chromosomes. When two haploid gametes join in fertilization, the normal diploid number of chromosomes is restored. In this way, meiosis allows the chromosome number of a species to remain the same from generation to generation.

In animals and certain algae, meiosis usually gives rise to haploid eggs and sperm cells directly. In plants, however, meiosis results in haploid **spores**, which are reproductive cells that give rise to new individuals without first fusing with another cell.

HAPLOID CELLS ARE PRODUCED FROM DIPLOID CELLS DURING MEIOSIS

Remember that in mitosis each daughter cell receives exactly the same number and kind of chromosomes that the parent cell had. In contrast, in meiosis the number of chromosomes is reduced by half. Thus, each new cell that results from meiosis is a haploid cell; it has n, not $2n$, chromosomes.

Unlike mitosis, in which pairs of homologous chromosomes end up in *each* daughter cell, the process of meiosis *separates* the members of each homologous pair of chromosomes. As a result, any contrasting genetic traits on each homologous pair of chromosomes are separated and distributed independently to different gametes or spores. The outcome of this distribution method is the likelihood that no two offspring of the same parents will be exactly alike.

Meiosis consists of two cell divisions, logically named the first and second meiotic divisions, or simply meiosis I and meiosis II (Figure 13-5). During **meiosis I**, the members of each homologous pair of chromosomes separate and are distributed into separate nuclei in two daughter cells. The separation of homologous chromosomes that occurs during meiosis is like shuffling a deck of cards—one of each pair is randomly "dealt" to each new cell. These chromosomes were duplicated prior to meiosis I, so each consists of two chromatids. In **meiosis II**, the chromatids separate into individual chromosomes and are distributed into different haploid daughter cells. Thus, what began as one diploid cell gives rise after meiosis to four haploid cells. Meiosis I and meiosis II each include stages comparable to the four stages of mitosis, namely prophase, metaphase, anaphase, and telophase.

Since it is easier to follow these events in an organism that possesses only a few chromosomes, we will discuss a plant with a diploid number of only four chromosomes (two homologous pairs).

Homologous Chromosomes Segregate into Different Daughter Cells During Meiosis I

As in mitosis, the chromosomes are duplicated during interphase before meiosis actually begins. Recall that when a chromosome is duplicated, it consists of two chromatids joined by their centromeres.

During prophase I (prophase of the first meiotic division), while the chromatids are still elongated and thin, the homologous chromosomes arrange themselves lengthwise side by side (Figure 13-5a). This pairing of homologous chromosomes is called **synapsis**. Since the diploid number in our example is four chromosomes, at synapsis we would see two homologous pairs. One of each pair, the maternal chromosome, was originally contributed by the egg, whereas the other member of each pair, the paternal chromosome, was contributed by the sperm cell. Since each chromosome is duplicated (in effect, doubled) at this time so that it actually consists of two chromatids, synapsis results in the coming together of *four* chromatids (2 homologous chromosomes × 2 chromatids per chromosome = 4 chromatids).

All the genes on a particular chromosome are said to be *linked* and tend to be inherited together. However, this tendency for linked genes to stay together is not absolute. During synapsis, genetic material may be exchanged between chromatids of paired homologous chromosomes, a process called **crossing over**. During crossing over, pieces of the maternal and paternal chromatids break off and are then precisely rejoined to the opposite chromatid (Figure 13-6). The exchange of genetic material between homologous chromosomes is a form of genetic **recombination**. The new combinations of genes that result from recombination greatly enhance the variety among offspring of sexual partners.

While synapsis and crossing over are occurring, other events also take place. During prophase I, a spindle forms and the nuclear envelope and nucleolus disappear.

During metaphase I, homologous chromosomes line up *in pairs* along the equator of the spindle (Figure 13-5b). During anaphase I, each pair of homologous chromosomes separates, with one chromosome moving toward one pole and its homolog moving toward the other pole (Figure 13-5c). The chromatids of each doubled chromosome are still united at their centromere regions.

In telophase I in this example, there would be two doubled chromosomes, one of each homologous pair, at each pole (Figure 13-5d). During telophase I, the nuclei often reorganize, the chromatids elongate, and cytokinesis generally takes place. Note that the haploid number of chromosomes ($n = 2$) has been established, although each chromosome is still doubled (that is, still consists of two chromatids).

During Meiosis II, Sister Chromatids (Now Called Chromosomes) Separate and Move to Opposite Poles

During the interphase-like stage that follows meiosis I, no further DNA replication or chromosome duplication takes place. In most organisms this period is very brief, and in some organisms it is absent.

Since the chromatids do not completely elongate between meiotic divisions, prophase II is also brief. In prophase II there

(Text continues on page 230)

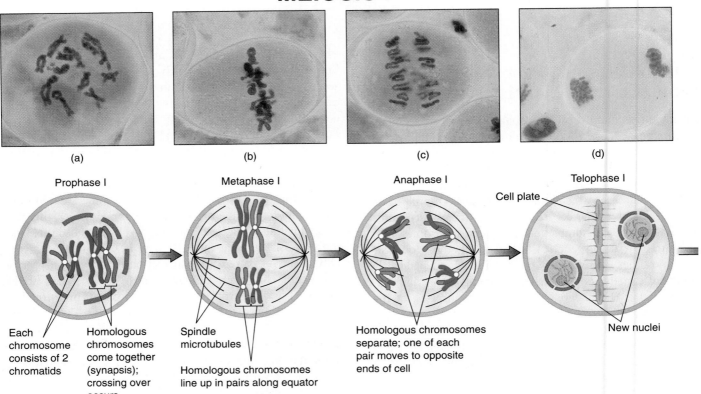

Figure 13-5   The stages of meiosis in Easter lily (*Lilium longiflorum*) cells prepared with stains and flattened on slides. The diploid number for the cells shown in the diagrams is four. (*Clare A. Hasenkampf/Biological Photo Service*)

Figure 13-6   A pair of homologous chromosomes during ▶ **late prophase I of meiosis. (a)** Note the four chromatids (Ch) that make up the paired homologous chromosomes in this photomicrograph. The centromeres are visible at Cm. Crossing over produces the configurations shown at each X. **(b)** Crossing over results in the exchange of DNA between homologous chromosomes. (*Courtesy of J. Keezer*)

MEIOSIS II

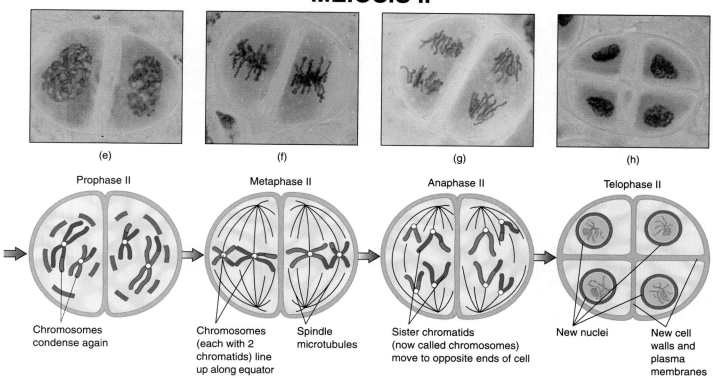

(e)   (f)   (g)   (h)

Prophase II

Chromosomes condense again

Metaphase II

Chromosomes (each with 2 chromatids) line up along equator

Spindle microtubules

Anaphase II

Sister chromatids (now called chromosomes) move to opposite ends of cell

Telophase II

New nuclei

New cell walls and plasma membranes

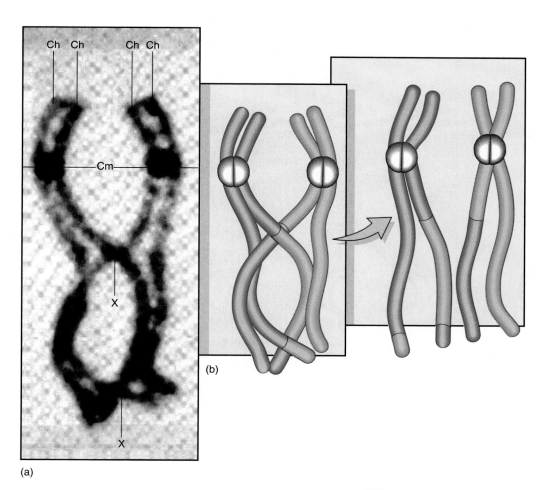

Ch   Ch   Ch   Ch

Cm

X

X

(a)

(b)

is no pairing of homologous chromosomes, because only one chromosome of each pair is in each new daughter cell; also no crossing over occurs (Figure 13-5e). New spindle fibers form, and the nuclear envelope, if it reformed during telophase I, breaks down.

During metaphase II, the chromosomes again line up on the equator, each doubled chromosome consisting of two sister chromatids (Figure 13-5f). Metaphase I and II can be distinguished from each other because in metaphase I there are four chromatids (of two homologous chromosomes), whereas in metaphase II there are two chromatids of a single chromosome.

During anaphase II, the centromeres that join the sister chromatids split (Figure 13-5g). The sister chromatids, each now considered a chromosome, separate and move to opposite poles.

Thus, in telophase II there is one member of each homologous chromosome pair at each pole of the cell, and each chromosome is in an unduplicated state. Nuclear envelopes

then form around each set of chromosomes, the chromosomes gradually elongate into threadlike chromatin, and cytokinesis occurs (Figure 13-5h). The two divisions of meiosis result in four haploid daughter cells from a single diploid parent cell.

A COMPARISON OF MITOSIS AND MEIOSIS REVEALS IMPORTANT DIFFERENCES

Although the events of mitosis and meiosis are somewhat similar, there are several important differences (Figure 13-7). Mitosis and meiosis differ in the number of divisions. In mitosis there is only one division, typically yielding two daughter cells. In meiosis there are two successive divisions, typically producing a total of four cells.

Each daughter cell produced in mitosis contains the diploid number of chromosomes. (An exception to this occurs in the plant life cycle, as we will discuss shortly.) Each of the

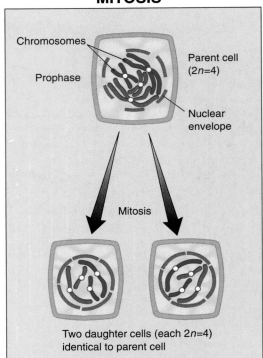

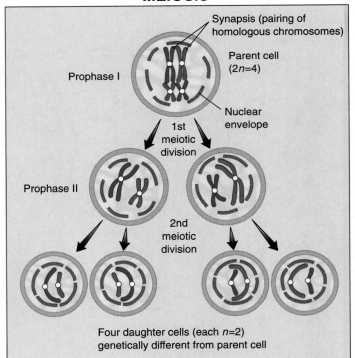

Figure 13-7 Mitosis compared with meiosis. The diploid number for each cell is four. **(a)** Mitosis. Note that the two daughter cells have identical sets of four chromosomes (two pairs), which is the diploid number. (The chromosomes in the daughter cells are in an unduplicated state, whereas those in the original parent cell are duplicated.) **(b)** Meiosis. Two divisions take place, giving rise to four daughter cells. Each daughter cell is haploid and has only two chromosomes, one of each pair.

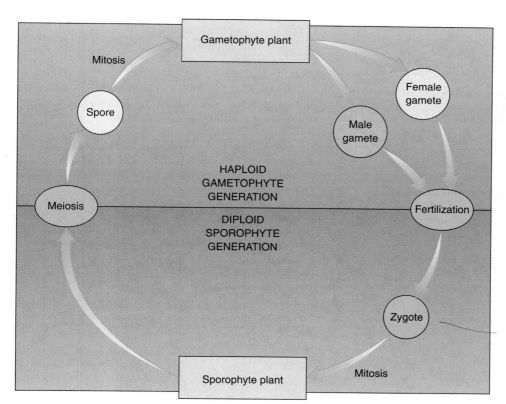

Figure 13-8 The life cycle of plants involves an alternation between diploid and haploid stages.

(handwritten annotation: one-celled stage)

four cells produced in meiosis contains the haploid number of chromosomes—that is, only one member of each homologous pair.

After mitosis, each daughter cell contains a set of chromosomes that is identical in every way to that of the parent cell. In contrast, the homologous chromosomes containing genetic information from each parent (that is, from egg and sperm cell) are thoroughly shuffled during meiosis, and one chromosome of each pair is randomly distributed to each new cell. The resulting haploid cells possess new combinations of chromosomes and therefore unique combinations of genes.

In mitosis there is little opportunity for an exchange of genetic material between homologous chromosomes (called crossing over). Crossing over, however, is an important part of meiosis, which means that the genes originally located together on one chromosome do not always stay together. Thus, crossing over further increases the genetic shuffling of meiosis.

PLANTS HAVE A LIFE CYCLE THAT INVOLVES AN ALTERNATION OF HAPLOID AND DIPLOID GENERATIONS

Plants and other organisms, such as certain algae and fungi, have a life cycle in which the diploid plant known as a **sporophyte** produces haploid spores by meiosis (Figure 13-8). Each spore divides by mitosis to give rise to a multicellular plant known as a **gametophyte**, whose cells *all* contain the haploid number of chromosomes. This haploid organism then produces haploid gametes (eggs and sperm cells) by *mitosis*. The gametes of plants are haploid, just as animal gametes are, but unlike animal gametes, they are not the immediate products of meiosis. After the gametes unite during fertilization, a new sporophyte is produced. We say more about alternation of generations in Chapters 22 to 25.

STUDY OUTLINE

I. Each somatic cell of every organism of a given species has the same characteristic number of chromosomes.
  A. In the somatic cells of diploid organisms, chromosomes are present in pairs, called homologous chromosomes.
  B. A cell with chromosomes occurring in pairs is diploid, or 2*n*. A cell with only one member of each pair of homologous chromosomes is haploid, or *n*.

II. The cell cycle is the period from the beginning of one cell division to the beginning of the next division.
  A. During interphase, the cell grows and prepares for the next division, and DNA replicates.
  B. Interphase can be divided into the first gap phase (G_1), the synthesis phase (S), and the second gap phase (G_2).

III. During mitosis, a complete set of chromosomes identical to the parent set is distributed to each daughter cell.
   A. During prophase, chromatin condenses into chromosomes, the nucleolus disappears, the nuclear envelope breaks down, and the mitotic spindle begins to form. At the end of prophase, each doubled chromosome is composed of two sister chromatids.
   B. During metaphase, the doubled chromosomes line up along the equator of the cell.
   C. During anaphase, the sister chromatids separate and move to opposite poles of the cell; each chromatid is now considered a separate chromosome.
   D. During telophase, a nuclear envelope forms around each set of chromosomes, nucleoli reappear, the chromosomes lengthen and become chromatin, and the spindle disappears.
   E. During cytokinesis, which generally takes place in telophase and therefore overlaps mitosis, the cytoplasm divides and two individual cells form.
IV. In meiosis, two successive cell divisions give rise to four haploid cells. Meiosis must occur at some time in the life of a sexually reproducing organism if gametes (eggs and sperm cells) are to be haploid.

   A. During prophase I, each pair of homologous chromosomes undergoes synapsis (pairing up) and crossing over, during which segments of homologous chromosomes are exchanged.
   B. Homologous chromosomes line up in pairs along the equator during metaphase I, separate during anaphase I, and are distributed to different daughter cells during telophase I.
   C. During the second meiotic division, the two chromatids of each doubled chromosome separate, and one is distributed to each daughter cell.
   D. As a result of meiosis, four haploid cells form.
   E. In sexual reproduction, a haploid set of chromosomes from a sperm cell and a haploid set of chromosomes from an egg unite to form a diploid zygote.
V. Plants alternate haploid and diploid generations.
   A. A diploid sporophyte plant forms haploid spores by meiosis.
   B. A spore divides mitotically to form a haploid gametophyte plant, which produces haploid gametes by mitosis.
   C. Two gametes then fuse to form a diploid zygote, which divides mitotically to produce a diploid sporophyte.

SELECTED KEY TERMS

cell cycle, p. 222
cell plate, p. 225
centromere, p. 224
chromatid, p. 224
crossing over, p. 227
cytokinesis, p. 222

diploid (2n), p. 222
gamete, p. 226
gametophyte, p. 231
haploid (n), p. 222
homologous chromosomes, p. 222

meiosis, p. 226
mitosis, p. 222
polyploidy, p. 221
recombination, p. 227
sexual reproduction, p. 226

spindle, p. 224
spore, p. 226
sporophyte, p. 231
synapsis, p. 227
zygote, p. 226

REVIEW QUESTIONS

1. Distinguish among chromatin, chromosome, and chromatid.
2. What is the relationship between genes and chromosomes?
3. Briefly summarize what occurs during the three phases of interphase (G_1, S, and G_2).
4. Define the following terms: diploid, haploid, and homologous chromosomes.
5. Draw and label the stages in mitosis.

6. Define cytokinesis, and distinguish between mitosis and cytokinesis.
7. Describe the stages in meiosis. Indicate at which stages synapsis, crossing over, and separation of homologous chromosomes occur.
8. What is the consequence of crossing over?
9. How does meiosis differ from mitosis?
10. What is the *immediate* product of meiosis in plants? In animals?

THOUGHT QUESTIONS

1. When during the life of a cell are chromosomes visible? Why are they visible at this time?
2. Explain why two very different organisms might have cells with the same number of chromosomes.
3. Why does mitosis take so much longer than cytokinesis?
4. How does genetic recombination aid in the long-term survival of a species?

5. The developing egg cell of a plant with a diploid chromosome number of 20 was treated with colchicine. This egg was subsequently fertilized by a normal sperm cell. How many chromosomes would you expect to find in the zygote?

SUGGESTED READINGS

McIntosh, J.R., and K.L. McDonald, "The Mitotic Spindle," *Scientific American*, October 1989. A review of what is known about how chromosomes separate during mitosis.

Murray, A.W., and M.W. Kirschner, "What Controls the Cell Cycle," *Scientific American*, March 1991. Discusses the universal regulators involved in controlling the cell cycle.

CHAPTER 14

PATTERNS OF INHERITANCE

Our modern knowledge of genetics, the branch of biology that deals with inheritance, is based on the work of a 19th-century Augustinian monk, Gregor Johann Mendel (1822–1884). Mendel lived in the town of Brünn, Austria, which is now Brno in the Czech Republic. In Mendel's time, inheritance was thought to be the result of a blending of two parents' traits, and there was no clear-cut understanding of the special role of reproductive cells in inheritance.

Mendel changed all that with a series of elegant experiments involving the garden pea (*Pisum sativum*). Native to the Mediterranean region, the garden pea is an annual plant cultivated during cool seasons for its edible seeds. It grows best when the temperature is 13°C to 18°C (55°F to 64°F). The garden pea has a climbing stem, which grows as tall as 1.8 meters (6 feet) and bears fruits that are pods, each containing two to ten seeds.

There are many varieties of garden peas that exhibit differences in height, flower color, seed coat color, and seed shape. In his experiments Mendel studied seven traits, or characteristics: (1) seed shape (round or wrinkled), (2) seed color (yellow or green), (3) flower color (red or white), (4) pod shape (inflated or constricted), (5) pod

Mendel's experiments with the garden pea (*Pisum sativum*) laid the foundations of heredity.
(Carlyn Iverson)

color (green or yellow), (6) position of flower and fruit on the stem (axial or terminal), and (7) stem length (tall or short).

Mendel bred his garden peas for many generations, carefully counting the number of plants that possessed each trait in each generation. His conclusions revolutionized the field of biology and laid the foundation for understanding inheritance in all sexually reproducing organisms, from garden peas to humans.

Mendel's work indicated that inheritance of traits was not due to a blending of genetic information but to the transmission of specific units of inheritance, which are now called *genes*. Mendel predicted not only the existence of genes but that genes normally occur in pairs. His results indicated that, during the formation of gametes (haploid reproductive cells), the two members of a gene separate from one another so that each gamete contains only one of each pair. During sexual reproduction, the pairs are restored when an egg containing one member of a gene combines with a sperm cell containing another member. Thus, Mendel's research also predicted meiosis and the existence of haploid and diploid cells.

LEARNING OBJECTIVES

After reading this chapter, you should be able to:

1. *Define the following terms relating to genetic inheritance: dominant and recessive; homozygous and heterozygous; genotype and phenotype.*
2. *Distinguish among chromosomes, genes, and alleles.*
3. *Solve simple genetics problems involving monohybrid and dihybrid crosses.*
4. *Explain how a test cross is used to determine the genotype of an individual exhibiting a dominant phenotype.*
5. *Define and give an example of linkage.*
6. *Describe how incomplete dominance and polygenes differ, despite the fact that the F_1 (first-generation off-spring) results associated with them may be similar.*

GREGOR MENDEL STUDIED INHERITANCE IN THE GARDEN PEA

Before describing Gregor Mendel's experiments and findings, it is helpful to briefly review the process and structures involved in fertilization in flowering plants, which were discussed in Chapter 9. Like other flowering plants, garden peas produce pollen grains, each of which contains haploid male gametes called sperm cells. When a pollen grain lands on the stigma of a carpel (the female flower part of a garden pea plant), it develops a long tube that grows through the carpel, eventually reaching the haploid egg. The egg and the sperm cell fuse, and an embryonic plant develops from the zygote (fertilized egg). As development continues, a seed forms, which contains a miniature plant and the nutrients it needs to get a start in life.

Unlike most flowering plants, garden peas normally self-fertilize—that is, the male and female gametes that fuse during sexual reproduction are from the same flower. Therefore, the simple surgical removal of the male flower parts makes the plant incapable of being fertilized except by artificial means (Figure 14-1). In this way, the cross fertilization (or, more simply, crossing) of different varieties can be closely controlled.

In a typical experiment, Mendel crossed a tall-growing variety of garden pea with a short, or bush, variety (Figure 14-2). The offspring were not of some intermediate height but instead resembled the tall parent. When two of these offspring were crossed, or when one offspring was allowed to fertilize itself, some of the offspring of the second generation were short. The second generation occurred in a 3:1 ratio—that is, about three tall plants developed for every short one. No matter what trait Mendel studied in garden peas, he obtained the same results. The first generation of offspring always resembled one of the parents, and in the second generation, plants with both traits always appeared in a 3:1 ratio.

These carefully recorded and repeated results led Mendel to conclude that some traits can persist "silently" even though they are not expressed visibly. Mendel therefore concluded

that an individual must possess two sets of instructions for a particular trait, such as height; that is, the instructions must occur in pairs. The 3:1 ratio in the second generation indicated to Mendel that these pairs of instructions combine and recombine in accordance with the rules of probability (probability being the likelihood that a given event will occur).

Mendel was active in the natural history society of Brünn and presented his findings in a series of research reports, which were published in the society's journal. His research was revolutionary, but its significance was not understood or appreciated until the early 20th century, when several biologists independently recognized most of the principles of inheritance and then rediscovered Mendel's papers describing them.

Because the behavior of chromosomes in meiosis and the role of deoxyribonucleic acid (DNA) in cells would not be determined until many decades after Mendel's death, Mendel had no clear conception of the physical basis of inheritance. From his experimental results he inferred that units of inheritance existed, but he could not have said what they were or how they influenced an organism's traits. To Mendel, the unit of inheritance was a mathematical abstraction whose behavior could be described by equations. Perhaps the significance of Mendel's work was not appreciated because most biologists of his time lacked his mathematical background and were not used to thinking in quantitative terms.

INHERITANCE IS GOVERNED BY RULES OF CHANCE

Mendel's breeding experiments led him to certain conclusions about the mechanisms of heredity. Later scholars restated some of those conclusions as **Mendel's principles of inheritance**:

1. Inherited traits are transmitted by genes, which occur in pairs called **alleles**.
2. **Principle of dominance**: When two alternative forms of the same gene (that is, two different alleles) are present

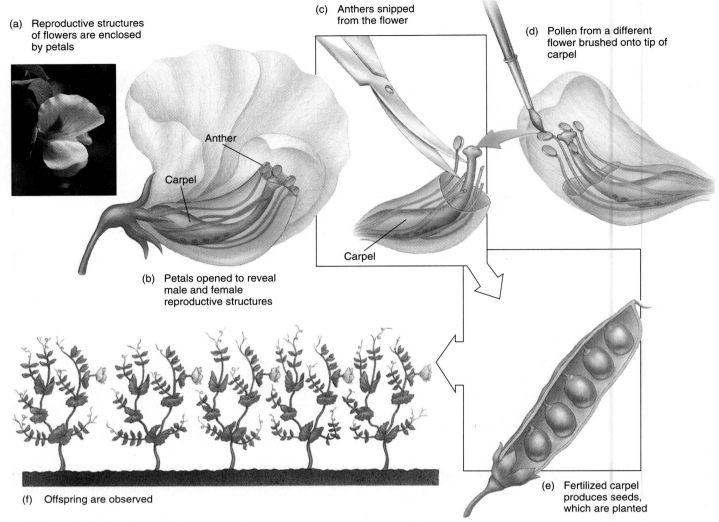

(a) Reproductive structures of flowers are enclosed by petals

Anther

Carpel

(b) Petals opened to reveal male and female reproductive structures

(c) Anthers snipped from the flower

(d) Pollen from a different flower brushed onto tip of carpel

Carpel

(e) Fertilized carpel produces seeds, which are planted

(f) Offspring are observed

Figure 14-1 How garden peas are cross-pollinated. (a) Male and female reproductive structures are enclosed by petals. **(b)** The open flower reveals the pollen-producing anthers and the tip of the carpel, the female part that receives the pollen. Since the petals completely enclose these parts, there is little chance of natural cross pollination between separate flowers. **(c,d)** After the anthers are removed from the flower, pollen from another plant is dusted onto the tip of the carpel. **(e,f)** The seeds produced from this procedure are planted, and the offspring are observed. (*Carlyn Iverson*)

in an individual, often only one—the dominant allele—is expressed.

3. **Principle of segregation**: When gametes form in meiosis, the two alleles of each gene, which are found on homologous chromosomes, *segregate* (separate) from each other, and each gamete receives only one allele for each gene.[1]

4. **Principle of independent assortment:** When two or more traits are examined in a single cross, each trait is inherited without relation to the other traits. This occurs because the alleles for each trait assort into the gametes independently of one another. All possible combinations of genes thus occur in the gametes.

---

[1]Mendel's principle of segregation is the result of the behavior of chromosomes during meiosis.

GENES OCCUR IN PAIRS CALLED ALLELES AND ARE INHERITED AS PARTS OF CHROMOSOMES

A typical gene is now viewed as a region of DNA that contains the information necessary to manufacture a specific polypeptide or protein. (Some DNA, particularly that involved in the control of genes, may not code for protein directly; this DNA usually affects the function of protein-producing genes.) This concept of gene expression can be illustrated with one of the traits Mendel studied in the garden pea, seed shape. Pea seeds are either round or wrinkled, and this trait is related to the type of carbohydrate stored in the seed. Round seeds contain starch, and wrinkled seeds contain sugar. A high sugar concentration is thought to cause seeds to accumulate water dur-

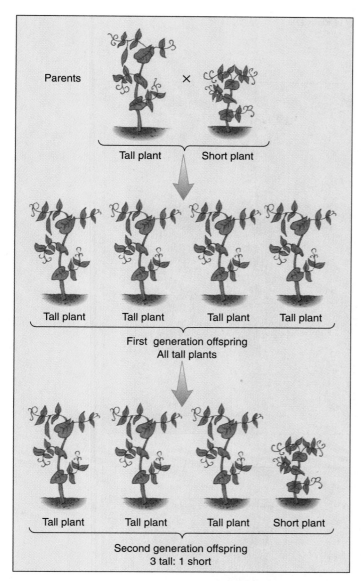

Figure 14-2 One of the crosses carried out by Gregor Mendel, using different varieties of garden pea. When a tall plant and a short plant were crossed, the offspring were all tall plants. When one of the offspring was self-fertilized or crossed with another F_1 plant, its offspring (the second generation) yielded three tall plants for every short one.

Labels within figure:

Parents ×

Tall plant   Short plant

Tall plant   Tall plant   Tall plant   Tall plant

First generation offspring
All tall plants

Tall plant   Tall plant   Tall plant   Short plant

Second generation offspring
3 tall: 1 short

ing development. Then, as they mature and dry out, these seeds become wrinkled.

The allele for round seed is the normal form of the gene. It codes for an enzyme that enables the seed to make starch from sugar. The allele for wrinkled seed codes for a defective enzyme, one that cannot make starch from sugar. Thus, the seed contains sugar and accumulates water, leading to its wrinkled state.

If an individual plant contained a defective wrinkled allele and a normal round allele, the seed would be round due to the presence of starch produced by the functional enzyme. Although the **genotype** (genetic makeup) of such a genetically mixed individual would contain a wrinkled allele, one would not know this from the seed's appearance. The individual's **phenotype**, or observable features (the portion of the genotype that is actually expressed), would be round.

A Dominant Allele Masks the Expression of a Recessive Allele

Recall from Chapter 13 that the two members of a pair of chromosomes are said to be *homologous*. Homologous chromosomes contain genes for similar traits arranged in similar order. The gene for each trait occurs at a particular site in the chromosome called a **locus** (pl. *loci*). In garden peas, for example, if one homologous chromosome contains a gene for pod color at a particular locus, so will the other chromosome of that pair.

Now that we understand that alleles of a gene occur at the same site on homologous chromosomes, we can define allele more precisely. The alternative forms of a gene that govern the same trait and that occupy corresponding loci on homologous chromosomes are **alleles** (Figure 14-3).

The definition of allele implies that there are at least two alternative forms of the gene that can occupy a specific locus on homologous chromosomes. For bookkeeping purposes, each of these forms is assigned a letter as its symbol. It is

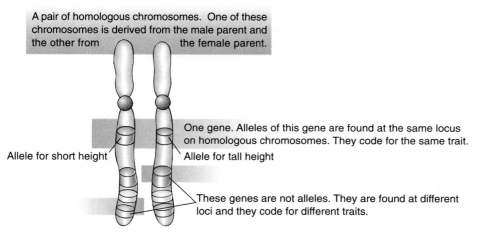

A pair of homologous chromosomes. One of these chromosomes is derived from the male parent and the other from the female parent.

One gene. Alleles of this gene are found at the same locus on homologous chromosomes. They code for the same trait.

Allele for short height   Allele for tall height

These genes are not alleles. They are found at different loci and they code for different traits.

Figure 14-3 Homologous chromosomes, genes, and alleles. Chromosomes occur in pairs in diploid cells. The members of a given pair correspond in shape, size, and type of genetic information and are referred to as homologous chromosomes. For purposes of illustration, each chromosome is shown in the unduplicated state. A gene is a region of DNA on a chromosome that contains the information necessary to manufacture a molecular cell product (that is, a polypeptide). Alleles are different forms of the same gene that are located on homologous chromosomes.

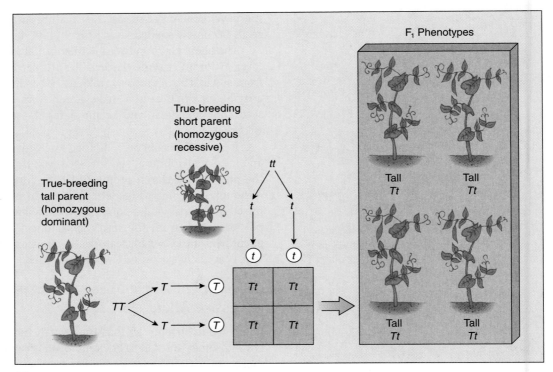

(a)

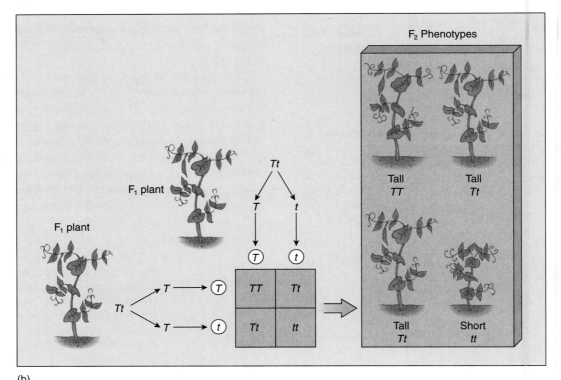

(b)

Figure 14-4   A monohybrid cross. A Punnett square is used to determine all possible genotypes in the offspring. **(a)** When a genetically pure tall garden pea is crossed with a genetically pure short garden pea, all the offspring are tall. **(b)** When two individuals of the F_1 generation are crossed, the F_2 generation is produced. The phenotypic ratio of the F_2 is 3:1.

customary to designate the **dominant** allele—the one that always manifests itself—with a capital letter, and the **recessive** allele—the allele that does not express itself in the presence of a dominant allele—with the same letter but lowercase. Thus, to specify the gene that determines height in garden peas, the letter *T* could be used to represent the allele for tall plant and the letter *t* the allele for short; the tall allele is dominant, whereas the short allele is recessive.

A Monohybrid Cross Involves a Single Pair of Alleles

Genetic terms and some of the basic rules of genetics can be illustrated by a simple **monohybrid cross**, which is a cross between two individuals in which only one trait is studied. Figure 14-4a shows the crossing of a genetically pure tall garden pea (designated *TT*) with a genetically pure short garden pea (designated *tt*). The two *tt* alleles separate during meiosis, so that each gamete produced by the short plant has only one *t* allele. In the formation of gametes in the tall plant, the *TT* alleles separate, so each gamete has only one *T* allele. The union of the *T*-bearing gamete with a *t*-bearing gamete during sexual reproduction results in offspring with the genotype *Tt*, that is, with one allele for tall and one allele for short. As you would expect, these offspring are all tall.

Suppose that two tall plants, each having alleles for both tall and short (*Tt*), are crossed (Figure 14-4b). After meiosis occurs, half of the gametes produced by each plant have a single allele for tall (*T*), and the other half have a single allele for short (*t*).

The probable combinations of gametes may be represented in a diagram called a **Punnett square** (Figure 14-4). The Punnett square shows all possible combinations of gametes to form offspring. In a Punnett square, the types of gametes from one individual are written across the top of the grid, and the types of gametes from the other individual appear along the left side. The squares, which are filled in with the resulting combinations of gametes, indicate the genotypes of all possible offspring from this cross.

The generation with which a particular genetic experiment is begun is called the **parental generation**, or **P**. Offspring of this generation are the **first filial generation**, or **F₁**. When two F₁ individuals are bred, or one self-fertilizes, the offspring constitute the **second filial generation**, or **F₂**.

A Genotype May Be Either Homozygous or Heterozygous

If both alleles specify short (*tt*), then the individual is said to be **homozygous**—that is, it has identical alleles for the same gene. In another garden pea, both alleles may specify tall (*TT*). This second plant is also homozygous. It often happens, however, that one allele carries instructions for tall while the other

carries instructions for short (*Tt*). In this case, the individual is said to be **heterozygous** for height—it contains two different alleles for the same gene. In the heterozygous condition, only the dominant allele is expressed; thus, a garden pea with a genotype designated *Tt* is tall.

To summarize:

1. When an organism is homozygous for the dominant allele, its phenotype reflects the dominant allele. (A *TT* plant is tall.)
2. When an organism is homozygous for the recessive allele, its phenotype reflects the recessive allele. (A *tt* plant is short.)
3. When an organism is heterozygous, the dominant allele is expressed in the phenotype just as it would be if the organism were homozygous for the dominant allele. (A *Tt* plant is tall.) In such cases, one cannot readily tell a heterozygous individual (*Tt*) from one that is homozygous for the dominant allele (*TT*).

All this information permits us to define *dominant* and *recessive* with greater precision. When one allele tends to dominate the other completely, so that it alone is expressed in the heterozygous condition, it is said to be **dominant**. If an allele is expressed *only* when homozygous, it is **recessive**.

A Test Cross Determines Whether an Individual with a Dominant Phenotype Is Homozygous or Heterozygous

There remains the practical question of how we can determine the genotype of an individual that displays a dominant phenotype, since it could be homozygous for the dominant allele (*TT*) or heterozygous (*Tt*). One way to discover the answer would be with a **test cross**, an experimental cross with an individual that is homozygous recessive (*tt*). Suppose, for instance, that a tall garden pea is heterozygous (*Tt*). If one were to cross it with a short plant (which can *only* be homozygous for the recessive short trait, or *tt*), at least some of the offspring should display the recessive trait and be short (Figure 14-5). On the other hand, if the tall plant is homozygous for tall (*TT*), all offspring will be tall.

GENES LOCATED ON DIFFERENT CHROMOSOMES ARE INHERITED INDEPENDENTLY

We now know that Mendel's principle of independent assortment applies only to traits carried on non-homologous chromosomes. If different genes are carried on the same chromosome, they tend to be inherited together (discussed shortly). If genes occur on non-homologous chromosomes, they are inherited independently.

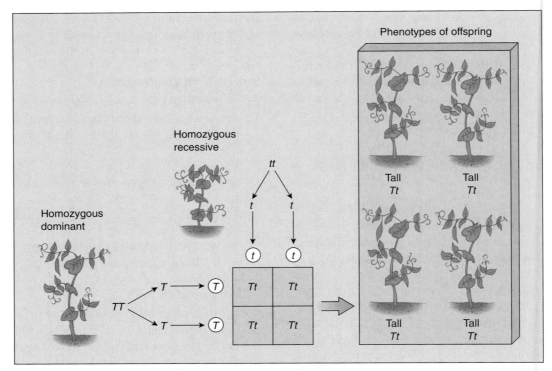

(a)

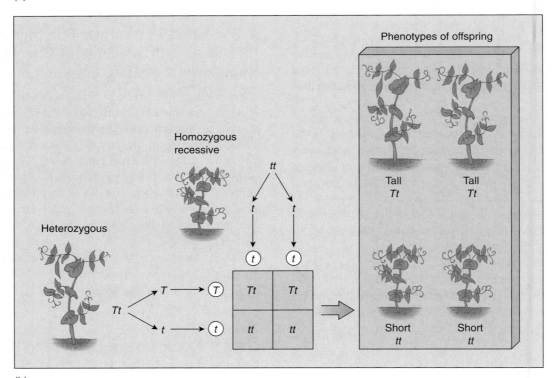

(b)

Figure 14-5   A test cross to determine the genotype of a tall garden pea. (a) If a
homozygous tall garden pea is crossed with a short one, all the offspring are certain to
be tall. **(b)** If any of the offspring are short, the tall garden pea must be heterozygous.

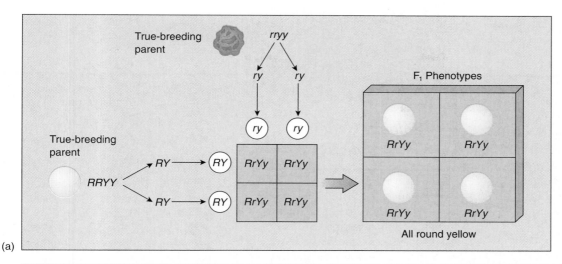

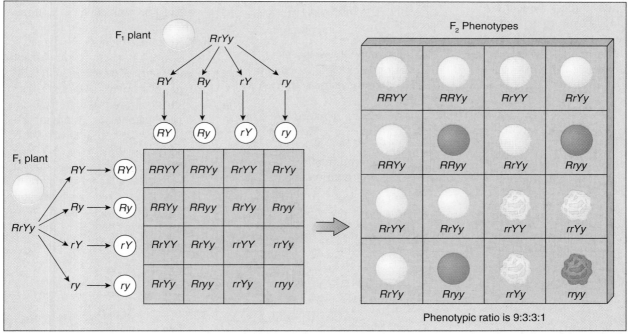

Figure 14-6   Independent assortment is illustrated in this dihybrid cross. In garden peas, the gene for round seed (*R*) is dominant over its allele for wrinkled (*r*), and the gene for yellow seed (*Y*) is dominant over its allele for green (*y*). **(a)** When Mendel crossed a true-breeding plant having round and yellow seeds with a true-breeding plant having wrinkled and green seeds, the seeds produced by the F₁ plants were all round and yellow. **(b)** When two heterozygous individuals are crossed, the ratio of phenotypes is 9:3:3:1 (9 round yellow:3 round green:3 wrinkled yellow:1 wrinkled green).

To illustrate the principle of independent assortment, consider two different garden pea traits, seed shape and seed color. A single gene controls each trait. Round seed is determined by a dominant allele, whereas wrinkled seed is recessive. The allele that results in yellow seed is dominant, and the allele for green seed is recessive.

A cross that involves individuals differing in *two* traits is referred to as a **dihybrid cross**. If a plant homozygous for round seed (*RR*) is crossed with a plant homozygous for wrinkled seed (*rr*), all their offspring (*Rr*) may be expected to have round seeds, since round is dominant and the offspring are heterozygous (*Rr*). Similarly (but quite unrelated), if one of the plants is homozygous for yellow seed (*YY*) and the other is homozygous for green seed (*yy*), the offspring will have yellow seeds (*Yy*). The offspring's genotype for both sets of traits is designated *RrYy* because the offspring are heterozygous for both traits (Figure 14-6a). Since the two traits are the results of genes found on separate chromosome

Focus On
Solving Genetics Problems

In teaching general botany for a number of years, it has been my experience that some students intuitively understand the logic behind solving genetics problems, while many others have great difficulty. This box teaches a methodical approach to solving genetics problems.

Step 1. Read the problem. If it gives you information about the parents and asks you to determine the offspring, go to Step 2. If it gives you information about the offspring and asks you to work backwards to determine the parents, go to Step 3.

Example 1. In garden peas, the allele for inflated pod is dominant over the allele for constricted pod, and green pod color is dominant over yellow pod. The two traits are inherited independently. Determine the genotypes and phenotypes of the offspring produced from a cross between an individual that is heterozygous for both traits and one that is homozygous recessive for both traits.

Example 2. In squash, the allele for disk-shaped fruit is dominant over the allele for round fruit. Two squash plants are crossed, and of their 110 F_1 offspring, 54 have disk-shaped fruits and 56 have round fruits. What are the genotypes and phenotypes of each parent?

In Example 1, you are given information about the parents and asked to determine the offspring, so go to Step 2.
In Example 2, you are asked to work backwards, so go to Step 3.

Step 2. To determine offspring when given information about the parents:

1. Make a key, using capital and lowercase letters to represent the various alleles.

I = Allele for inflated pod
i = Allele for constricted pod
G = Allele for green pod
g = Allele for yellow pod

2. Determine the genotypes of the parents. One parent is heterozygous for both traits—*IiGg;* the other parent is homozygous recessive for both traits—*iigg.* (Remember that each individual has two alleles for each trait.) Therefore, the parents are *IiGg* × *iigg.*

3. Determine the possible gametes produced by each parent. Individual *IiGg* produces four different types of gametes: *IG, Ig, iG, ig.* Individual *iigg* produces only one type of gamete: *ig.* (Remember that gametes contain only one allele for each gene.)

4. Set up a Punnett square. Put all possible gametes for one parent across the top of the square and all possible gametes for the other parent down the side of the square (see below).

	IG	Ig	iG	ig
ig				

pairs, they are inherited (or assorted) independently of each other.

Let us suppose that some of these offspring self-fertilize (*RrYy* × *RrYy*). As you see in Figure 14-6b, four kinds of gametes are possible because each gamete contains one allele for each gene. If, by chance, the chromosome bearing *R* is sorted during meiosis into the same gamete as the chromosome bearing *y*, the resulting gamete will have the genotype *Ry.* Similarly, if the chromosomes bearing *r* and *Y* are assorted together, the gamete will have the genotype *rY.* Other gametes will contain *RY,* and still others *ry. There are no possible com-* *binations other than these four.* Since alleles are always borne on homologous chromosomes that are *separated* from one another in meiosis, a gamete can ordinarily have no more than *one* copy of each allele (for example, *R* or *r* but not both; *Y* or *y* but not both).

Since two parents produce four possible kinds of gametes, working out a cross of that kind requires a Punnett square of 16 boxes (16 different combinations of offspring are possible when these gametes come together randomly). Count the individuals exhibiting each of the four possible phenotypes (round yellow seeds, round green seeds, wrinkled yellow

5. Perform the cross by filling in the boxes to represent the offspring. These four boxes show the genotypes of the offspring. (To keep things straight, place the *I*'s together and the *G*'s together, and put capital letters before lowercase letters.)

6. Finish the problem by listing the phenotypes for the offspring.

	IG	Ig	iG	ig
ig	IiGg	Iigg	iiGg	iigg

	IG	Ig	iG	ig
ig	IiGg	Iigg	iiGg	iigg
	Inflated	Inflated	Constricted	Constricted
	green pods	yellow pods	green pods	yellow pods

Step 3. To determine parents when given information about the offspring:

1. Make a key, using capital and lowercase letters to represent the various alleles.

   D = Allele for disk-shaped fruit

   d = Allele for round fruit

2. List what is known about the genotypes of the parents. Put a blank for each unknown allele. (Remember that each individual has two alleles for each trait.)

   _ _ × _ _

3. Work backwards by filling in the blanks. Because approximately half of the offspring exhibit the dominant trait, you know that at least one of the parents possesses at least one allele *D*.

   D _ × _ _

4. Continue this type of reasoning. Because half of the offspring exhibit the recessive trait, *each* parent must possess at least one allele *d*. Fill in the blanks.

   Dd × _ *d*

5. Calculate a phenotypic ratio for the offspring if it is not given in the original problem. Do this by dividing each number of offspring by the largest number given.

   $$\frac{56}{56} = 1$$

   $$\frac{54}{56} = 0.96$$

   When rounded, the ratio is 1:1; that is, half of the offspring have disk-shaped fruits and half are round (one round to one disk-shaped).

6. Determine which cross would give a 1:1 ratio. Because you already know that the parental genotypes are *Dd* × _ *d*, there are only two possibilities: *Dd* × *Dd* and *Dd* × *dd*. You may be able to determine the answer by simply thinking about the two possibilities. If not, then perform both crosses and calculate the phenotypic ratio expected in each case. The phenotypic ratio for the offspring of *Dd* × *Dd* is 3:1; the phenotypic ratio for the offspring of *Dd* × *dd* is 1:1.

Therefore, the genotypes of the parents are *Dd* × *dd*. Their phenotypes are disk-shaped fruit (for the *Dd* parent) and round fruit (for the *dd* parent).

Additional genetics problems (and their answers) are given at the end of the chapter.

seeds, and wrinkled green seeds). The phenotypic ratio among the F$_2$ offspring is 9:3:3:1; that is, nine offspring produce round yellow seeds, three produce round green seeds, three produce wrinkled yellow seeds, and one produces wrinkled green seeds.

When comparing a monohybrid cross with a dihybrid cross, it becomes clear that, as the number of traits being considered increases, the number of possible genotypes in the offspring multiplies rapidly. Imagine, then, how many different genotypes are possible among the offspring of two parents who differ in hundreds of traits! (Focus On: Solving Genetics Problems provides step-by-step directions for working genetics problems.)

Genetic Linkage Can Be Deduced from the Way in Which Genes Are Inherited

We have seen that genes that occur on non-homologous chromosomes are assorted (and inherited) independently. The inheritance of multiple genes borne on a single pair of homologous chromosomes, however, is very much like that of a *single* gene, if the genes are located close together. The ten-

Figure 14-7   Linkage and crossing over. (a) Initially, genes *A* and *B* are linked on a single chromosome, whereas *a* and *b* are linked on a separate (but homologous) chromosome. **(b)** Crossing over results in the exchange of segments between chromatids of homologous chromosomes. **(c)** Crossing over permits the formation of new combinations of genes in the gametes. For example, note that genes *A* and *b* are now linked in one of the gametes, as are genes *a* and *B*. These linkages would not have been possible without crossing over.

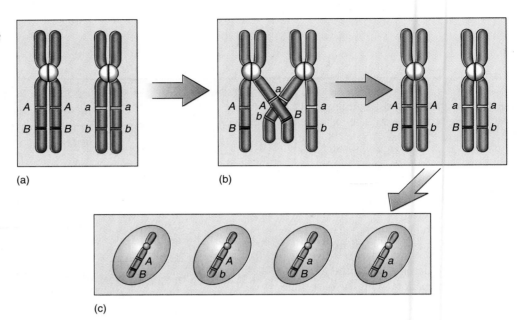

(a)   (b)

(c)

dency for a group of genes on the same chromosome to be inherited together is known as **linkage**. Linkage occurs because genes that happen to occur on the same chromosome tend to remain together during meiosis and thus tend to be inherited together.

Sometimes linked genes are not inherited together, however. By watching the behavior of chromosomes in meiosis, researchers determined that the explanation for the failure of linked genes to always stay together lay in the crossing over of segments of homologous chromosomes. In meiosis, homologous chromosomes, each consisting of two chromatids, come together during prophase I and exchange pieces of themselves during crossing over (Figure 14-7). Crossing over results in new combinations of genes in gametes. Without crossing over, all gametes are either *AB* or *ab*. With crossing over, there are four kinds of gametes (*AB, Ab, aB,* and *ab*). Greater variation in gametes gives rise to greater variation in traits among the offspring of this parent. As you will see in Chapter 16, the evolutionary significance of greater variation is that some of these combinations of genes may improve the chances of survival in the individuals that possess them. (Also see Plants and People: Protecting Genetic Diversity in Crop Plants.)

Genetic Maps Give the Order and Relative Distances Between All Known Genes on the Chromosome

By observing a large number of crosses involving linked genes, one can determine how frequently linked genes stay together in their original combination and how frequently they cross over. This information can be used to estimate the relative distances between linked genes: the farther from one another the loci of linked genes are on the chromosomes, the

more frequently they tend to cross over. Thus, maps of chromosomes can be constructed that show the physical locations of their known genes.

Why is genetic mapping important? In humans, genetic mapping is helping to locate the genes responsible for genetic disorders. This knowledge will assist physicians in their diagnosis and treatment. Genetic mapping of plants is important because precise genetic maps facilitate the development of new varieties of agricultural plants by genetic engineering.[2]

INHERITANCE IS COMPLICATED BY OTHER FORMS OF GENETIC VARIATION

Some traits are not inherited by simple Mendelian rules. Geneticists know of many examples in which an organism's appearance (its phenotype) is the result of complex interactions among genes. These interactions include incomplete dominance and polygenes.

Dominance Can Be Incomplete

Not every gene consists of a dominant allele and a recessive allele. For example, when red-flowered (R^1R^1) and white-flowered (R^2R^2) four-o'clocks are crossed, the offspring do not have red or white flowers (Figure 14-8). Instead, all the F_1 offspring have pink flowers. Does this result negate Mendel's assumptions that inheritance is not a blending phenomenon

---

[2]The mapping of the entire *Arabidopsis* plant **genome** (the complete genetic material carried by an *Arabidopsis* cell) is now under way to determine the chromosome locations and structures of the thousands of genes that this important research plant possesses. Mapping of important agricultural plants, such as corn and tomatoes, is also well advanced.

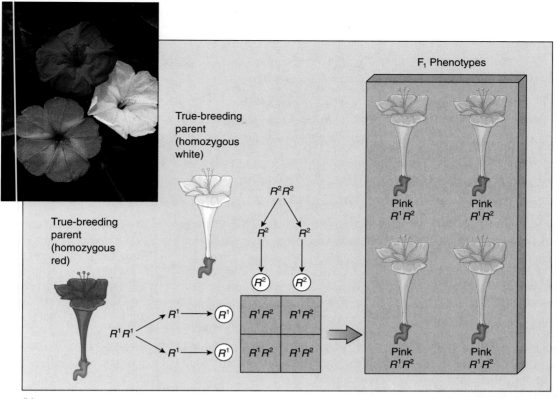

(b)

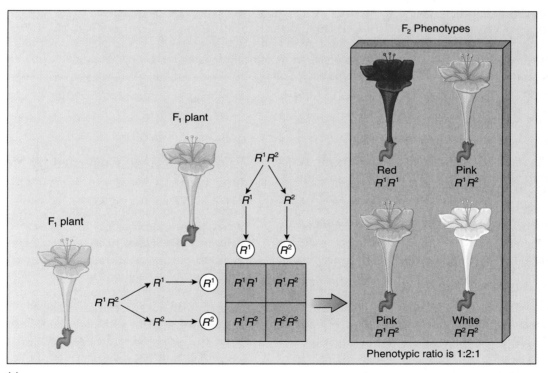

(c)

Figure 14-8   Four o'clocks (*Mirabilis jalapa*). (a) Red, pink, and white four o'clock flowers. The plant gets its common name from the fact that the flowers open in late afternoon. **(b)** Flower color in four o'clocks is controlled by a single gene with two alleles, one for red (R^1) and one for white (R^2) flower color. Red is incompletely dominant to white. A plant with the genotype R^1R^2 has pink flowers. Note that there are alleles only for red and white flowers, not for pink. **(c)** When two pink-flowered plants are crossed, approximately a fourth of the offspring have red flowers, half have pink flowers, and a fourth have white flowers. (*R. Calentine/CBR Images*)

245

PLANTS AND People

Protecting Genetic Diversity in Crop Plants

A plant that is cultivated is protected from natural competition with other plants and from plant-eating animals. The farmer selects the traits that are desirable in a cultivated plant (such as high yield, good flavor, or appealing color) and encourages the transmission of those traits to future generations by selecting the seeds to save for planting. Over time, such selection brings about genetic changes in the plants, and they may become so altered from their original ancestors that it is doubtful that they could survive and compete successfully in the wild. Such plants are said to be **domesticated**.

Plants that exist in the wild often exhibit **genetic diversity**, which means that great genetic variety exists among individuals of a single species. Genetic diversity contributes to a species' long-term survival by providing the variation that enables a population to adapt to changing environmental conditions. When plants become domesticated, much of this genetic diversity is usually lost, because the farmer artificially selects for propagation only those plants with certain marketable characteristics. At the same time, the farmer selects against other traits that may not be of obvious value in

the domesticated plant. As a result, an individual crop variety generally contains a small fraction of the genetic variation inherent in the species.

Despite the fact that domestication reduces genetic diversity, the many farmers who have been selecting for specific traits worldwide have produced an enormous number of varieties of each domesticated plant. Each variety—which is the legacy of hundreds of farmers who have developed it over thousands of years—is adapted to the specific climate where it was bred and contains a unique combination of traits conferred on it by its unique combination of genes.

and that some traits persist silently? Quite the contrary, for when two of these pink-flowered plants are crossed, the F_2 offspring appear in the ratio of one red-flowered to two pink-flowered to one white-flowered plant. If inheritance were a blending phenomenon, all offspring of two pink-flowered plants would be pink.

In this instance, as in other aspects of science, results that differ from those predicted simply prompt scientists to reexamine and modify their assumptions to account for the new experimental results. The pink-flowered plants are clearly the heterozygous individuals in which neither the red allele nor the white allele is completely dominant. When the heterozygote has a phenotype that is intermediate between those of its two parents, the responsible alleles are said to show **incomplete dominance**.

Incomplete dominance is not unique to four-o'clocks. Red- and white-flowered snapdragons also produce pink-flowered plants when crossed. The reason is that the single "red" allele in these plants is unable to code for the production of enough red pigment to make the petals look red.

Some Traits Are Governed by More Than One Gene

Some traits are governed in their expression by many related genes, known as **polygenes**. Because each individual gene in polygenic inheritance makes a small contribution to the organism's phenotype, a trait affected by polygenes exhibits a wide range of variability. The color of wheat kernels is an example of polygenic inheritance, as are height, fruit size, and several other traits in certain plants.

If a wheat plant with dark red kernels is crossed with a plant that produces white kernels, their offspring are light red—that is, intermediate in color. If that were all there was to it, this would be an obvious case of incomplete dominance. To test whether or not kernel color is caused by incomplete dominance, you would need to examine F_2 plants that are the offspring of two plants with intermediate kernel color. If kernel color were the result of simple incomplete dominance, the F_2 plants would exhibit a 1:2:1 ratio among their offspring. In actuality, however, the offspring exhibit a complex

The trend worldwide is to replace the many unique local varieties of a given crop with a few improved varieties. When local farmers stop planting their traditional varieties in favor of the more modern ones, the traditional varieties frequently become extinct. This represents a loss in genetic diversity, because each variety's unique combination of genes gives it a distinctive nutritional value, size, color, flavor, resistance to disease, and adaptability to a specific climate and soil type. The gene combinations in traditional varieties are potentially valuable to plant breeders because they can be transferred to other varieties either by traditional breeding methods or by genetic engineering.

The conservation of plant genetic resources—both wild relatives and locally adapted varieties of important crop plants—is clearly important because the world's food supply depends on it. Plant breeding today is an ongoing process, and breeders are continually developing new varieties of crops in response to changing conditions, such as new insect pests, rapidly evolving disease organisms, changing climates, and new agricultural techniques. Fortunately, many crop varieties are being conserved by agricultural research centers and seed banks (see Chapter 9, Plants and People: Seed Banks). An international effort is also under way to set aside land reserves to preserve the wild relatives of some important crop species where they exist in the natural environment.

distribution of kernel colors. Although the majority of offspring are intermediate in color, other offspring range from white to dark red, and the distribution of phenotypes is not easily expressed by any ratio.

The explanation for this outcome is that kernel color in wheat is governed by at least two genes, each located on a different chromosome. Thus, a crossing between a plant with dark red kernels and one with white kernels may be represented as *AABB* × *aabb* (Figure 14-9). The F$_1$ offspring can be only *AaBb*. What about the F$_2$? The genotypes of the F$_2$ have been worked out for you in the illustration. It is easy to determine their phenotypes: an individual with the genotype *AABB* would have dark red kernels, whereas an individual with the genotype *aabb* would have white kernels; intermediate kernel colors are predicted by counting the *number* of capital letters (represented visually by darkened circles in Figure 14-9) in each genotype (Figure 14-10). For example, an *AABb* would be darker than an *aaBB*.

THE ENVIRONMENT ALSO AFFECTS AN ORGANISM'S PHENOTYPE

We have seen some of the complex ways in which genes interact to determine an organism's phenotype. In addition, the environment in which an individual lives exerts a strong influence over phenotype. For example, consider ear length in corn, which is genetically determined by many genes (that is, by polygenic inheritance). A corn plant may possess a genotype for very long ears, but if it does not have access to adequate light, water, and essential minerals during growth, it will not be able to produce ears as long as those called for by its genotype. Certain aquatic plants provide another example of how the external environment affects gene expression. The water starwort (*Callitriche heterophylla*) produces two kinds of leaves, depending on whether they are underwater or aerial (Figure 14-11). The submerged leaves are long and narrow, whereas the leaves surrounded by air are broad. Thus, environmental conditions affect the expression of the genes that determine leaf shape.

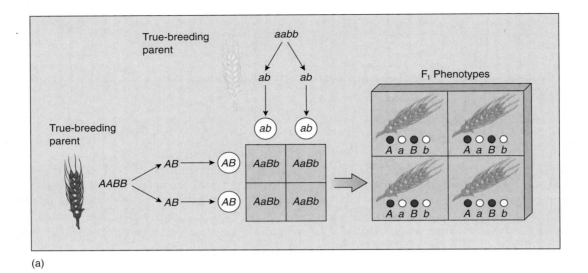

(a)

(b)

Figure 14-9   Wheat kernel color is an example of polygenic inheritance, in which several different genes affect a single trait. Wheat kernel color involves the interaction of at least two separate genes. Here, *A* and *B* represent the genes for dark kernels, and *a* and *b* represent their respective light alleles. **(a)** When a wheat plant with dark kernels is crossed with a plant with white kernels, the offspring are all intermediate in color. **(b)** The expected results in the F_2 generation. The dark dots (signifying the number of dark genes) are counted to determine the kernel colors in the F_2, as shown in Figure 14-10.

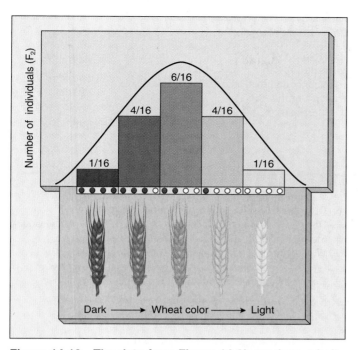

Figure 14-10 The data from Figure 14-9b produce a bell-shaped curve, which demonstrates continuous variation in wheat kernel color, from dark to white, in the F₂. The bell-shaped curve is typical of the distribution of phenotypes in a trait governed by polygenes.

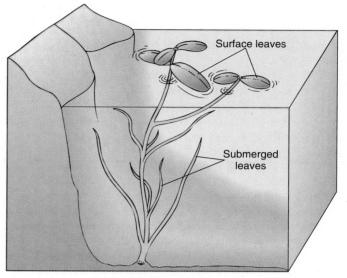

Figure 14-11 Water starwort (*Callitriche heterophylla*) leaves vary depending on the environment to which they are exposed.

STUDY OUTLINE

I. In the 19th century, Gregor Mendel used the garden pea to demonstrate important principles of inheritance. His research founded the science of genetics.
   A. Most inherited traits are governed by one or more genes that are present in each cell. The two members of a gene in a diploid cell are called alleles. Alleles may be alike (homozygous) or unlike (heterozygous).
   B. An allele that is expressed in the heterozygous condition is said to be dominant. An allele that is expressed only when homozygous is said to be recessive.
   C. When gametes form during meiosis, alleles separate, and only one allele for each gene goes to each gamete.
      1. Alleles are carried on homologous chromosomes.
      2. The behavior of alleles in meiosis is similar to the behavior of homologous chromosomes.
      3. Each offspring receives one allele for each gene from each parent.
   D. The member of a pair of homologous chromosomes (and its genes) that a gamete receives is governed entirely by chance. Thus, each trait is inherited without relation to the other traits.
II. A monohybrid cross is a cross between two individuals in which only one trait is being studied.
   A. A cross between two individuals, one homozygous dominant

and the other homozygous recessive, yields F₁ offspring that are all heterozygous.
   B. When two heterozygous individuals are crossed, a 3:1 phenotypic ratio occurs in the F₂ offspring.
III. A dihybrid cross is a cross between two individuals in which two traits are being studied.
   A. A cross between two individuals, one homozygous dominant for both traits and the other homozygous recessive for both traits, yields F₁ offspring that are heterozygous for both traits.
   B. When two heterozygous individuals are crossed, a 9:3:3:1 phenotypic ratio occurs in the F₂ offspring.
IV. Some traits are not inherited by simple Mendelian rules.
   A. Two genes on the same chromosome are said to be linked and tend to be inherited together.
   B. In incomplete dominance, neither allele is completely dominant. A heterozygote exhibits a phenotype intermediate between those of its two parents.
   C. The inheritance of a trait affected by many related genes (that is, by polygenes), in which each individual gene makes a small contribution to the organism's phenotype, is known as polygenic inheritance.
   D. The environment interacts with an organism's genotype to produce its phenotype.

SELECTED KEY TERMS

allele, p. 235, 237
dihybrid cross, p. 241
dominant, p. 239
genotype, p. 237
heterozygous, p. 239

homozygous, p. 239
incomplete dominance, p. 246
linkage, p. 244
locus, p. 237
monohybrid cross, p. 239

phenotype, p. 237
polygenes, p. 246
principle of dominance, p. 235
principle of independent
   assortment, p. 236

principle of segregation,
   p. 236
recessive, p. 239
test cross, p. 239

REVIEW QUESTIONS

1. Distinguish between genes and alleles.
2. State Mendel's principle of dominance.
3. Why can an individual receive only one member of a homologous pair of chromosomes from each parent?
4. State Mendel's principle of segregation.
5. Distinguish between homozygous and heterozygous individuals.

6. Distinguish between dominance and recessiveness.
7. Distinguish between the genotype and phenotype of an individual.
8. State Mendel's principle of independent assortment.
9. Would you need to perform a test cross to determine the genotype of four o'clocks that have pink flowers? Why or why not?
10. Define incomplete dominance and polygenic inheritance.

THOUGHT QUESTIONS

1. Mendel originated the principle of independent assortment on the basis of several crosses involving genes that today are known to be linked. How could the linked genes that he studied have appeared to be inherited independently?
2. If you want to determine whether or not an organism exhibiting a dominant trait is homozygous, you are best advised to cross the unknown organism with an individual homozygous for the recessive allele rather than with one known to be heterozygous. Why?

3. A new species of plant is discovered growing in the mountains of Tibet. Some individuals have blue flowers, and others have red. When a biologist crosses blue-flowered and red-flowered plants, the offspring all have purple flowers. What are the two most likely explanations? How would you determine which explanation is correct?

GENETICS PROBLEMS

1–3. Most of the individuals of a certain wildflower population have white flowers, although a few are purple-flowered. Crosses have demonstrated that the allele for white flower color (W) is dominant over the allele for purple (w).
   1. Give the genotype of a purple-flowered plant, and show the gametes that it would produce as a result of meiosis.
   2. If two heterozygous white-flowered plants are crossed, what fraction of the offspring do you expect to be purple?
   3. If two purple-flowered plants are crossed, what fraction of their offspring do you expect to be white?
4. In one study, Gregor Mendel crossed a yellow-seeded, tall garden pea with a green-seeded, short garden pea. The F_1 offspring were all yellow-seeded and tall. Assuming independent assortment of these two genes, what phenotypes and proportions did he find among the F_2 offspring when the F_1 garden peas were allowed to fertilize themselves?
5–6. Thorn apples produce either purple (P) or white (p) flowers, and their fruits are either spiny (S) or smooth (s). These two genes are located on non-homologous chromosomes.
   5. What is/are the phenotype(s) and genotype(s) of the offspring that result from the cross $PPss \times PpSs$?

6. If the cross $PpSS \times ppSs$ is made, which of the following is/are *not* represented in the offspring?

   (1) *PpSS*   (2) *PpSs*   (3) *ppSS*   (4) *ppSs*   (5) *PPSs*

7–8. In garden peas, the gene for tall (T) is dominant over its allele for short (t), and round seed (R) is dominant over wrinkled seed (r). A homozygous tall, wrinkle-seeded plant is crossed with a homozygous short, round-seeded plant.
   7. What is/are the phenotype(s) of the F_1? The genotype(s) of the F_1?
   8. What is the phenotypic ratio of the F_2 generation?
9. In corn, a single gene controls height, with the allele for normal height (N) being dominant over the allele for short (n). A normal corn plant was crossed with a short plant, but the genotypes of the two plants were not known. Of the offspring, 150 were normal height and 153 were short. Based on the information given, determine the genotypes of the parent plants.
10. Sesame plants produce seed pods with one chamber (O) or three chambers (o), and leaves that are normal (N) or wrinkled (n). The two traits are inherited independently. Determine the

genotypes and phenotypes of the two parents that produced the following offspring: 304 with one-chamber pods and normal leaves; 100 with one-chamber pods and wrinkled leaves; 298 with three-chamber pods and normal leaves; 103 with three-chamber pods and wrinkled leaves.

SUGGESTED READINGS

Diamond, J., "How to Tame a Wild Plant," *Discover,* September 1994. The fascinating story of how our ancestors may have first domesticated wild plants for human consumption.

Heim, W.G., "What Is a Recessive Allele?" *The American Biology Teacher,* February 1991. What is known at the molecular level about the traits that Mendel studied in garden peas.

Kemper, S., "He 'Wrestles' Wild Plants Until They Release His Dreams," *Smithsonian,* August 1994. The interesting account of how a plant breeder (Ronald Parker) develops new varieties of garden plants.

Rhoades, R.E., "The World's Food Supply at Risk," *National Geographic* 179:4, April 1991. Documents the loss of genetic diversity in the world's food crops.

KEY TO GENETICS PROBLEMS

1. *ww;* all gametes would have *w*
2. 1/4 purple
3. None
4. 9:3:3:1 (9 yellow, tall:3 green, tall:3 yellow, short:1 green, short)
5. *PPSs* (purple, spiny), *PPss* (purple, smooth), *PpSS* (purple, spiny), and *Ppss* (purple, smooth)
6. (5) *PPSs*
7. All tall, round (all *TtRr*)
8. 9:3:3:1
9. *Nn × nn*
10. *OoNn × ooNn*

CHAPTER

GENETIC FRONTIERS

One of the world's most important crops is corn (*Zea mays*), a grain crop that was first cultivated thousands of years ago somewhere in South or Central America. Corn is a tall plant with a fibrous root system and long narrow leaves. Because it is an annual, it grows from seed, reproduces, and dies in one growing season. Corn produces male and female flowers on separate parts of the plant. The male flowers are clustered together at the top of the plant (the tassel), and the female flowers are clustered together along the sides of the plant (the ears). Each corn kernel that develops in the ear is a fruit that contains a single seed.

Many different varieties of corn exist. Most of the corn grown for livestock is a starchy feed type that is known as *dent corn* because a dent forms in the top of the grain when the kernel dries. Some of the varieties of corn consumed directly by humans include sweet corn, in which the grains contain a high percentage of sugar, and popcorn, whose small, hard grains explode into soft white masses when heated.

Today corn is cultivated in many countries, including the United States, Canada, Russia, Chile, Australia, and South Africa. It is the world's third most important food crop; only rice and wheat are produced in greater quantities. Much of the corn grown in the United States is cultivated in the Corn

A developing ear of corn (*Zea mays*). Because corn is such an important crop plant, it is the focus of much scientific research, including genetic engineering. (*Carlyn Iverson*)

Belt—Ohio, Illinois, Indiana, Missouri, Nebraska, Iowa, North and South Dakota, and Minnesota.

The cultivation of corn removes many essential minerals from the soil, particularly nitrogen. For every 2500 kilograms (5500 pounds) of grain harvested from a field, 80 kilograms (175 pounds) of nitrogen are removed. To prevent a loss in soil fertility and an accompanying decline in crop productivity, fertilizer must be added to the soil, including *extra* nitrogen fertilizer, because some nitrogen leaches into surface water and groundwater and is lost from the soil.

The expense of applying such large amounts of fertilizer is prohibitive, and farmers in poorer nations cannot afford it. Added to the monetary costs are the water pollution problems associated with applying fertilizer to farmlands. Fertilizer runoff from agricultural land into streams, rivers, and lakes causes excessive growth of algae and aquatic plants, which disrupts the natural balance between aquatic organisms and causes other environmental problems.

The high nitrate levels caused by inorganic fertilizers are dangerous in drinking water, particularly for infants and small children. When nitrates get into the body, they are converted to compounds that reduce the body's ability to transport oxygen. This condition is one of the causes of cyanosis ("blue baby syndrome"), a serious disorder in very young children.

Given the need for fertilizing to grow corn and the problems associated with fertilizers, a variety of corn that made its own fertilizer would be a wonderful boon to humanity. Some food crops, including peas and soybeans, make use of bacteria to fertilize themselves. **Nitrogen-fixing bacteria** live in their roots and convert atmospheric nitrogen to a form that the plants can use. However, attempts to get nitrogen-fixing bacteria to colonize the roots of other crops, such as corn, have not been successful.

Research is under way in many laboratories to transfer genes from a plant such as soybean into corn, with the hope that the gene will be expressed in the corn plant and enable it to provide a suitable environment in its roots for the nitrogen-fixing bacteria. Other scientists are taking a different approach; they want to transfer certain genes from nitrogen-fixing bacteria into corn so that corn can make its own fertilizer directly. The technical challenges involved in such gene transfers are formidable and were thought insurmountable just a few years ago. However, new genetic techniques developed during the last two decades make the dream of self-fertilizing corn seem attainable. This chapter is concerned with these techniques.

LEARNING OBJECTIVES

After reading this chapter, you should be able to:

1. *Outline the primary techniques used in recombinant DNA technology, and explain the actions and importance of restriction enzymes and ligase.*

2. *Identify the role of biological vectors in recombinant DNA technology, and describe a biological vector and a nonbiological method used to introduce genes into plant cells.*

3. *Describe the polymerase chain reaction, and give several potential uses of this technique.*

4. *List at least five potential or realized applications of recombinant DNA technology.*

5. *Summarize the concerns that have been raised about recombinant DNA research and the guidelines that have been formulated to address these concerns.*

HUMANS HAVE BEEN MODIFYING ORGANISMS FOR THOUSANDS OF YEARS

Selective breeding, the breeding by humans of plants with desired traits, has been used for thousands of years to develop better food crops such as corn, apples, and beans. The first farmer who ever decided which seeds to save for the following year's crop was practicing selective breeding. However, selective breeding requires a long time to produce results and does not always have the desired outcome. Corn with the protein content of soybeans, for example, is beyond the capability of selective breeding.

Using organisms to produce products that benefit humanity is known as **biotechnology.** Selective breeding of plants is an example of biotechnology, as is the use of yeast (a single-celled fungus) to cause bread to rise during baking. Recent advances in genetics promise unprecedented control of heredity. When people talk about biotechnology today, they are often using the word in a narrower sense, referring to the manipulation of the genetic material in cells so that new strains of organisms with new characteristics can be constructed (Figure 15-1). Such modification of the DNA of an organism to produce new characteristics is also known as **genetic engineering.**

Figure 15-1   In biotechnology, new combinations of genes produce novel traits in organisms, such as crop resistance to insects. A gene from a common soil bacterium (*Bacillus thuringiensis*) was engineered into cotton (*Gossypium* sp); this gene codes for the production of a protein that acts as a natural insecticide. In a field test, cotton plants were subjected to an insect infestation. The genetically engineered plant (*left*) received little damage and, as a result of being healthier, was able to produce larger bolls of cotton. The unaltered plant (*right*) was weakened by insects and produced much smaller bolls. (*Courtesy of Monsanto Company*)

Genetic engineering differs from selective breeding in several respects. For one thing, genetic engineering enables new strains of organisms to be developed in a fraction of the time required for selective breeding. For example, with traditional breeding methods, it might take 15 years or more to incorporate genes for disease resistance into a particular crop plant. Genetic engineering has the potential to accomplish the same goal in a fraction of that time. Moreover, genetic engineering differs from traditional breeding methods in that it can make use of desirable genes from *any* organism, not just those from the species of plant being improved. For example, if a gene for disease resistance in petunias would be beneficial in tomatoes, the genetic engineer can splice the petunia gene into the tomato plant's DNA to produce a disease-resistant tomato plant. This process could never be carried out by traditional breeding methods, because petunias and tomatoes are different species and cannot interbreed.

THE METHODS USED IN GENETIC ENGINEERING ARE KNOWN AS RECOMBINANT DNA TECHNOLOGY

Recombinant DNA technology, a field of biology that started in the mid-1970s, permits the formation of new combinations of genes by isolating genes from one organism and introducing them into either a similar or an unrelated organism. Using recombinant DNA technology, geneticists can insert DNA not only into simple single-celled bacteria but also into cells of complex organisms. Once inside the cell, this foreign DNA may be expressed. The genetic composition in bacteria may be altered, for example, so that they produce novel proteins such as human growth hormone.

Scientists are not the only ones who practice recombinant DNA technology. Organisms have natural mechanisms for exchanging genes, and such gene transfers have appar-

ently always occurred in nature. For example, bacteria sometimes transfer DNA through a special tube that extends from one cell to another. The transfer of genetic material from one bacterium to another by means of a bacterial virus is another example of natural genetic exchange. However, these natural "experiments" have occurred randomly, and certainly not to satisfy human goals and desires.

Inserting a foreign gene into a cell requires several steps. The gene that is being introduced into a different cell must be (1) isolated from its original cell; (2) used to construct a recombinant DNA molecule; (3) transferred into a new (that is, foreign) cell; (4) identified as having been taken up by the cell; and (5) expressed in that cell. We consider each of these steps in detail.

A Specific Region of DNA Is Isolated from the Cell's DNA

To begin the process of genetic engineering, the investigator must isolate the gene of interest. This is done by first breaking up the cell's DNA into more manageable fragments, using bacterial enzymes known as **restriction enzymes**, which cut DNA molecules at specific base sequences. Restriction enzymes generally recognize a specific DNA sequence four to six base pairs in length. For example, one restriction enzyme recognizes and cuts DNA whenever it encounters the base sequence -AAGCTT-, while another cuts DNA at the sequence -GAATTC-.

Many of these base sequences that are recognized by restriction enzymes read the same from left to right as from right to left; that is, the base sequence of one strand reads the same as the bases on its complementary strand read in the opposite direction. For example, note that the DNA base-pair sequence -AAGCTT- reads the same as the reverse of its complementary sequence, -TTCGAA- (Figure 15-2). When the particular restriction enzyme recognizes the -AAGCTT- sequence, it cleaves the double-stranded DNA between A and

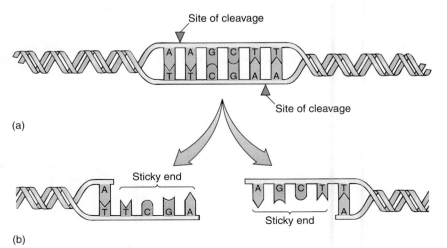

Figure 15-2 Restriction enzymes cleave DNA at specific sites. (a) A portion of a DNA molecule (unwound for simplification) contains the recognition site for a restriction enzyme that recognizes the DNA base sequence -AAGCTT- and its complementary base pairs. Note that -AAGCTT- and its complement, -TTCGAA-, have the same base sequence in opposite directions. **(b)** The cleavage of DNA by this restriction enzyme results in two DNA fragments that have short single-stranded stubs—that is, sticky ends.

A on *each* strand. This action leaves one strand longer than the other on each DNA fragment; the single-stranded stub that extends from the end of each fragment of DNA is called a **sticky end**.

DNA segments from entirely different organisms are potentially able to recombine with one another if they've been broken up by the same restriction enzyme. Each restriction enzyme always cuts the same base sequence, no matter what kinds of DNA it cleaves. For this reason, the sticky ends of the DNA are complementary and can be joined, or spliced, by an enzyme known as **ligase**.

Recombinant DNA Molecules Contain DNA from Two Different Sources

Restriction enzymes split the potential donor DNA into many sticky-ended fragments. The next step is to construct recombinant DNA molecules by combining these isolated DNA fragments with DNA from another source, which acts as a **vector**, or DNA carrier. The vector carries the gene from one organism into the cells of a second organism, where the gene is incorporated into that organism's existing DNA. The most common vector in recombinant DNA technology comes from bacteria.

In addition to having one large circular DNA molecule that contains a bacterium's genetic information, some bacteria have small circular DNA molecules, known as **plasmids**, that carry one or two genes (Figure 15-3). Any plasmids that are present within a bacterial cell are replicated along with the bacterial chromosome, and the copies are distributed to daughter cells.

When plasmids are used as genetic vectors, desired genes are inserted into the plasmids, which are then incorporated into bacterial cells. Each plasmid and its genes (its own genes plus the inserted gene) are replicated every time the bacterial cells divide. As a result, the entire bacterial population descended from the original bacterial cells contains the novel genes in its plasmids.

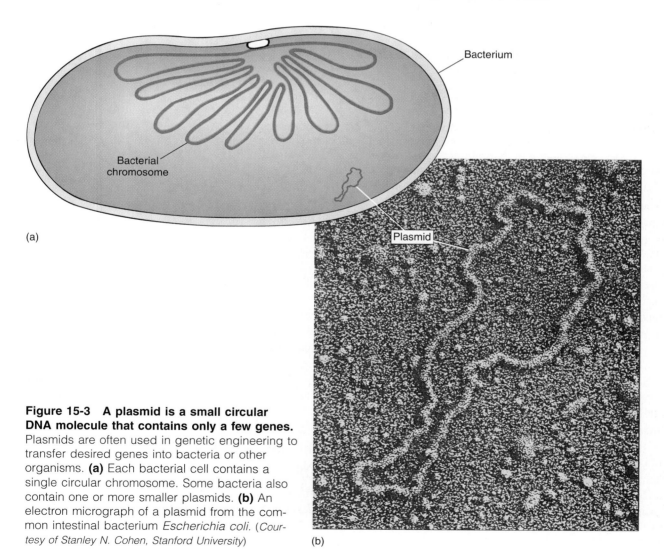

(a)

Figure 15-3 A plasmid is a small circular DNA molecule that contains only a few genes. Plasmids are often used in genetic engineering to transfer desired genes into bacteria or other organisms. **(a)** Each bacterial cell contains a single circular chromosome. Some bacteria also contain one or more smaller plasmids. **(b)** An electron micrograph of a plasmid from the common intestinal bacterium *Escherichia coli*. (*Courtesy of Stanley N. Cohen, Stanford University*)

(b)

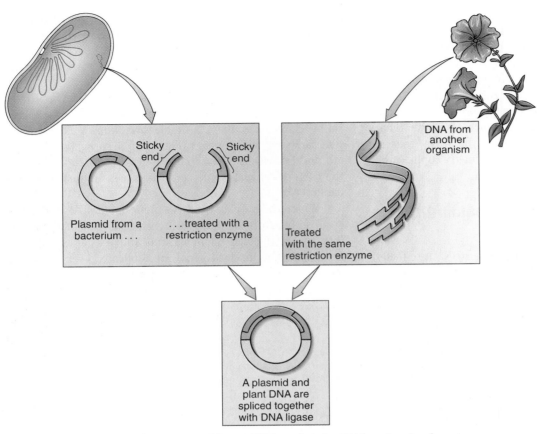

Figure 15-4   Constructing a recombinant DNA molecule. DNA molecules from two different sources are cut with the same restriction enzyme, which results in complementary sticky ends. In this example, one of the DNA molecules is a circular plasmid from a bacterium, and the other is a segment of DNA from a common petunia (*Petunia X hybrida*). When mixed, their sticky ends form base pairs, and ligase splices the two together.

To construct recombinant DNA molecules in which a plasmid is the vector, the investigator treats plasmids isolated from bacterial cells with a restriction enzyme (Figure 15-4). Because the restriction enzyme cuts the plasmids at the base sequence it recognizes, the formerly circular plasmids now become linear molecules of DNA with sticky ends. These plasmid strands are then mixed with sticky DNA fragments from the donor cell. Then, using ligase, the sticky ends of the donor and plasmid DNA fragments are joined into a single circular plasmid.

The result is a mixture of recombinant plasmids, with each recombinant plasmid containing a different fragment of donor DNA. Although the vast majority of plasmids in the mixture do *not* contain the desired gene, a few can be expected to possess exactly what is needed. To identify the plasmid that contains the gene of interest, all the plasmids must be *amplified;* that is, many copies of each must be made so that there will be millions of copies to work with. Amplification takes place inside bacterial cells.

Viruses can also be used to introduce recombinant DNA into cells, and they are commonly used as vectors in mammalian cells, including human cells. Other methods used to introduce foreign DNA into a cell do not involve a biological vector, that is, a bacterial plasmid or virus. One of the most intriguing of these methods is a **gene gun**, which shoots tiny, DNA-coated metallic "bullets" into living cells (Figure 15-5). The gene gun was first used successfully to introduce genes into algal and yeast cells, but it is now used routinely in plant cells and recently has been used successfully in animal cells.

Recombinant DNA Molecules Are Transferred into a New Host Organism's Cell

The recombinant DNA molecules (now existing as plasmids) are next taken into bacterial cells. Under normal conditions, bacteria do not take in plasmids; however, some bacteria receive plasmids under certain conditions, such as when the bacte-

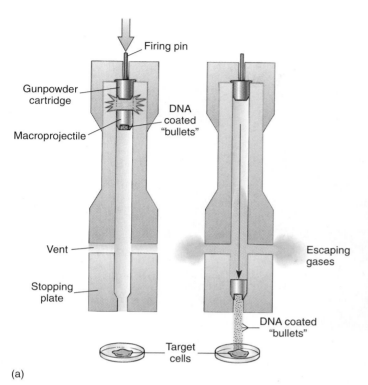

(a)   (b)

Figure 15-5   How a gene gun works. (a) A gene gun fires tiny metallic "bullets" coated with DNA into cells. The macroprojectile that holds the bullets is powered by gas (helium) or by gunpowder. The stopping plate stops the macroprojectile, but the bullets enter the target cells from the momentum. **(b)** A researcher places a dish containing plant cells into the gene gun. (*Courtesy of Dr. John Sanford, Cornell University*)

ria are treated with heat and calcium. Bacteria that take in plasmids are said to be **transformed**.

Because only a small proportion of the bacteria take up the recombinant plasmids even under these special conditions, there has to be a way to determine which bacteria contain the genetically altered plasmids. Fortunately, plasmids often carry genetic information that makes the bacterium resistant to certain antibiotics, such as penicillin and tetracycline, that kill or inhibit bacteria and other microorganisms. If the plasmid used as a vector has a gene for resistance to a particular antibiotic, then investigators can grow, or *culture,* the bacterial population in the presence of that antibiotic. Only bacteria that have taken up the genetically altered plasmids resist the antibiotic and survive; the rest conveniently die (Figure 15-6).

To select the cells with the desired gene from among all other genetically altered cells, the researcher spreads a sample of the bacterial culture on plates of nutrient agar, a solid growth medium. If the sample is dilute enough, each bacterial cell is separated from other cells on the plate. As each cell reproduces, it gives rise to a **colony**, a cluster of millions of genetically identical cells; each bacterium in the colony con-

tains the same recombinant plasmid. Because there may be as many as several thousand colonies, each containing a different portion of foreign DNA, researchers must then identify which colony contains the gene of interest.

Cells That Have Taken up the Gene of Interest Are Identified

A needle could be found in a haystack of any size if one had a magnet or, better, a metal detector. The genetic equivalent of a metal detector can be used to find just the desired DNA base sequence among the vast number of genes in the bacterium's chromosomes. This device, called a **genetic probe**, is a radioactively labeled segment of RNA or single-stranded DNA that is complementary to the target gene.

For example, suppose we want to identify the gene that codes for the protein nitrogenase, an enzyme involved in nitrogen fixation. Because we know the amino acid sequence of nitrogenase, we can synthesize the mRNA that produces it. This mRNA, which is the probe nucleic acid, attaches by complementary base pairing to the DNA that contains the base sequence needed for production of nitrogenase. Because the

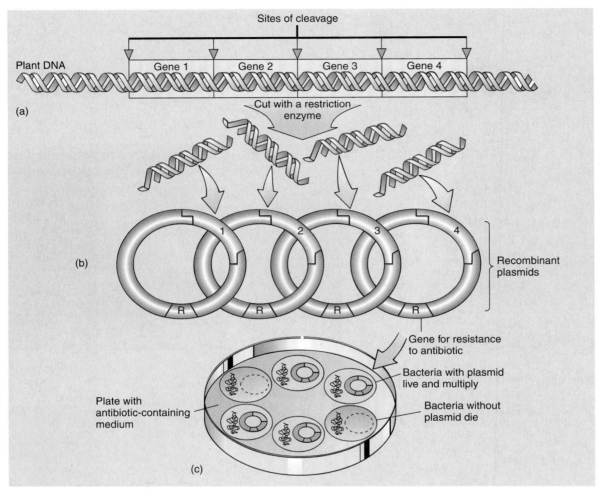

Figure 15-6   Identifying bacteria that have taken up genetically altered plasmids.
(a) DNA from plant cells is cut into multiple fragments with a restriction enzyme.
(b) Recombinant plasmids are formed by cutting plasmids with the same restriction
enzyme, mixing the plasmids with the segments of plant DNA, and treating with ligase.
(c) Because the recombinant plasmids contain a gene for resistance to an antibiotic
(R), bacterial cells that contain the plasmids are resistant to that antibiotic. The bacteria
are then grown on an antibiotic-containing nutrient medium, and only those that contain
the recombinant plasmid survive. (Bacteria and plasmids are not drawn to scale.)

synthesized mRNA is radioactively labeled, it can be detected
by x-ray film. If, after this treatment, any radioactive DNA
is detected in a particular colony of bacterial cells, that is the
colony that contains the gene for the production of the desired protein—in this case, nitrogenase (Figure 15-7).

The Gene Is Expressed Inside the New Cell

Even though a gene has now been identified as being incorporated into certain bacteria, growing the bacterial strain
would not necessarily produce the desired protein. The gene
of interest is not transcribed (and subsequently translated into
a protein) unless it is associated with a set of regulatory genes

that, in effect, turn the gene "on." When these genes are added
to plasmids, the bacterial cells can then transcribe and trans-
late the gene of interest.

Gene Insertion in Plants Is Technically Challenging

Although several vectors for the introduction of genes into
plant cells have been used, the most widely used vector is the
crown gall bacterium (*Agrobacterium tumefaciens*), which
produces plant tumors (Figure 15-8). This bacterium causes
disease by inserting DNA from its plasmid into the plant chromosome. The inserted DNA instructs the plant cells to pro-

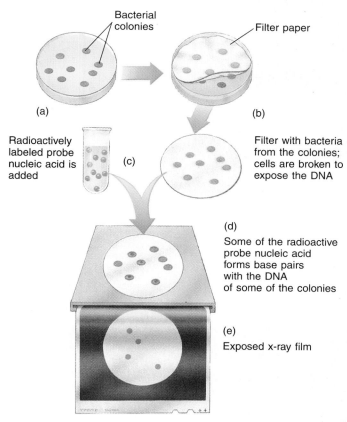

Bacterial colonies

Filter paper

(a)

(b)

Radioactively labeled probe nucleic acid is added

(c)

Filter with bacteria from the colonies; cells are broken to expose the DNA

(d)
Some of the radioactive probe nucleic acid forms base pairs with the DNA of some of the colonies

(e)
Exposed x-ray film

Figure 15-7   Genetic probes can be used to determine which cells have taken up the gene of interest. (a) Bacterial cells are spread on solid nutrient medium so that only one cell is found in each location. Each cell multiplies to give rise to genetically identical descendants that form a colony on the medium. **(b)** A few cells from each colony are transferred to special filters. **(c)** Radioactively labeled probe nucleic acid (in this case, mRNA) is added to the filter; the probe nucleic acid contains a sequence of nucleotides complementary to the gene of interest. **(d)** Some of the probe nucleic acid forms base pairs with the DNA of some of the colonies. **(e)** DNA from bacterial cells that contain the DNA sequence complementary to the radioactive probe becomes radioactive and can be detected by x-ray film. The pattern of spots on the film enables the researcher to identify which colonies from the original plate contain the correct plasmid.

Figure 15-8   Crown gall tumors on the stem of a tomato (*Lycopersicon lycopersicum*) plant. A plasmid carried by the crown gall bacterium (*Agrobacterium tumefaciens*) induces the growth of these tumors. The strain of *A. tumefaciens* used in genetic engineering has been modified so that it contains a plasmid that does not cause tumors. (*Visuals Unlimited/Jack M. Bostrak*)

duce abnormal quantities of plant growth hormones, which result in excessive growth and tumor formation.

It is possible, however, to disarm the plasmid so that it does not induce tumor formation and to incorporate desirable genes in the *A. tumefaciens* plasmid for insertion into a plant host (Figure 15-9). The plant cells into which such genetically altered plasmids are introduced are essentially normal except for the new genes that have been inserted. Genes placed in the plant cells in this fashion are transmitted to the next generation by sexual reproduction (that is, in seeds) or asexual reproduction.

There are several problems with using *A. tumefaciens* plasmids to introduce foreign genes into plants. First, only dicots (the larger of the two classes of flowering plants) are susceptible to this bacterium; important monocots (the smaller class of flowering plants) such as corn, wheat, and rice are outside the host range of the crown gall bacterium. The gene gun (described earlier) is an alternative to using *A. tumefaciens*.

A second problem with gene insertion in plants is that it must use plant tissue culture to regenerate whole plants after individual cells have been genetically transformed. Tissue culture techniques have not been developed for all plants, although more than 1000 species have been regenerated using tissue culture techniques, and the number increases yearly. Recombinant DNA techniques cannot be used for plants in which tissue culture is a problem.

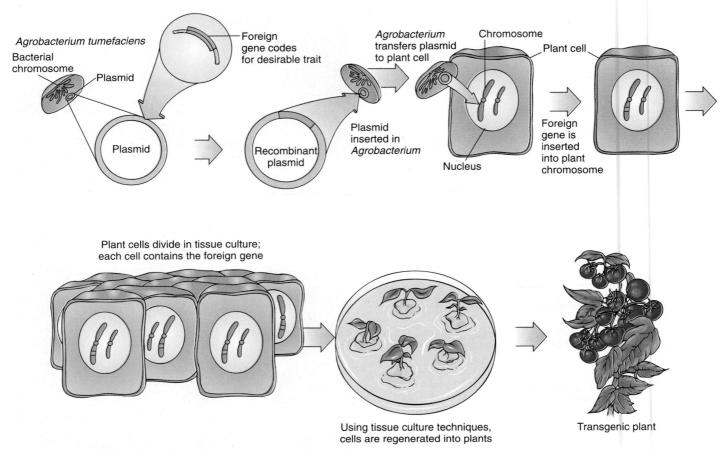

Figure 15-9   The plasmid of *Agrobacterium tumefaciens* can be used to introduce desirable genes from another organism into a plant. A restriction enzyme and the enzyme ligase are used to splice the foreign DNA into the crown gall plasmid. The recombinant plasmid is inserted into *A. tumefaciens,* which is then used to infect plant cells in culture. The foreign gene is inserted in the plant's chromosome. Genetically engineered plants are produced from the cultured plant cells through the use of plant tissue culture techniques.

An additional complication of plant genetic engineering is that some important plant genes are located *in the chloroplasts* instead of the nucleus. Recall from Chapter 3 that chloroplasts are essential in photosynthesis, the process that is the basis of plant productivity. Chloroplast genetic engineering is currently the focus of intense research interest.

DNA CAN BE AMPLIFIED BY THE POLYMERASE CHAIN REACTION

The methods just described involve making large amounts of DNA inside living cells. These techniques are time-consuming and require an adequate starting amount of DNA. Another method, called the **polymerase chain reaction (PCR)**, enables researchers to produce millions of copies of DNA from a tiny sample in just a few hours. PCR has hundreds of applications, such as amplifying small amounts of DNA found at crime scenes, analyzing DNA from fossil leaves, and diagnosing disease, including HIV, the virus that causes AIDS (see Focus On: Using PCR To Identify Mycorrhizae).

The polymerase chain reaction involves subjecting DNA to alternating treatments of heat and enzymatic action. First, a double-stranded DNA molecule is heated to separate it into two single strands. Then the enzyme DNA polymerase assembles nucleotides along each strand, causing the formation of *two* double-stranded DNA molecules. The two double-stranded molecules are then heated, which causes them to separate into four single strands, and a second cycle of replication with DNA polymerase results in *four* double-stranded DNA molecules. After the next cycle of heating and replication, there

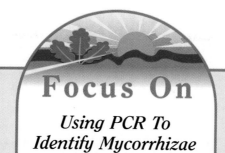

Focus On

Using PCR To Identify Mycorrhizae

The polymerase chain reaction (PCR) has the potential for many applications in basic biological research in addition to genetic engineering. For example, ecologists can use PCR to help survey the individual partners in symbiotic relationships (intimate associations between members of different species). One such relationship, known as a mycorrhiza (pl. *mycorrhizae*), occurs between the roots of forest trees and fungi. Mycorrhizae are widespread in nature. The roots of most, if not all, trees are colonized by such fungi. This association is beneficial for both organisms; the fungus supplies the plant with essential minerals that it has absorbed from the soil, and the plant provides the fungus with carbohydrates and other organic molecules.

Identifying the exact species and strains of fungi found in mycorrhizal associations is often difficult. However, biologists successfully used PCR techniques to amplify DNA from a single mycorrhizal fungus colonizing jack pine roots. They demonstrated that the fungus was a different species from the one with which the jack pine had been thought to form an association.

Knowing exactly which species or strain of fungus forms mycorrhizae with which tree species has a practical value. Because mycorrhizal associations enhance tree growth, deliberately inoculating the roots of young seedlings with the specific fungus that colonizes them would cause forest and nursery trees to grow faster.

---

are *eight* DNA molecules, and so on, with the number of DNA molecules doubling with each cycle.

GENETIC ENGINEERING OF PLANTS HOLDS GREAT PROMISE

The genetic engineering of plants is still in its infancy, and only a few applications have been fully developed by research laboratories and private companies. Nonetheless, the genetic engineering of plants has the potential to transform all aspects of agriculture, making it more efficient and more profitable (see Plants and People: More Food for a Hungry World: Is Genetic Engineering the Final Answer?). Among other applications, genetic engineers are currently working to improve pest and disease control and extend the shelf life of post-harvest crops.

Plants Resistant to Insect Pests Are Being Developed

Plant genetic engineering is typically used to introduce into a crop plant a gene that targets a single insect pest. One method that has met with some success is to introduce a gene for a natural pesticide from the soil bacterium *Bacillus thuringiensis,* known as *Bt* for short. There are many different varieties of the *Bt* bacterium, and each variety produces a specific protein that is toxic to only a small group of insects; none is harmful to humans. For example, the *Bt* variety that works against corn borers would not be effective against Colorado potato beetles. Thus, scientists must identify the *Bt* gene that is most effective against the insect being targeted. Scientists have inserted *Bt* genes into crops such as tomato and cotton plants, making **transgenic** (genetically engineered) plants that are more resistant to pests such as the tomato pinworm and the cotton bollworm, respectively (see Figure 15-1). Such transgenic plants require fewer applications of insecticide, which saves the farmer money and helps the environment. For example, beneficial insects that are killed by insecticides are not affected by the transgenic plants.

Plants Resistant to Viral Diseases Are Being Developed

Traditionally, the only way farmers could control viral diseases in plants was by spraying insecticides to kill the insects that transmit the disease. Because cures are not known for most plant viruses, the development of virus-resistant crops is being actively studied. One approach that has been used successfully is to insert the gene that codes for the virus's coat protein (that is, its outer covering) into the plant cell. For reasons not completely understood, when a plant cell produces the viral protein, the virus cannot infect the cell. When the gene that codes for the tobacco mosaic virus coat protein was inserted into tobacco plants, for example, those genetically altered plants were resistant to the virus. The Chinese have grown such transgenic tobacco plants, with a resistance

PLANTS AND People

More Food for a Hungry World: Is Genetic Engineering the Final Answer?

Haunting photographs of starving African children in the news became commonplace during the 1980s and early 1990s, and most people are upset by the human suffering and misery that such pictures reveal. Positive news about hunger relief did occasionally appear, such as when medical treatment and balanced nutrition resulted in a reversal of nutritional deficiency diseases in certain populations. But it is clear that the specter of famine, which has been with humanity for thousands of years, has not disappeared. Despite advances made in many fields related to food production, numerous people—adults and children—still do not get enough food.

Producing enough food to feed the world's people is the largest single challenge in agriculture today, and the challenge grows more difficult each year as the human population expands. Currently, 1.4 billion metric tons (1.5 billion tons) of grains such as wheat, rice, corn, and barley are required to feed the world's population of 5.8 billion* for 1 year. Each year, an *additional* 28 million metric tons of grain must be produced to account for that year's increase in population.

The food production challenge confronting agriculture is being met in a variety of ways, with genetic engineering being touted by some as the high-technology answer to agriculture's problems. Although it is true that genetic engineering has the potential to revolutionize agri-

culture, the changes will not occur overnight. A great deal of research must be done before most of the envisioned benefits from genetic engineering are realized.

Furthermore, genetic engineering is not the final solution to the challenge of producing enough food. The production of adequate food for the world's people will be an impossible goal until population growth is brought under control. Thus, the ultimate solution to world hunger is related to achieving a stable population in each nation at a level that it can support.

*Estimated human population as of mid-1996.

to the tobacco mosaic virus, commercially since 1993. Similar results have been obtained with a number of plant species, such as grapes (against the grape fan-leaf virus) and the yellow crookneck squash (against the zucchini yellow mosaic virus and the watermelon mosaic virus 2).

Genetic Engineering Has the Potential To Extend the Shelf Life of Harvested Crops

As anyone who has ever gardened will tell you, vine-ripened tomatoes have much more flavor than those that are picked while green, treated with ethylene, and shipped to supermarkets. The reason supermarket tomatoes are not vine-ripened is that the ripening process softens the fruits so that it is difficult to harvest and handle them without damaging them. Some scientists have addressed this problem by trying to suppress fruit softening, usually by blocking the expression of a gene that codes for one of the enzymes responsible for fruit softening. As a result, fruits on these transgenic plants,

Figure 15-10 Genetic engineering delays softening in FlavrSavr™ tomatoes, developed by Calgene, Inc. (*Courtesy of Calgene, Inc.*)

mainpts

Environment

Field Testing of Genetically Engineered Plants

An increasing number of field tests are being conducted on transgenic crops; as of mid-1995 the U.S. Department of Agriculture (USDA) had issued more than 2200 permits for field tests of plants as diverse as alfalfa, cantaloupe, rice, strawberry, and walnut. These and other crops have been engineered to resist insects or diseases or to be more nutritious. Many of the plants in these field trials performed so well that they have demonstrable commercial value. The next step in marketing transgenic plants is deregulation, which means that the USDA will allow them to be grown as any traditional crop is grown. Deregulation of a particular transgenic crop will be approved only if it is established that the crop poses no greater risk than the traditional crop from which it was developed.

One of the main concerns with the development of transgenic plants is the possible risk that they could become invasive, that is, become established in places where we do not want them to grow. Although scientists widely recognized the need for studies to ascertain whether a particular transgenic plant could become an environmental disaster, no such studies were performed until recently. In 1993 a landmark paper appeared in the journal *Nature,** reporting a rigorous field analysis of the invasiveness of a genetically engineered form of oilseed rape, a plant whose seeds yield a food oil (Canola). Populations of transgenic oilseed rape and normal oilseed rape (as a control) were grown for a period of 3 years in 12 very different environments. These habitats included the presence and absence of the following variables: a moist habitat, direct sunlight, grazing animals, plant-eating insects, other vegetation, and disease-causing fungi. This experiment, executed in the United Kingdom, is considered one of the most comprehensive studies ever performed in ecology; it required the collaboration of ecologists with many different specialties.

The results of this experiment demonstrated conclusively that the transgenic oilseed rape poses no risk of invasiveness; it did not grow differently from the control plants under any of the experimental conditions. Although this is good news for the people who wish to market the genetically engineered oilseed rape, the study did much more than provide a green light for the deregulation of transgenic oilseed rape. This study is significant because it shows how such tests should be designed and conducted to determine whether the deregulation of a particular transgenic plant is appropriate.

---

*Crawley, M.J., et al., "Ecology of Transgenic Oilseed Rape in Natural Habitats," *Nature,* Vol. 363, June 17, 1993.

called FlavrSavr tomatoes, can stay on the vine until they turn red and therefore develop more flavor before being picked (Figure 15-10). FlavrSavr tomatoes have been available in U.S. supermarkets since 1994.

SAFETY GUIDELINES HAVE BEEN DEVELOPED FOR RECOMBINANT DNA TECHNOLOGY

Even while acknowledging that the potential uses of recombinant DNA techniques were important and beneficial, many scientists were initially concerned about potential misuses or unpredictable side effects. The fear was that someone might accidentally engineer an organism that would have undesirable ecological or other effects. Such totally new strains of bacteria or other organisms, with which other living things would have no previous experience, might be difficult to control. This possibility was recognized by those who developed the recombinant DNA methods and led them to propose stringent guidelines for making the new technology safe.

Recent history has failed to bear out such worries. Millions of experiments have demonstrated that recombinant DNA experiments can be carried out safely, and no reports of such experiments causing human illness or damage to the environment have surfaced so far.

Scientists recognize the practical importance of recombinant DNA technology and generally agree that the perceived threat to humans and to the environment was overestimated. For example, the engineering of crops resistant to pests is widely perceived as posing a minimal threat to the environment.

In fact, such genetically engineered crops *benefit* the environment because they reduce the use of chemical pesticides.

Although many regulations have been relaxed, strict guidelines still exist in areas of recombinant DNA research in which there are unanswered questions about possible effects on the environment. Much research is currently being conducted to assess the effects of introducing genetically engineered organisms into natural environments. For example, could any of the new gene sequences ("new" in the sense that they are novel to the organism into which they've been engineered) possibly be transferred by natural reproductive processes to closely related species living in the environment? If so, could this cause a previously harmless species to become dangerous? What if genetically altered organisms escaped human control and became established in the natural environment? Could they become "weedy" and overwhelm native organisms? In the next few years, as more data accumulate on the environmental effects of the release of genetically engineered organisms, we will know more clearly whether such effects will be dangerous (see Plants and the Environment: Field Testing of Genetically Engineered Plants on page 263).

STUDY OUTLINE

I. Genetic engineering is the genetic manipulation of cells so that desirable new characteristics can be developed in organisms. Recombinant DNA technology encompasses the methods employed in genetic engineering.

II. Recombinant DNA technology enables investigators to isolate, identify, and manipulate genes.
A. Restriction enzymes cleave DNA at specific sites to break it into more manageable fragments. Each fragment cleaved by a restriction enzyme has sticky ends.
B. Segments of DNA from different sources can be joined by an enzyme called ligase.
C. Many bacteria contain small, accessory, circular DNA molecules called plasmids. These plasmids may be used as vectors (DNA carriers) to transfer genes from one organism to another.
D. The recombinant DNA molecules are taken into a host cell. The cells that take up the specific gene of interest are identified through the use of a genetic probe.
E. The gene may be transcribed and translated within the host cell, leading to the production of a protein not previously produced by the host organism.

III. The polymerase chain reaction (PCR) provides a way to quickly and easily amplify small amounts of DNA and has hundreds of applications.

IV. Recombinant DNA technology is being used to improve crops in a variety of ways—for example, by lessening our need for insecticides to control insect pests, by imparting resistance to viral diseases, and by extending the shelf lives of post-harvest crops.

V. Concerns of scientists who developed recombinant DNA technology led to safety guidelines to make the new technology safe.
A. Millions of recombinant DNA experiments have been performed without any adverse effects on human health or the environment.
B. Strict restrictions still exist for the introduction of genetically engineered organisms into the environment, because many unanswered questions remain.

SELECTED KEY TERMS

biotechnology, p. 253
colony, p. 257
gene gun, p. 256
genetic engineering, p. 253
genetic probe, p. 257

ligase, p. 255
plasmid, p. 255
polymerase chain reaction, p. 260

recombinant DNA technology, p. 254
restriction enzymes, p. 254
selective breeding, p. 253

sticky end, p. 255
transformed, p. 257
transgenic, p. 261
vector, p. 255

REVIEW QUESTIONS

1. What is genetic engineering?
2. What are restriction enzymes? Ligases? How are these enzymes employed in recombinant DNA research?
3. Explain how recombinant DNA molecules are usually constructed.
4. Describe how a foreign gene is implanted in a bacterial cell. What might be gained by this procedure?
5. Explain how a foreign gene might be inserted into a plant cell.
6. What is a genetic probe? How is it used to identify genetically modified cells that contain a specific gene?
7. Briefly describe the PCR method of amplifying DNA.
8. What are some of the benefits of plant genetic engineering?
9. Why does research involving recombinant DNA technology have strict safety guidelines.

THOUGHT QUESTIONS

1. Would genetic engineering be possible if restriction enzymes did not exist? Explain.
2. Think of at least two potential benefits of plant genetic engineering *not* discussed in this chapter.
3. What safeguards could be employed in recombinant DNA technology to guard against potential harm and yet not excessively hamper research?

SUGGESTED READINGS

Arnheim, N., T. White, and W. Rainey, "Application of PCR: Organismal and Population Biology," *BioScience,* Vol. 40, No. 3, March 1990. Some of the biological applications of the polymerase chain reaction.

Diamond, J., "How to Tame a Wild Plant," *Discover,* September 1994. The fascinating story of how our ancestors may have first altered wild plants for human consumption.

Gasser, C., and R. Fraley, "Transgenic Crops," *Scientific American,* June 1992. Methods used to genetically engineer crops.

Schmidt, K., "Whatever Happened to the Gene Revolution?" *New Scientist,* January 7, 1995. Although genetic engineering has gotten off to a slow start, its future is promising.

Stone, R., "Large Plots Are Next Test for Transgenic Crop Safety," *Science,* Vol. 266, December 2, 1994. Large-scale field trials are required for a complete picture of any possible environmental hazards of transgenic plants.

CHAPTER 16

CONTINUITY THROUGH EVOLUTION

Cabbage, broccoli, brussels sprouts, cauliflower, collard greens, kale, and kohlrabi are distinct vegetable crops that are all members of the same species, *Brassica oleracea*. Cabbage, which forms a large head of overlapping leaves around a small terminal bud, is thought to have originated along the northwestern coast of Europe. It is the oldest vegetable in the *Brassica oleracea* group and has been cultivated for centuries.

The edible parts of the broccoli plant are the highly nutritious green flower buds, which form dense clusters on the ends of succulent stems. If not harvested in the immature state, the flower buds of broccoli develop into yellow flowers. Broccoli is probably native to Asia and the eastern Mediterranean region.

Brussels sprouts grow tiny cabbage-like lateral buds along a main unbranched stem. Although named for the city of Brussels, Belgium, where they became popular, brussels sprouts are thought to have evolved from cabbage in the Mediterranean region.

The edible parts of the cauliflower plant are the dense clusters of short-stemmed, immature flower buds. Cauli-

Artificial selection has resulted in many varieties of the same species, *Brassica oleracea*. Shown are cabbage, broccoli, brussels sprouts, cauliflower, collards, kale, and kohlrabi. (*John Arnaldi*)

flower differs from broccoli in that the flowers of cauliflower are mostly sterile. The evolutionary history of cauliflower is unknown, but it probably evolved from broccoli in northwestern Europe.

Collards and kale are closely related leafy plants which are native to Europe. Their edible leaves do not overlap to form dense heads as those of cabbage do.

Kohlrabi is a lesser-known vegetable with a mild flavor reminiscent of turnips. It has an enlarged, fleshy aboveground storage stem that can be eaten cooked or raw. Kohlrabi's origin is unknown. It is not cultivated to a great extent in the United States.

It is appropriate to begin our study of evolution with the seven different vegetable crops of *Brassica oleracea*. All seven were produced by selective breeding of the colewort, or wild cabbage, a leafy plant native to Europe and Asia. In selective breeding (also known as artificial selection), plants with certain desirable characteristics are selected for propagation by humans. Beginning more than 4000 years ago, some breeders artificially selected wild cabbage plants that formed overlapping leaves; over time, these leaves were so emphasized that the plants, which resembled modern-day cabbages, became recognized as separate and distinct from their wild cabbage ancestor. Other breeders emphasized different parts of the wild cabbage, giving rise to the other modifications. For example, kohlrabi developed after selection for an enlarged storage stem. Thus, humans are responsible for the evolution of *Brassica oleracea* into seven distinct vegetable crops.

LEARNING OBJECTIVES

After reading this chapter, you should be able to:

1. *Explain the four premises of evolution by natural selection as envisioned by Darwin.*
2. *Explain how the synthetic theory of evolution differs from Darwin's original theory of evolution.*
3. *Summarize the evidence for evolution from the fossil record, comparative anatomy, biogeography, and biochemistry and molecular biology.*
4. *Describe how scientists make inferences about evolutionary relationships among organisms from the sequence of amino acids in specific proteins or the sequence of nucleotides in particular genes.*
5. *Discuss the opposing views on the pace of evolution: gradualism and punctuated equilibrium.*

THE CONCEPT OF EVOLUTION ORIGINATED BEFORE CHARLES DARWIN

All the organisms on our planet, from microscopic bacteria to coastal redwoods, from tropical tree frogs to desert cacti, evolved from one or a few simple kinds of organisms. This vast diversity of species developed from earlier species by a process Charles Darwin—a 19th-century naturalist considered one of the greatest biologists of all time—called "descent with modification," or evolution. In biology, **evolution** is defined as a genetic change in a population of organisms. (A population is a group of individuals of the same species that live in a particular place at a specific time.) The world "evolution" refers not to the changes that occur to an individual organism during its lifetime but to the changes in populations over many generations.

Although Charles Darwin is universally associated with evolution, ideas concerning the gradual evolution of life predate Darwin by centuries. Aristotle (384–322 B.C.) saw much evidence of design and purpose in nature and arranged all the organisms that he knew in a "Scale of Nature" that extended from the very simple to the most complex. He visualized the Earth's organisms as being imperfect but moving toward a more perfect state. Some have interpreted this idea as a rudimentary concept of evolution, but Aristotle is very vague on the nature of life's "movement toward the more perfect state" and certainly did not propose any notion of evolution by natural selection.

Long before Darwin, fossils—old fragments resembling bones, teeth, leaves, and shells—had been discovered embedded in rocks all over the world. Some of these fossils corresponded to parts of familiar living organisms, but others were strangely unlike any known form. Fossils of marine invertebrates, such as clams and corals, were found in rocks high on mountains. In the 15th century, Leonardo da Vinci correctly interpreted such finds as the remains of organisms that had existed in previous ages but had become extinct.

The most thoroughly considered view of evolution before Charles Darwin was expressed by Jean Baptiste de Lamarck in 1809. Like many biologists of his time, Lamarck thought that all organisms were endowed with a vital force that drove them to evolve toward greater complexity. He also thought that organisms could pass on traits they had acquired during their lifetimes to their offspring. As an example of this line of reasoning, Lamarck suggested that the long neck of the giraffe had evolved when a short-necked ancestor took to browsing on the leaves of trees instead of grass (Figure 16-1). Lamarck reasoned that the ancestral giraffe, in reaching up, stretched and elongated its neck. Its offspring, after inheriting the longer neck, stretched still farther. The process, repeated over many generations, supposedly produced the very long neck of the modern giraffe.

The mechanism for Lamarckian evolution was an "inner drive" for self-improvement. Other scientists discredited this concept when they discovered the mechanisms of heredity. Lamarck's contribution to science is important, however, because he was the first to propose that organisms undergo change over time as a result of some natural phenomenon rather than divine intervention. It remained for Charles Darwin to discover the actual mechanism of evolution—natural selection.

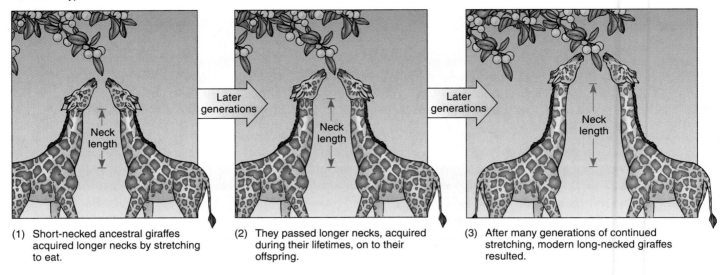

(1) Short-necked ancestral giraffes acquired longer necks by stretching to eat.

(2) They passed longer necks, acquired during their lifetimes, on to their offspring.

(3) After many generations of continued stretching, modern long-necked giraffes resulted.

Darwin's hypothesis

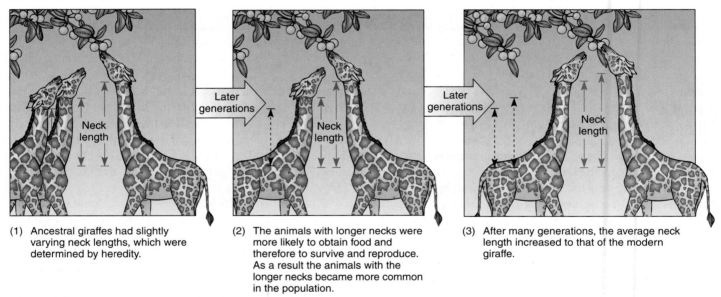

(1) Ancestral giraffes had slightly varying neck lengths, which were determined by heredity.

(2) The animals with longer necks were more likely to obtain food and therefore to survive and reproduce. As a result the animals with the longer necks became more common in the population.

(3) After many generations, the average neck length increased to that of the modern giraffe.

Figure 16-1 How did the giraffe get its long neck? A comparison of Lamarck's **(a)** and Darwin's **(b)** hypotheses on the evolution of giraffes. Scientific evidence supports Darwin's hypothesis. Although Lamarck's mechanism of evolution through the inheritance of acquired characteristics was incorrect, he was the first scientist to propose that organisms undergo evolution by natural means.

DARWIN GATHERED MUCH EVIDENCE FOR THE THEORY THAT ORGANISMS EVOLVE BY NATURAL SELECTION

As a young man, Charles Darwin (1809–1882) was appointed naturalist on the ship HMS *Beagle,* which was taking a 5-year voyage around the world to prepare navigational charts for the British navy (Figure 16-2). The *Beagle* sailed from Plymouth, England, in 1831 and cruised slowly down the east coast and up the west coast of South America. While other members of the company mapped the coasts and harbors, Darwin studied the animals, plants, fossils, and geological formations of both coastal and inland regions, areas that had not been extensively explored. He observed firsthand the diversity of the plants and animals of these regions. During the *Beagle*'s voyage, Darwin collected and catalogued thousands of specimens of plants and animals and kept notes of his observations.

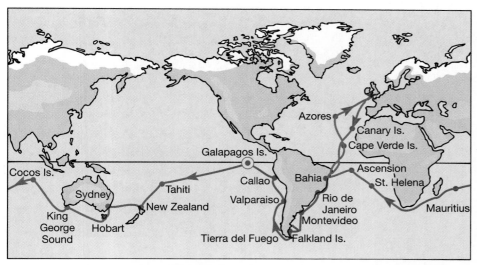

Figure 16-2 The voyage of the HMS *Beagle*. The 5-year voyage began in Plymouth, England (*red star*), in 1831. Darwin's observations of the Galapagos Islands were the basis for his theory of evolution by natural selection.

In 1832 the *Beagle* stopped for a time at the Galapagos Islands, 965 kilometers (600 miles) west of Ecuador. There Darwin continued his observations and collections (Figure 16-3). He compared the plants and animals of the Galapagos with those of the South American mainland, which he had observed the previous year. He was particularly impressed by their similarities and wondered why, for example, the plants and animals of the Galapagos should resemble those from South America more than they resembled those from other islands, where the environmental conditions were more similar. Moreover, although there were similarities between Galapagos and South American species, there were also distinct differences. Birds and reptiles even differed from one island in the Galapagos to the next! Darwin pondered these observations and tried to develop an adequate explanation for the distribution of the plants and animals.

Europeans in the mid-1800s generally believed that organisms were the results of divine creation and therefore did not change significantly over time. They thought each species looked the same as it had the day it was created. True, there were some obvious exceptions to this premise. For one thing, breeders could produce a great deal of modification in domesticated plants and animals in just a few generations. Breeders accomplished such change by breeding individuals that possessed desired traits, through the process of artificial selection. Fossils discovered during the mid-19th century also contradicted the accepted views, because many fossils did not resemble any living organism.

Geological evidence suggested that the Earth was far older than had previously been thought. During the early 19th century, geologists such as Charles Lyell advanced the idea that mountains, valleys, and other physical features of the Earth's surface had not been created in their present forms. Instead, they had been formed over long periods of time by the slow geological processes of volcanic activity, uplift, erosion, and glaciation, processes that still occur today. The slow pace of these geological processes indicated that the Earth was very, very old.

The ideas of Thomas Malthus, a clergyman and economist, were another important influence on Darwin. Malthus noted that populations increase in size until checked by factors in the environment. In the case of humans, Malthus suggested that wars, famine, and pestilence served as the inevitable and necessary brakes on the growth of human populations.

Darwin's years of observing the habits of plants and animals had introduced him to the struggle for existence described by Malthus. It occurred to Darwin that, in this struggle, favorable variations would tend to be preserved and unfavorable ones eliminated. As a result, a population would adapt to the environment, and eventually, enough modifications would give rise to a new species. Time was all that was required for new species to evolve, and the geologists of the era, including Lyell, had supplied evidence that Earth was indeed old enough to provide an adequate amount of time for such changes to have occurred.

Darwin had at last conceived of a working theory of evolution—that of evolution by natural selection. He spent the next 20 years accumulating an immense body of evidence to demonstrate that evolution had occurred and formulating his arguments for natural selection.

As Darwin was pondering his ideas, Alfred Russel Wallace, who was studying the plants and animals of Malaysia and

Figure 16-3   The plants and animals of the Galapagos Islands. (a) Giant prickly pear trees (*Opuntia echios*). Other *Opuntia* species in the Galapagos are not trees. **(b)** A lava cactus (*Brachycereus nesioticus*) that grows on recent lava flows is found only in the Galapagos Islands. **(c)** A land iguana (*Conolophus subcristatus*) is also found nowhere in the world except the Galapagos. **(d)** A red-footed booby (*Sula sula*). Unlike other boobies, which nest on the ground, the red-footed booby nests in shrubs and trees. Why its ducklike feet are bright red is not known. (a, *Stan Osolinski/Dembinsky Photo Associates*; b, *Patti Murray*; c,d, *Ed Reschke*)

Indonesia, was similarly struck by the diversity of organisms and the peculiarities of their distribution. Wallace, too, arrived at the conclusion that evolution occurred by natural selection. In 1858 he sent a brief essay to Darwin, who was by then a world-renowned biologist, asking his opinion.

Darwin's friends persuaded him to have Wallace's paper presented along with an abstract of his own views, which he had prepared and circulated to a few friends several years earlier. Both papers were presented at a meeting of the Linnaean Society in London in July 1858. Darwin's classic book *On the* *Origin of Species by Means of Natural Selection* was published in 1859.

Natural Selection Has Four Premises

Natural selection, the process in nature that causes evolution, is based on the tendency of organisms that possess favorable adaptations to their environments to survive and become the parents of the next generation. Natural selection consists of four observations about the natural world: overproduction, variation, competition for limited resources, and survival to reproduce.

1. **Overproduction**. Each species produces more offspring than will survive to maturity. Natural populations have the reproductive potential to continuously increase their numbers over time. For example, each individual puffball (see Figure 21-4c) produces millions of spores, and if each of those spores germinated and in turn gave rise to millions of spores, the land would be covered with puffballs in a very short time. Yet puffballs have not overrun the planet.

2. **Variation**. The individuals in a population exhibit variation in their traits. Some of those traits improve the chances of an individual's survival and reproductive success, whereas others do not. It is important to remember that the variation necessary for evolution by natural selection must be *heritable,* that is, able to be passed on to offspring.

3. **Competition for limited resources**, or **struggle for existence**. Only so much food, water, light, space for growing, and other resources are available in the environment, and organisms must compete with one another for the limited resources available to them. Because there are more individuals than the environment can support, not all will survive to reproductive age.

4. **Survival to reproduce**, or "**survival of the fittest**." The individuals that possess the most favorable combination of characteristics are most likely to survive and reproduce, passing their heritable traits on to the next generation.

Natural selection thus causes an increase in the frequency of appearance of favorable traits and a decrease in the frequency of appearance of unfavorable traits within a population. Over succeeding generations, the population becomes better adapted to local conditions. Successful reproduction is the key to natural selection: the "fittest" individuals are those that reproduce most successfully.

Over time, changes may accumulate in geographically separated populations of a species (often with slightly different environments), making them increasingly different from one another. Eventually these changes may be significant enough that one of the populations becomes a new species.

THE SYNTHETIC THEORY OF EVOLUTION COMBINES DARWIN'S THEORY WITH MENDELIAN GENETICS

One of the premises on which Darwin based his theory of evolution by natural selection is that individuals pass traits on to the next generation. However, Darwin was unable to explain *how* traits were passed from one generation to another. He was also unable to explain *why* individuals vary within a population. Darwin was a contemporary of Gregor Mendel, who worked out the basic patterns of inheritance. However,

Darwin was not aware of Mendel's work, since it wasn't to be recognized by the scientific establishment until the early part of the 20th century.

Over fifty years ago, biologists combined Mendelian genetics with Darwin's theory to formulate a comprehensive explanation of evolution, which is known as **neo-Darwinism** or, more commonly, the **synthetic theory of evolution**. (*Synthesis* in this context refers to the putting together of parts of several previous theories to form a whole.) The synthetic theory of evolution explains Darwin's observation of variation among offspring in terms of genetic mutation and recombination. This theory has dominated the thinking and research of many biologists since it was formulated and has resulted in an enormous accumulation of scientific evidence that supports evolution.

Although almost all biologists accept the basic principles of the synthetic theory of evolution, they may disagree over certain mechanisms of evolution. For example, what is the role of chance in determining the direction of evolution? How rapidly do new species evolve? These questions have arisen in part from a reevaluation of the fossil record and in part from discoveries about molecular aspects of inheritance. Debates are an integral part of the scientific process; they stimulate additional observation, experimentation, and reanalysis of older evidence. Science is a continuing process, and information obtained in the future may require us to modify certain parts of the synthetic theory of evolution.

MANY TYPES OF SCIENTIFIC EVIDENCE SUPPORT EVOLUTION

Evolution by natural selection is supported by a huge body of scientific observations and experiments. In this book we can report only a small fraction of this wealth of evidence, which comes from both living organisms and the fossil record. Although biologists still do not agree completely on some mechanisms by which evolutionary changes occur, the concept that evolution by natural selection has taken place is now well documented and accepted.

Comparative Anatomy of Related Species Demonstrates Similarities in Their Structures

Comparison of the structural details of any given organ found in different but related organisms reveals a basic similarity of form with some variation from one group to another. For example, a cactus spine and a pea tendril, although quite different in appearance, are similar in structure and development, because both are modified leaves (Figure 16-4; also see Figure 8-16b). The spine protects the succulent stem tissue of the cactus, whereas the tendril, which winds around a small

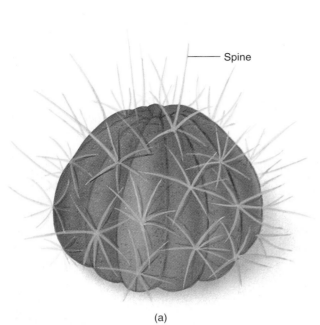

Spine

(a)

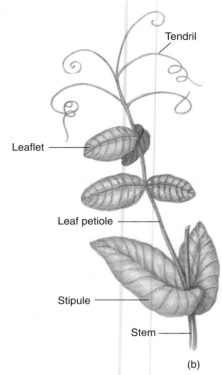

Tendril

Leaflet

Leaf petiole

Stipule

Stem

(b)

Figure 16-4   Spines and tendrils are homologous organs. (a) The spines of the fish-hook cactus (*Ferocactus wislizenii*) are modified leaves, as are **(b)** the tendrils of the garden pea (*Pisum sativum*). Leaves of the garden pea are compound, and the terminal leaflets are modified into tendrils that are frequently branched. Note the leafy stipules at the base of the leaf; the stipules in garden pea are often larger than the leaflets.

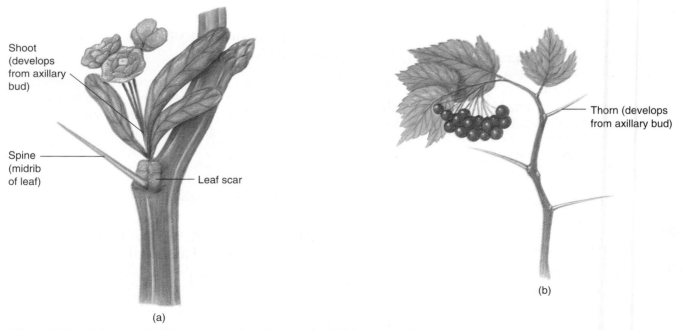

Shoot (develops from axillary bud)

Spine (midrib of leaf)

Leaf scar

(a)

Thorn (develops from axillary bud)

(b)

Figure 16-5   Spines and thorns are analogous organs. (a) A spine of Japanese barberry (*Berberis thunbergii*) is a modified leaf. (In this example, the spine is actually the midrib of the original leaf, which has been shed.) **(b)** Thorns of downy hawthorn (*Crataegus mollis*) are modified stems that develop from axillary buds formed during the previous year's growth.

object once it makes contact, helps support the climbing stem of the pea plant. Darwin pointed out that such basic similarities in organs used in different ways are precisely the expected outcome of evolution. Organs that are similar in underlying form in different organisms owing to a common evolutionary origin are termed **homologous**. Spines and tendrils are homologous organs.

Organs that are not homologous but simply have similar functions in different organisms are termed **analogous** organs. For example, spines, which are modified leaves, and thorns, which are modified stems, are analogous structures that have evolved over time to meet, in different ways, the common need for protection (Figure 16-5). Hawthorns and honey locusts are examples of plants that produce thorns. Tendrils are an additional example of analogous organs in plants. Some tendrils, such as those of grape and Boston ivy, are modified shoots and are therefore analogous to pea tendrils, which are modified leaves.

Like homologous organs, analogous organs offer crucial proof of evolution. Analogous organs are of evolutionary interest because they show how unrelated groups may adapt in

similar ways to common situations. The independent evolution of similarities in two or more unrelated organisms as a result of adaptation to similar environmental conditions is known as **convergent evolution**. For example, convergent evolution has resulted in structural similarities in two unrelated plant families, the cactus family and the spurge family (Figure 16-6). These plants, which evolved in similar desert environments in different parts of the world, are similar to each other in appearance even though they are not closely related.

Comparative anatomy also demonstrates the existence of **vestigial organs**. Many organisms contain organs or parts of organs that seem to serve no function. In plants, the greatly reduced petals of maple flowers are considered vestigial (Figure 16-7).

Darwin was interested in vestigial organs because their presence conflicted with the prevailing view of creation. He wondered how organisms that were the products of a "perfect creation" could have useless parts. Evolution, however, offered an easy explanation. The occasional presence of a vestigial organ is to be expected as a species evolves and adapts to a different mode of life. Organs that become less impor-

(a)   (b)

Figure 16-6   Convergent evolution results in structural similarities in members of two unrelated plant families that evolved in similar desert environments in different parts of the world. (a) *Euphorbia ingens,* a member of the spurge family, is native to Africa. **(b)** *Cereus hankianus,* a member of the cactus family, is native to North America. (*Dennis Drenner*)

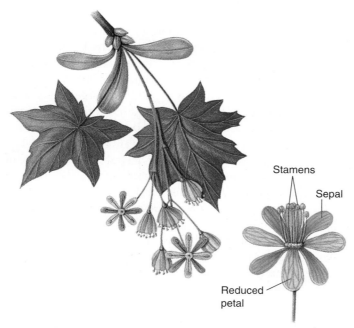

Figure 16-7 Vestigial organs in plants. The petals of wind-pollinated maple (*Acer* sp) flowers are reduced in size or totally absent. Because the petals do not serve a known function, they can be interpreted as vestigial organs. A unisexual maple flower that contains stamens but not carpels is shown.

tant for an organism's survival may end up as vestiges. When an organ loses much or all of its function, it no longer has any selective advantage. If its presence does not hurt the organism, however, the vestigial organ tends to remain.

Mimicry Provides an Evolutionary Advantage

Mimicry is a subtle example of how natural selection improves an organism's chances of surviving and producing offspring in a particular environment. **Mimicry** is the resemblance of one organism to another organism or to an inanimate object. The flowers of certain orchids mimic female bees. For example, the yellow bee orchid, (*Ophrys lutea*), contains three yellow-green sepals and two yellow petals, which form a lip (Figure 16-8a). Note the shiny purple-black center of the lip; it is what looks like a female bee to the male bees. Two clumps of pollen grains are hidden from view under the overhanging middle sepal. When the male bee attempts to copulate with the flower, the pollen clumps stick to his back. He carries them to another flower, which is pollinated when he tries to copulate again (a phenomenon called *pseudocopulation*). In this example, mimicry serves to attract pollinators.

Mimicry can also make an organism difficult to find. **Protective coloration**, coloration that permits an organism to blend into its surroundings, is a type of mimicry. Protective coloration can either benefit a prey organism by screening it

from its predators or benefit a predator by keeping the prey from noticing it until it is too late. Many examples of protective form and coloration involve animals resembling plants. Some caterpillars resemble twigs so closely that you would never guess they are animals—until they start to move (Figure 16-8b). Natural selection has accentuated and preserved such protective coloration.

The Distribution of Plants and Animals Supports Evolution

The study of the distribution of plants and animals is called **biogeography**. One of its basic tenets is that each species evolved only once. The particular place where a species originated is known as its **center of origin**. The center of origin is not a single point but the range of the population when the new species formed. From its center of origin, each species spreads out until halted by a barrier of some kind. The barrier can be physical, such as an ocean, a desert, or a mountain; environmental, such as an unfavorable climate; or ecological, such as the presence of other species that compete for food or living space. (See Plants and the Environment: Islands and the Problem of Exotic Species for discussion of a serious environmental threat that is directly related to some of the principles of biogeography.)

Biogeography examines why species live where they do. Palm species, for example, are common in tropical rain forests of South America but almost nonexistent in comparable forests of central Africa. Similarly, South America has balsa trees (*Ochroma* sp), snakewood trees (*Cecropia* sp), sloths, and tapirs, whereas central Africa, with a similar climate and environmental conditions, has none of these. These organisms evolved in tropical America and could not expand their range into Africa because the Atlantic Ocean was an impassable barrier. Likewise, central Africa has umbrella trees (*Musanga* sp), guapiruvu trees (*Schizolobium* sp), chimpanzees, and elephants, none of which is found in tropical America. Thus, the natural distribution of organisms on Earth seems understandable only on the basis of where they evolved.

Molecules Contain a Record of Evolutionary Change

Similarities and differences in the biochemistry and molecular biology of various organisms provide evidence for evolutionary relationships. Indeed, if we established evolutionary lines of descent based solely on biochemical and molecular data instead of the usual structural and fossil evidence, there would be many close resemblances.

Organisms owe their characteristics to the types of proteins they possess, which are determined by the order of nucleotides in their DNA. Evidence that all life is related comes from the fact that all organisms use a genetic code that is virtually

(a)

(b)

Figure 16-8   Examples of mimicry. (a) The yellow bee orchid (*Ophrys lutea*) flowers resemble a certain species of female bee. The highly specialized mimicry of bee orchids is not just visual. These plants also secrete a scent similar to that produced by female bees, and the males are irresistibly attracted to it. **(b)** Geometrid larvae are caterpillars that resemble twigs when resting on a branch. Can you find the caterpillar? (a, *Phil Gates/Biological Photo Service*; b, *James L. Castner*)

identical. The genetic code specifies a triplet in DNA that codes for a particular codon (a sequence of three nucleotides in RNA), which codes for a particular amino acid in a polypeptide chain. For example, "AAA" in DNA codes for "UUU" in mRNA, which codes for the amino acid phenylalanine in organisms as diverse as shrimp, humans, bacteria, and tulips. In fact, "AAA" codes for phenylalanine in *all* organisms examined to date.

The universality of the genetic code—no other code has been found in any organism—is compelling evidence that organisms arose from a common ancestor. The genetic code has been passed along through all the branches of the evolutionary tree since its origin in some extremely early form of life.

Proteins Contain a Record of Evolutionary Change

Investigations of the sequences of amino acids in the proteins that play the same roles in different species have revealed both great similarities and specific differences. Even organisms that are related only remotely, such as humans, oaks, and bacteria, have some proteins such as cytochrome *c* in common. To survive, all aerobic organisms need a respiratory protein with the same basic structure and function as the cytochrome *c* of their common ancestor. Consequently, not all of the amino acids that give the protein those features are free to change. In the course of the long, independent evolution of different organisms, however, mutations have resulted in the

substitution of many amino acids at less important locations in the cytochrome *c* protein. The longer it has been since two organisms diverged, or took separate evolutionary pathways, the greater the differences in the amino acid sequences of their cytochrome *c* molecules.

DNA Contains a Record of Evolutionary Change

Since DNA codes for proteins, the differences in amino acid sequences indirectly demonstrate the nature and number of underlying DNA base pair changes that must have occurred during evolution. Such molecular information is obtained directly by **DNA sequencing**, the determination of the order of nucleotide bases in DNA, that codes for a gene shared by several organisms. Generally, the more closely species are thought to be related on the basis of other scientific evidence, the greater is the proportion of nucleotide sequences that their DNA molecules usually have in common.

Fossils Indicate That Organisms Evolved in the Past

Perhaps the most direct evidence for evolution comes from the discovery, identification, and interpretation of fossils. **Fossils** (Latin *fossilis*, "something dug up") are the remains or traces of organisms left in layers of rock (Figure 16-9). Fossils provide a record of plants and animals that lived earlier,

Environment

Islands and the Problem of Exotic Species

Biotic pollution, the introduction of a foreign, or exotic, species into an area, often upsets the balance among the organisms living in that area. The foreign species may compete with native species for food or habitat or may prey on them. Generally, an introduced competitor or predator has a greater negative effect on local organisms than do native competitors or predators. Although exotic species may be introduced into new areas by natural means, humans are usually responsible for such introductions, either knowingly or accidentally. The water hyacinth, for example, was deliberately brought from South America to the United States because it has lovely flowers. Today it has become a nuisance in waterways across the entire southeastern United States, clogging them so that boats cannot move easily and

crowding out native aquatic plants (see Figure *a*).

Islands are particularly vulnerable to the introduction of exotic species. Many plant and animal species living on islands are *endemic*—that is, they are not found anywhere else in the world. Endemism is common on islands because they are so isolated. During an island's early history, only a few kinds of plant and animal colonists make it there from across the water. Over many generations, the colonists' descendants adapt to the unique environmental characteristics of that island, gradually evolving into completely new species. Because they have evolved in isolation from competitors, predators, and disease organisms, island species have few defenses when such organisms are then introduced to their habitat. This makes island species particu-

larly vulnerable to extinction.

For example, more than two thirds of the plant species endemic to the Galapagos Islands are at risk, largely due to the introduction of goats, pigs, and dogs. In Hawaii, the introduction of sheep has imperiled the mamane tree (because the sheep eat it) and the honeycreeper, an endemic bird (because it relies on the mamane tree for food). When goats and sheep were introduced to Santa Catalina Island, California, they overgrazed the Catalina mahogany until only seven trees remained (see Figure *b*).

In conclusion, islands were important centers of evolution in the past and are major centers of extinction today. Their isolation has resulted in a rich evolutionary history as well as a vulnerability to biotic pollution.

(a)

(b)

(a) The water hyacinth (*Eichhornia crassipes*), introduced into the southeastern United States from tropical South America, has become a pest. **(b)** Closeup of a branch of the Catalina mahogany (*Cercocarpus traskiae*). This species was overgrazed by sheep and goats introduced by humans to Santa Catalina Island, California. Today the Catalina mahogany is one of the rarest trees in the United States and has a total population of seven small trees. (a, *Visuals Unlimited/William J. Weber;* b, *courtesy of Robert Thorne, Rancho Santa Ana Botanic Garden*)

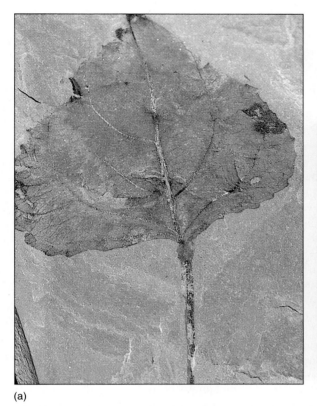

(a)   (b)

Figure 16-9   Plant fossils. (a) An impression fossil of a cottonwood (*Populus* sp) leaf that formed about 50 million years ago. **(b)** Fossilized wood from the Petrified Forest National Park in Arizona. (a, *Carlyn Iverson;* b, *B. Gruliow*)

some understanding of where and when they lived, and an idea of the lifestyles they had. By studying the fossils of organisms of different geological ages, we can trace the lines of evolution that gave rise to those organisms. Thus, the fossil record, although incomplete, provides a history of life on Earth.

The formation and preservation of a fossil require that the organism be buried under conditions that slow the process of decay. Slow decay is most likely if the remains are covered quickly by fine particles of soil suspended in water, which envelop the organism with sediment. Remains of aquatic organisms may be buried in bogs, mud flats, sand bars, or river deltas. Remains of terrestrial organisms may also be covered by waterborne sediments in a flood plain or by windblown sand in an arid region.

Because of the conditions required for preservation, the fossil record is not a random sample of all past life. The record is biased toward aquatic organisms and those living in the few terrestrial habitats that are conducive to fossil formation, such as deserts, areas with tar pits, and arctic regions. For example, relatively few fossils of tropical rainforest organisms have been found, because plant and animal remains decay very rapidly on the forest floor, before fossils can form. Another reason for bias is that organisms with hard body parts, such as wood, bones, or shells, are more likely to form fossils than

are those with soft tissues, such as flowers, mushrooms, or worms.

To be interpreted, the sedimentary layers containing fossils must be arranged in chronological order (see Focus On: Dating Fossils). The layers of sedimentary rock, if they have not been disturbed, occur in the sequence of their deposition, with the more recent layers on top of the older, earlier ones.

EVOLUTION IS GRADUAL, OCCURS IN SPURTS, OR IS A COMBINATION OF BOTH PROCESSES

Biologists have long recognized that the fossil record lacks many transitional forms; that is, the starting points and end points are present, but the intermediate stages in the evolution from one species to another may be absent. This observation was traditionally blamed on the incompleteness of the fossil record, and biologists attempted to fill in the missing parts much as a writer might fill in the middle of a novel after composing the beginning and end.

Beginning in the 1970s, however, many biologists have begun to question whether the fossil record really is incomplete. The **punctuated equilibrium** model proposes that the fossil record reflects evolution accurately, with long periods

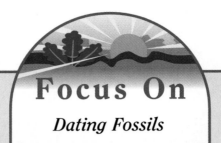

Focus On
Dating Fossils

To interpret fossil evidence of the history of life, the rocks in which fossils appear must be dated. The radioactive isotopes that are present in a rock give us an accurate measure of its age. Recall from Chapter 2 that elements exist as different isotopes. Some isotopes (those with excess neutrons) are unstable and tend to break down, or decay, to a more stable isotope, usually of a different element. Such isotopes, termed **radioactive isotopes**, emit powerful, invisible radiation when they decay. As a radioactive isotope emits radiation in the process called **radioactive decay**, its atomic nucleus changes into the nucleus of a different element. For example, the nucleus of one radioactive isotope of uranium (U-235) decays over time into lead (Pb-207).

Each radioactive isotope has its own characteristic rate of decay. The period of time required for one half of a radioactive isotope to change into a different material is known as its **half-life** (see figure). There is an enormous variation in the half-lives of different radioactive isotopes. For example, the half-life of iodine (I-132) is only 2.4 hours, whereas the half-life of uranium (U-235) is 704 million years. The half-life of a particular radioactive isotope is constant and never varies; it is not influenced by temperature, pressure, or any other environmental factor.

Scientists estimate the age of a fossil by measuring the proportion of the original radioactive isotope and its decay product. For example, the half-life of a radioactive isotope of potassium (K-40) is 1.25 billion years, meaning that in 1.25 billion years half of the radioactive potassium will have decayed into its decay product , argon (Ar-40). If the ratio of potassium (K-40) to argon (Ar-40) in the rock being tested is 1:1, the rock is 1.25 billion years old. (The radioactive clock begins ticking when the rock solidifies. The rock initially contains some potassium but no argon.)

Several different radioactive isotopes are used in dating fossils, including potassium (K-40; half-life 1.25 billion years), uranium (U-235; half-life 704 million years), and carbon (C-14; half-life 5730 years). Carbon-14 is used to date the carbon remains of anything that was once living, such as wood, bones, and shells; the other isotopes are used to date the rock in which fossils are found. Because of its relatively short half-life, carbon-14 is useful for dating fossils that are 50,000 years old or less. In contrast, potassium-40, with its long half-life, can be used to date fossils that are hundreds of millions of years old. Wherever possible, the age of a fossil is verified independently, using two or more different radioactive isotopes.

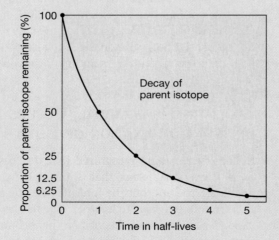

The decay of radioactive isotopes. At time zero, the radioactive clock begins ticking. At this point a sample is composed entirely of the radioactive isotope. After one half-life, only 50 percent of the original radioactive isotope remains.

of little or no evolutionary change punctuated, or interrupted, by short periods of rapid evolution. Thus, in punctuated equilibrium, active evolution normally proceeds in relatively short "spurts" interspersed with long periods of inactivity. During each spurt, new species form, and many old ones are outcompeted and become extinct.

With punctuated equilibrium, the evolution of a species can occur in a relatively short period of time. It is important to realize, however, that a "short" amount of time for the evolution of a new species may mean anything from a few years to thousands of years. Such a period of time is short when compared to the several million years that a species typically exists.

Punctuated equilibrium accounts for the abrupt appearance of a new species in the fossil record with little or no record of intermediate forms. That is, proponents think that there are few transitional forms in the fossil record because most species evolved so rapidly that they did not produce transitional forms.

In contrast to punctuated equilibrium, the traditional Darwinian view of evolution is embodied in the **gradualism** model, which reasons that evolution proceeds at a more or less steady rate, but it is not observed in the fossil record because the record is incomplete. Occasionally, a complete fossil record of transitional forms *is* discovered and is cited as a strong case for gradualism. The gradualism model maintains that populations diverge from one another slowly by the gradual accumulation of adaptive characteristics within each population. These adaptive characteristics accumulate as a result of differing selective pressures on the populations living in different environments.

The fact that there is abundant evidence in the fossil record of long periods with little or no change in a species seems to argue against gradualism. Gradualists, however, point out that little or no change in fossils is deceptive because fossils do not show all aspects of evolutionary change. Fossils typically show changes in hard body parts and external structures but do not reveal such characteristics as changes in cell physiology, behavior, and ecological roles, which also represent evolution.

Thus, while evolutionary biologists generally agree that natural selection is the main mechanism responsible for evolution, there is disagreement as to the timing, or pace, of evolutionary change during a species' existence. Many biologists are of the opinion that *both* types of evolution may occur; that is, the pace of evolution may be steady and gradual in certain instances and abrupt in others.

STUDY OUTLINE

I. Evolution can be defined as a genetic change in a population of organisms over time.
II. Charles Darwin proposed the theory of evolution by natural selection on the basis of four observations.
   A. **Overproduction.** Each species produces more offspring than will survive to maturity.
   B. **Variation.** The individuals in a population exhibit heritable variation in their traits.
   C. **Competition for limited resources, or struggle for existence.** Organisms compete with one another for the food, space, water, light, and other resources needed for life.
   D. **Survival to reproduce.** The offspring with the most favorable combination of characteristics are most likely to survive and reproduce, passing those genetic characteristics to the next generation.
III. The synthetic theory of evolution combines Darwin's theory of evolution by natural selection with the genetic mechanisms for explaining evolution (Mendelian genetics).
IV. That evolution occurs is now well documented scientifically.
   A. Evidence supporting evolution is derived from comparative anatomy.
      1. Homologous organs have basic structural similarities, even though the organs may be used in different ways. Homologous organs indicate evolutionary ties between the organisms possessing them.
      2. Analogous organs have similar functions but are not structurally similar and do not indicate close evolutionary ties.
      3. The occasional presence of a vestigial organ is to be expected as an ancestral species adapts to a different mode of life.
   B. Mimicry, in which an organism evolves to resemble another organism or an inanimate object, improves an organism's chances of surviving and producing offspring in a particular environment.
   C. Biogeography, the distribution of plants and animals, supports evolution.
      1. Areas that have been isolated from the rest of the world for a long time have plants and animals unique to those areas.
      2. Each species originated only once, at its center of origin. From its center of origin, each species spread out until halted by a barrier of some kind.
   D. Biochemistry and molecular biology provide compelling evidence for evolution.
      1. The sequence of amino acids in common proteins such as cytochrome *c* reveals greater similarities in closely related species than in remotely related species.
      2. A greater proportion of the sequence of nucleotides in DNA is identical in closely related organisms than in remotely related organisms.
   E. Perhaps the most direct evidence for evolution comes from fossils, the remains or traces of ancient plants and animals.

V. Scientists are debating the pace of evolutionary change.
   A. According to proponents of gradualism, populations diverge from one another slowly by the gradual accumulation of adaptive characteristics within a population.

B. According to proponents of punctuated equilibrium, evolution proceeds in spurts. Short periods of active evolution are followed by much longer periods of inactivity.

SELECTED KEY TERMS

analogous organs, p. 273
biogeography, p. 274
biotic pollution, p. 276
center of origin, p. 274
convergent evolution, p. 273

DNA sequencing, p. 275
evolution, p. 267
fossil, p. 275
gradualism, p. 279

homologous organs, p. 273
mimicry, p. 274
natural selection, p. 270
protective coloration, p. 274

punctuated equilibrium, p. 277
synthetic theory of evolution
   (neo-Darwinism), p. 271
vestigial organ, p. 273

REVIEW QUESTIONS

1. Explain briefly the four premises of evolution by natural selection.
2. In what ways does Lamarck's idea of evolution not agree with present scientific evidence?
3. Consider the giraffe's long neck. Use Lamarck's theory of evolution to explain how this came about. Then use Darwin's mechanism of natural selection to explain the giraffe.
4. Why are only inherited variations important in the evolutionary process?
5. What part of his theory was Darwin unable to explain? How does the synthetic theory of evolution explain it?

6. Discuss at least four different types of evidence used to determine evolutionary relationships among organisms.
7. Give an example of a plant that mimics an animal and of an animal that mimics a plant.
8. Explain how cacti and spurges demonstrate convergent evolution.
9. Distinguish between gradualism and punctuated equilibrium.
10. Explain why many island species are found nowhere else.

THOUGHT QUESTIONS

1. In what ways are the study of evolution and human history similar? How are they different?
2. How can you account for the fact that both Darwin and Wallace independently and almost simultaneously proposed essentially identical theories of evolution by natural selection?

3. Explain the following statement. Natural selection picks from among the individuals in a population those that are best suited to current environmental conditions. It does not select based on some view of "best design" or on "possible future need."

SUGGESTED READINGS

Amos, W.H., "Hawaii's Volcanic Cradle of Life," *National Geographic,* July 1990. Depicts the colonization of volcanic lava by a few hardy species and describes some of the endemic species that have evolved in the Hawaiian Islands.

Darwin, C.R., *On the Origin of Species by Means of Natural Selection, or the Preservation of the Favored Races in the Struggle for Life,* New York, Cambridge University Press, 1975. A readily obtainable reprint of one of the most important books of all time. Darwin's long essay is still of great significance to modern readers.

Lemonick, M.D., "Can the Galapagos Survive?" *Time,* October 30, 1995. The unique flora and fauna of the Galapagos are increasingly threatened by a growing human population, the introduction of foreign species, poaching, and other environmental ills.

Moore, J.A., *Science as a Way of Knowing: The Foundations of Modern Biology,* Cambridge, Mass., Harvard University Press, 1993.

Contains an excellent introduction to the history of evolutionary thought.

Porter, D.M., and P.W. Graham, *The Portable Darwin,* New York, Penguin Books, 1993. This collection of Darwin's writings reveals the diverse interests of the man who profoundly changed the intellectual climate of the 19th and 20th centuries.

Royte, E., "On the Brink: Hawaii's Vanishing Species," *National Geographic,* September 1995. Hawaii has more endangered species than any other state in the United States. In many cases, native species are threatened by exotic ones.

Thompson, G.R., and J. Turk, *Modern Physical Geology,* 2nd ed., Philadelphia, Saunders College Publishing, 1996. Contains informative material on radioactive dating and fossils.

Wallace, A.R., *The Malay Archipelago,* New York, Oxford University Press, 1987. A reprint of Wallace's classic investigation.

THE EVOLUTION OF POPULATIONS AND SPECIES

Primroses (*Primula* sp) are attractive perennial herbs that are popular as house and garden plants. They produce clusters of trumpet-shaped flowers at the end of a long stem; the leaves are generally located at the base of the stem. About 400 primrose species exist, and these exhibit a wide variety of heights (30 to 120 centimeters, or 1 to 4 feet), flower colors (white, pink, red, yellow, blue, and purple), and flower arrangements on the stem.

Each primrose species produces two different kinds of flowers. Some primroses have flowers that contain long ("male") stamens and short ("female") carpels, that is, carpels located deep inside the flower. Other plants have flowers with short stamens and long carpels. These differing flower structures increase the likelihood of cross pollination by insect pollinators such as bees and butterflies. Each type of insect enters primrose flowers in the same manner and penetrates the trumpet-shaped flowers the same distance each time. Thus, one kind of insect may carry pollen from short stamens of one flower to short carpels of another,

The kew primrose (*Primula kewensis*) has bright yellow flowers. (*Visuals Unlimited/Dick Keen*)

and another insect may carry pollen from long stamens to long carpels in a different flower.

Primroses are a perfect plant with which to begin this chapter on evolution, because a new species of primrose, the kew primrose (*Primula kewensis*), originated in 1898 at the Royal Botanic Gardens at Kew, England. The specific epithet *kewensis* was assigned in recognition of the plant's place of origin. Details about the evolution of the kew primrose are discussed shortly.

In this chapter we study heredity at the population level. A **population** consists of all the individuals of the same species that live in a particular place at a specific time. (Recall from Chapter 1 that a **species** is a group of similar organisms that interbreed in their natural environment.) First we examine genetic variation in populations, including how the relative abundances of different alleles (forms of each gene) change during the course of evolution. Then we consider how evolutionary changes within populations can ultimately lead to the evolution of new species.

After reading this chapter, you should be able to:

1. *Define microevolution and distinguish between the gene pool of a population and the genotype of an individual.*
2. *Discuss the significance of the Hardy-Weinberg law as it relates to evolution, and explain how each of the following alters allele frequencies in populations: mutation, genetic drift, gene flow, natural selection.*
3. *Define species, and describe some of the limitations of your definition.*
4. *Distinguish among different reproductive isolating mechanisms, and give an example of each.*
5. *Distinguish between allopatric and sympatric speciation.*
6. *Discuss the evolutionary significance of adaptive radiation and extinction.*

INDIVIDUALS IN A POPULATION EXHIBIT GENETIC VARIATION

Individual organisms within a population exhibit genetic variation for the traits characteristic of the population. A field of garden peas may all possess the same trait, such as a fruit that is a pod, but vary from one individual to another in pod color (green or yellow) or pod shape (inflated or constricted).

Each population possesses a **gene pool**, the total genetic material of all the individuals that make up that population at a given time. The gene pool includes all possible alleles of all genes present in the population (Figure 17-1). An individual organism possesses only a small fraction of the alleles present in a population's gene pool. In diploid (*2n*) organisms, each somatic cell (body cell) contains only two alleles for each gene, one on each of the homologous chromosomes (members of a chromosome pair). The genetic variation that is evident among individuals in a given population indicates that each individual has a different combination of the alleles in the gene pool.

Evolution occurs in populations, not individuals (Table 17-1). Although natural selection acts on individuals by determining which of them will survive to reproduce, individuals do not evolve during their lifetimes. Populations, however, undergo evolutionary change over many generations. Evolutionary change, which includes modifications in structure, physiology, ecology, and behavior, is inherited from one generation to the next. Although Darwin recognized that evolution occurs in populations, he did not understand how traits are passed to successive generations. One of the most significant advances in biology since Darwin's time has been the demonstration of the genetic basis of evolution.

If a population is not undergoing evolutionary change, the frequency of each allele in the gene pool remains constant from generation to generation. Changes in allele frequencies over a few successive generations indicate that evolution has occurred. This type of evolution is sometimes referred to as **microevolution**, because it involves relatively small or minor changes that take place *within* a population.

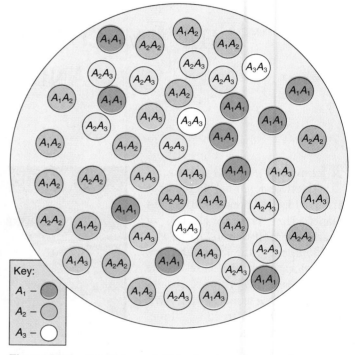

Figure 17-1   A gene pool. This drawing shows only one gene (*A*), with three different possible alleles (A_1, A_2, and A_3). Because each individual plant (*represented by the small circles*) is diploid, it possesses only two alleles for each gene.

THE HARDY-WEINBERG LAW DEMONSTRATES THAT ALLELE FREQUENCIES DO NOT CHANGE IN A POPULATION THAT IS NOT EVOLVING

Suppose we are studying a population of well-adapted plants growing in an environment that is relatively constant from year to year. Such a population would be in genetic equilibrium—that is, not undergoing evolutionary change. If we were to compare the numbers of red-flowered plants and white-flowered plants in the population, for example, we

TABLE 17-1   A Comparison of Individuals and Populations

	Individual	Population
Period of existence	One generation	Many generations
Genetic description	Genotype	Gene pool
Genetic variation	None	Substantial
Ability to evolve	No	Yes (change in allele frequencies)

might find 1820 plants with red petals and 180 with white petals, which is a ratio of 9 to 1. If we came back to the same location the following summer, we would expect to find a population essentially the same as the previous one, with roughly nine red-flowered plants to every white-flowered plant. If we continued the study for several generations and always got the same result, the population would clearly be in genetic equilibrium.

In 1908 G.H. Hardy, an English mathematician, and W. Weinberg, a German physician, independently developed the explanation for this stability of successive generations of populations in genetic equilibrium. They pointed out that the frequencies of various genotypes (genetic makeups) in a population can be described mathematically.

The principles of inheritance worked out by Mendel describe the frequencies of genotypes among offspring of a single mating pair (see Chapter 14). In contrast, the **Hardy-Weinberg law** describes the frequencies of alleles in the genotypes of an *entire breeding population*. The Hardy-Weinberg law shows that the process of inheritance by itself does not cause changes in allele frequencies. The proportions of alleles in successive generations will always be the same, provided that the following criteria are met.

1. **Random mating.** The individuals represented by the genotypes *AA, Aa,* and *aa* must mate with one another at random and must not select their mates on the basis of genotype. That is, matings between genotypes must occur in proportion to the frequencies of the genotypes.
2. **No mutations.** There must be no mutations of *A* or *a*. (Recall from Chapter 12 that a **mutation** is an alteration in a cell's genetic material, that is, its DNA.)
3. **Large population size.** The population of individuals must be large enough that the laws of probability function. Chance events do not alter the frequency of the alleles in a large population as they do in a small population.
4. **No migration.** There can be no exchange of genes with other populations that might have different allele frequencies. In other words, there can be no movement of individuals (in animals) or of pollen or seeds (in plants) into or out of a population.
5. **No natural selection.** There can be no natural selection, because then certain genotypes would be favored over others, and the allele frequencies would change.

EVOLUTION OCCURS WHEN THERE ARE CHANGES IN ALLELE FREQUENCIES IN A GENE POOL

In studies of populations in nature or in the laboratory, the Hardy-Weinberg law is used to test whether evolutionary changes are taking place. *Evolution, stated in its simplest terms, represents a departure from the allele frequencies predicted by the Hardy-Weinberg law of genetic equilibrium.* Such changes in the gene pool of a population result from four microevolutionary processes: mutation, genetic drift (an especially important factor in small populations), migration, and natural selection. When one or more of these processes is acting on the population, allele frequencies undergo change from one generation to the next.

Mutation Increases Variation in the Gene Pool

Genetic variation is introduced into a gene pool through mutation. Mutations arise from one of three occurrences: a change in the nucleotide base pairs of the gene; a rearrangement of genes within chromosomes so that the interactions among genes produce different effects; or a change in the number of chromosomes. Mutation is the source of all new alleles (Figure 17-2).

Mutations occurring in somatic cells are not passed on to the next generation. When an individual with a somatic mutation dies, the mutation dies with it. Some mutations, however, alter the DNA in reproductive cells, and these mutations may or may not affect the offspring. If a mutation occurs in the DNA that codes for a protein, it may still have little effect in terms of altering the structure or function of that protein. However, when the protein undergoes a change great enough to affect its functioning, the mutation is usually harmful. For instance, a mutation in a gene that codes for an enzyme involved in producing blue petals might result in a plant with nonblue petals that would not attract its normal insect pollinators.

Mutations do not determine the *direction* of evolutionary change. Consider, for example, a plant population living in an environment that is becoming increasingly arid. A muta-

(a)

(b)

Figure 17-2 The ancestors of cacti were flowering trees and shrubs. Mutations provided the raw material that led to the evolution of the cactus family. **(a)** The leaf cactus (*Pereskia sacharosa*) still retains many of the features of its ancestors, including broad leaves and a "normal" stem. This plant's evolution was apparently the result of relatively few mutations. **(b)** The fishhook cactus (*Ferocactus wislizenii*) is very different from its ancestors. Mutations have resulted in much-reduced leaves (the spines) and a rounded, much-thickened succulent stem. These traits have enabled the cactus to exist in arid environments where its ancestors could not survive. (a, *James Mauseth*; b, *Stan Osolinski/Dembinsky Photo Associates*)

tion that produces a new allele that helps an individual adapt to dry conditions is no more likely to occur than one for adapting to wet conditions or one that has no relationship to the changing environment. The production of new mutations simply increases the genetic variability within a population.

Genetic Drift Causes Changes in Allele Frequencies Through Random Events

The size of a population affects allele frequencies because random events, or chance, tend to cause changes in a small population. For example, consider two populations, one with 10,000 individuals and one with 10 individuals. If an uncommon allele occurs at a frequency of 10 percent, or 0.10, then 1900 individuals in the larger population possess the allele. That same allele frequency, 10 percent, in the smaller population means that no more than two individuals possess the allele. From this example, it is easy to see that the uncommon allele is more likely to be lost from the smaller population than from the larger population. For example, purely by chance, herbivores might eat the only two individuals of the smaller population that possess the uncommon allele.

The production of random evolutionary changes in small breeding populations, known as **genetic drift**, results in changes in the gene pool of a population from one generation to the next. An allele may be eliminated from the population by chance, regardless of whether the allele is beneficial, harmful, or of no particular advantage or disadvantage (Figure 17-3). Thus, genetic drift can *decrease* genetic variation *within* a population, although it tends to *increase* the genetic differences *between* populations.

Organisms in small populations that have limited gene pools are particularly vulnerable to genetic drift, because the loss of genetic variation reduces their ability to adapt to stressful changes in their environment. For many rare and endangered species whose populations have been reduced in size, genetic drift tends to increase their chances of extinction.

Gene Flow in Plants Is Caused by the Migration of Pollen or Seeds

The geographical distribution of a particular species is its **range**. Members of a species are not usually distributed evenly over their entire range but instead tend to be found in local populations that are more or less genetically isolated from

284

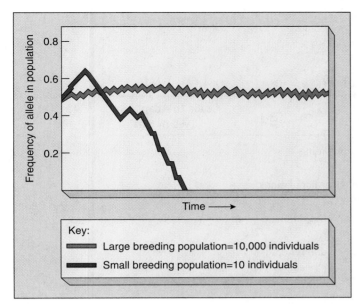

Figure 17-3 Genetic drift. The frequency of a given allele is shown in two hypothetical populations, one large and one small. Note how the frequency of the allele in the large population varies only slightly due to chance, but in the small population, as a result of chance, the allele disappears in a short period of time.

one another. Two populations of wind-pollinated plants, for example, are isolated genetically when the distance separating them is greater than the maximum distance their pollen is blown by the wind.

When individuals of one population reproduce with members of another population, they contribute their alleles to that population's gene pool. Thus, migration causes a corresponding movement of alleles, or **gene flow**, between populations. Gene flow can have significant evolutionary consequences. As alleles "flow" from one population to another, they usually increase the amount of variability within the population that receives them. If the gene flow between two populations is great enough, these populations become more similar genetically. Because gene flow has a tendency to reduce the amount of genetic variation between two populations, it tends to counteract the effects of natural selection (discussed shortly) and genetic drift, both of which cause individual populations to become increasingly distinct.

Natural Selection Changes Allele Frequencies in a Way That Leads to Adaptation to the Environment

Natural selection is the mechanism of evolution first proposed by Darwin, in which members of a population that possess better adaptations to the environment are more likely to survive and reproduce. As a result, these members, as parents of the next generation, increase the proportion of their alleles in the population.

Natural selection functions to preserve individuals with favorable phenotypes (and therefore favorable genotypes) and to eliminate individuals with unfavorable genotypes. Individuals have a selective advantage if they can survive and produce fertile offspring. Natural environmental pressures, such as competition for light or water or living space, select the individuals that survive to reproduce.

Natural selection not only explains why organisms are well adapted to the environments in which they live but accounts for the astonishing diversity of life. Natural selection enables populations to adapt to different environments and different ways of life. In time, these evolving populations may become separate and distinct species.

The mechanism of natural selection does not lead to the development of a "perfect" organism. Rather, natural selection weeds out organisms with phenotypes that are less adapted to environmental challenges so that those that are better adapted survive and pass their alleles on to their offspring. Natural selection is the only process known that brings genetic variation into harmony with the environment and leads to adaptation. By reducing or eliminating alleles that result in the expression of less favorable traits, natural selection changes the composition of the gene pool in a direction that favors the population's survival (Figure 17-4). These

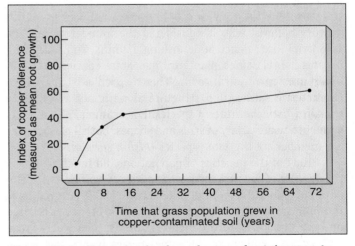

Figure 17-4 Natural selection of genes for tolerance to high levels of copper in creeping bent grass (*Agrostis stolonifera*). Individuals were taken from grass populations growing in areas known to have been contaminated by copper for 0, 4, 8, 14, and 70 years. Year 0 represents the control. All plants were grown in a nutrient solution containing a high level of copper. Root growth was measured to give an index of copper tolerance: 0 = no root growth (complete inhibition), and 100 = maximum root growth (no inhibition). Note that the longer the grass populations had been exposed to copper-contaminated soil, the greater their tolerance for copper. This clearly shows natural selection occurring in a direction that favors the population's survival.

285

TABLE 17-2 Processes of Microevolution

Process	Effect on Gene Pool	Leads to Adaptation (yes or no)
Mutation	Changes allele frequencies by providing inheritable variation	No
Genetic drift	Changes allele frequencies in small populations due to random sampling error	No
Gene flow (migration)	Changes allele frequencies due to migration of individuals between populations	No
Natural selection	Changes allele frequencies due to differential survival and reproduction	Yes

changes in the gene pool increase the probability that the favorable alleles responsible for an adaptation will come together in the offspring.

Table 17-2 summarizes the four microevolutionary processes just discussed.

THE BIOLOGICAL CONCEPT OF SPECIES IS BASED ON REPRODUCTIVE ISOLATION

The concept of distinct groups of organisms, known as species, is not new. However, every definition of exactly what constitutes a species has some sort of limitation. Linnaeus, an 18th century botanist who is considered the founder of modern taxonomy (the science of describing, naming, and classifying organisms), classified plants into separate species based on differences in physical form. This method is still used to characterize species, but structure alone is not adequate to explain what constitutes a species. For example, garden peas come in a wide variety of sizes and shapes, but all garden peas are members of the same species (*Pisum sativum*).

Study of the genetics of populations did much to clarify the concept of species, and now a species is defined as a group of organisms with a common gene pool. Often referred to as the **biolocial species concept**, this definition of species is based on reproductive isolation. Members of a species freely breed with other members of the same species to produce fertile offspring and do not interbreed with—that is, are **reproductively isolated** from—members of other species. In other words, each species has a gene pool that is isolated from the gene pools of other species, and each species is restricted by reproductive barriers from mixing genetically with other species.

One of the problems with the biological species concept in relation to plants is that it applies only to sexually reproducing organisms. Plants often reproduce asexually or by self-pollination, so the concept of reproductive isolation plays no role in defining them as species. The general concept of species is, however, still valid for these organisms; they are classified into species on the basis of structural and biochemical characteristics.

Another problem with the biological species concept is that two populations that are widely separated geographically may be so much alike that they are classified as the same species, but it is impossible to test whether they will interbreed in nature. For example, the quaking aspen (*Populus tremuloides*) has the largest range of any tree in North America, from northern Alaska and Canada to Mexico. It is not known whether quaking aspens from an extremely northern population could successfully interbreed with those from an extremely southern population. In the absence of this information, quaking aspens are considered a single species.

Conversely, organisms assigned to different species may interbreed if they are brought together in a greenhouse, zoo, aquarium, or laboratory. Therefore, we usually include in our definition of species that the members of a species do not normally interbreed with members of other species *in nature*.

To summarize, a species is a group of organisms that have a common gene pool and are capable of interbreeding with one another in nature but are reproductively isolated from other species. This definition is far from perfect, and the biological concept of species has limitations.

Species Have Various Mechanisms That Are Responsible for Reproductive Isolation

Certain mechanisms prevent interbreeding between different species, thus maintaining reproductive isolation. These mechanisms preserve the integrity of the gene pool of each species by preventing gene flow between different species. Sometimes timing prevents genetic exchange between two groups, as when they flower at different times of the day, season, or year. For example, black sage and white sage, two very similar species, live in the same area of southern California but do not interbreed because black sage flowers in early spring and white sage blooms in late spring and early summer (Figure 17-5).

(a)   (b)

**Figure 17-5   Reproductive isolation due to timing of re-
production. (a)** The black sage (*Salvia mellifera*) flowers in
early spring. **(b)** The white sage (*Salvia apiana*), photographed
at the same time of year, has unopened flower buds.
(*Courtesy of Robert Thorne, Rancho Santa Ana Botanic Garden*)

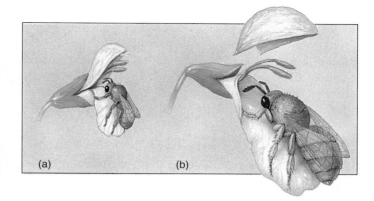

(a)   (b)

**Figure 17-6   Reproductive isolation occurs between
black sage (*Salvia mellifera*) and white sage (*Salvia
apiana*) as a result of the differences in floral structure,
which evolved to make use of different insect pollinators.
(a)** The petal of the black sage is shaped as a landing platform for small carpenter bees. Larger bees cannot use this
platform. **(b)** The larger landing platform and longer stamens
of the white sage allow pollination by larger carpenter bees
(a different species). If smaller bees land on white sage, they
do not accomplish pollination because their bodies do not
brush against the pollen-bearing stamens. (The upper part of
the white sage flower has been removed.)

Many flowering plants have physically distinct flower
parts that help them maintain their reproductive isolation
from one another. Black sage, which is pollinated by small
bees, has a floral structure different from that of white sage,
which is pollinated by large carpenter bees (Figure 17-6). The
differences in floral structures prevent insects from cross-
pollinating the two species, should they happen to flower at
the same time.

Sometimes, despite these reproductive isolating mechanisms, fertilization occurs between gametes of two species.
However, other reproductive isolating mechanisms may still
ensure reproductive failure. Often the embryo of such a union
is aborted. A plant's development from fertilized egg to a
complete multicellular organism is a complex process that
requires the precise interaction and coordination of many
genes. Apparently, the genes from gametes of different species
do not interact properly to regulate normal embryonic development. If an **interspecific hybrid** (a hybrid formed between
two species) does live, it still may not be able to reproduce,
often because of chromosomal differences.

THE KEY TO SPECIATION IS THE EVOLUTION OF REPRODUCTIVE ISOLATING MECHANISMS

With our understanding of reproductive isolation and its role
in defining what a species is, we are now ready to consider
speciation, the formation of a new species. Speciation occurs
when a population diverges, or splits apart, from the rest of
the species. A required step during speciation is the repro-

ductive isolation of that population from other members of
the species. When a population has become sufficiently different from its ancestral species so that no genetic exchange
can occur between them, even if the two populations meet,
speciation has occurred. Such a situation is thought to arise in
one of two ways, through allopatric or sympatric speciation.

Long Geographical Isolation and Different Selective Pressures Result in Allopatric Speciation

Speciation that occurs when one population becomes geographically separated from the rest of the species and subsequently evolves is known as **allopatric speciation** (*allo,* "different," and *patri,* "fatherland").

The geographical isolation required for allopatric speciation may occur in several ways. The Earth's surface is in a
constant state of change: rivers shift their courses; glaciers
migrate; mountain ranges form; land bridges develop, separating previously united aquatic populations; large lakes dimin-
ish into several smaller, geographically separated pools.

What might be an imposing geographical barrier to one
species may be of no consequence to another. For example,
as a lake subsides into smaller pools, fish are usually unable
to cross the land barriers between the pools and so become
reproductively isolated. On the other hand, plants such as
cattails, which disperse their fruits by air currents, would not
be isolated by this barrier.

Allopatric speciation also occurs when a small population migrates and colonizes a new area away from the original species' range. This colony is geographically isolated from its parent species. The Galapagos Islands and the Hawaiian Islands were colonized by individuals of a few species of plants and animals. From these original colonizers, the distinctive species of plants and animals unique to each island evolved.

New Plant Species May Evolve in the Same Geographical Region as the Parent Species

Although geographical isolation is an important factor in many cases of evolution, it is not an absolute requirement. When a population forms a new species within the same geographical region as its parent species, **sympatric speciation** (*sym,* "together," and *patri,* "fatherland") has occurred. Sympatric speciation is especially common in plants.

We have seen that interspecific hybrids, which form from the union of gametes from two species, rarely produce offspring and that, when they do, the offspring are usually sterile. This is because the chromosomes of an interspecific hybrid are not homologous and therefore cannot properly be parceled into gametes during meiosis.

However, if the number of chromosomes doubles *before* meiosis takes place, then the chromosomes occur in pairs, and meiosis can occur successfully. This spontaneous doubling of chromosomes has been well documented in plants. It produces nuclei with multiple sets of chromosomes.

Polyploidy, the possession of more than two sets of chromosomes, is a major factor in plant evolution. When polyploidy occurs in conjunction with sexual reproduction between individuals of two species, it is known as **allopolyploidy** and can produce a fertile interspecific hybrid (Figure 17-7). Because allopolyploidy provides the homologous chromosomes necessary for meiosis, gametes may be viable. Plants that are allopolyploids can reproduce with themselves (self-fertilize) or with similar individuals. However, they are reproductively isolated from both parent species because the allopolyploid has a different number of chromosomes from either of its parents.

Allopolyploidy has been a significant factor in the evolution of flowering plants. As many as 80 percent of flowering plants are polyploids, and most of these are allopolyploids. Moreover, allopolyploidy provides a mechanism for extremely rapid speciation. A single generation is all that is needed to form a new, reproductively isolated species. Allopolyploidy helps explain the rapid appearance of flowering plants in the fossil record and the remarkable diversity (more than 250,000 species) in flowering plants today.

The kew primrose (*Primula kewensis*), discussed in the chapter introduction, is an example of sympatric speciation (Figure 17-8). The interspecific hybrid of two primrose species, *P. floribunda* (2n = 18) and *P. verticillata* (2n = 18), *P. kewensis* had a chromosome number of 18 but was sterile. Then, at three different times, it was reported to have spon-

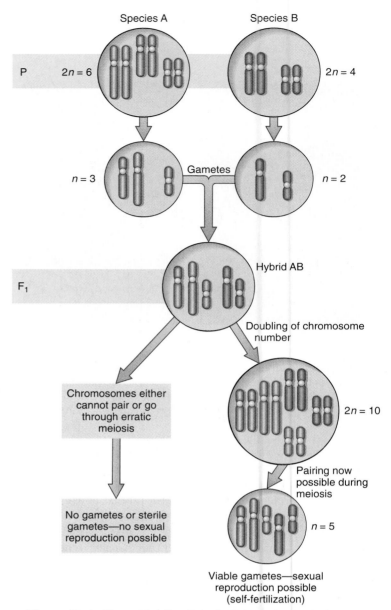

Figure 17-7   How a fertile allopolyploid is formed. When two species (designated the *P* generation) successfully interbreed, the interspecific hybrid offspring (the *F₁* generation) are almost always sterile (*bottom left*). If the chromosomes double before meiosis occurs, the interspecific hybrid is able to undergo meiosis and is fertile (*bottom right*).

taneously formed a fertile branch, which was an allopolyploid (2n = 36) that produced viable seeds of *P. kewensis.*

ADAPTIVE RADIATION IS THE DIVERSIFICATION OF AN ANCESTRAL SPECIES INTO MANY SPECIES

In **adaptive radiation,** an ancestral organism evolves in a relatively short period of time into many new species that fill a variety of ecological niches—that is, roles or ways of living in a community.

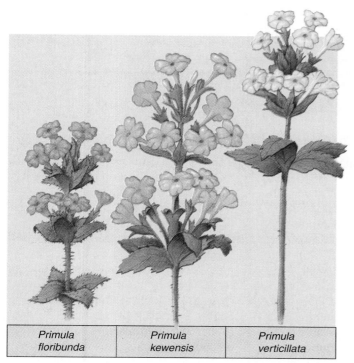

(a)

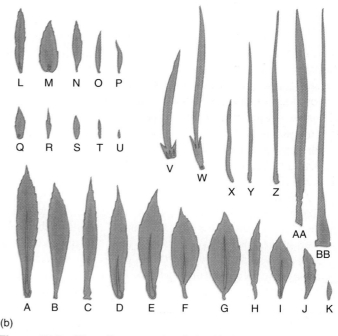

(b)

Figure 17-8 An allopolyploid primrose (*Primula kewensis*) arose during the early part of the 20th century in a greenhouse at the Royal Botanic Gardens at Kew, England. The F₁ hybrid of *P. floribunda* ($2n = 18$) and *P. verticillata* ($2n = 18$) was a sterile perennial with 18 chromosomes. Three different times this sterile hybrid spontaneously formed a fertile branch, which was *P. kewensis*, a fertile allopolyploid ($2n = 36$) that produced seeds. Today *P. kewensis* is a popular houseplant.

Each ecological niche can be occupied by only one type of organism. A newly evolved species can take over an ecological niche even if that niche is already occupied by another species, provided the new species has features that make it competitively superior to the original occupant. Alternatively, if a number of ecological niches are empty, they may be exploited by way of adaptive radiation. For example, consider the Hawaiian silverswords, 28 species of closely related plants found only on the Hawaiian Islands. When the silversword ancestor, a California plant related to daisies, reached the Hawaiian Islands, there were many diverse environments, such as exposed lava flows, dry woodlands, moist forests, and bogs. The succeeding generations of silverswords quickly diversified to occupy the many ecological niches available to them. The diversity in their leaves, which changed during the course of natural selection to enable different populations to adapt to various levels of light and moisture, is a particularly good illustration of adaptive radiation (Figure 17-9). For example, leaves of silverswords that are adapted to shady, moist forests are large, whereas those of silverswords living in exposed dry areas are small. Leaves of silverswords living on volcanic slopes, which receive

(Text continues on page 292)

Figure 17-9 The silverswords of the Hawaiian Islands are a dramatic example of adaptive radiation. Their ancestor was a California plant similar to the daisy. **(a)** Shown is a silversword (*Argyroxyphium sandwicense*) that grows only in Haleakala National Park in Hawaii. **(b)** Leaf shapes of the 28 Hawaiian silverswords, all at a scale of ×.25. Leaf color, which varies considerably among the species, is not shown. (a, *H. Gordon Morris, University of Tennessee*; b, *From Carr et al. 1989*)

PLANTS AND THE Environment

Extinction Today—Endangered and Threatened Species

Currently the Earth's organisms are disappearing at an alarming rate. Conservation biologists estimate that species are becoming extinct at a rate approximately 10,000 times the rate of background extinctions. As much as one fifth of all species may become extinct within the next 30 years. The Center for Plant Conservation estimates that about 4000 native plant species are of conservation concern in the United States; this number represents about 20 percent of U.S. plant species (see figure).

A species is **endangered** when its numbers are so severely reduced that it is in danger of becoming extinct. When extinction is less imminent but the population of a particular species is still quite low, the species is said to be **threatened**. The genetic variability of an endangered or threatened species is severely diminished. Because long-term survival and evolution depend on genetic diversity, endangered and threatened species are at greater risk of extinction than are species with more genetic variability.

Endangered species share certain characteristics that seem to make them more vulnerable to extinction than other species. These include (1) an extremely small, localized range; (2) an extended territory that has been greatly modified by humans; (3) an island habitat; and (4) low reproductive success, which is usu-

ally the result of a small population size.

Many endangered species have very limited natural ranges, which makes them particularly prone to extinction if their habitats are altered. The Tiburon mariposa lily (*Calochortus tiburonensis*), for example, is found nowhere in nature except on a single hilltop near San Francisco. Similarly, the Florida stinking cedar (*Torreya taxifolia*) grows only on limestone banks along Florida's Apalachicola River. Development of either area would almost certainly cause the extinction of the species that lives there.

Many species that are endemic to certain islands (that is, are not found anywhere else in the world) are endangered. These organisms often have small populations that cannot be replaced should their numbers be destroyed. For example, by 1986 only two trees of the highly endangered *Banara vanderbiltii* remained in Puerto Rico. Fortunately, botanists obtained a few cuttings that are now growing in a tropical garden in Miami.

In order for a species to survive, its members must be present in large enough numbers in their range for successful reproduction. The minimum population density and size that ensure reproductive success vary from one type of organism to another. However, for all organisms,

if the population density and size fall below a critical minimum level, the population decreases, making it susceptible to extinction.

Today the most serious threat to the survival of many species is habitat disruption caused by human activities. We alter habitats when we build roads, parking lots, and buildings; clear forests to grow crops or graze domestic animals; or log forests for timber. We drain marshes to build on aquatic habitats, thus making them terrestrial, and we flood terrestrial habitats when we build dikes and dams, making them aquatic. Because most organisms are utterly dependent on a particular type of environment, habitat destruction reduces their biological range and their ability to survive.

Even habitats left undisturbed and in their natural state are modified by human activities that produce acid precipitation and other forms of *pollution*. Acid precipitation is thought to have contributed to the decline of large stands of forest trees and to the biological death of many freshwater lakes. The production of other types of pollutants also adversely affects wildlife. Such pollutants include industrial and agricultural chemicals, organic pollutants from sewage, acid wastes seeping from mines, and thermal pollution from the heated waste water of industrial plants.

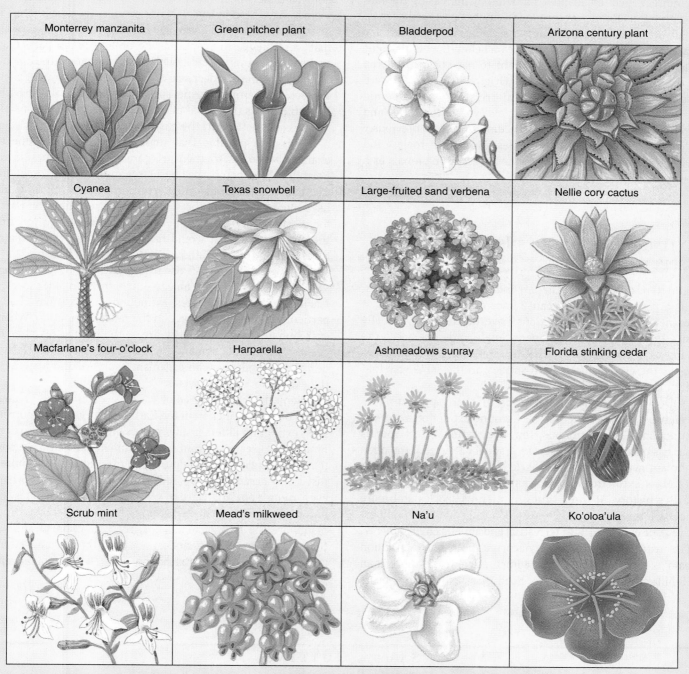

Sixteen of the 4000 native plant species of conservation concern in the United States.

intense ultraviolet radiation, are covered with dense silvery hairs that may reflect some of the radiation off the plant.

Adaptive radiation appears to be more common during periods of major environmental change, but it is difficult to determine whether such change actually triggers adaptive radiation. It is possible that major environmental change has an indirect effect on adaptive radiation by increasing the rate of extinction. Extinction produces empty ecological niches, which are then available for adaptive radiation. Mammals, for example, had evolved millions of years before they underwent adaptive radiation, which is thought to have been triggered by the extinction of the dinosaurs. Originally, mammals were small animals that ate insects. In a relatively short period after the dinosaurs' demise, mammals evolved to occupy and exploit a variety of roles that the dinosaurs had previously filled. Flying bats, running gazelles, burrowing moles, and swimming whales evolved from the ancestral mammals as a result of adaptive radiation.

EXTINCTION OF SPECIES IS AN IMPORTANT ASPECT OF EVOLUTION

Extinction occurs when the last individual of a species dies. It is a permanent loss, because once a species is extinct, it can never reappear. Extinctions have occurred continually since the origin of life. By one estimate, only one species is living today for every 2000 that have become extinct. Extinction is the eventual fate of all species, in the same way that death is the eventual fate of all individuals.

During the course of life on Earth, there appear to have been two types of extinction, background extinction and mass extinction. The continuous, low-level extinction of species that has occurred in response to gradual changes in the environment over time is called **background extinction.**

Mass extinctions have occurred five or six times during Earth's history. At these times, numerous species and more inclusive taxonomic categories (such as genera and families) have gone extinct in both terrestrial and aquatic environments. Mass extinctions may have occurred over several million years, a relatively short time compared with the more than 3.5-billion-year history of life. Each mass extinction appears to have been indiscriminate in terms of which species survived and which became extinct. Each episode of mass extinction was followed by a period during which the surviving species evolved rapidly to fill many of the ecological niches vacated by the extinct species.

The causes of extinction, particularly mass extinction, are not well understood, but environmental and biological factors seem to have been involved. Major changes in the climate, for instance, would have adversely affected plants and animals that were unable to adapt. Marine organisms, in particular, are adapted to a very steady, unchanging climate. If the Earth's temperature were to decrease or increase overall by just a few degrees, numerous marine species would probably perish. Many biologists think that climatic changes were responsible for mass extinctions in the past.

It is also possible that mass extinctions were due to changes in the environment triggered by catastrophes. If the Earth had been bombarded by a large meteorite or comet, for example, the dust going into the atmosphere on impact could have blocked much of the sunlight from reaching the Earth for at least several months. In addition to killing many plants, the atmospheric dust would have lowered Earth's temperature, leading to the deaths of many marine organisms.

Biological factors can also trigger extinction. When a new species evolves, it may be able to outcompete an older species, leading to the older species' extinction. Humans have had a particularly profound impact on the rate of extinction. Whenever humans move into new areas, their activities alter or destroy the habitats of many plant and animal species. Habitat destruction endangers an organism's survival and may cause the extinction of that species.

Some biologists fear that we have entered the greatest period of mass extinction in the Earth's history, but the current mass extinction differs from previous ones in several respects. First, it is directly attributable to human activities. Second, it is occurring in a tremendously compressed period of time— just a few decades as opposed to millions of years. Perhaps even more sobering, larger numbers of plant species are becoming extinct than in previous mass extinctions. Since plants are the base of food webs, the extinction of animals that depend on those plants cannot be far behind (see Plants and the Environment: Extinction Today—Endangered and Threatened Species).

STUDY OUTLINE

I. Microevolution is a change in allele frequencies in the gene pool of a population. Each individual within a population contains only a portion of the alleles in the gene pool

    A. The Hardy-Weinberg law states that allele frequencies in a population tend to remain constant in successive generations unless certain factors are operating.

B. Allele frequencies may be changed by mutation, genetic drift, gene flow, or natural selection.
  1. The source of new alleles in a gene pool is mutation.
  2. Genetic drift is the random change in allele frequencies of a small breeding population. The changes are usually not adaptive.
  3. The migration of individuals between local populations results in a corresponding movement of alleles, or gene flow, that can cause changes in allele frequencies.
  4. Changes in allele frequencies that lead to adaptation to the environment are caused by natural selection, in which members of a population that possess better adaptations to the environment are more likely to survive and reproduce, thereby increasing the proportion of their alleles in the population.

II. A biological species is a group of organisms with a common gene pool that have the potential to breed with one another in nature but not with members of other species.

III. Reproductive isolating mechanisms restrict the gene flow between species.
  A. Reproductive isolation occurs when two species reproduce at different times of the day, season, or year.

B. Reproductive isolation also occurs as a result of structural differences in the reproductive organs of species.
C. Reproductive failure is common even when fertilization has taken place between gametes of two species.
  1. The interspecific hybrid usually dies at an early stage of embryonic development.
  2. If an interspecific hybrid survives to adulthood, it usually cannot reproduce successfully.

IV. Speciation is the evolution of a new species from an ancestral population.
  A. Allopatric speciation occurs when one population becomes geographically isolated from the rest of the species and subsequently evolves.
  B. Sympatric speciation does not require geographical isolation. In plants it occurs as a result of allopolyploidy.

V. Adaptive radiation is the process of evolutionary diversification of an ancestral species into many species.

VI. Extinction is the termination of a species. Once a species is extinct, it can never reappear.

SELECTED KEY TERMS

adaptive radiation, p. 288
allopatric speciation, p. 287
allopolyploidy, p. 288
extinction, p. 292

gene flow, p. 285
gene pool, p. 282
genetic drift, p. 284
Hardy-Weinberg law, p. 283

interspecific hybrid, p. 287
microevolution, p. 282
mutation, p. 283
natural selection, p. 285

population, p. 281
reproductive isolation, p. 286
species, p. 281, 286
sympatric speciation, p. 288

REVIEW QUESTIONS

1. Define population, species, and gene pool. How did population genetics help clarify the definition of a species?
2. Explain the effect of each of the following on genetic variation: (a) natural selection, (b) mutation, (c) gene flow, (d) genetic drift.
3. If a mutation occurs in a somatic cell, can the new allele become established in a population? Explain why or why not.
4. Why are mutations almost always neutral or harmful?
5. Give several examples of reproductive isolation.
6. Describe why interspecific reproduction usually fails even when fertilization has taken place.

7. Identify at least five geographical barriers that may lead to allopatric speciation.
8. Why is allopatric speciation more likely to occur if the original isolated population is small?
9. Explain how allopolyploidy can cause a new plant species to form in as little time as one generation.
10. What role does extinction play in evolution?

THOUGHT QUESTIONS

1. Given that the Hardy-Weinberg equilibrium occurs only under conditions that populations in nature seldom, if ever, experience, why is it important?
2. Explain why we discuss evolution in terms of the selective advantage that a particular *genotype* confers on an individual, yet natural selection acts on an organism's *phenotype*.

3. Why is it necessary to understand the biological concept of species in order to understand evolution?
4. How far apart must two populations of insect-pollinated plants be to be genetically isolated from each other?

5. Examine the maps on the right, which show the continents as they appeared 240 million years ago (*a*), when they were joined together, and today (*b*). How might continental drift—the separation and drifting apart of once-joined landmasses—have affected the evolution of organisms on those landmasses? Organisms in the ocean?

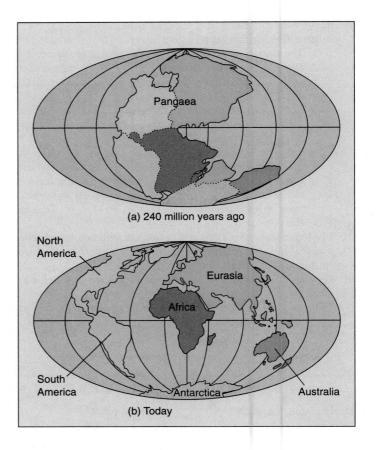

(a) 240 million years ago

(b) Today

SUGGESTED READINGS

Arnold, M.L., "Natural Hybridization and Louisiana Irises: Defining a Major Factor in Plant Evolution," *BioScience,* Vol. 43, No. 3, March 1994. Interspecific hybridization appears to be a significant factor in the evolution of the Louisiana irises.

Kerr, R.A., "When Climate Twitches, Evolution Takes Great Leaps," *Science,* Vol. 257, September 18, 1992. How climate changes have driven the evolutionary process.

Rennie, J., "Are Species Specious?" *Scientific American,* November 1991. A brief essay on the problems biologists still have with the definition of species.

Thompson, J.D., "The Biology of an Invasive Plant," *BioScience,* Vol. 41, No. 6, June 1991. A new species of cordgrass that evolved approximately 100 years ago has become common in the salt marshes of Great Britain.

DIVERSITY

CHAPTER

18

THE CLASSIFICATION OF PLANTS AND OTHER ORGANISMS

The European white water lily (*Nymphaea alba*), which is found in Europe, North Africa, and parts of Asia, is an herbaceous, aquatic flowering plant that thrives in shallow, still water. About 35 temperate and tropical species of water lilies belong to the genus *Nymphaea* worldwide. All possess underground stems (either tubers or rhizomes) from which leaves arise. Each leaf consists of a long petiole and a large round or heart-shaped blade that usually floats on the water's surface. European white water lilies have large, attractive white flowers that also float on the water's surface, although the fruit that develops from the flower ripens underwater. Like some other water lilies, the European white water lily is widely cultivated in aquatic gardens.

The genus of water lilies—*Nymphaea*—is named after water nymphs, the beautiful fairy-like goddesses in Greek mythology. Linnaeus, the 18th-century scientist who described and named these plants, was familiar with classical literature that depicted nymphs. Because water lilies live in ponds, lakes, and

The European white water lily has more than 200 different common names but only one scientific name: *Nymphaea alba*. (*Richard H. Gross*)

slow-moving waterways, they reminded Linnaeus of the mythical water nymphs.

The scientific name of the European white water lily, *Nymphaea alba*, consists of two words. One tells its genus, *Nymphaea*, which is common to all water lilies, and the other its species (or, more accurately, its specific epithet), *alba*, which is Latin for "white."

Although scientific names are often incomprehensible to lay people, most of them, like *Nymphaea alba*, are full of meaning. Such names are also a scientific necessity. The European white water lily has over 200 local, or common, names in at least four languages (English, French, German, and Dutch); in English it is called, among others, water lily, white water lily, European white water lily, water nymph, and platter dock. Clearly, biologists from different parts of the world who are interested in *Nymphaea alba* need to know that they are all talking about the same species. Scientific nomenclature enables biologists of all nationalities to be precise when they discuss and share data on organisms.

TAXONOMY IS THE SCIENCE OF DESCRIBING, NAMING, AND CLASSIFYING ORGANISMS

How would you use what you already know about plants if you wanted to assign them to groups? Would you place the red Christmas cactus, red rose, and red hibiscus in one group because they all have the same flower color? Or would you classify plants according to their uses, placing edible plants in one group and poisonous plants in another?

Each of these schemes might be valid, depending on your purpose. Similar classification methods have been used throughout history. Theophrastus, a Greek philosopher and biologist who lived during the third century B.C., classified several hundred different plants into such groups as herbs, shrubs, and trees. His system of classification persisted for many centuries.

After the invention of a printing press with movable metal type in 1448, early botanical works known as **herbals** were printed. These volumes contained drawings and descriptions of plants, particularly those with medicinal uses (Figure 18-1). Some herbals also contained attempts at classification. European explorers of the 16th and 17th centuries introduced to Europe hundreds of new plants gathered from other continents, compounding the need for a simple method of classification (see Focus On: Plant Exploration—Then and Now).

The classification system designed in the mid-18th century by Carolus Linnaeus, a Swedish botanist, has survived, with some modification, to the present day (Figure 18-2). Linnaeus' Swedish name was Carl von Linné. Like other scholars of his time, he used the Latinized form of his name. In *Species Plantarum*, which was published in 1753, Linnaeus described all the plants known in his time—some 7300 species—and provided each with a binomial name. Linnaeus based his system of classification in *Species Plantarum* on visual observations of flower parts. He noted, for example, that all flowers

of the same kind of plant contain the same number of stamens (pollen-producing structures).

Linnaeus probably intended to design an unchanging system of classification, for he carried out his work long before Darwin's theory of evolution made common ancestry the basis for classification. Neither Linnaeus nor his colleagues had any concept of the vast number of living and extinct organisms that would later be discovered. Yet it is remarkable how flexible and adaptable to new biological knowledge and theory his

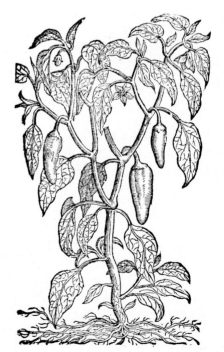

Figure 18-1   A drawing of chili pepper (*Capsicum annuum*), taken from the *Lobel Herbal,* published in 1581. Scholarly books of plants, known as herbals, were printed during the 15th, 16th, and 17th centuries. (*The Field Museum, Chicago, IL, neg #72370*)

(Text continues on p. 300)

OUI

Focus On

Plant Exploration— Then and Now

Humans have continuously introduced wild plants into their gardens since they began cultivating food crops. The deliberate introduction of plants from other regions has occurred throughout history as people explored or invaded new areas. When the Spanish invaded South America, for example, they introduced potatoes, peppers, tomatoes, and many other plants to Europe.

The first expeditions that were organized exclusively to collect plants are thought to have occurred during the 18th century, when plant explorers were typically employed by kings, wealthy men, or nurseries. Plant explorers faced many dangers, including hostile natives and deadly diseases. During the 19th century, for example, seven of the 22 plant explorers hired by the Veitch nursery in England died overseas from disease, and one had to have his leg amputated.

Transporting the plants home frequently involved long sea journeys. Live plants that were kept on deck to obtain sunlight often succumbed to the salt spray. Plants that were transported as seeds, bulbs, corms, or tubers were more likely to survive. However, they had to be completely dried out to avoid mold, no easy feat if the collecting occurred in the humid tropics. To keep rats on board from eating them, the dried plant parts were packed in wooden cases that were filled with broken glass. If a turbulent storm threatened the ship, the plant material was often tossed overboard to lighten the load.

The experiences of two important plant explorers—Robert Fortune and David Fairchild—highlight the important contributions of the many plant explorers who introduced plants from the wild. Robert Fortune (1812–1880) was a Scottish plant explorer who collected plants in China during the 1840s and 1850s (see Figure *a*). Fortune, who worked for the British East India Company and the Horticultural Society of London, smuggled tea, along with knowledge of its cultivation and processing after harvest, out of China to the British. He discovered that green tea and black tea were produced by the same plant, *Camellia sinensis*. Black tea is made from "fermented" leaves, whereas green tea is "unfermented." Fortune also introduced winter jasmine, forsythia, Chinese holly, and many other beautiful ornamental plants to Europe.

Fortune smuggled plants out of China despite the fact that China was little known by outsiders and was politically unstable. To obtain access to areas declared off limits to foreigners, Fortune sometimes disguised himself as a native. Once he even had to fight and kill pirates who were attempting to board his boat on the South China Sea.

David Fairchild (1869–1954) was a famous American plant explorer (see Figure *b*). Working for the U.S. Department of Agriculture, Fairchild searched all over the world for economically useful plants. He caught

(a)

(b)

(a) Robert Fortune was a 19th-century plant explorer. (b) David Fairchild was a famous American plant explorer. This photo shows him when he was in his early 80s. He is holding his favorite tropical fruit, the mangosteen, which is native to Malaysia and requires a hot, humid climate. Mangosteens are not well known outside the tropics because they do not ship or store well. (a, *Courtesy of Hunt Institute for Botanical Documentation, Carnegie Mellon University, Pittsburgh, Pennsylvania*; b, *Courtesy of Fairchild Tropical Garden*)

typhoid in Ceylon (now Sri Lanka) but continued his travels after a brief recovery period. Fairchild introduced many thousands of fruits, vegetables, grains, and ornamental plants to the United States. Some of the plants he collected include varieties of the water chestnut from China; cotton, grape, carob, barley, and dates from Egypt; pistachios from Greece; blood oranges from Malta; Jordan almonds from Spain; wheat from Russia; alfalfa from Peru; and rice and soybeans from Japan.

Plant exploration still goes on but only rarely involves the dangers of former times—although a botanist sent to Colombia by the Missouri Botanical Garden in the 1980s was beaten and held captive by Indians until the military intervened. Plant exploration today is generally focused on the collection of a particular plant—for example, cotton plants to breed with existing varieties. This focus replaces the "Bring back whatever you find that is interesting" approach of past plant exploration.

Usually a team of two botanists is sent to a foreign country to collect the plants, but first they do extensive research to determine where they should look for their target plants. For example, plant explorers first examine a collection of dried, pressed, and carefully labeled plant specimens collected in the past, which are stored in an **herbarium** (see Figure c). Some herbarium specimens more than 200 years old are still in use. The specimens brought back from the Lewis and Clark expedition (1804–1806), for example, are still preserved and are included in the Smithsonian National Herbarium. Often the labels on herbarium specimens tell plant

(c)

No 454

HERBARIUM OF ORLAND E. WHITE ARBORETUM
BOYCE, VIRGINIA

HIBISCUS SYRIACUS

COMMON NAME: SHRUBBY ALTHEA
FAMILY: MALVACEAE
TREE #: 4654
LOCATION: 121
COLLECTED BY: ALICIA GRAYSON
COLLECTION DATE: 29 JULY 1992
IDENTIFIED BY: ALICIA GRAYSON

(d)

(c,d) An herbarium specimen of shrubby althaea (*Hibiscus syriacus*), collected in 1992. (c) The entire herbarium sheet. (d) A closeup of the label. Location 121 refers the researcher to a list that indicates exactly where this specimen was collected, on the University of Virginia's Blandy Experimental Farm in Boyce. (*Marion Lobstein*)

explorers precisely where the plants were collected—for example, "in a field 15 km south of the city, along the western banks of the river."

Plant exploration occurs in many places worldwide. For example, a team of Australian plant biologists might be sent to North Africa to collect plants to be screened for their potential as sources of new drugs to fight diseases such as cancer. Or a team of U.S. botanists might be sent to the Middle East to collect seeds of forage grasses (grasses grown for animal fodder) that will be used in selective breeding experiments to improve the rangeland grasses of western North America. Tropical areas of Central and South America and Asia are common sites

for plant exploration, in part because most of the world's fruits and vegetables originated in the tropics. Locating the wild ancestors of important crops might enable us to introduce beneficial genes from these wild plants into existing crops.

The transport of plant material from one country or continent to another is more successful now than it once was, because the plants or seeds are wrapped in polyethylene, refrigerated, and transported quickly by air rather than ship. Plant exploration today relies on international cooperation, and special care is taken to avoid the accidental introduction of devastating pests along with the plant material.

Figure 18-2   Carolus Linnaeus (1707–1778), a Swedish botanist and physician, is considered the father of modern taxonomy because of his contributions to the classification of plants and other organisms. He standardized scientific terminology by establishing the use of the binomial system of nomenclature to assign scientific names. (*The Linnaean Society of London*)

system has proved to be. Very few other 18th-century inventions survive today in a form that their originators would recognize. Linnaeus' system of classification provided the foundation on which modern biology was built.

Species Are Classified with the Binomial System of Nomenclature

Prior to Linnaeus, scholars used Latin sentences up to 12 words long to describe each type of plant. For example, the spiderwort, a popular garden plant, was scientifically described as *Tradescantia ephemerum phalangoides tripetalum non repens Virginianum gramineum* (Figure 18-3). Translated, this name means "the annual upright Tradescantia from Virginia with a grasslike habit, three petals, and stamens with hairs like spider legs." Linnaeus used this lengthy scientific description as well, but he added in the margin a simplified scientific name for this plant, *Tradescantia virginiana*.

Thus, Linnaeus' most significant contribution to biology is his **binomial system of nomenclature**, a system of naming organisms based on a unique two-part name for each. The first part of the name designates the genus, and the addition of the second part, the specific epithet, designates the species. The specific epithet is usually a word that describes some particular quality of the organism.

The generic (genus) name can be used alone to designate all species in the genus, but the specific epithet can never be used alone; it must always be preceded by the full or abbreviated generic name. In each scientific name, the generic name appears first, and its first letter is always capitalized; the specific epithet is given second and is not capitalized. Both names must be underlined or italicized. Corn, for example, is assigned to the genus *Zea;* its specific epithet is *mays*. The proper scientific name for corn is *Zea mays*.

Although any given specific epithet can be used for only one species within each genus, it can be used in more than one genus. For example, *Quercus alba* is the scientific name for the white oak, and no other oak species can have *alba* as its specific epithet. However, *Nymphaea alba* is the scientific name for the European white water lily, and *Salix alba* for

Figure 18-3   Spiderworts (*Tradescantia virginiana*), found in meadows and along roadsides and the edges of woods, are also popular garden plants. **(a)** A spiderwort plant. **(b)** A closeup of the flower reveals six golden stamens with hairy filaments. (a,b, *Marion Lobstein*)

(a)

(b)

TABLE 18-1   Derivations of Selected Plant Genera Mentioned in This Chapter

Genus	Derivation
Acer	Latin name for maple
Camellia	For George Kamel (Camellus was his Latinized name) (1661–1706), a Moravian Jesuit who studied Asian plants
Gazania	For Theodore of Gaza (1398–1478), who translated the botanical works of Theophrastus
Gilia	Perhaps for Felipe Gil, an 18th-century Spanish botanist
Hedera	Latin name for ivy
Hibiscus	Greek name for mallow (a kind of plant)
Hymenocallis	From the Greek words for "membrane" and "beauty," in reference to the stamens, which are joined by a membrane
Lupinus	Latin name for lupine
Mirabilis	From the Latin word for "wonderful"
Nymphaea	After Nymphe, a water nymph
Picea	Latin name for pitch, a resin produced by spruce
Quercus	Latin name for oak
Salix	Latin name for willow
Tradescantia	For John Tradescant (1570–1638), English gardener and botanist

the white willow. (Recall that *alba* is a Latin word meaning "white.") Both parts of an organism's scientific name must be used to accurately identify the species.

Scientific names are generally composed from Greek or Latin roots or from Latinized versions of the names of persons, places, or characteristics (Table 18-1). *Hedera,* the generic name for the ivy *Hedera canariensis,* for example, is based on the Latin word for ivy. The specific epithet *canariensis* tells us that this plant is native to the Canary Islands.

The use of Latin, a dead language that does not change, rather than a modern language in naming organisms is a carry-over from the days when Latin was the language of European scholars. Why do we continue to use Latin rather than common names for plants and animals? Why call a sugar maple *Acer saccharum?* (*Acer* is derived from the Latin word meaning "maple," and *saccharum* from the Greek for "sugar.") The main reason is to be accurate and to avoid confusion, for in some parts of the United States this same tree is called hard maple or rock maple. The plant *Mirabilis jalapa* is generally called four-o'clock, but some people call it marvel-of-Peru and others call it beauty-of-the-night. Because of the many instances of confusing common names, exact scientific names are important for the accurate identification of organisms. In addition, many scientifically important organisms lack common names. Since each organism can have only one scientific name, a researcher in Canada reading a study published by a German knows exactly which organisms were used.

The Species Is the Basic Unit of Classification

Recall from Chapter 17 that a **species** is a group of organisms that have a common gene pool and can breed with one another under natural conditions but are reproductively isolated from other organisms.[1] Members of a species share a common evolutionary ancestry.

The species is the basic unit of classification but not the smallest taxonomic group in use. Geographically distinct populations within a species often display certain characteristics that distinguish them from other populations of the same species. If they can interbreed, however, they are not truly separate species but are known as **subspecies** or **varieties**. The full scientific name of a subspecies or variety consists of three names—the generic name, the specific epithet, and the subspecies or variety name—instead of two. For example, peaches are *Prunus persica* var. *persica,* and nectarines, a smooth-skinned variety of peaches, are *Prunus persica* var. *nucipersica.*

Experts can usually distinguish subspecies or varieties from one another. However, subspecies may grade into each other at the borders of their geographical ranges, where there is opportunity to interbreed. Some of these subspecies or

---

[1]Chapter 17 discussed some of the shortcomings of the definition of species. One shortcoming is that different plant species can sometimes interbreed to form fertile hybrids. Thus, reproductive isolation is not always characteristic of plant species.

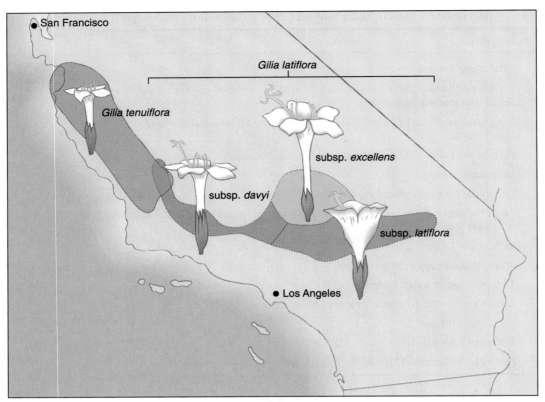

Figure 18-4   The ranges and distinguishing features of flowers of the subspecies of the California wildflower *Gilia latiflora* and of a closely related species, *G. tenuiflora*. From their similarities, it seems probable that *G. tenuiflora* was originally a subspecies of *G. latiflora*. Because it now overlaps *G. latiflora* geographically without interbreeding, *G. tenuiflora* is considered a separate species. For simplicity, three other subspecies of *G. latiflora* have been omitted.

varieties appear to be in the process of becoming reproductively isolated and may, over time, become separate species (Figure 18-4). Thus, some subspecies or varieties provide biologists with the opportunity to study evolution in progress.

Each Taxonomic Level Is More General Than the One Below

Classification is hierarchical; the narrowest category in the Linnaean system is the species, and the broadest is the kingdom. Closely related species are grouped together in the next higher level of classification, the **genus** (pl. *genera*). One or more related genera are assigned to the same **family**, and families are grouped into **orders**, orders into **classes**, classes into **phyla**,[2] and phyla into **kingdoms** (Table 18-2; also see Figure 1-11). These groupings can also be separated into subgroupings—for example, subphylum or subclass.

Each taxonomic level is broader (more inclusive) than the level below. For example, the family Zingiberaceae (the ginger family) contains about 1300 species in 49 genera, including genus *Zingiber* (the ginger genus), genus *Curcuma* (the turmeric genus), and genus *Elettaria* (the cardamom genus) (Figure 18-5). The family Zingiberaceae, along with four other families, is placed in the order Zingiberales. Order Zingiberales and 18 other orders belong to class Monocotyledones (the monocots)—one of two classes of flowering plants. The monocots produce seeds that contain one cotyledon (seed leaf) and possess flower parts (for example, petals, stamens, and carpels) in threes or multiples of three. Class Monocotyledones (monocots) and class Dicotyledones (dicots) are grouped into phylum Anthophyta,[3] which, along with phylum Coniferophyta, phylum Gnetophyta, and several others, belongs to kingdom Plantae.

A **taxon** (pl. *taxa*) is a taxonomic grouping at any level, such as a species, a genus, or a phylum. For example, the

---

[2]Plants have traditionally been classified into divisions rather than phyla. The International Botanical Congress, however, recently approved the use of the phylum designation for plants.

---

[3]Some botanists prefer phylum Magnoliophyta instead of phylum Anthophyta, class Magnoliopsida instead of class Dicotyledones, and class Liliopsida instead of class Monocotyledones. The International Botanical Congress has not yet standardized phylum and class names, although the matter is currently under study.

TABLE 18-2   Classification of Texas Bluebonnet, Alligator Lily, and Blue Spruce

	Texas Bluebonnet	Alligator Lily	Blue Spruce
Kingdom	Plantae	Plantae	Plantae
Phylum	Anthophyta	Anthophyta	Coniferophyta
Class	Dicotyledones	Monocotyledones	Coniferopsida
Order	Fabales	Liliales	Coniferales
Family	Fabaceae	Liliaceae	Pinaceae
Genus	*Lupinus*	*Hymenocallis*	*Picea*
Species	*Lupinus texensis*	*Hymenocallis palmeri*	*Picea pungens*

phylum <u>Anthophyta</u> is a taxon that contains two classes, <u>Dicotyledones and Monocotyledones</u>. Similarly, class Monocotyledones is a taxon that <u>includes</u> many different orders.

MODERN CLASSIFICATION IS BASED ON EVOLUTIONARY RELATIONSHIPS

The classification of plants and other organisms into groups determined by their evolutionary relationships is called **systematics**. A systematist seeks to reconstruct the evolutionary history, or **phylogeny**, of organisms. Once these relationships are defined, the classification of organisms can be based on common ancestry.

If all the plants within a taxon share the same common ancestor, the taxon is referred to as **monophyletic** (one branch). Monophyletic taxa are *natural* groupings, as they represent true evolutionary relationships, and they include all close relatives.

Many taxa are **polyphyletic**, consisting of several evolutionary lines and not including a common ancestor. Polyphyletic taxa may misrepresent evolutionary relationships. For this reason, taxonomists try to avoid constructing polyphyletic taxa.

Cladistics Emphasizes Evolutionary Relationships

An approach to systematics that has been widely adopted in the past several decades is **cladistics**. Cladistics emphasizes phylogeny by focusing on when evolutionary lineages (lines of descent) divide into two different branches. Cladists (taxon-

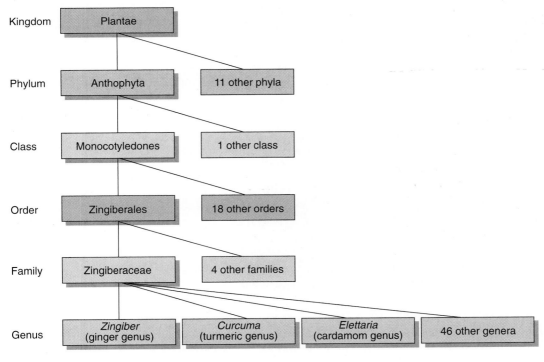

Kingdom — Plantae

Phylum — Anthophyta | 11 other phyla

Class — Monocotyledones | 1 other class

Order — Zingiberales | 18 other orders

Family — Zingiberaceae | 4 other families

Genus — *Zingiber* (ginger genus) | *Curcuma* (turmeric genus) | *Elettaria* (cardamom genus) | 46 other genera

Figure 18-5   Each taxonomic level is more inclusive than the one below it. For example, the order Zingiberales consists of five families. The family Zingiberaceae contains 49 genera and a total of about 1300 species.

Species

Could use Key.

ch 1

Focus On

Using a Dichotomous Key To Identify Plants

Most taxonomists identify organisms primarily on the basis of structural features. Using such features, taxonomists have developed special guides known as dichotomous keys to aid in the identification of unknown plants. A **dichotomous key** consists of a series of two contrasting statements made about plants. The person trying to identify the unknown plant chooses which of two statements is correct for the plant in question and is then directed to another set of paired, contrasting statements. By working through the statements in the key, the investigator can eventually identify and classify the unknown organism.

　　To understand how such a key works, try to identify the plants in the illustration by using the following simplified, abbreviated key to ten common woody plants. The entry in the right-hand column opposite the appropriate structural feature identifies the plant or tells you which number to go to next. Be aware that students often have difficulty using such keys because they are unfamiliar with the terminology or because they make a "wrong turn" in haste.

Key to Ten Common Woody Plants in the Eastern United States

1. a. Needles (elongated, thin leaves) . . . . Pine
   b. Leaves with broad, flat blades 2
2. a. Tree (one main stem) 3
   b. Shrub (multiple stems) or vine 8
3. a. Simple leaves (undivided leaves) 4
   b. Compound leaves (each leaf composed
   of two or more leaflets) 7
4. a. Palmate venation (veins radiating
   from a common point 5
   b. Pinnate venation (veins arranged like a feather) 6
5. a. Opposite leaf arrangement on stem . . . . Maple
   b. Alternate leaf arrangement on stem . . . . Sweet gum
6. a. Narrow blade (1 centimeter or less) . . . . Willow
   b. Broad blade . . . . Tulip poplar
7. a. Leaflets sharply tapered at tip . . . . Hickory
   b. Leaflets blunt at tip . . . . Black locust
8. a. Simple (undivided) leaves . . . . Mountain laurel
   b. Compound leaves (each leaf composed
   of two or more leaflets) 9
9. a. Leaves palmately compound (leaflets
   radiating from a common point) . . . . Virginia creeper
   b. Leaves pinnately compound (leaflets
   arranged like a feather) . . . . Sumac

(a)

(b)

TABLE 18-3   The Six Kingdoms of Life

3
domains

Kingdom	Examples	Characteristics
Eubacteria	Most bacteria belong to this group, including the nitrogen-fixing bacteria, cyanobacteria, lactic acid bacteria (in yogurt), and enterobacteria (in the intestines of humans and other animals)	Prokaryotic cells (lack distinct nuclei and other membranous organelles); single-celled; microscopic; cell walls composed of peptidoglycan
Archaebacteria	Halophiles (live in salt ponds); thermoacidophiles (live in hot sulfur springs); methanogens (live in swamps and in the digestive tracts of humans and other animals)	Prokaryotic cells; single-celled; microscopic; cell walls without peptidoglycan; very different biochemically from eubacteria; adapted to extreme environments, such as hot springs and undersea thermal vents
Protista	Algae; slime molds; water molds; protozoa	Eukaryotic cells (possess distinct nuclei and other membranous organelles); single-celled or simple multicellular; varied modes of nutrition—some are photosynthetic (like plants), some consume food (like animals), and some absorb nutrients (like fungi)
Fungi	Mushrooms; puffballs; mildews; yeasts	Eukaryotic cells; most have a threadlike, multicellular body; nonphotosynthetic; absorb nutrients; cell walls of chitin
Plantae	Mosses; ferns; pines; flowering plants	Eukaryotic cells; multicellular; photosynthetic; possess multicellular reproductive organs; cell walls of cellulose
Animalia	Jellyfish; clams; grasshoppers; frogs; mammals	Eukaryotic cells; multicellular; no cell walls; nonphotosynthetic; most obtain nutrients by consuming other organisms; most move about by muscular contraction; specialized nervous tissue to coordinate responses

omists who practice cladistics) insist that taxa be monophyletic and use carefully defined objective criteria to determine branch points. Thus, common ancestry is the basis for classification. A cladist would say that daisies are classified with flowering plants rather than with ferns because daisies and other flowering plants share a more recent common ancestor than do daisies and ferns. Cladists develop branching diagrams called **cladograms** (see Focus On: Building and Interpreting Cladograms).

DNA Studies Support a Six-Kingdom Scheme of Classification

The broadest level of taxonomic classification is the kingdom. From the time of Aristotle to the mid-19th century, biologists divided the living world into two kingdoms, plants and animals. With the development of microscopes, it became increasingly obvious that many organisms could not easily be assigned to either the plant or the animal kingdom.

More than a century ago, a German biologist, Ernst Haeckel, suggested that a third kingdom be established, the kingdom Protista. Simple and ambiguous organisms, such as bacteria and most microorganisms, many of which were formerly considered members of the plant kingdom, were to be classified in the kingdom Protista.

In the 1960s advances in electron microscopy and biochemical techniques revealed basic cellular differences that inspired many new proposals for classifying organisms. In 1969 R.H. Whittaker proposed a five-kingdom classification (Monera, Protista, Plantae, Fungi, and Animalia) based on cell structure and ways of obtaining energy (that is, modes of nutrition). Kingdom Monera was established to accommodate the bacteria, which are fundamentally different from all other organisms in that they are prokaryotic and lack distinct nuclei and other membranous organelles. Whittaker also suggested that fungi be classified as a separate kingdom, kingdom Fungi, rather than as part of the plant kingdom, because fungi are nonphotosynthetic and obtain energy by absorbing nutrients. Fungi also differ from plants in the composition of their cell walls, in body structure, and in reproduction.

Most biologists accepted the five kingdoms proposed by Whittaker until DNA studies in the late 1980s revealed that there are two very different groups of bacteria. These two groups, the more familiar Eubacteria and the very distinctive Archaebacteria (Chapter 19), are so profoundly different that most biologists think each merits its own kingdom. As a result, a six-kingdom scheme has been proposed that takes into account these evolutionary relationships as well as cell structure and ways of obtaining energy.

The six-kingdom system of classification recognized by many biologists today is described above in Table 18-3 and represented in Figure 18-6 on page 308. The six kingdoms are

305

(Text continues on page 309)

Focus On

Building and Interpreting Cladograms

The general goal of cladistics is to reconstruct phylogenies using an analysis of evolutionary changes in specific traits, or characters. The kinds of characters used for the analysis can be structural, physiological, or molecular. The characters must be homologous (identical by common descent) and must have evolved independently of each other.

The first step in constructing a cladogram is to select the taxa. Here we use a representative group of four plant taxa: moss, fern, pine, and daisy. The next step is to select the homologous characters to be analyzed. In our example we use three characters: vascular tissue, seeds, and flowers (see Table 18-A). For each character, we must define all of the different conditions. For simplicity, we will consider our characters to have only two different states—present and absent.

The last step in preparing the data is to organize the character states into their correct evolutionary order. The most common method of doing this is by outgroup analysis. An **outgroup** is a taxon that is con-

sidered to have diverged earlier than the other taxa under investigation and thus may represent an approximation of the ancestral condition. In our example, moss is the chosen outgroup, and all of its character states are considered to be ancestral. Therefore, the character state "absent" is the ancestral condition, and the character state "present" is the derived (evolved) condition for each of the characters.

Examine the table, and notice that all taxa except the outgroup share a derived trait, vascular tissue. We may therefore conclude that these three taxa—fern, pine, and daisy—form a valid monophyletic group, and, using these data, we may construct a preliminary cladogram (Figure *a*). The base of the diagram represents the common plant ancestor, and the branch points (referred to as *nodes*) represent the divergence

of the ancestral lineage into two lineages. Thus, node 1 represents the divergence of the outgroup (moss) and the common ancestor of the three vascular plant taxa.

Similarly, node 2 represents a subsequent divergence of the fern and the ancestor of plants with seeds (Figure *b*). The branching process is continued with the data from the table (Figure *c*). In Figure *c*, note that pine and daisy are more similar to each other than to any other taxon. This relationship is indicated by the presence of a common ancestor at node 3. The relative time of divergence is indicated by the distance from the base of the tree. For example, the farther a node is up the tree, the more recent the time of divergence. In our example, node 3 represents the most recent divergence, and node 1 represents the most ancient divergence. Thus, the daisy is more closely related to the pine and more distantly related to the fern.

Adapted from Dr. John Beneski, Department of Biology, West Chester University, West Chester, Pennsylvania

TABLE 18-A   A Comparison of Four Plants

Taxon	Characters		
	Vascular Tissues	*Seeds*	*Flowers*
Moss	A	A	A
Fern	P	A	A
Pine	P	P	A
Daisy	P	P	P

A = absent. P = present.

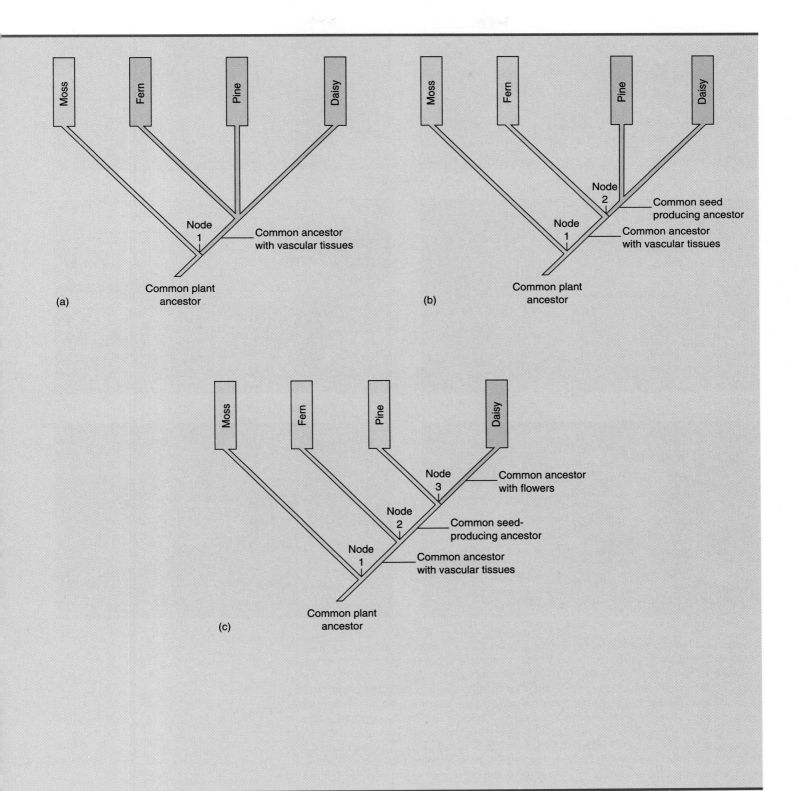

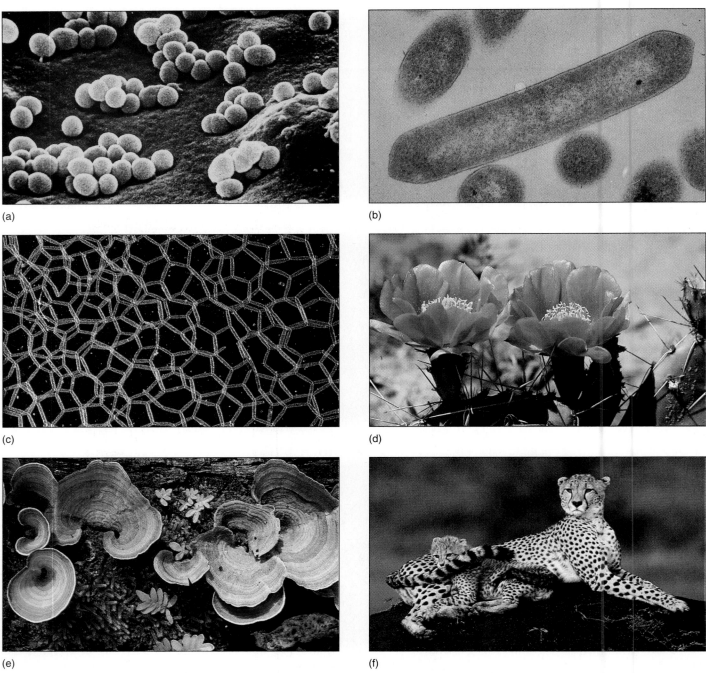

(a)

(b)

(c)

(d)

(e)

(f)

Figure 18-6   A survey of life. (a) Bacteria in the genus *Micrococcus* are classified in the kingdom Eubacteria. Species of *Micrococcus* occur widely in the soil and air as well as on human skin. (×21,000) **(b)** *Methanobacterium formicum* is a member of kingdom Archaebacteria. It is found in flooded soils and other anaerobic environments. (×111,000) **(c)** Water net (*Hydrodictyon* sp), which consists of a tube of interconnected cells, is found in lakes and slow-moving streams. It is a member of kingdom Protista. **(d)** Kingdom Plantae is represented by prickly pear (*Opuntia* sp). This specimen was photographed at Colorado National Monument. **(e)** This turkey-tail (*Trametes versicolor*) on a decaying log is a member of kingdom Fungi. **(f)** A cheetah (*Acinonyx jubatus*) mother and cub in East Africa. The cheetah is one of the swiftest members of kingdom Animalia. (a, *Visuals Unlimited/David M. Phillips;* b, *Visuals Unlimited/T.J. Beveridge;* c, *Dwight Kuhn;* d,e *John Arnaldi;* f, *Fritz Polking/Dembinsky Photo Associates*)

(1) Eubacteria (most bacteria), (2) Archaebacteria (some bacteria found in extreme environments), (3) Protista (algae, protozoa, water molds, and slime molds), (4) Fungi (mushrooms and other fungi), (5) Plantae (plants), and (6) Animalia (animals).

Protists Do Not Fit Together Well in a Single Kingdom

Ideally, all members of a kingdom should have a common ancestor. There seems to be no common ancestor for the members of the kingdom Protista, because it appears that eukaryotic cells arose several different times during the course of evolution. Some biologists want to resolve this problem by dividing the protist kingdom into several natural groupings, each of which would be another kingdom. However, most biologists would rather deal with the limitations of the six-kingdom system than with several additional kingdoms.

VIRUSES ARE NOT ASSIGNED TO ANY OF THE SIX KINGDOMS

Viruses are not cellular, cannot move about on their own, and cannot carry on metabolic activities independently. All cellular forms of life contain both DNA and RNA, but a virus contains *either* DNA *or* RNA, not both. A typical virus consists of a core of nucleic acid (DNA or RNA) surrounded by a protein coat (Figure 18-7a). Viruses lack ribosomes and the enzymes necessary for protein synthesis. They can reproduce, but only within the complex environment of the living cells that they infect.

Because they are not cellular and cannot carry on metabolic activities on their own, viruses are not classified in any of the six kingdoms. Furthermore, no system of viral classification has yet been agreed on, because so little is known about how viruses evolved.

Some biologists have suggested that viruses, because of their simplicity, represent a primitive form of life. However, because all viruses depend totally on living cells to reproduce, other biologists think that viruses are descendants of cellular organisms that have become highly specialized as parasites. These biologists propose that, during the course of their evolution, viruses lost all their cellular components except their genetic material and a few components needed for replication and infection.

The hypothesis currently thought to be most likely is that viruses were originally fragments of DNA or RNA that broke off from the nucleic acids of cellular organisms. According to this view, some viruses may trace their origins to animal cells, others to plant cells, and still others to bacterial cells. These multiple origins may explain why each virus usually infects only certain species—that is, a virus may infect a species that

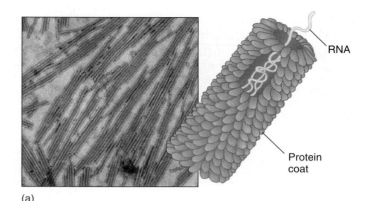

(a)

(b)

Figure 18-7   Some viruses cause disease in plants.
(a) Tobacco mosaic virus is an RNA-containing plant virus that appears rod-shaped. It consists of a core of RNA surrounded by a protein coat. **(b)** A tobacco leaf infected with tobacco mosaic virus. The virus produces a yellow and green mottling, or mosaic pattern, on a variety of plants. The disease tends to reduce crop yields rather than killing the plants outright. (a, *Visuals Unlimited/K.G. Murti;* b, *Kenneth M. Corbett*)

is the same as or closely related to the organism in which it originated.

Some Viruses Infect Plants

Animals are not the only organisms to be infected by viruses. In 1892 a Russian botanist found that he could transmit tobacco mosaic disease—so called because the infected tobacco leaves have a spotted, mosaic appearance—to healthy plants by dabbing their leaves with the sap of diseased plants (Figure 18-7b). The sap was infective even after it had been passed through filters fine enough to remove all bacteria. Tobacco mosaic disease was the first evidence of a viral disease in plants.

Viral diseases are spread among plants mostly by insects. They are also inherited by way of infected seeds or by asexual propagation. Once a plant is infected, the virus can spread throughout the plant body.

Plant viruses cause serious agricultural losses. Because no cures for viral diseases of plants are known, it is common to burn plants that have been infected. Some biologists are focusing their efforts on prevention of viral disease by developing virus-resistant strains of important crop plants.

STUDY OUTLINE

I. Organisms are named using the binomial system of nomenclature, which was first used consistently by Linnaeus.
   A. In this system, the basic unit of classification is the species.
   B. The scientific name of each species has two parts: the generic name (genus) and the specific epithet. For example, the scientific name for corn is *Zea mays*.
II. The hierarchical system of classification includes (from most inclusive to least inclusive): kingdom, phylum, class, order, family, genus, and species.
III. Modern classification is based on evolutionary relationships, or phylogeny.
   A. All of the plants in a monophyletic taxon have a common ancestor; the plants in a polyphyletic taxon evolved from different ancestors.
   B. Homologous structures indicate evolution from a common ancestor.

   C. The cladistic approach to taxonomy insists that taxa be monophyletic. Each taxon consists of a common ancestor and all its descendants.
IV. The six-kingdom classification recognizes the kingdoms Eubacteria, Archaebacteria, Protista, Fungi, Plantae, and Animalia. Protists apparently do not share a common ancestor, and some biologists think they should not be grouped together in a single kingdom.
V. Viruses do not fall into any of the six kingdoms.
   A. Viruses are not cellular.
   B. A typical virus consists of a core of nucleic acid (DNA or RNA) surrounded by a protein coat.
   C. Plant viruses, which are spread among plants by insects, cause serious agricultural losses.

SELECTED KEY TERMS

binomial system of nomenclature, p. 300
cladistics, p. 303
class, p. 302
dichotomous key, p. 304
family, p. 302

genus, p. 302
herbal, p. 297
herbarium, p. 299
kingdom, p. 302
monophyletic, p. 303

order, p. 302
phylogeny, p. 303
phylum, p. 302
polyphyletic, p. 303
species, p. 301

subspecies, p. 301
systematics, p. 303
taxon, p. 302
variety, p. 301
virus, p. 309

REVIEW QUESTIONS

1. Distinguish between an organism's common names and its scientific name, and explain why scientific names are used in scientific work.
2. How did scientific names change as a result of Linnaeus' contributions?
3. What is a species? fertile offspring / mate.
4. List the taxonomic levels in the hierarchical system of classification, beginning with species.

5. What is cladistics?
6. Why have biologists proposed a six-kingdom system to replace the five-kingdom system?
7. Explain why the kingdom Protista is not a natural kingdom, as compared with the other kingdoms. trouble to trace back
8. Why is it difficult to assign viruses to one of the six kingdoms? no cells, protein coat + DNA center

THOUGHT QUESTIONS

1. Biologists think that there may be millions of species that have not yet been scientifically described or named. Discuss why these organisms have not yet been studied, where most of them probably live, and to which kingdom or kingdoms they probably belong.

2. All fields of biology use taxonomy. List at least five different disciplines within plant biology (see Chapter 1 for the disciplines), and tell why taxonomy would be important to each.

SUGGESTED READINGS

Fairchild, D., *The World Was My Garden: Travels of a Plant Explorer,* New York, Scribner, 1938. David Fairchild's fascinating account of his long career with the U.S. Department of Agriculture.

Gould, S.J., "The First Unmasking of Nature," *Natural History,* April 1993. Darwin's contributions were built on the foundations of biology established by Linnaeus.

Healey, B., *The Plant Hunters,* New York, Scribner, 1975. The adventures and accomplishments of plant collectors.

May, R.M., "How Many Species Inhabit the Earth?" *Scientific American,* Vol. 267, No. 4, October 1992. An argument for the importance of identifying and classifying the organisms that inhabit our planet. This information has an impact on environmental issues.

Savage, J.M., "Systematics and the Biodiversity Crisis," *BioScience,* Vol. 45, No. 10, November 1995. Systematics has a crucial role in the conservation of plants and other organisms.

Schiebinger, L., "The Loves of the Plants," *Scientific American,* February 1996. A fascinating account of Linnaeus' taxonomic use of the sex organs of plants.

KINGDOMS EUBACTERIA AND ARCHAEBACTERIA

Anabaena is the generic name for several species of unobtrusive photosynthetic organisms whose bodies consist of hairlike filaments, or unbranched chains, of prokaryotic cells. Biologists consider *Anabaena* to be a special type of bacteria known as **cyanobacteria**. Formerly, however, *Anabaena* species were considered a primitive type of alga, called blue-green algae; thus, these organisms were originally classified in the plant kingdom. Because cyanobacteria and other bacteria were traditionally considered plants, they are usually studied in plant biology courses. Despite their prokaryotic cell structure, *Anabaena* and other cyanobacteria are remarkably like plants and algae in many ways. For example, photosynthesis is essentially identical in cyanobacteria, algae, and plants, and all three kinds of organisms contain the green pigment, chlorophyll *a.* Cyanobacteria are found worldwide in both freshwater and marine environments as well as in the soil.

Although most students have never seen or thought much about it, *Anabaena* is an important organism to learn

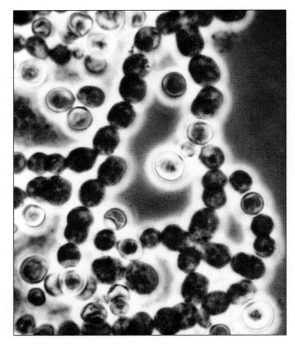

Anabaena sp, a filamentous cyanobacterium that fixes nitrogen. Nitrogen fixation in *Anabaena* occurs in the swollen, rounded cells, which are called heterocysts. (*Visuals Unlimited/S. Thomson*)

about because it and other cyanobacteria play crucial roles in the living world. *Anabaena* is photosynthetic and, along with algae, forms the base of the food web in aquatic environments. Without producers such as *Anabaena,* aquatic animals could not exist because they would have nothing to eat. Furthermore, the oxygen that *Anabaena* generates during photosynthesis helps to oxygenate the water, which benefits aquatic organisms.

Even if cyanobacteria were nonphotosynthetic, they would still be ecologically important. Many cyanobacteria, including *Anabaena,* can take nitrogen from the air and convert it to ammonia (NH_3), a form of nitrogen that plants can use to synthesize proteins and other nitrogen-containing organic compounds. The conversion of atmospheric nitrogen to ammonia is called **nitrogen fixation**. Although nitrogen is an essential component of such biologically important molecules as proteins and nucleic acids, only a few prokaryotic organisms have the ability to fix nitrogen. The nitrogenous compounds that *Anabaena* and other cyanobacteria produce are used by countless other organisms in the environment.

Anabaena sometimes forms a close association with an aquatic fern known as *Azolla*. The *Azolla-Anabaena* symbiosis, or partnership, is common in flooded rice paddies, where it provides needed nitrogen "fertilizer" to the rice. Although rice has been grown in Asia for thousands of years, the rice fields there do not exhibit a nitrogen deficiency. In contrast, crops grown in fields without populations of *Azolla-Anabaena* require the application of nitrogen fertilizer annually.

LEARNING OBJECTIVES

After reading this chapter, you should be able to:

1. *Label the basic parts of a bacterial cell and give their functions.*
2. *Define the following terms: endospore, plasmid, capsule, pilus.*
3. *Characterize the metabolic diversity of bacteria using the following terms: autotrophy and heterotrophy; aerobes, facultative anaerobes, and obligate anaerobes.*
4. *Distinguish between archaebacteria and eubacteria.*
5. *Explain why saprotrophic bacteria and nitrogen-fixing bacteria (including cyanobacteria) are so crucial to the environment.*

BACTERIA ARE PLACED IN KINGDOMS EUBACTERIA AND ARCHAEBACTERIA

All prokaryotes—the bacteria—are assigned to two kingdoms, Eubacteria and Archaebacteria. Biologists consider prokaryotic cell structure to be a fundamental difference between bacteria and other organisms, and as a result, all bacteria were grouped in a single kingdom until recently. Molecular work in the 1980s, however, verified that there are two groups of bacteria and that these two groups differ from each other as much as they differ from the eukaryotes. Thus, many biologists now recognize two kingdoms of prokaryotes.

Prokaryotic Cell Structure Is Simpler Than Eukaryotic Cell Structure

Prokaryotic cells lack the membrane-bounded organelles that are typical of eukaryotic cells: they have no nuclei, no mitochondria, no chloroplasts, no endoplasmic reticulum, and no Golgi complex. Most prokaryotes are unicellular organisms, but some form colonies or filaments that contain specialized cells. Figure 19-1 shows the structure of a typical bacterium.

Almost all bacterial cells are tiny. Their cell volume is typically about one-thousandth that of a small eukaryotic cell, and their length only about one-tenth. Most prokaryotic cells vary from 1 to 10 micrometers in length.[1] (A bacterial cell 1 micrometer long is far too short to be seen with the unaided eye, because a micrometer is one-millionth of a meter or one thousandth of a millimeter.)

Some species of bacteria are covered by a **capsule**, or slime layer, that may prevent the bacterial cell from drying out under adverse conditions. The capsule may also serve for defense, as it appears to provide the cell with added protection against being engulfed by other microorganisms in the environment (or by white blood cells in the human body).

Most prokaryotic cells have a cell wall just inside the capsule. The composition of bacterial cell walls differs from those of eukaryotic cell walls in that most bacterial cell walls are

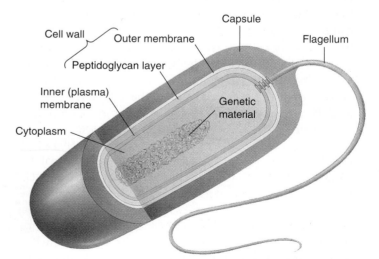

Figure 19-1   The structure of a typical bacterium. Shown is a gram-negative bacterium, which has an outer membrane exterior to the peptidoglycan layer (discussed later in the chapter). Other bacteria lack the outer membrane. Note the absence of a nuclear envelope surrounding the genetic material.

---

[1]At least one bacterium is quite large—about a million times larger than a typical bacterium. This giant bacterium (*Epulopsicium fishelsoni*), reported in the journal *Nature* in 1993, was discovered living in the intestines of surgeonfish in the Red Sea and off the Great Barrier Reef in Australia.

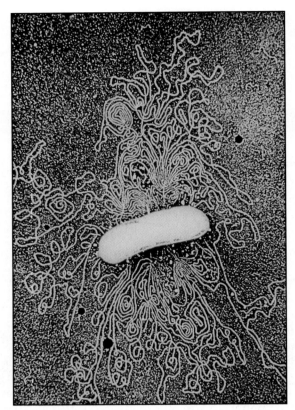

Figure 19-2   DNA spilling out of a ruptured bacterial cell. The bacterial chromosome, which is about 1000 times the length of the bacterium, is tightly coiled inside the cell. (*Visuals Unlimited/K.G. Murti*)

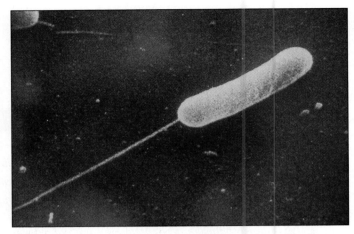

Figure 19-3   An SEM of *Escherichia coli,* showing its single flagellum. Other bacteria have a tuft of flagella at one end of the cell, and some have numerous flagella that project from many surfaces of the cell. (*Visuals Unlimited/Science VU*)

composed of **peptidoglycan**, a macromolecule that confers strength and rigidity. The bacterial cell wall provides a rigid framework that supports the cell, maintains its shape, and keeps it from bursting. Normally, bacteria cannot survive without their cell walls.

Inside the cell wall is the bacteria's plasma membrane, the active barrier between the cell and its external environment. The plasma membrane governs the passage of molecules into and out of the cell. Most of the functions performed by elaborate systems of internal membranes in eukaryotes are carried out by the plasma membranes of bacteria.

Within the bacterial cell's cytoplasm are the ribosomes, where protein synthesis occurs, and the genetic material (DNA). Bacterial DNA is found mainly in a single long, circular molecule, often referred to as a bacterial chromosome, that lies in the cytoplasm and is not surrounded by a nuclear envelope. No proteins are associated with the bacterial chromosome as there are with eukaryotic chromosomes, which have histones and other proteins as part of the chromosome structure. When stretched out to its full length, the bacterial chromosome is about 1000 times longer than the bacterial cell itself (Figure 19-2). In addition to the bacterial chromosome, a small amount of genetic information may be present as smaller DNA loops, called **plasmids**, which replicate inde-

pendently of the bacterial chromosome. Bacterial plasmids often bear genes that provide resistance to antibiotics, which are drugs intended to harm or kill bacteria and other microorganisms.

Some Prokaryotes Have Specialized Structures

Some prokaryotes have flagella, but their structure differs from that of eukaryotic flagella in that each bacterial flagellum has only a single filament (Figure 19-3). At the base of a bacterial flagellum is a complex structure that rotates the flagellum and pushes the cell forward much as a propeller pushes a ship.

Some bacteria have hundreds of straight, hairlike appendages known as **pili** (sing. *pilus*) that help the bacteria adhere to certain surfaces, such as the cells that they infect. Some pili are involved in the transmission of genetic material between related *strains* (varieties of the same species) of bacteria.

When the environment of a bacterium becomes unfavorable—very dry, for instance—many species become dormant. The cell loses water, shrinks slightly, and remains dormant until water is available again. Other species form extremely durable resting structures called **endospores** that can survive when environmental conditions are extremely dry, hot, or cold or when food is scarce (Figure 19-4). Some endospores are so resistant that they can survive an hour or more of boiling or centuries of freezing.[2] When environmental conditions are again suitable for growth, the endospore absorbs water,

---

[2]In 1995 in the journal *Science,* scientists reported that 25-million-year-old endospores preserved inside a bee trapped in amber had survived.

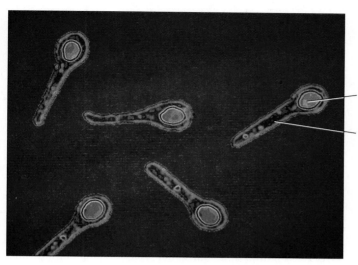

— Endospore

— Rod-shaped bacterial cell

Figure 19-4   An endospore within a cell of *Clostridium tetani,* the bacterium that causes tetanus. Each bacterial cell contains only one endospore, which is a resistant, dehydrated remnant of the original cell. (*Alfred Pasieka/Peter Arnold, Inc.*)

breaks out of its inner wall, and becomes an active, growing bacterial cell again.

Bacteria Have Diverse Methods of Obtaining Organic Molecules

Most bacteria are **heterotrophs** (from the Greek *hetero,* "other," and *tropho,* "nourishment") and must obtain organic molecules from other organisms. The majority of heterotrophic bacteria are free-living **saprotrophs**, organisms that get their nourishment from dead organic matter. Other heterotrophic bacteria live in or on other organisms. These bacteria may be **commensals,** which neither help nor harm their hosts, or they may be **parasites**, which live at the expense of their hosts and can cause disease. Other bacteria form **mutualistic** associations with another organism, in which both the bacterium and its partner derive benefits from the association.

Some bacteria are **autotrophs** (from the Greek *auto,* "self," and *tropho,* "nourishment") and can manufacture their own organic molecules from carbon dioxide. Autotrophic bacteria are either photosynthetic or chemosynthetic. **Photosynthetic bacteria** obtain energy to manufacture organic compounds from light, whereas **chemosynthetic bacteria** obtain energy from chemical reactions, primarily the oxidation of (addition of oxygen to) inorganic compounds that contain ammonia, iron, or sulfur. Five groups of photosynthetic bacteria exist: cyanobacteria, green sulfur bacteria, purple sulfur bacteria, green nonsulfur bacteria, and purple nonsulfur bacteria.

Cyanobacteria photosynthesize in the same way algae and plants do. Photosynthesis in other photosynthetic bacteria differs from "standard" photosynthesis in two important ways. First, bacterial chlorophyll, which is different from chlorophyll *a,* absorbs light most strongly in the near-infrared portion of the electromagnetic spectrum, which means that these bacteria can carry on photosynthesis in red light at wavelengths that would not work well for plants, algae, or cyanobacteria. Second, photosynthesis by bacteria other than cyanobacteria does not produce oxygen, because water is not used as a hydrogen donor. Instead, hydrogen sulfide (H_2S) is the hydrogen donor, and sulfur (S) is the byproduct.

Bacteria Differ in Their Needs for Oxygen

Whether they are heterotrophs or autotrophs, most bacterial cells are **aerobic**; that is, they require atmospheric oxygen for cellular respiration. Plant and animal cells are also aerobic. Some bacteria are **facultative anaerobes**, meaning that they can use oxygen for cellular respiration if it is available but will respire anaerobically when oxygen is absent. Other bacteria are **obligate anaerobes** and can carry on cellular respiration only in the absence of oxygen. Some obligate anaerobes are actually poisoned by even low concentrations of oxygen.

Bacteria Reproduce by Binary Fission

Bacteria generally reproduce asexually by **binary fission**, in which one cell divides into two similar cells. After the circular bacterial chromosome has been replicated, the cell splits into two cells. A transverse wall forms between the two new cells by an ingrowth of both the plasma membrane and the cell wall. Binary fission occurs with remarkable speed; under ideal conditions, some species divide every 20 to 30 minutes! At this rate, if nothing interfered, one bacterium could give rise to more than 130,000 bacteria within 6 hours. This explains why the entrance of only a few pathogenic (disease-producing) bacteria into a human can result so quickly in the symptoms of disease. Fortunately, bacteria cannot reproduce at this rate for very long, because they are soon checked by lack of food or by the accumulation of waste products.

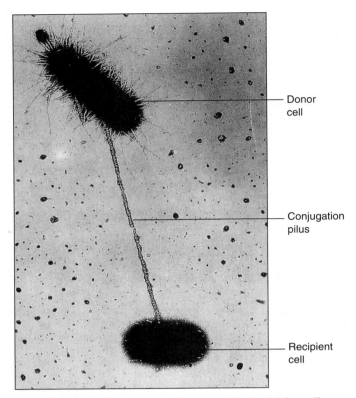

Figure 19-5   A pilus connecting two *Escherichia coli* cells. Plasmid DNA is transferred from donor cell to recipient cell during conjugation. (*Courtesy of Charles C. Brinton, Jr., and Judith Carnahan*)

Labels in figure: Donor cell; Conjugation pilus; Recipient cell

Although sexual reproduction involving the fusion of gametes does not occur in bacteria, genetic material is sometimes exchanged between individuals. In **conjugation**, two cells of different mating types come together, and genetic material is transferred through pili from one cell (the donor) to the other (the recipient) (Figure 19-5). In addition, fragments of DNA released by a broken cell are sometimes taken in by another bacterial cell (**transformation**). Alternatively, genetic material may be carried from one bacterial cell to another by a virus (**transduction**). These three methods of genetic exchange allow genes for resistance to antibiotics to be transferred from one bacterium to another (see Focus On: Multidrug-Resistant Tuberculosis on page 319).

THERE ARE SIGNIFICANT DIFFERENCES BETWEEN ARCHAEBACTERIA AND EUBACTERIA

Under a microscope, all bacteria appear to be similar. Their structural similarities, along with a scant fossil record, have made it difficult to determine evolutionary relationships

among them. However, evidence from molecular biology has helped biologists conclude that ancient prokaryotes split into two lineages very early in the history of life. The modern descendants of these two ancient lines of bacteria are the archaebacteria, which include a few genera of prokaryotes able to live in extreme environments, and the eubacteria, which comprise all other groups of prokaryotes. The differences between archaebacteria and eubacteria are considered great enough to warrant their classification in separate kingdoms.

Three Groups of Archaebacteria Are Recognized

Biochemically, the 200 or so species of **kingdom Archaebacteria** are very different from other bacteria. One of their most distinguishing features is the absence of peptidoglycan in the cell wall. There are also other important (though quite technical) differences that set the archaebacteria apart from other bacteria.

The biochemical differences between the archaebacteria and other bacteria suggest that these groups may have diverged from each other long ago—relatively early in the history of life. Many of the extreme environments to which the modern archaebacteria are adapted resemble conditions that are thought to have been common on early Earth billions of years ago but that are somewhat rare today. Examples include hot springs and thermal vents at the bottom of the ocean, both of whose temperatures may exceed 100°C.

The archaebacteria include three groups: the halophiles, the methanogens, and the thermoacidophiles. The **halophiles** live only in extremely salty environments such as salt ponds (Figure 19-6a). The most common archaebacteria are the **methanogens**, which are obligate anaerobes that produce methane from carbon dioxide and hydrogen. They inhabit sewage and swamp sediments and are common in the digestive tracts of humans and other animals. The **thermoacidophiles** normally grow in hot, acidic environments (Figure 19-6b). For example, one species that is found in the hot sulfur springs of Yellowstone National Park flourishes at temperatures near 60°C and pH values of 1 to 2 (the pH of concentrated sulfuric acid).

Interestingly, the archaebacteria appear to be more closely related to eukaryotes, such as plants and humans, than do the eubacteria (Figure 19-7). Evidence that supports this relationship comes from molecular data, including the fact that archaebacteria and eukaryotes share certain genes that are not found in eubacteria. The relationship between archaebacteria and eukaryotes is a distant one, however; the line of descent leading to present-day archaebacteria is thought to have diverged from the eukaryote line some 3 billion years ago!

(a)

(b)

Figure 19-6   Many archaebacteria live in extreme environments. (a) Seawater evaporating ponds near San Francisco Bay are colored pink, orange, and yellow from the many extreme halophiles (salt-loving archaebacteria) growing in them. (Algae also add color to the ponds.) The salt that remains after the water has evaporated has commercial value. **(b)** Yellowstone National Park's Grand Prismatic Spring, the world's third largest hot spring, teems with thermophilic archaebacteria. The rings around the perimeter, where the water is cooler, get their distinctive colors from the various kinds of bacteria living there. (a, *Helen E. Carr/ Biological Photo Service;* b, ©*1995* Discover *Magazine*)

The Eubacteria Are the Most Familiar Prokaryotes

Some 3000 species of bacteria belong to **kingdom Eubacteria**, or "true bacteria," and when bacteria are mentioned casually, eubacteria are usually what are meant. Early biologists quickly determined that eubacteria are present almost universally, being abundant in air, in soil, in water, and in and on the bodies of living and dead organisms. In fact, relatively few places in the world are completely devoid of eubacteria, for these organisms can be found in fresh and salt water, as far down as several meters deep in the soil, in deep underground water supplies, in the ice of glaciers, and even in oil deposits far underground.

Eubacteria have three main shapes: sphere, rod, and helical (Figure 19-8). Spherical bacteria, known as **cocci** (sing. *coccus*), occur singly in some species, in groups of two in others (diplococci), in long chains (streptococci), or in irregular clumps that look like bunches of grapes (staphylococci). Rod-shaped bacteria, called **bacilli** (sing. *bacillus*), may occur

Eukaryotes

Archaebacteria

Eubacteria

First ancestral cells

Figure 19-7   This phylogenetic tree illustrates probable relationships among eubacteria, archaebacteria, and eukaryotes. Eukaryotes appear to be more closely related to archaebacteria than they are to eubacteria.

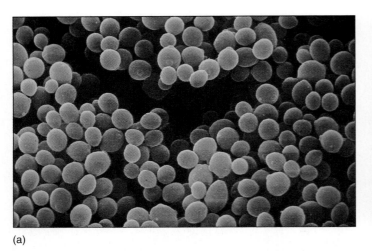

(a)

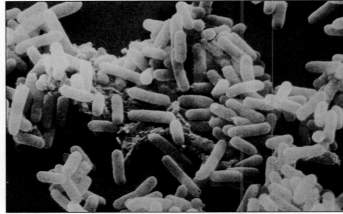

(b)

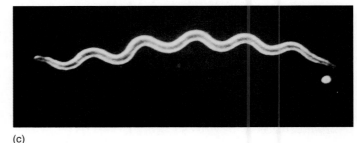

(c)

Figure 19-8   Three characteristic bacterial shapes.
(a) *Micrococcus* sp are cocci, or spherical bacteria.
(b) *Escherichia coli* are bacilli, or rod-shaped bacteria.
(c) *Borrelia* sp are spirilla, or helical bacteria. (a,b, *Visuals Unlimited/David M. Phillips; c, Visuals Unlimited/Science VU-Charles W. Stratton*)

as single rods or as long chains of rods. Helical bacteria are known as **spirilla** (sing. *spirillum*).

Eubacteria may be grouped on the basis of their staining properties

In 1884 the Danish physician Hans Christian Gram developed the **Gram staining procedure**, which divides the eubacteria into two groups. Bacteria that absorb and retain the primary stain during the procedure are referred to as **gram-positive**, whereas those that do not retain the stain are **gram-negative** (Figure 19-9). The cell walls of gram-positive bacteria are very thick and consist primarily of peptidoglycan, which retains the Gram stain. The cell wall of a gram-negative bacterium consists of two layers, a thin peptidoglycan wall on the inside and an outer membrane that does not retain the stain.

The differences in composition of the cell walls of gram-positive and gram-negative bacteria are of great practical importance. The antibiotic penicillin, for example, interferes with peptidoglycan synthesis and ultimately results in a cell wall so fragile that it cannot effectively protect the cell or keep it from bursting. Penicillin works most effectively against gram-positive bacteria due to the greater abundance of peptidoglycan in their cell walls.

Microbiologists recognize three main groups of eubacteria on the basis of differences in their cell walls: the wall-less bacteria, the gram-negative bacteria, and the gram-positive bacteria (Table 19-1).

TABLE 19-1   Examples of Common Eubacteria

Wall-less Bacteria

Mycoplasmas—extremely small cells that lack cell walls

Gram-Negative Bacteria

Nitrogen-fixing bacteria—aerobic bacteria that fix nitrogen

Enterobacteria—a large group of diverse bacteria; facultatively anaerobic, heterotrophic

Spirochetes—spiral-shaped, flexible cell walls

Cyanobacteria—photosynthetic autotrophs, occur mainly as colonial masses or filaments; some fix nitrogen

Rickettsias—obligate parasites; a few cause diseases that are transmitted to humans by arthropods

Chlamydias—obligate parasites; are not transmitted by arthropods

Myxobacteria—unicellular bacilli; move by gliding

Gram-Positive Bacteria

Lactic acid bacteria—fermenting bacteria that can ferment the sugar in milk

Streptococci—fermenting bacteria

Staphylococci—aerobic bacteria

Clostridia—fermenting bacteria; anaerobic

Focus On
Multidrug-Resistant Tuberculosis

Since the late 1980s, the U.S. Centers for Disease Control in Atlanta have documented an alarming increase in tuberculosis (TB). The number of cases of TB had declined in the United States during the prior 30 or so years, largely as the result of successful treatments with antibiotics. One of the reasons for the current outbreak of TB is the recent development and spread of drug-resistant strains of the bacteria that cause TB. These strains are unaffected by one or more of the antibiotics traditionally used to treat TB. Drug-resistant TB is deadly: 80 percent of the people infected with multidrug-resistant TB (MDR-TB) die within 2 months of diagnosis, even with medical care.

How has bacterial resistance to antibiotics come about? When physicians began to use antibiotics to treat human and animal infections, it was thought that the drugs would eliminate bacterial diseases. This has not taken place, however. Bacteria are constantly evolving, even inside the lungs of human hosts. Each time an antibiotic is used to treat a bacterial infection, some bacteria survive. The survivors, because of certain genes they have acquired,* are genetically resistant to the antibiotic, and they pass that trait on to subsequent generations. As a result, the bacterial population contains a larger percentage of drug-resistant bacteria than it did before.

Drug resistance is usually found in patients who have been treated for TB, and quite often human behavior is a factor in the development of drug resistance. A person with TB must take three to ten pills each day for at least 6 months. After the first week or two of treatment, the patient usually feels better; many patients decide to quit taking their pills at that point. Then the TB bacteria still lurking in the body—those with a resistance to the discontinued antibiotic—rally. The evolution of a strain of bacteria resistant to several drugs is a worst-case scenario. MDR-TB is extremely difficult to treat effectively and, as mentioned, is often fatal.

---

*Recall that bacteria can acquire new genes, including those for resistance to antibiotics, by conjugation, transformation, and transduction.

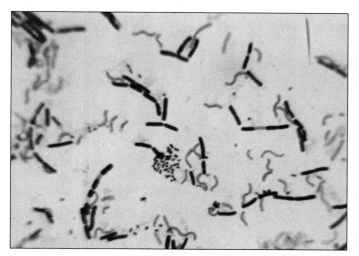

Figure 19-9 A variety of Gram-stained bacteria as viewed under a light microscope. Gram-positive cells are purple, and gram-negative cells are pink. (*Visuals Unlimited/Jack M. Bostrack*)

Mycoplasmas lack a rigid cell wall

A **mycoplasma** is a tiny bacterium bounded by a plasma membrane but lacking a typical bacterial cell wall. Some mycoplasmas are so small that, like viruses, they pass through bacteriological filters. In fact, mycoplasmas are smaller than some viruses. Mycoplasmas may be the simplest form of cellular life. Some live in soil, some live in sewage, and others are parasitic on plants or animals. Many plant diseases that were originally attributed to viruses are now thought to be caused by mycoplasmas (Figure 19-10).

Gram-negative bacteria have thinner cell walls

The gram-negative bacteria exhibit a lot of variation in shape, structure, and metabolism. Among them are the **nitrogen-fixing bacteria**, which have the ability to fix atmospheric nitrogen into inorganic compounds that can be used by plants. These gram-negative rods, most of which are flagellated, are common in soil and aquatic environments worldwide. Some, such as *Azotobacter,* are free-living, whereas others, such as *Rhizobium,* form symbiotic relationships with plants (see Chapter 26).

Sewage is wastewater carried off from toilets, washing machines, and showers by drains or sewers. Because sewage contains human wastes, its release into bodies of water (for example, rivers, lakes, and oceans) causes two pollution problems that are related to bacteria. First, untreated municipal wastewater usually contains many disease-causing bacteria as well as other pathogenic microorganisms such as viruses, protozoa, and parasitic worms. Typhoid, cholera, bacterial dysentery, and enteritis are some of the more common diseases caused by bacteria that are transmissible through contaminated water. Thus, water polluted by sewage poses a threat to public health.

The bacteria in sewage cause a second serious environmental problem in water: oxygen demand. The action of microorganisms, particularly bacteria, decomposes sewage and other organic materials into carbon dioxide, water, and similar inoffensive materials. This degradation process, known as cellular respiration, requires the presence of oxygen. Dissolved oxygen is also used by most organisms living in healthy aquatic ecosystems, including fish and aquatic plants. But oxygen has a limited ability to dissolve in water, and when an aquatic ecosystem contains high levels of sewage or other organic material, the decomposing bacteria use up most of the dissolved oxygen, leaving little for fish or other aquatic organisms. At extremely low oxygen levels, fish die off in large numbers.

Sewage and other organic wastes are measured in terms of their **biological oxygen demand**, or **biochemical oxygen demand (BOD)**, the amount of oxygen needed by bacteria to decompose the wastes. A large amount of sewage in water creates a high BOD, which robs the water of dissolved oxygen. When dissolved oxygen levels are low, anaerobic bacteria also produce compounds that have very unpleasant odors, further deteriorating water quality.

Because sewage-contaminated water is a threat to public health, periodic tests are conducted for the presence of sewage in our water supplies. Although many different microorganisms thrive in sewage, the common intestinal bacterium *Escherichia coli* is typically used as an indication of the amount of sewage present in water and as an indirect measure of the presence of disease-causing agents. Although **coliform bacteria** (bacteria, such as *E. coli,* that inhabit the large intestine) themselves do not cause disease, their presence indicates the likely presence of disease-causing agents in water. *Escherichia coli* is perfect for monitoring sewage because it is not present in the environment except in human and animal feces, where it is found in large numbers.

The **fecal coliform test** is performed to test for the presence of *E. coli* in water (see figures). A small sample of water is passed through a filter to trap all bacteria. The filter is then transferred to a petri dish that contains nutrients. After an incubation period, the number of greenish colonies present indicates the number of *E. coli*. Safe drinking water should contain no more than 1 coliform bacterium per 100 milliliters (about 1/2 cup) of water, safe swimming water should have no more than 200 per 100 milliliters of water, and general recreational water (for boating) should have no more than 2000 per 100 milliliters. In contrast, raw sewage may contain several million coliform bacteria per 100 milliliters of water.

The fecal coliform test is used to indicate the likely presence of disease-causing agents in water. A water sample is first passed through a filtering apparatus. Then, for a period of 24 hours, the filter disk is placed on a culture medium that supports coliform bacteria (Figure *a*). After incubation, the number of bacterial colonies is counted (Figure *b*). Each colony arose from a single coliform bacterium in the original water sample. (a,b, *Courtesy of Millipore Corp.*)

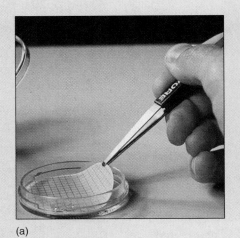

(a)

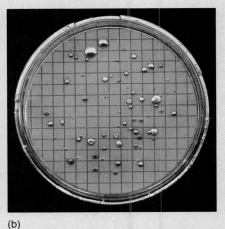

(b)

Figure 19-10   Yellow blight in palm is a disease caused by a mycoplasma. Originally, a virus was thought to be the causative agent of this disease (© *Peter B. Kaplan, The National Audubon Society Collection/Photo Researchers, Inc.*)

Many important pathogens are gram-negative eubacteria. About one half of the almost 200 species of bacteria that cause disease in plants are gram-negative rods in a single genus, *Pseudomonas.* Different species of *Pseudomonas* cause a variety of diseases in plants, such as leaf spots, cankers (local decay of bark and wood), wilts, and seedling death. In humans, the gram-negative *Neisseria gonorrhoeae* causes the sexually transmitted disease gonorrhea.

Enterobacteria are a group of gram-negative bacteria that include free-living saprotrophs, plant pathogens, and some bacteria that inhabit humans. One enterobacterium, *Escherichia coli,* lives in the intestines of humans and other animals as part of the normal microbial population (see Focus On: Problems Associated with Bacteria in Wastewater). Some strains of the enterobacterium *Salmonella* cause food poisoning.

Spirochetes are corkscrew-shaped, gram-negative bacteria with flexible cell walls. Some spirochetes are free-living and inhabit freshwater and marine habitats, whereas some form close associations with other organisms, including a few that are parasitic. The spirochetes of greatest medical importance cause Lyme disease, which is transmitted by ticks, and the sexually transmitted disease syphilis.

Cyanobacteria (such as *Anabaena,* mentioned in the chapter introduction) are gram-negative bacteria found in ponds, lakes, swimming pools, and moist soil as well as on logs and the bark of trees. Some cyanobacteria also occur in the ocean, and a few species inhabit hot springs. Several cyanobacteria are unicellular, but most species occur as round colonies or as long filaments (Figure 19-11). Most cyanobacteria are photosynthetic autotrophs and contain chlorophyll *a,* which is also found in plants and algae. Cyanobacteria have several accessory pigments, including **phycocyanin** (a blue pigment) and **phycoerythrin** (a red pigment).

Rickettsias are gram-negative bacteria that must live within cells as parasites in order to survive. Most rickettsias parasitize arthropods such as fleas, lice, ticks, and mites without causing specific diseases in them. Diseases caused by the few rickettsias known to be pathogenic to humans are transmitted by arthropods through bites or contact with their excretions. Among diseases caused by rickettsias are typhus, transmitted by fleas and lice, and Rocky Mountain spotted fever, transmitted by ticks.

Chlamydias also must live within cells as parasites in order to survive; unlike rickettsias, however, chlamydias do not depend on arthropods for transmission. Studies indicate that these gram-negative bacteria infect almost every species of bird and mammal. Perhaps 10 to 20 percent of the human population worldwide is infected, but interestingly, individuals may be infected for many years without apparent harm. However, chlamydias sometimes cause acute infectious diseases. For example, trachoma, the leading cause of blindness in the world, is caused by a strain of *Chlamydia,* and the most common sexually transmitted disease in the United States is an infection of the genitourinary tract caused by chlamydias.

Myxobacteria are gram-negative bacteria that excrete slime. When they are cultured in a petri dish, their growth is marked by a spreading layer of slime on which they glide or creep along. Most myxobacteria are saprotrophs that break down organic matter in the soil, manure, or rotting wood that is their habitat. A few myxobacteria prey on other bacteria.

Gram-positive bacteria have thicker cell walls

Gram-positive bacteria include the lactic acid bacteria, streptococci, staphylococci, and clostridia. **Lactic acid bacteria** are gram-positive bacteria that produce lactic acid as the main end product of their fermentation of sugars. Lactic acid bacteria may be found in decomposing plant material and in milk, yogurt, and other dairy products. They are commonly present in animals and are among the normal inhabitants of the human mouth and vagina.

Streptococci, spherical gram-positive bacteria that occur in chains, are found in the human mouth as well as the digestive tract. Among the harmful species of streptococci are those that cause "strep throat," scarlet fever, and other infections.

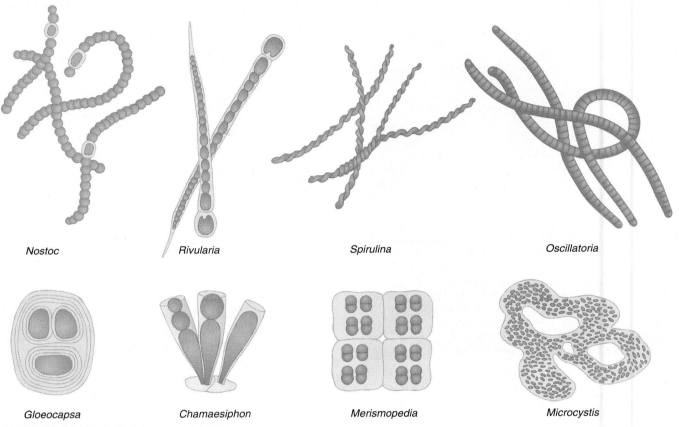

Nostoc   Rivularia   Spirulina   Oscillatoria

Gloeocapsa   Chamaesiphon   Merismopedia   Microcystis

Figure 19-11   Variation in cyanobacteria.

Staphylococci, spherical gram-positive bacteria that occur in irregular clusters, normally live in the human nose and on the skin. They are opportunistic, which means that they can cause disease when the immunity of the host is lowered. *Staphylococcus aureus* causes boils and skin infections and may infect wounds. Certain strains of *S. aureus* cause food poisoning, and some are thought to cause toxic shock syndrome.

Clostridia are a notorious group of anaerobic gram-positive bacteria. One species causes tetanus, or lockjaw; another causes gas gangrene; and a third causes botulism, a potentially fatal type of food poisoning.

BACTERIA ARE ECOLOGICALLY IMPORTANT

Although it is easy to overlook bacteria in the world of living things or to associate them exclusively with diseases, bacteria perform important environmental services. Indeed, bacteria are essential for life, and there are many more useful species than harmful ones. The contribution of some bacteria is to alter atmospheric nitrogen to a form that can be used by plants. Nitrogen fixation enables plants (and animals, because they eat plants) to manufacture essential nitrogen-containing compounds such as proteins and nucleic acids. The importance of nitrogen fixation to life cannot be overstated. If there were no prokaryotic nitrogen fixers, there would be no biological way to convert atmospheric nitrogen to usable nitrogen, and life would cease to exist.

Other bacteria play an essential role in the biosphere as decomposers, breaking down organic molecules into their simpler components. Bacteria, along with fungi, are nature's recyclers. Without bacteria and fungi, all available carbon, nitrogen, phosphorus, and sulfur would eventually be tied up in the wastes and dead bodies of plants and animals. Life would soon cease because of the lack of raw materials for the synthesis of new cellular components.

Bacteria are also used to clean up hazardous waste sites produced by humans. In a process known as **bioremediation**, a contaminated site is exposed to bacteria or other microorganisms that produce enzymes that break down the poisons, leaving behind harmless byproducts such as carbon dioxide and chlorides. To date, more than 1000 species of bacteria and fungi have been used to clean up various forms of pollution.

STUDY OUTLINE

I. Kingdoms Eubacteria and Archaebacteria contain bacteria—organisms that have a prokaryotic cell structure.
   A. Prokaryotic cells lack membrane-bounded organelles such as nuclei and mitochondria.
   B. The genetic material of a prokaryote is a single circular DNA molecule called a bacterial chromosome. In addition to the bacterial chromosome, a small amount of genetic information may be present as smaller DNA loops (plasmids).
   C. Most bacteria have cell walls composed of peptidoglycan. Some bacteria also produce a capsule that surrounds the cell wall.
   D. Bacterial flagella are structurally different from eukaryotic flagella.
   E. Some bacteria have pili that help the bacteria adhere to certain surfaces and that are involved in the transmission of genetic material between related strains of bacteria.
II. Bacteria are metabolically diverse.
   A. Some bacteria are heterotrophic and must obtain organic molecules from other organisms. Heterotrophic bacteria include saprotrophs (organisms that get their nourishment from dead organic matter) and parasites (organisms that get their nourishment from other organisms).
   B. Some bacteria are autotrophic and can make their own organic molecules. Autotrophic bacteria may be photosynthetic or chemosynthetic, depending on whether the energy they use to manufacture organic compounds comes from light or inorganic compounds.
   C. Bacteria may be aerobes (organisms that require oxygen to respire); facultative anaerobes (organisms that can respire with or without oxygen); or obligate anaerobes (organisms that can respire only in the absence of oxygen).

III. Bacteria reproduce asexually by binary fission (one cell divides into two similar cells). Genetic recombination can take place in several ways.
   A. In conjugation, genetic material is transferred through pili from one cell to another.
   B. In transformation, fragments of DNA released by a broken cell are taken in by another bacterial cell.
   C. In transduction, genetic material is carried from one bacterial cell to another by a virus.
IV. Bacteria are divided into two kingdoms, Archaebacteria and Eubacteria.
   A. Archaebacteria are anaerobic, have cell walls of unusual chemical composition, and are often adapted to harsh conditions. The three groups of archaebacteria are halophiles, methanogens, and thermoacidophiles.
   B. The remaining bacteria, collectively known as the eubacteria, are divided into three groups on the basis of cell wall composition.
      1. Mycoplasmas are bacteria that lack cell walls.
      2. Gram-negative bacteria have thin cell walls surrounded by an outer membrane. Gram-negative bacteria include nitrogen-fixing bacteria, enterobacteria, spirochetes, cyanobacteria, rickettsias, chlamydias, and myxobacteria.
      3. Gram-positive bacteria have thick cell walls of peptidoglycan. Gram-positive bacteria include the lactic acid bacteria, streptococci, staphylococci, and clostridia.
V. Bacteria play an important role in the biosphere, because many of them are decomposers and break down dead organic materials, allowing them to be recycled in the biosphere. Some bacteria are important in nitrogen fixation, the conversion of atmospheric nitrogen to a form that other organisms can use.

SELECTED KEY TERMS

aerobic, p. 315
autotroph, p. 315
bacillus, p. 317
binary fission, p. 315
bioremediation, p. 322
capsule, p. 313
coccus, p. 317

cyanobacteria, p. 312
endospore, p. 314
facultative anaerobe, p. 315
fecal coliform test, p. 320
Gram staining procedure,
   p. 318

heterotroph, p. 315
kingdom Archaebacteria,
   p. 316
kingdom Eubacteria, p. 317
nitrogen-fixing bacteria,
   p. 319

obligate anaerobe, p. 315
peptidoglycan, p. 314
pili, p. 314
plasmid, p. 314
saprotroph, p. 315
spirillum, p. 318

REVIEW QUESTIONS

1. What are the differences between archaebacteria and eubacteria? To which group are plants and other eukaryotes more closely related?
2. How does the cell wall of a gram-positive bacterium differ from that of a gram-negative bacterium?
3. Distinguish between autotrophs and heterotrophs. What is a chemosynthetic autotroph?

4. Describe one way in which bacteria might exchange genetic material.
5. Draw and label the three major shapes of bacterial cells.
6. What is an endospore?
7. Explain how saprotrophic bacteria help recycle materials in the environment.

8. Label the following diagram of a bacterial cell.

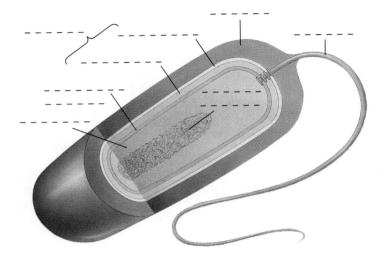

THOUGHT QUESTIONS

1. Imagine that you discover a new microorganism. After careful study, you determine that it should be classified in the kingdom Eubacteria as a cyanobacterium. What characteristics might lead you to this classification?
2. The bacterium *Clostridium tetani*, which causes tetanus (lockjaw), is an obligate anaerobe. Which type of wound would be more likely to develop tetanus, a surface cut or a deep puncture or gunshot wound? Why?
3. Why are bacteria classified primarily on the basis of chemical and metabolic features rather than their shapes?
4. Why is it important to take the entire prescribed dose of an antibiotic when being treated for a bacterial infection?

SUGGESTED READINGS

Angert, E.R., K.D. Clements, and N.R. Pace, "The Largest Bacterium," *Nature,* Vol. 362, March 18, 1993. The discovery of a novel bacterium so large that it can be seen *without* a microscope!

Caldwell, M., "Prokaryotes at the Gate," *Discover,* August 1994. How bacteria counteract the human arsenal (such as antibiotics) aimed against them.

Canby, T.Y., "Bacteria: Teaching Old Bugs New Tricks," *National Geographic,* August 1993. Because they are so diverse metabolically, bacteria have an enormous potential to perform such important tasks as degrading pollutants and making industrial chemicals.

Dixon, B., *Power Unseen: How Microbes Rule the World,* Salt Lake City, W.H. Freeman, 1994. This collection of 75 short essays, many of which discuss disease-causing bacteria, is entertaining and easy to read.

Gould, S.J., "Microcosmos," *Natural History,* March 1996. Some bacteria, such as those living in deep-sea thermal vents, are totally independent of the sun and of photosynthesis.

Hively, W., "Life Beyond Boiling," *Discover,* May 1993. Research discoveries about bacteria that thrive at temperatures well above 100°C. Most of these bacteria live in thermal vents at the bottom of the ocean.

KINGDOM PROTISTA

A large area of the North Atlantic Ocean between Africa and the West Indies (between 20 and 35° N latitude) is known as the Sargasso Sea. This expanse of warm water, about 2 million square miles in area, is not greatly affected by surface ocean currents, which rotate clockwise around the margins of the North Atlantic. Early European explorers feared sailing into the Sargasso Sea, because their ships would drift aimlessly through the comparatively still water for days or even weeks.

The Sargasso Sea gets its name from *Sargassum,* a gold and olive-colored seaweed that floats there in great abundance. (The genus name *Sargassum* is derived from *sargazo,* the Spanish word for "seaweed.") Although most species of sargassum grow attached to rocks in shallow ocean waters along rocky coastlines, the sargassum in the Sargasso Sea has adapted to the open ocean. Sargassum looks for all the world like a plant instead of an alga; it grows a meter or more in length and possesses stemlike and leaflike parts that produce round, gas-filled bladders, which keep it buoyant.

The sargassum living in the Sargasso Sea supports a variety of organisms that interact with one another to form a complex community. Among them is the sargassum fish (*Histrio histrio*), a minnow-sized fish with a fishing lure (a fleshy bulb with a short "pole") on its snout to attract small prey. The mottled brown and golden coloration of the sargassum fish enables it to blend almost perfectly into the sargassum. Although its closest relatives are all bottom-dwelling fish found in shallow seas, the sargassum fish swims in the open ocean among the sargassum or even crawls over the sargassum on its fins.

Sargassum fish (*Histrio histrio*) resembles the floating sargassum (*Sargassum* sp) on and around which it lives. (*Visuals Unlimited/David B. Fleetham*)

The floating meadows of sargassum offer shelter from the dangers of the open ocean and supply food to a complex community in the middle of the Atlantic Ocean. The sargassum fish and many other animals spend their entire lives among the tangled branches of sargassum, hiding from larger predators that lurk nearby and preying on smaller fish and invertebrates that consume sargassum. Other fish, including dolphinfish, flying fish, and eels, journey to the Sargasso Sea to breed. Eels migrate spectacular distances from European and American rivers to reproduce in the Sargasso Sea; the young return to the rivers to live out their adult lives.

Sargassum is one of many plantlike organisms placed in the kingdom Protista. In this chapter we examine a variety of protists—some plantlike and others fungus-like—that were originally classified in the plant kingdom. These organisms are usually studied in plant biology classes because they were traditionally considered plants.

After reading this chapter, you should be able to:

1. *Describe the common features of members of the kingdom Protista, and discuss in general terms the diversity within this kingdom, including modes of nutrition, body forms, and methods of reproduction.*

2. *Explain why the algae are classified in kingdom Protista rather than kingdom Plantae, and distinguish among six phyla of algae: dinoflagellates, diatoms, euglenoids, green algae, red algae, and brown algae.*

3. *Explain why slime molds and water molds are classified in kingdom Protista rather than kingdom Fungi, and briefly discuss three phyla of fungus-like protists: plasmodial slime molds, cellular slime molds, and water molds.*

4. *Define the endosymbiont theory on the origin of eukaryotic cells.*

PROTISTS ARE THE "SIMPLE" EUKARYOTES

When Robert Whittaker proposed the five-kingdom system of classification in 1969, only unicellular eukaryotic organisms were placed in kingdom Protista (from the Greek *protos,* "the very first"). The boundaries of this kingdom have been expanding since that time, although there is no universal agreement among biologists about what constitutes a protist. Today most biologists interpret the protist kingdom broadly, to include not only unicellular eukaryotes but eukaryotes of multicellular yet relatively simple organization. Kingdom Protista encompasses animal-like protists (protozoa), plantlike protists (algae, including seaweeds), and fungus-like protists (slime molds and water molds). Although these groups superficially resemble animals, plants, or fungi in certain respects, they are not considered merely simple kinds of animals, plants, or fungi.

Kingdom Protista is not a natural grouping of organisms, and organisms are placed within it for taxonomic convenience. If natural relationships based on a shared evolutionary history were the sole means of classifying organisms into kingdoms, there would be many more than six kingdoms. Some biologists think as many as 50 phyla would be needed to recognize natural relationships within the protists alone.

The kingdom Protista consists of a vast assortment of eukaryotic organisms whose diversity makes them difficult to characterize (Figure 20-1). Biologists estimate that there are as many as 200,000 species of protists living today. Their salient feature, eukaryotic cell structure characterized by the presence of nuclei and other membrane-bounded organelles, is shared with organisms from three other kingdoms—animals, plants, and fungi. There is, however, a distinct separation between eukaryotic protists and the organisms in kingdoms Eubacteria and Archaebacteria, all of which are prokaryotic.

Size varies considerably within the protist kingdom, from microscopic algae and protozoa to giant kelps, brown algae that can reach 60 meters (almost 200 feet) in length. Although most protists are unicellular organisms, some form **colonies** (loosely connected groups of cells), some are **coenocytic** (a multinucleate mass of cytoplasm), and some are multicellular. Multicellular protists have relatively simple body forms without specialized tissues.

Methods of obtaining nutrients in the kingdom Protista are varied. Most of the algae are autotrophic and photosynthesize as plants do. Some heterotrophic protists obtain their nutrients by absorption, as fungi do, whereas others resemble animals and ingest food. Some protists switch their mode of nutrition and are autotrophic at certain times and heterotrophic at others.

Although many protists are free-living, others form symbiotic relationships with various organisms. These intimate associations range from mutualism, a more-or-less equal partnership in which both partners benefit, to parasitism, in which one partner lives on or in another and is metabolically dependent on it. Some parasitic protists are important pathogens (disease-causing agents) of plants or animals. Specific examples of symbiotic associations are described throughout this chapter.

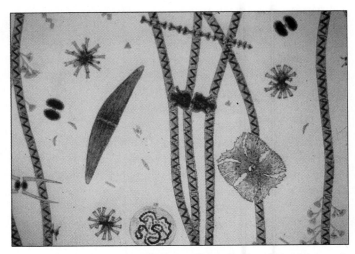

Figure 20-1   Various photosynthetic protists in a drop of pond water. The protists are an extremely diverse kingdom of organisms. (*Visuals Unlimited/Science VU*)

TABLE 20-1   A Comparison of Representative Algae and Fungus-like Members of the Kingdom Protista

Common Name	Phylum	Body Form	Locomotion	Photosynthetic Pigments	Special Features
Dinoflagellates	Pyrrhophyta	Single cell, some colonial	Two flagella	Chlorophylls *a* and *c*; carotenoids, including fucoxanthin	Many covered with cellulose plates; store oils and polysaccharides
Diatoms	Chrysophyta	Single cell, some colonial	Most nonmotile; some move by gliding over secreted slime	Chlorophylls *a* and *c*; carotenoids, including fucoxanthin	Silica in shell, store oils and carbohydrates
Euglenoids	Euglenophyta	Single cell	Two flagella (one of them very short)	Chlorophylls *a* and *b*; carotenoids	Flexible outer covering; store paramylon
Green algae	Chlorophyta	Single cell, colonial, siphonous, multicellular	Most flagellated at some stage in life; some nonmotile	Chlorophylls *a* and *b*; carotenoids	Reproduction highly variable; store starch
Red algae	Rhodophyta	Most multicellular, some single cell	None	Chlorophyll *a*; carotenoids; phycocyanin; phycoerythrin	Some reef builders; store floridean starch
Brown algae	Phaeophyta	Multicellular	Two flagella on reproductive cells	Chlorophylls *a* and *c*; carotenoids, including fucoxanthin	Differentiation of body into blade, stipe, and holdfast; store laminarin
Plasmodial slime molds	Myxomycota	Multinucleate plasmodium	Streaming cytoplasm, flagellated or amoeboid reproductive cells	—	Reproduce by spores formed in sporangia
Cellular slime molds	Acrasiomycota	Vegetative form—single cell	Amoeboid (for single cells)	—	Aggregation of cells signaled by chemical
		Reproductive form—multicellular (slug)	Cytoplasmic streaming (for multicellular)		
Water molds	Oomycota	Coenocytic mycelium	Biflagellate zoospores	—	Cellulose and/or chitin in cell walls

Most protists are aquatic and live in oceans or freshwater ponds, lakes, and streams. They make up most of the **plankton**, the floating, often microscopic organisms that are the base of the food web in aquatic ecosystems. Other aquatic protists attach to rocks and other surfaces in the water. Terrestrial (land-dwelling) protists are restricted to damp places such as soil, bark, and leaf litter. Even the parasitic protists are aquatic because they live in the wet environments of other organisms' body fluids.

Reproduction is varied in the kingdom Protista. Almost all protists reproduce asexually, and many can also reproduce sexually, often by **syngamy**, the union of gametes. Most protists, however, do not develop multicellular reproductive organs, nor do they form embryos the way more complex organisms do.

Protists, most of which are motile at some point in their life cycles, have various means of locomotion. They may move by pushing out cytoplasmic extensions as an amoeba does, by flexing individual cells, by waving cilia (short, hairlike structures), or by lashing flagella (long, whiplike structures). Some protists use a combination of two or more means of locomotion.

Protozoa, such as paramecia and amoebae, were traditionally classified as animals and are not usually studied in botany. In this chapter we discuss a number of representative algal and fungus-like phyla of the kingdom Protista (Table 20-1).

ALGAE ARE PLANTLIKE PROTISTS

The algae (from the Latin word for "seaweed") are a diverse group of organisms that are mostly photosynthetic. They range in size from unicellular microscopic forms to large multicellular seaweeds. Although they are photosynthetic, algae are not plants. Unlike a plant, an alga lacks a *cuticle,* the waxy protective coating that covers the aerial parts of plants and reduces water loss. Because of this, algae are restricted to damp or wet environments, such as oceans; freshwater ponds; lakes and streams; hot springs; polar ice; moist soil, trees, and rocks; and the bodies of certain animals, including sloths, sea anemones, corals, and worms. Also, most algae lack multicellular **gametangia** (sing. *gametangium*), reproductive organs in which gametes are produced. Algal gametangia are formed from single cells, whereas plant gametangia are multicellular (see Figure 22-2).

In addition to chlorophyll *a* and yellow and orange carotenoids, which are photosynthetic pigments found in all algae as well as all plants, different groups of algae possess a variety of other pigments that are also important in photosynthesis. Classification of algae into phyla is largely based on their pigment composition and what kinds of materials they produce to store energy reserves (fuel molecules). Other characteristics used to classify algae include their cell wall composition, number and placement of flagella, and chloroplast structures. We will consider six algal phyla: dinoflagellates, diatoms, euglenoids, green algae, red algae, and brown algae.

Most Dinoflagellates Are a Part of Marine Plankton

About 2100 species of **dinoflagellates** (*phylum Pyrrhophyta*) exist. Most of these unusual organisms are unicellular, although a few are colonies. Their cells are often covered with shells of interlocking cellulose plates, some of which contain silicates (minerals that contain silicon dioxide) that impart strength (Figure 20-2). Each dinoflagellate has two flagella; one is

wrapped like a belt around a transverse groove in the center of the cell, and the other is located in a longitudinal groove (at a right angle to the transverse groove) and projects behind the cell. The undulation of these flagella propels the dinoflagellate through the water like a spinning top. Indeed, the dinoflagellates' name is derived from the Greek *dinos*, meaning "whirling."

Most dinoflagellates are photosynthetic and possess the pigments chlorophyll *a*, chlorophyll *c*, and carotenoids. A special yellow-brown carotenoid, **fucoxanthin**, is found only in dinoflagellates and a few other groups of algae (the diatoms and brown algae). However, other dinoflagellates are colorless (and therefore nonphotosynthetic) and ingest other microorganisms for food. Dinoflagellates usually store energy reserves as oils or polysaccharides.

Many dinoflagellates reside (as endosymbionts) in the bodies of marine invertebrates such as jellyfish, corals, and mollusks. These symbiotic dinoflagellates lack cellulose plates and flagella and are called **zooxanthellae**. Zooxanthellae photosynthesize and provide carbohydrates for their invertebrate partners. The contribution of zooxanthellae to the productivity of coral reefs is substantial. Other dinoflagellates that reside in invertebrates lack pigments and do not photosynthesize; these heterotrophs are parasites that live off their hosts.

Reproduction in the dinoflagellates is primarily asexual, by cell division, although a few species have been reported to reproduce sexually. The dinoflagellate nucleus is unusual because the chromosomes are permanently condensed and always evident. Meiosis and mitosis are unique because the nuclear envelope remains intact throughout cell division and the spindle is located *outside* the nucleus. Based on the

Chlorophyll a = universal

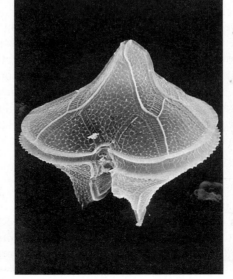
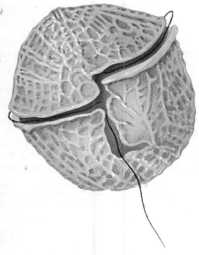

Figure 20-2   Representative dinoflagellates.
(a) A scanning electron micrograph of *Protoperidium* sp. Note the plates that encase the unicellular body. **(b)** A species of *Gonyaulax*, showing the two flagella located in grooves. (*Courtesy of T.K. Maugel, University of Maryland*)

(a)   (b)

Figure 20-3   A red tide in Mexico. The orange cloudiness in the water is produced by billions of dinoflagellates. (*Kevin Schafer/Peter Arnold, Inc.*)

Humans sometimes get paralytic shellfish poisoning by eating oysters, mussels, or clams that have fed on certain dinoflagellates. Paralytic shellfish poisoning induces respiratory failure and can result in death. (The dinoflagellates do not appear to harm the shellfish.)

Diatoms Are Nonflagellated and Have Silicon in Their Cell Walls

Although three groups of algae—the **diatoms**, yellow-green algae, and golden algae—are classified in the *phylum Chrysophyta*, we restrict our discussion to the diatoms because of their ecological and economic importance. There are about 5600 known species of diatoms. Most are unicellular, although a few exist as filaments or colonies. Diatoms are protected by a shell that is composed of two halves that overlap where they fit together, much like the two halves of a petri dish. The shells are impregnated with silica, a glasslike material laid down in striking, intricate patterns, which are used to classify the species (Figure 20-4).

Diatom cells have one of two shapes: *radial symmetry* (wheel-shaped) and *bilateral symmetry* (boat-shaped or needle-shaped). Although most diatoms are part of the floating plankton, some diatoms live on rocks and other surfaces, where they move by gliding. This gliding movement is facilitated by the secretion of a slimy material from a small groove along the shell.

Most diatoms are photosynthetic and contain the pigments chlorophyll *a*, chlorophyll *c*, and carotenoids, including the yellow-brown fucoxanthin; their pigment composition gives them a yellow or brown color. Energy reserves are stored as oils or carbohydrates.

uniqueness of their chromosome structure and mitosis, the dinoflagellates are thought to have no close living relatives.

In terms of ecological contributions, the dinoflagellates are one of the most important groups of producers in marine ecosystems. A few dinoflagellates are known to have occasional population explosions, or **blooms**. These blooms frequently color the water orange, red, or brown and are known as **red tides** (Figure 20-3). It is not known what environmental conditions initiate dinoflagellate blooms, but they are more common in the warm waters of late summer and early fall. The number of red tides has been increasing in recent years, and many experts think that human-produced coastal pollution triggers them, presumably by providing nutrients to the dinoflagellates. Some of the dinoflagellate species that form red tides produce a toxin that attacks the nervous systems of fishes, leading to massive fish kills. When airborne, this toxin can irritate eyes and make breathing difficult for coastal residents and beachgoers.

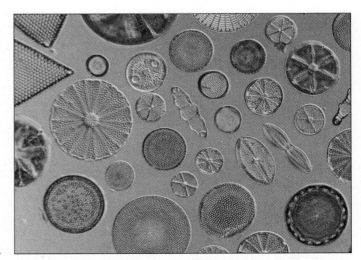

Figure 20-4   Diatoms have remarkably beautiful patterns on their symmetrical shells. (*The Stock Market/Phillip Harrington*)

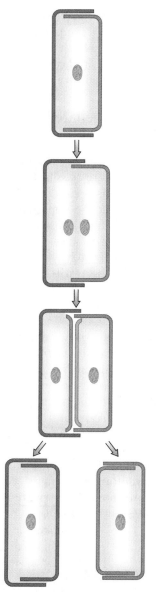

Figure 20-5   Asexual reproduction in diatoms. As cell division occurs, each new cell retains half of the original shell. The other half of the shell is synthesized, always to fit *inside* the original half. As a result, one of the new cells is smaller than the other.

Diatoms most often reproduce asexually by cell division (Figure 20-5). When a diatom divides, the two halves of its shell separate, and each becomes the larger half of the shell for a new diatom cell. Therefore, the diatom cells that receive the smaller halves of the shells get progressively smaller with each succeeding generation. When diatom descendants reach a certain fraction of the original size of the ancestral individual, sexual reproduction is triggered, and the small diatoms produce shell-less gametes. Sexual reproduction restores the

diatoms to the original size, because each resulting zygote (fertilized egg) grows substantially before producing a new shell.

Diatoms are common in fresh water and ocean water, but they are especially abundant in cooler marine waters. They are major producers in aquatic ecosystems because of their extremely large numbers. When diatoms die, their shells, which do not decompose, trickle down and accumulate in layers of what eventually becomes sedimentary rock. After millions of years, geological upheaval exposes some of these deposits, called **diatomaceous earth**, on land. For example, huge deposits of diatomaceous earth are found in the White Hills of Lompoc, California. Diatomaceous earth is mined and used as a filtering, insulating, and soundproofing material. As a filtering agent, it is used to refine raw sugar and to process vegetable oils. Because of its abrasive properties, diatomaceous earth is a common ingredient in scouring powders and metal polishes. It is no longer added to most toothpastes because it is too abrasive for tooth enamel.

Most Euglenoids Are Freshwater Unicellular Flagellates

There are about 1000 species of **euglenoids** (*phylum Euglenophyta*). Most are unicellular flagellates that possess two flagella, one long and whiplike and one so short that it doesn't protrude outside the cell (Figure 20-6). A euglenoid changes shape continually as it moves through the water, because its outer covering, called a *pellicle,* is flexible rather than rigid. The red to orange *eyespot,* a carotenoid-containing region of the cell, is thought to help the euglenoid perceive the direction of light. Euglenoids reproduce asexually by cell division; none has ever been observed to reproduce sexually.

Most euglenoids contain chloroplasts and photosynthesize. They have chlorophyll *a,* chlorophyll *b,* and carotenoids, the same pigments found in green algae and plants. (Although the euglenoids have the same pigmentation as the green algae and plants, they are not thought to be closely related to either group.) Energy reserves are stored as a polysaccharide called paramylon. Some photosynthetic euglenoids lose their chlorophyll when grown in the dark and turn into heterotrophs that obtain their nutrients by ingesting organic matter. Some species of euglenoids are always colorless and heterotrophic.

Euglenoids are both plantlike and animal-like and have often been used as an example of one of the dilemmas of classification. When biologists recognized two kingdoms—plants and animals—euglenoids were classified in the plant kingdom (with the algae) because many are photosynthetic. At the same time, other biologists classified them in the animal kingdom because they move, and some of them feed.

Most euglenoids inhabit freshwater ponds and puddles, particularly those with large amounts of organic material. For that reason their numbers are used to detect organic pollution. If a body of water has a large population of euglenoids, it is

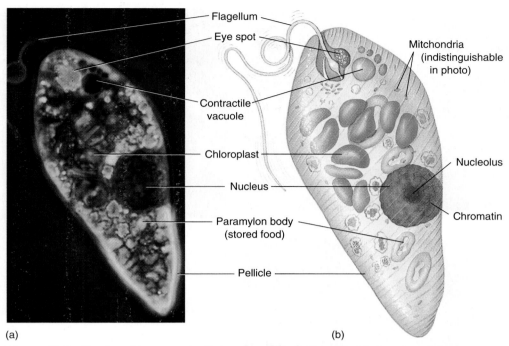

Figure 20-6   **Euglenoids are unicellular, flagellated algae that have at various times been classified in the plant kingdom (with the algae) and in the animal kingdom (when protozoa were considered animals). (a)** A living species of *Euglena*. **(b)** *Euglena* has a complex cellular structure. The eyespot is a light-sensitive organelle that helps it react to light. *Euglena's* outer covering, called a pellicle, is flexible and enables the organism to change shape easily. (*Biophoto Associates*)

probably polluted with excess organic matter. Some euglenoids are found in marine waters and mud flats.

The Green Algae Exhibit a Great Deal of Diversity

If one had to pick a single word to describe the 7000 species of **green algae** (*phylum Chlorophyta*), it would have to be "variety." Green algae exhibit many different body forms and methods of reproduction. Their body forms range from single cells to colonies to coenocytic, **siphonous** (tubular) algae to multicellular filaments and sheets (Figure 20-7). The multicellular forms do not have cells differentiated into specialized tissues, however. Most green algae are flagellated during at least part of their life history, although there are a few that are totally nonmotile.

Although the green algae are structurally diverse, they are biochemically very uniform. Green algae are photosynthetic, with chlorophyll *a*, chlorophyll *b,* and carotenoids present in chloroplasts of a wide variety of shapes. Starch (a polysaccharide) is the main energy reserve. Most green algae possess cell walls that contain cellulose, although some lack

walls and some are covered with scales. Green algae have a significant number of characteristics in common with plants, including pigments, storage products (energy reserves), and cell wall composition. Because of these and other similarities, it is generally accepted that plants evolved from green alga–like ancestors.

Reproduction in the green algae is as varied as structure, and both sexual and asexual reproduction occur. Asexual reproduction is by cell division in unicellular forms or by fragmentation in multicellular forms; in fragmentation, pieces that break off from the original alga give rise to new individuals. Many green algae reproduce asexually by forming reproductive cells called **spores** by mitosis; if these spores are flagellated and motile, they are called **zoospores**. Each spore is capable of developing into a new individual without fusing with another cell as gametes do.

Sexual reproduction in the green algae involves the formation of gametes in unicellular gametangia. Three types of sexual reproduction are recognized in green algae: isogamy, anisogamy, and oogamy (Figure 20-8). If the two gametes that fuse are identical in size and appearance, sexual reproduction is said to be **isogamous** (Figure 20-9). **Anisogamous** sexual reproduction involves the fusion of two gametes that differ in

(*Text continues on p. 335*)

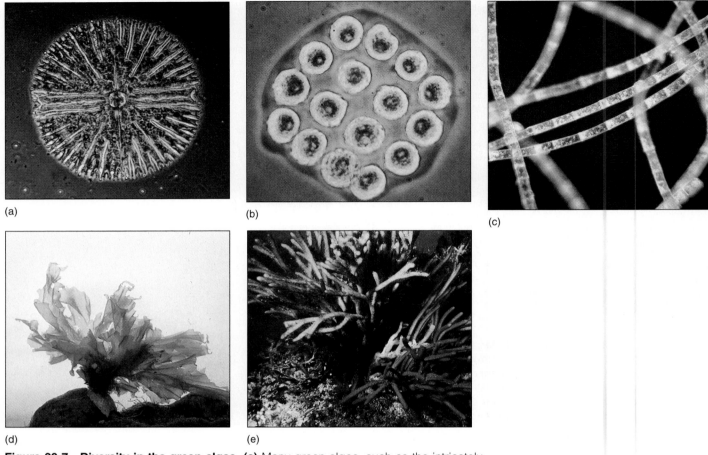

(a)

(b)

(c)

(d)

(e)

Figure 20-7   Diversity in the green algae. (a) Many green algae, such as the intricately beautiful *Micrasterias* sp, are single cells. **(b)** *Platydorina caudata* is a colonial green alga that consists of 16 to 32 cells. **(c)** *Spirogyra* sp is a multicellular green alga with a filamentous body form. **(d)** Some multicellular green algae are sheetlike. The thin, leaflike form has given *Ulva* sp its common name, "sea lettuce." **(e)** Siphonous green algae such as dead man's fingers (*Codium fragile*) are coenocytic, which means that each individual's body is composed of one giant cell with multiple nuclei. (a, *Ronald W. Hoham;* b, *Visuals Unlimited/Cabisco;* c, *Jack M. Bostrack/CBR Images;* d, *J.M. Kingsbury;* e, *Visuals Unlimited/William C. Jorgensen*)

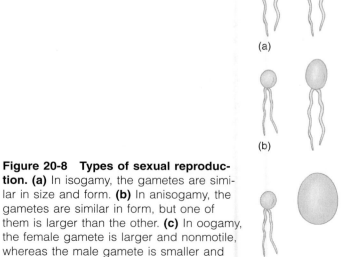

(a)

(b)

(c)

Figure 20-8   Types of sexual reproduction. (a) In isogamy, the gametes are similar in size and form. **(b)** In anisogamy, the gametes are similar in form, but one of them is larger than the other. **(c)** In oogamy, the female gamete is larger and nonmotile, whereas the male gamete is smaller and usually motile.

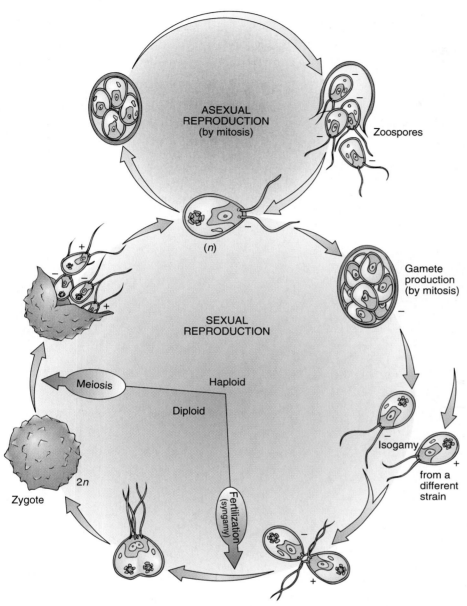

Figure 20-9   Isogamous sexual reproduction occurs during the life cycle of *Chlamydomonas* sp. *Chlamydomonas* is a haploid green alga with two strains, (+) and (−), that are visually indistinguishable. Both strains reproduce asexually by mitosis. During sexual reproduction, a (+) gamete fuses with a (−) gamete, forming a diploid zygote. Meiosis occurs, and four haploid cells emerge, two (+) and two (−).

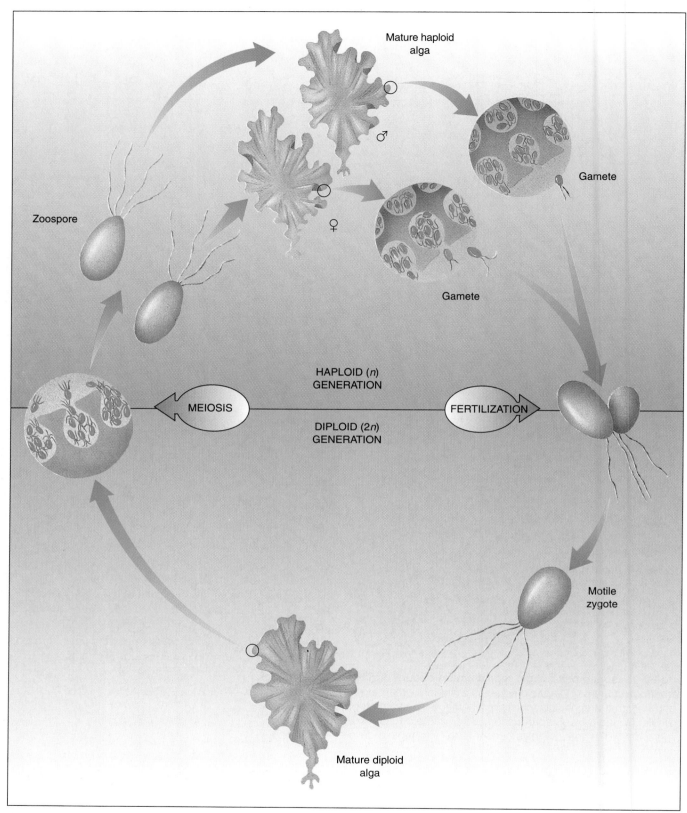

Figure 20-10 Anisogamous sexual reproduction occurs during the life cycle of ***Ulva* sp.** *Ulva* alternates between haploid and diploid multicellular phases, which are identical in overall appearance. The male and female haploid algae give rise to anisogamous gametes that fuse and subsequently develop into the diploid alga. Special cells in the diploid alga undergo meiosis to form haploid zoospores that develop directly into haploid algae, and the cycle continues.

Labels within figure:

Mature haploid alga

Gamete

Zoospore

Gamete

HAPLOID (*n*) GENERATION

MEIOSIS

FERTILIZATION

DIPLOID (2*n*) GENERATION

Motile zygote

Mature diploid alga

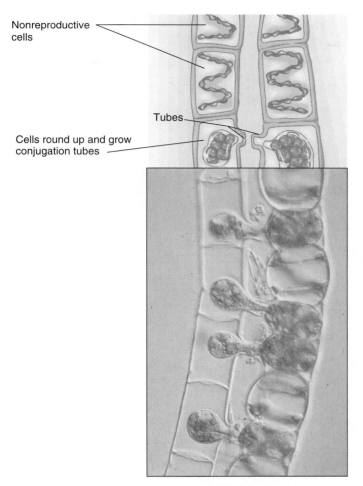

Nonreproductive cells

Tubes

Cells round up and grow conjugation tubes

Figure 20-11 *Spirogyra* **sp, a haploid organism, has a sexual phenomenon known as conjugation.** (*Top*) Filaments of two different strains line up, and conjugation tubes form between cells of the two filaments. (*Micrograph on bottom*) The contents of one cell pass into the other through the conjugation tube. The two cells then fuse and form a diploid zygote. Following a period of dormancy, the zygote undergoes meiosis, restoring the haploid condition. (*Biophoto Associates*)

size and/or motility (Figure 20-10). Some green algae are **oogamous** and produce large nonmotile female gametes (eggs) and small flagellated male gametes (sperm cells). Instead of sexual reproduction by the fusion of gametes, some green algae join together temporarily and exchange genetic information through **conjugation**, a process in which the genetic material of one cell passes into and fuses with the genetic material of a recipient cell (Figure 20-11).

Both aquatic and terrestrial green algae exist. Aquatic green algae primarily inhabit fresh water, although marine species also occur. Green algae that inhabit the land are restricted to damp soil, cracks in tree bark, and other moist places. Some green algae live as endosymbionts in the bodies of animals such as freshwater sponges, mollusks, and flat-

worms; others grow together with fungi as a "dual organism" called a lichen (see Chapter 21).

Regardless of where they live, green algae are ecologically important. Because of their photosynthetic activity, they are at the bases of many food webs, particularly in freshwater habitats. Green algae also help to oxygenate the water during daylight hours (recall that oxygen is a byproduct of photosynthesis). Thus, aquatic animals depend on green algae to provide them with both food and oxygen.

Most Red Algae Are Multicellular Organisms

There are about 4000 species of **red algae** (*phylum Rhodophyta*). The vast majority are multicellular organisms, although a few unicellular species exist. The multicellular body of a red alga is commonly composed of complex, interwoven filaments. Red algae are sometimes delicate and feathery, although a few species are flattened sheets of cells (Figure 20-12). Most multicellular red algae attach to rocks or other solid materials with a rootlike **holdfast**.

The chloroplasts of red algae contain **phycoerythrin**, a red pigment, and **phycocyanin**, a blue pigment, in addition to chlorophyll *a* and carotenoids. The cyanobacteria are the only other organisms that possess the same pigment composition as the red algae. Red algae store their energy reserves as a polysaccharide (floridean starch) similar to glycogen.

The cell walls of red algae often contain thick, sticky polysaccharides that are of commercial value. For example, **agar** is a polysaccharide extracted from certain red algae and used as a food thickener and culture medium (a substance on which to grow microorganisms and propagate some plants, such as orchids). Another polysaccharide extracted from red algae, **carrageenan**, is a food additive used to stabilize chocolate milk and to provide a thick, creamy texture to ice creams and other soft, processed foods. (A stabilizer keeps the texture uniform by preventing different ingredients from settling out.) Carrageenan is also used to stabilize cosmetics and paints. Red algae are a source of vitamins (particularly A and C) and minerals for humans, particularly in Japan and other East Asian countries where they are eaten fresh, dried, or toasted in such traditional foods as sushi and nori.

Reproduction in the red algae has been studied in detail for only a few species, but it is remarkably complex, with an alternation of sexual and asexual stages. Although sexual reproduction is common, at no stage in the life history of red algae are there any flagellated cells.

The red algae are primarily found in warm tropical oceans, although a few species occur in fresh water and in soil. Some red algae incorporate calcium carbonate into their cell walls from the ocean waters (Figure 20-12c). These coralline red algae are central in building "coral" reefs and are possibly more important than coral animals in this process.

(a)

(b)

(c)

Figure 20-12   Diversity in the red algae. Most red algae are multicellular, many with complex, highly branched, filamentous bodies. **(a)** *Rhodymenia* sp, which is common in the Northern Hemisphere, has a flattened, branching body. **(b)** *Polysiphonia* sp, which is widely distributed throughout the world, has a highly branched body. **(c)** *Bosiella* sp, which is found in the Pacific Ocean, is a coralline red alga with a hard, brittle body. (a, *Richard H. Gross;* b, *Visuals Unlimited/Philip Sze;* c, *Visuals Unlimited/D. Gotshall*)

Brown Algae Are the Giants of the Protist Kingdom

About 1500 species of **brown algae** (*phylum Phaeophyta*) exist. All brown algae are multicellular. Their bodies, which may be tufts, ropes, or thick, flattened branches, range in size from several centimeters to approximately 60 meters in length. The largest brown algae, called kelps, are tough and leathery in appearance and exhibit considerable differentiation into leaflike **blades**, stemlike **stipes**, and anchoring holdfasts (Figure 20-13). They often have gas-filled floats to increase buoyancy (recall sargassum in the chapter introduction). It is important to remember that the blades, stipes, and holdfasts of brown algae are not homologous with the leaves, stems, and roots of plants. Brown algae and plants are thought to have evolved from different unicellular ancestors.

Brown algae are photosynthetic and possess chlorophyll *a,* chlorophyll *c,* and carotenoids, including the yellow-brown fucoxanthin, in their chloroplasts. The main energy reserve in brown algae is a carbohydrate called laminarin.

Brown algae are commercially important for several reasons. Their cell walls contain a polysaccharide, **algin**, which possibly helps cement the cell walls of adjacent cells together. Algin is used as a thickening and stabilizing agent in ice creams, marshmallows, toothpastes, shaving creams, hair sprays, and hand lotions. Brown algae are eaten by humans, particularly in East Asian countries, and they are a rich source of certain vitamins and of minerals such as iodine. Brown algae are one source of the antiseptic tincture of iodine.

Reproduction is varied and complex in the brown algae. They reproduce sexually, and most spend a portion of their lives as haploid organisms and a portion as diploid organisms. Their reproductive cells, both asexual zoospores and sexual gametes, are flagellated.

Brown algae are common in cooler marine waters, especially along rocky coastlines, where they are found mainly in the intertidal zone or in relatively shallow waters. Kelps form extensive underwater "forests" and are essential in that ecosystem as the primary producer of food (Figure 20-14). Kelp beds also provide habitats for many marine invertebrates,

Figure 20-13   *Laminaria* sp, a typical brown alga. Note the blade, stipe, and holdfast. (*J. Robert Waaland, University of Washington/Biological Photo Service*)

fishes, and mammals such as sea otters. The diversity of life in these underwater forests is astounding, and all of it is supported by the brown algae.

SLIME MOLDS AND WATER MOLDS ARE FUNGUS-LIKE PROTISTS

The fungus-like protists superficially resemble fungi in that they are nonphotosynthetic and their body forms often consist of threadlike filaments called *hyphae.* However, fungus-like protists are not fungi, for several reasons. Many fungus-like protists produce flagellated reproductive cells, which true fungi never produce. In addition, the cell walls of many fungus-like protists contain cellulose, while the cell walls of fungi contain chitin, the same substance found in the exoskeletons of insects.

The Plasmodial Slime Molds Have an Unusual Feeding Stage

There are about 500 species of **plasmodial slime molds** (*phylum Myxomycota*). In the feeding stage of the plasmodial slime mold life cycle, a **plasmodium** streams over damp, decaying logs and leaf litter, ingesting bacteria, yeasts, spores, and

Figure 20-14   **A kelp bed off the coast of California.** These underwater "forests" are ecologically important, supporting large numbers of aquatic organisms. (*Beverly Factor*)

decaying organic matter (Figure 20-15a). The plasmodium, a wall-less amoeba-like mass that contains many diploid nuclei in its cytoplasm, changes shape as it moves, much as an amoeba does. Often brightly colored, it forms a network of channels to cover a larger surface area as it creeps along.

When the food supply dwindles or there is insufficient moisture, the plasmodium flows to a surface exposed to the air and initiates reproduction. Stalked structures of intricate complexity and beauty usually form from the drying plasmodium (Figure 20-15b). Within these structures, called **sporangia**, meiosis occurs to produce haploid nuclei. Later, a wall of cellulose and/or chitin develops around each nucleus, forming a spore. These spores are extremely resistant to adverse environmental conditions (drought and extremes of temperature). When conditions are favorable, however, the spores crack open, and a haploid reproductive cell emerges from each. This haploid cell may be flagellated or amoeba-like, depending on how wet the environment is; flagellated cells

(a)

(b)

Figure 20-15   Plasmodial slime molds are characterized by a creeping plasmodium and the production of spores. (a) The plasmodium of *Physarum polycephalum* is colored bright yellow. This naked mass of protoplasm, shown on a rotting log, is multinucleate and feeds on bacteria and other microorganisms. (b) The reproductive structures of *P. polycephalum* are stalked sporangia. (a,b, *R. Calentine/CBR Images*)

form in wet conditions. These two reproductive cells, the flagellated **swarm cell** and the amoeba-like **myxamoeba**, act as gametes. Eventually two gametes fuse to form a zygote with a diploid nucleus. The resultant diploid nucleus divides by mitosis, but the cytoplasm does not divide, so the result in this case is a multinucleate plasmodium.

The Cellular Slime Molds Aggregate into a "Slug"

Although the 70 or so species in the *phylum Acrasiomycota* are called **cellular slime molds**, their resemblance to the plasmodial slime molds is superficial. Indeed, cellular slime molds have much closer affinities with the amoeba. Each

cellular slime mold is a small individual amoeba-like cell that behaves as a separate, solitary organism; it moves over rotting logs and soil, ingesting bacteria and other particles of food. Each cell has a haploid nucleus.

When moisture or food becomes inadequate, the individual haploid cells send out a chemical signal that causes them to aggregate by the hundreds or thousands. During this stage the cells migrate as a single multicellular "organism," called a **pseudoplasmodium** or "slug." Each cell of the slug retains its plasma membrane and individual identity. Eventually, the slug settles and develops into a stalked structure. The cells in the front third of the slug form the stalk, and the rear portion of the slug forms a rounded structure at the top of the stalk, within which spores differentiate. Each spore grows into an individual amoeba-like cell, and the cycle repeats itself (Figure 20-16). This reproductive cycle is generally asexual, although sexual reproduction has been observed occasionally.

The Water Molds Produce Flagellated Zoospores

The 580 species of **water molds** (*phylum Oomycota*) were once classified as fungi because of their superficial structural resemblance; both a water mold and a fungus have a body, termed a **mycelium**, that grows over a food source, digests it with enzymes that it secretes, and then absorbs the predigested nutrients (Figure 20-17). The threadlike **hyphae** that make up the mycelium in a water mold are coenocytic; that is, the water mold body consists of one giant multinucleate "cell." The cell wall of a water mold may be composed of cellulose (as in a plant), chitin (as in a fungus), or both cellulose and chitin. Because a water mold produces flagellated cells at some point in its life cycle, whereas fungi never produce motile cells, most biologists classify the water molds as protists rather than fungi.

When food is plentiful and environmental conditions are good, water molds reproduce asexually (Figure 20-18). A hyphal tip swells and a cross wall is formed, separating the hyphal tip from the rest of the mycelium. Within this structure, called a *zoosporangium*, tiny biflagellate zoospores form, each of which can develop into a new mycelium. When environmental conditions worsen, water molds initiate sexual reproduction. After fusion of male and female nuclei, a thick-walled **oospore** develops from the zygote. Water molds often survive low temperatures in the winter as oospores.

Some water molds have played infamous roles in human history. The Irish potato famine of the 19th century, for example, was precipitated by the water mold that causes a disease in potatoes called late blight. As a result of several rainy, cool summers in Ireland during the 1840s, the late blight water mold, which thrives under such conditions, multiplied unchecked. Potato tubers rotted in the fields, and since pota-

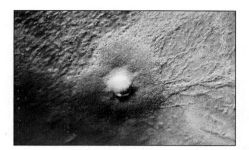

(a) Aggregation of amoeboid cells

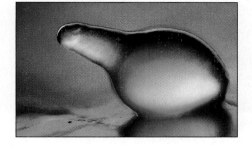

(b) Slug (migrating pseudoplasmodium)

Feeding
amoeboid cells

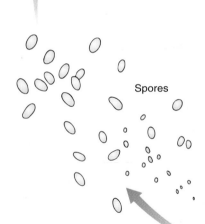

Spores

Developing
fruiting
body

(c) Fruiting body

**Figure 20-16 The life cycle of the cellular slime mold, _Dictyostelium discoideum,_
includes a unicellular stage and a multicellular pseudoplasmodium.** Unicellular
amoeba-like cells (*upper left*) ingest food, grow, and reproduce by cell division.
(a) After their food supply is depleted, the cells stream together. **(b)** The aggregation
organizes into a slug-shaped, multicellular pseudoplasmodium that migrates for a
period of time before forming a stalked fruiting body **(c)**. The mature fruiting body
releases spores, each of which opens to liberate an amoeba-like organism, and the
cycle continues. (a,b,c, *Visuals Unlimited/Cabisco*)

339

Figure 20-17   The water mold that has attacked this unhappy goldfish is a parasite that will soon kill it. Note the hairy mycelium surrounding the fish. (*Fred M. Rhoades*)

toes were the staple of the Irish peasants' diet, many people starved to death. Estimates of the number of deaths that resulted from the outbreak of this plant disease range from 250,000 to more than 1 million. The famine prompted a mass migration out of Ireland to other countries, such as the United States.

Late blight is still a problem today. In 1992 new strains of the water mold that causes it were reported in New York and Washington states; a year later, late blight was detected in 11 states. The new strains have also appeared in northern Europe, South America, Japan, South Korea, and parts of Africa. These water mold strains are resistant to the fungicide usually used to control the disease. Because today most people eat a varied diet, late blight is not expected to cause a famine, but it is costing potato growers millions of dollars in lost crops.

THE EARLIEST EUKARYOTES WERE PROTISTS

Protists are thought to have been the first eukaryotic cells to evolve from prokaryotes. They appeared in the fossil record some 2.1 billion years ago. The **endosymbiont theory**, popularized since the 1960s by Lynn Margulis, suggests that certain eukaryotic organelles (mitochondria, chloroplasts, and perhaps even cilia and flagella) may have originated from symbiotic relationships between larger prokaryotes and the small prokaryotes that lived within them (see Focus On: The Evolution of Eukaryotic Cells in Chapter 3). Chloroplasts are thought to have originated from photosynthetic prokaryotes (possibly cyanobacteria), and mitochondria from aerobic bacteria.

A few early protists with hardened shells—diatoms, for example, produced abundant fossils. However, most ancient protists did not leave extensive fossils because their bodies were too soft to leave permanent traces behind. Therefore, evolutionary studies of protists are based primarily on comparisons of present-day organisms, which contain many clues about their evolutionary history. Some of the most useful data for evolutionary interpretations come from electron microscopy, biochemistry, and molecular biology.

Multicellularity Evolved in the Protist Kingdom

The green algae, red algae, and brown algae are protistan phyla with multicellular species. However, multicellular green algae have more in common with unicellular green algae than they do with other multicellular protists. Similarly, multicellular forms of both red and brown algae have more in common with certain unicellular protists than they do with each other or with the multicellular green algae. In other words, the multicellular green algae, red algae, and brown algae probably had different unicellular ancestors, and multicellularity apparently evolved several times within the kingdom Protista.

Multicellular organisms may have evolved from colonial forms

Studying the protists living today provides clues about how multicellularity may have originated. Colonies, for example, may have been transitional stages between unicellular organisms and multicellular organisms. *Chlamydomonas* (Figure 20-19a) is a unicellular green alga that uses two flagella for motility. The green algae also include a number of colonies that consist of attached *Chlamydomonas*-like cells. For example, *Gonium* is a colony that consists of four *Chlamydomonas*-like cells, and *Pandorina* is a colonial organism composed of 16 to 32 *Chlamydomonas*-like cells. The largest colonies of *Chlamydomonas*-like cells—from 1000 to 50,000 cells—are in the genus *Volvox* (Figure 20-19b). As colonies increase in size and number of cells, specialization in cell structure and function occur. *Volvox* sp, which has some division of labor among the cells (for example, only some cells are capable of carrying out reproduction), approaches multicellularity.

The *volvocine line* (the series of *Chlamydomonas*-like organisms just discussed) is an evolutionary dead-end—that is, it did not give rise to multicellular green algae. However, the trend in increasing colony size and cell differentiation within the volvocine line indicates one possible way in which multicellularity may have originated: single cells → colonies → multicellular organisms.

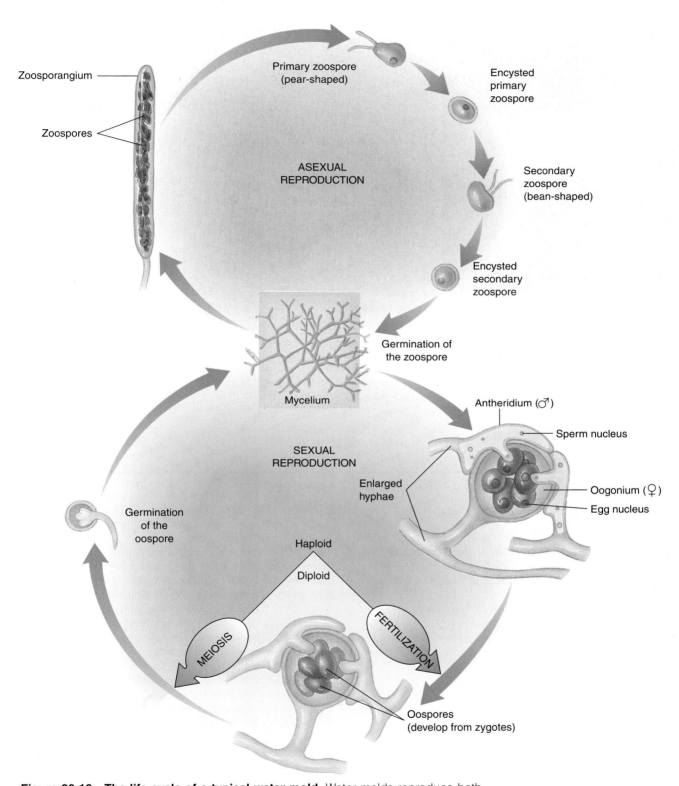

Figure 20-18   The life cycle of a typical water mold. Water molds reproduce both sexually (by oospores) and asexually (by zoospores).

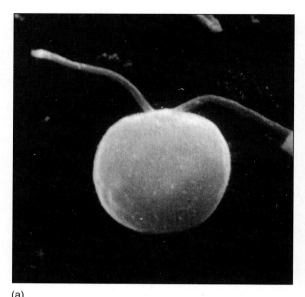

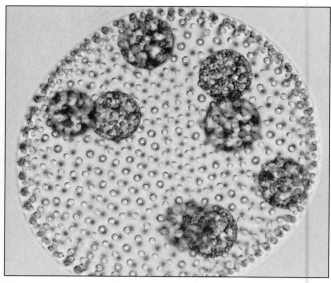

(a)   (b)

Figure 20-19   Multicellularity may have evolved in this sequence: unicellular organisms → colonial organisms → multicellular organisms. (a) A scanning electron micrograph of a species of *Chlamydomonas*, a biflagellate unicellular green alga. **(b)** A light micrograph of a species of *Volvox*, a colonial green alga composed of up to 1000 to 50,000 *Chlamydomonas*-like cells. New colonies can be observed inside the parental colony, which eventually breaks apart to release them. (a, *Courtesy of T.K. Maugel/University of Maryland*; b, *James W. Richardson/CBR Images*)

S T U D Y   O U T L I N E

I. The kingdom Protista is composed of "simple" eukaryotic organisms. Most protists are aquatic.

II. Kingdom Protista exhibits a remarkable diversity.
  A. Protists range in size from microscopic single cells to large multicellular organisms.
  B. Different protists obtain their nutrients photosynthetically or heterotrophically.
  C. Protists may be free-living or form symbiotic relationships, which range from mutualism to parasitism.
  D. Almost all protists reproduce asexually; many reproduce both sexually and asexually.
  E. Protists have various means of locomotion, although some are nonmotile.

III. Algae are plantlike protists.
  A. Dinoflagellates are mostly unicellular, biflagellate, photosynthetic organisms of great ecological importance.
  B. Diatoms are major producers in aquatic ecosystems. These are mostly unicellular protists, with silica-impregnated shells.
  C. Euglenoids are unicellular, flagellated protists with pigments identical to those found in green algae and plants. Euglenoids do not appear to be close relatives of either group, however.
  D. Green algae exhibit a wide diversity in size, complexity, and reproduction. It is generally accepted that plants evolved from green alga-like ancestors.

  E. Most red algae are multicellular and lack motile cells.
  F. All brown algae are multicellular and produce flagellated cells during reproduction.

IV. Fungus-like protists were originally classified with the fungi but have features, such as flagellated reproductive cells and cell walls that contain cellulose, that are clearly not fungal.
  A. The body of a plasmodial slime mold is a multinucleate plasmodium. Plasmodial slime molds reproduce by spores.
  B. The cellular slime molds live as individual amoeba-like cells, but they aggregate into a pseudoplasmodium that eventually forms a spore-producing reproductive structure.
  C. The water molds have a coenocytic mycelium, that is, a threadlike body that is multinucleate but not multicellular. They reproduce asexually by forming biflagellate zoospores and sexually by forming oospores.

V. The protists originated about 2.1 billion years ago and were the first eukaryotes.
  A. Compelling evidence exists that certain eukaryotic organelles (mitochondria and chloroplasts) evolved from endosymbiotic relationships between smaller prokaryotes and the larger prokaryotes in which they resided.
  B. Multicellularity evolved several times within the kingdom Protista, possibly by following a trend in greater complexity from single cells to colonies to multicellular organisms.

SELECTED KEY TERMS

brown alga, p. 336

cellular slime mold, p. 338

coenocytic, p. 326

colony, p. 326

conjugation, p. 335

diatom, p. 329

dinoflagellate, p. 328

endosymbiont theory, p. 340

euglenoid, p. 330

gametangium, p. 328

green alga, p. 331

holdfast, p. 335

hypha, p. 338

mycelium, p. 338

plankton, p. 327

plasmodial slime mold, p. 337

plasmodium, p. 337

pseudoplasmodium, p. 338

red alga, p. 335

red tide, p. 329

syngamy, p. 327

water mold, p. 338

zoospore, p. 331

zooxanthellae, p. 328

REVIEW QUESTIONS

1. What are the characteristics of a typical protist? Why are protists so difficult to characterize?
2. In what way does a unicellular protist differ from a bacterium?
3. How are the protists important to humans?
4. How are the protists important ecologically?
5. Some biologists still classify the algae as plants. Why could algae be considered "simple" plants? Why do many biologists classify algae as protists rather than plants?
6. What features are used to classify the algae into phyla?
7. Name the three parts of a brown alga's body.
8. Some biologists still classify the water molds as fungi. Why could water molds be considered fungi? Why do many biologists classify them as protists rather than fungi?

THOUGHT QUESTIONS

1. How might *Sargassum* have first colonized the Sargasso Sea?
2. *Euglena* is a freshwater alga that lacks a cell wall. Why, then, is the presence of a contractile vacuole, which expels excess water, so important to *Euglena?*
3. Plasmodial slime molds reproduce sexually, whereas cellular slime molds do not. What advantages might there be for each organism to have the type of reproduction it does?
4. When food is plentiful and environmental conditions are good, water molds reproduce asexually, whereas when environmental conditions worsen, they reproduce sexually. The same is true for many algae and other organisms. Explain why it might be advantageous to reproduce asexually in favorable environments but sexually in unfavorable, changing environments.
5. What characteristics of the green algae indicate their close relationship to plants?

SUGGESTED READINGS

Boyle, R.H., "Phantom," *Natural History,* February 1996. A recently discovered dinoflagellate causes serious fish kills and human health problems.

Daly, D.C., "The Leaf That Launched a Thousand Ships," *Natural History,* January 1996. A fascinating account of the Irish potato famine in the 19th century.

Jacobs, W.P., "*Caulerpa,*" *Scientific American,* December 1994. Discusses a coenocytic green alga that is the world's largest "unicellular" organism.

Round, F.E., R.M. Crawford, and D.G. Mann, *The Diatoms: Biology and Morphology of the Genera,* Cambridge, England, Cambridge University Press, 1990. A beautifully illustrated reference book on the diatoms. Includes 5000 electron micrographs.

Sharnoff, S.D., "Beauties from a Beast: Woodland Jekyll and Hydes," *Smithsonian,* July 1991. Beautiful photographs of the sporangia of plasmodial slime molds.

KINGDOM FUNGI

Since ancient times edible mushrooms have been considered a delicacy largely reserved for the wealthy. Although mushrooms contain some vitamins and minerals, most are not particularly nutritious and do not contain much in the way of proteins, carbohydrates, or lipids (fats). Nonetheless, they are an important part of the world's major cuisines.

In the United States, more than 780 million pounds of mushrooms are produced annually, which is equivalent to about 3 pounds of mushrooms for every American man, woman, and child. The common mushroom (*Agaricus bisporus*) is the only fungal species grown extensively for food, although several exotic mushrooms (such as oyster, shiitake, portobello, and straw mushrooms) are becoming increasingly popular. More than half of the mushrooms produced commercially in the United States are grown in Pennsylvania, where mushrooms are the state's biggest cash crop.

Cultivating mushrooms requires very exacting conditions. First the microscopic spores germinate in a laboratory under sterile conditions and grow into a network of delicate threads called a *mycelium*. Tiny plugs of the mycelium are then placed on sterile millet seeds, which they

A mushroom bed at a commercial mushroom farm.
(*Courtesy of Paul Wuest, Pennsylvania State University*)

invade and digest. The seeds are then incubated and sent to a mushroom farm.

In windowless buildings at the mushroom farm, the farmer prepares beds of compost formed from the decomposed remains of straw and horse, cow, or chicken manure. The mycelium-containing millet seeds are then scattered across the compost, and the temperature and humidity are carefully regulated. After 2 to 3 weeks, the mycelium, which has formed a mat across the compost, is covered by peat moss. The temperature of the room is lowered to 10 to 15°C (50 to 60°F) and the humidity raised to 95 percent; these sudden changes in environmental conditions cause the mycelium to form tiny "buttons," immature mushrooms, in a week or two. Mushrooms can be harvested from the bed over the next 3 to 7 months.

Mushrooms are only one of many diverse types of organisms classified in kingdom Fungi. This chapter examines the biology and significance of fungi. Although they are not plants, fungi are usually studied in plant biology because they were traditionally considered plants when all organisms were classified into two or three kingdoms.

L E A R N I N G   O B J E C T I V E S

After reading this chapter, you should be able to:

1. *Describe the distinguishing characteristics of the kingdom Fungi.*
2. *Explain the fate of a fungal spore that lands on an appropriate food source, such as an overripe peach, and describe conditions that permit its growth.*
3. *Explain the ecological significance of fungi as saprotrophs and as mycorrhizae.*
4. *List distinguishing characteristics for each of the four phyla of fungi (zygomycetes, ascomycetes, basidiomycetes, and deuteromycetes), and give examples of each.*
5. *Characterize the unique nature of a lichen.*
6. *Describe the destructive nature of certain fungi, especially to crops.*

FUNGI ARE EUKARYOTES WITH A UNIQUE MODE OF NUTRITION

As unappetizing as this fact may be, mushrooms, morels, and truffles—the delights of gourmet cooks around the world—have much in common with the black mold that grows on stale bread and the mildew that appears on damp shower curtains. All of these organisms belong to the kingdom Fungi, a diverse group of more than 60,000 known species, most of which are terrestrial. All of the fungi are eukaryotes; their cells contain membrane-bounded organelles such as nuclei and mitochondria. Fungi, which vary strikingly in size and shape, were originally classified in the plant kingdom, but biologists today recognize that they are not plants. Because they are distinct from plants and other eukaryotes in many ways, fungi are accorded a separate kingdom.

Fungal cells, like plant cells, are enclosed by cell walls. However, fungal cell walls have a different chemical composition from plant cell walls. In most fungi the cell wall is composed in part of **chitin**, which is also a component of the external skeletons of insects and other arthropods. Chitin, a polysaccharide that consists of subunits of a nitrogen-containing sugar, is far more resistant to breakdown by microorganisms than is the cellulose that makes up plant cell walls.

Fungi lack chlorophyll and chloroplasts and therefore are nonphotosynthetic. Although they are **heterotrophs** (that is, they are unable to synthesize their own organic material), fungi do not ingest food as animals do. Instead, fungi secrete digestive enzymes onto living or dead organic material and then *absorb* small organic molecules of the predigested food through their cell walls and plasma membranes. Some fungi are **saprotrophs** (decomposers), and others are parasites. As saprotrophs, fungi obtain their nutrients from dead organic matter (animal carcasses, leaves, and so on); as parasites, they obtain nutrients from other organisms.

dead material   living

Fungi grow best in dark, moist habitats but are found wherever organic material is available. They require moisture to grow and can obtain water from the atmosphere as well as from the organic material on which they live. When the environment becomes very dry, fungi survive by going into a resting stage or by producing spores that are resistant to desiccation (drying out). Although the optimum pH for most species is slightly acidic (about 5.6), some fungi can tolerate and grow in environments where the pH ranges from 2 (very acidic) to 9 (alkaline). Many fungi are less sensitive to high osmotic pressures than bacteria and therefore can grow in concentrated salt or sugar solutions, such as jelly, that discourage or prevent bacterial growth. Because fungi can thrive over a wide temperature range, even refrigerated food is subject to spoilage by fungi.

Most Fungi Possess a Filamentous Body Plan

The body structures of fungi vary in complexity, ranging from the single-celled yeasts to the multicellular, filamentous molds (a term used loosely to include mildews, rusts and smuts, mushrooms, and many other fungi). **Yeasts** are unicellular fungi that reproduce asexually mainly by **budding**, in which a small bulge (bud) grows and eventually separates from the parent cell (see Figure 4-17). Each bud that separates from the parent cell is a new yeast cell that can grow and bud again. Yeasts also reproduce asexually by fission (the equal division of one cell into two cells) and sexually through spore formation (discussed shortly).

Most fungi are filamentous **molds**. A mold consists of long, branched threads (filaments) of cells called **hyphae** (sing. *hypha*). Hyphae form a tangled mass or tissue-like aggregation known as a **mycelium** (pl. *mycelia*) (Figure 21-1). The cobweb-like mold sometimes seen on bread is the mycelium of a fungus. What is not seen is the extensive part of the mycelium that grows down into the substance of the bread.

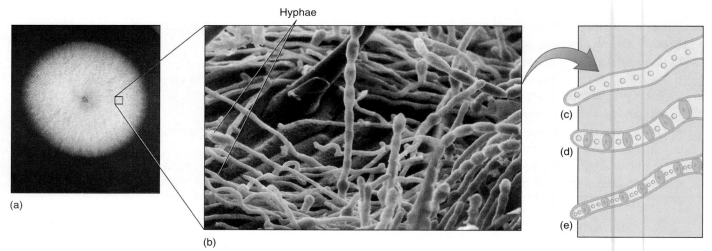

Hyphae

(a)

(b)

(c)

(d)

(e)

Figure 21-1   Molds are filamentous fungi. (a) A fungal mycelium growing on agar in a culture dish. In nature, fungal mycelia are rarely so symmetrical. **(b)** The mycelium consists of a mass of threadlike hyphae. **(c)** A coenocytic hypha. **(d)** A hypha divided into cells by septa; each cell has one nucleus (is monokaryotic). In some hyphae the septa are perforated, as shown here, permitting cytoplasm to stream from one cell to another. **(e)** A septate hypha in which each cell has two nuclei (is dikaryotic). (a, *Dennis Drenner;* b, *G.T. Cole/Biological Photo Service*)

Some hyphae are **coenocytic**; that is, they are not divided into individual cells but are like an elongated, multinucleated giant cell. Other hyphae are divided by cross walls, called **septa** (sing. *septum*), into individual cells that each contain one or more nuclei. The septa of septate fungi often contain large pores that permit nuclei and cytoplasm to flow from cell to cell, providing a system of internal transport.

Fungi Reproduce by Spores

Fungi reproduce by means of microscopic **spores**, nonmotile (nonflagellated) reproductive cells dispersed by wind or by animals. Spores are usually produced on aerial hyphae (hyphae that project up into the air). This arrangement permits the spores to be carried to new areas by air currents. In some fungi, such as mushrooms, the aerial hyphae form large complex reproductive structures, called **fruiting bodies**, in which spores are produced. The familiar part of a mushroom is a large fruiting body. We do not normally see the nonreproductive part of the fungus, a nearly invisible network of hyphae buried out of sight in the rotting material on which it grows.

Fungi can produce spores either asexually (by mitosis) or sexually (by meiosis). Fungal cells usually contain haploid nuclei. In sexual reproduction, the hyphae of two genetically different mating types come together and their nuclei fuse, forming a diploid zygote. Meiosis then occurs, forming haploid spores. During sexual reproduction in two phyla (the

ascomycetes and basidiomycetes), hyphae come together, but the two different nuclei do not fuse immediately; instead, they remain separate within the fungal cytoplasm. Hyphae that contain two genetically distinct nuclei within each cell are referred to as **dikaryotic**, which is described as $n + n$ rather than $2n$ (Figure 21-1e). Hyphae that contain only one nucleus per cell are said to be **monokaryotic** (Figure 21-1d). Fungal reproduction is discussed in greater detail later in the chapter.

When a fungal spore contacts an appropriate food source, perhaps an overripe peach that has fallen to the ground or has lain in the fruit basket in the kitchen too long, it germinates and begins to grow. A threadlike hypha emerges from the tiny spore and grows into the peach, branching frequently. Soon a tangled mat of hyphae infiltrates the peach. Cells of the hyphae secrete digestive enzymes into the peach, degrading its organic compounds to small molecules that the fungus can absorb. Later, other hyphae extend upward into the air to bear spores.

FUNGI ARE ECOLOGICALLY IMPORTANT

Fungi make important contributions to the ecological balance of our world. Like bacteria, most fungi are saprotrophs, decomposers that absorb nutrients from organic wastes and dead organisms. When fungi degrade wastes and dead organisms, carbon (as CO_2) and the mineral components of organic com-

TABLE 21-1   Phyla of Kingdom Fungi

Phylum	Common Types	Asexual Reproduction	Sexual Reproduction
Zygomycota	Black bread mold	Nonmotile spores form in a sporangium	Zygospores
Ascomycota (sac fungi)	Yeasts, powdery mildews, molds, morels, truffles	Conidia pinch off from conidiophores	Ascospores
Basidiomycota (club fungi)	Mushrooms, bracket fungi, puffballs, rusts, smuts	Uncommon	Basidiospores
Deuteromycota (imperfect fungi)	Molds, verticillium wilt, athlete's foot fungus	Conidia	Sexual stage not observed

Mildew

athletes foot

pounds are released into the environment, where they become available to plants and other organisms. Without this continuous decomposition, essential nutrients would remain locked up in huge mounds of dead animals, feces, branches, logs, and leaves. The nutrients would be unavailable for use by new generations of organisms, and life would cease.

Although most fungi are saprotrophs, others form symbiotic relationships—intimate relationships between organisms of two species. In some of these relationships, the fungus is a **parasite**, an organism that lives in or on the other organism, called the *host,* and harms it. Parasitic fungi absorb food from the living bodies of their hosts. Such fungi are important causes of diseases in plants and can also cause diseases in humans and other animals. Their activities cost billions of dollars in agricultural damage yearly. We discuss such disease-causing fungi later in this chapter.

Some types of fungi form mutualistic symbiotic relationships with other organisms. In **mutualism**, both partners benefit from the relationship. At the same time that a mutualistic fungus absorbs nutrients from its host, it makes some contribution to its host's well-being. **Mycorrhizae** (from Greek words meaning "fungus roots") are mutualistic relationships between fungi and the roots of plants (see Figure 6-14). Such relationships occur in more than 90 percent of all plant families. The fungus benefits the plant by decomposing organic material in the soil and providing the plant with water and minerals such as phosphorus. At the same time, the roots supply sugars, amino acids, and some other organic molecules to the fungus.

The importance of mycorrhizae first became evident when horticulturalists observed that orchids do not grow unless an appropriate fungus lives with them. Similarly, it has been shown that many forest trees die from mineral deficiencies when transplanted to nutrient-rich grassland soils that lack the appropriate mycorrhizal fungi. When forest soil that contains the appropriate fungi or their spores is added to the soil around these trees, they quickly assume normal growth.

FUNGI ARE CLASSIFIED INTO FOUR PHYLA

The classification of fungi is based mainly on the characteristics of their sexual spores and fruiting bodies. Biologists do not unanimously agree on how to classify these diverse organisms, but the current trend is to assign them to the phyla Zygomycota, Ascomycota, and Basidiomycota. In addition, fungi with no known sexual stage are assigned to a form phylum, Deuteromycota. Members of a **form phylum** are similar to one another in certain respects but probably do not share a common ancestry; they are classified together simply as a matter of convenience. Table 21-1 summarizes fungal classification.

Zygomycetes Reproduce Sexually by Forming Zygospores

The 765 or so species of *phylum Zygomycota* are referred to as **zygomycetes**. They produce sexual spores called **zygospores** that remain dormant for a time. Their hyphae are coenocytic—that is, they lack septa; however, septa do form to separate the hyphae from reproductive structures. Most zygomycetes are saprotrophs that live in the soil on decaying plant or animal matter, although some are parasites of plants and animals.

Perhaps the best known zygomycete is the black bread mold *Rhizopus nigricans,* a saprotroph that breaks down bread and other foods (Figure 21-2). Before preservatives were added to food, bread left at room temperature often became covered with a black, fuzzy growth in a few days. Bread becomes moldy when a spore falls on it and then germinates and grows into a tangled mass of hyphae, forming a mycelium. Hyphae penetrate the bread and absorb nutrients. Eventually, certain hyphae grow upward and develop **sporangia** (spore sacs) at their tips. Clusters of black asexual spores develop within each sporangium and are released when the delicate sporangium ruptures. The black spores give the black bread mold its characteristic color.

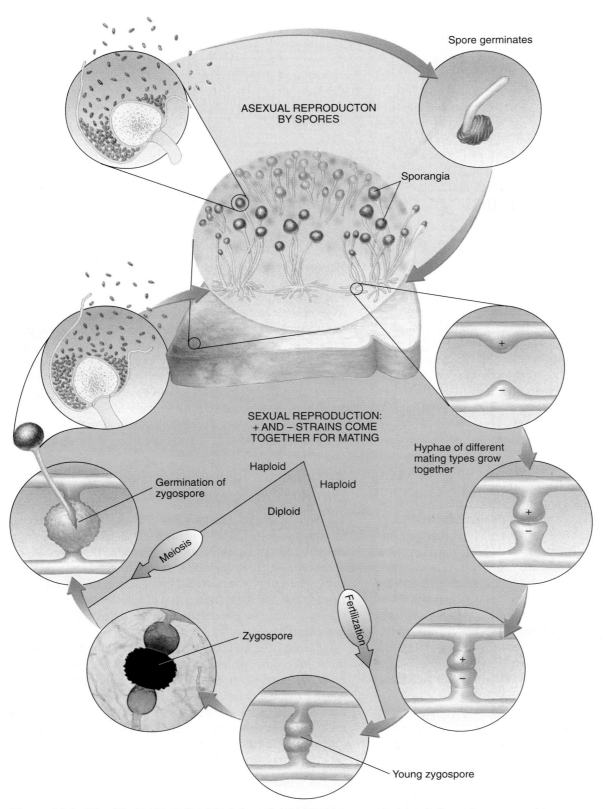

Figure 21-2   The life cycle of the black bread mold (*Rhizopus nigricans*). Sexual reproduction takes place only between genetically different mating types, designated + and −. (*Ed Reschke*)

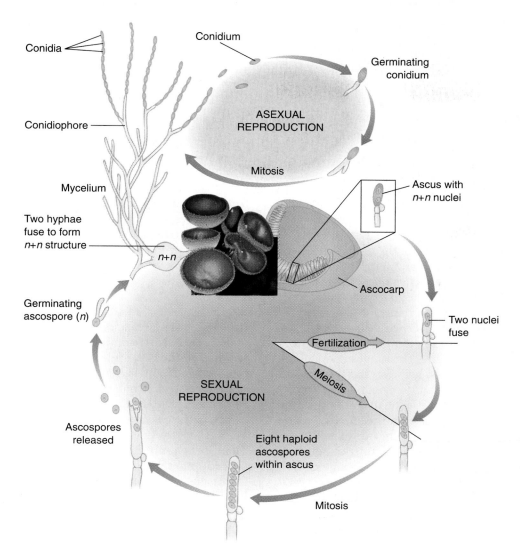

Conidia

Conidium

Germinating conidium

ASEXUAL REPRODUCTION

Mitosis

Conidiophore

Mycelium

Two hyphae fuse to form *n+n* structure

n+n

Germinating ascospore (*n*)

Ascus with *n+n* nuclei

Ascocarp

Two nuclei fuse

Fertilization

Meiosis

SEXUAL REPRODUCTION

Ascospores released

Eight haploid ascospores within ascus

Mitosis

Figure 21-3   The life cycle of ascomycetes. (*Top*) Asexual reproduction involves the formation of haploid conidia. (*Bottom*) Sexual reproduction involves the fusion of two haploid hyphae to form a dikaryotic (*n + n*) structure. Asci form from this structure; in each ascus the *n + n* nuclei fuse, followed by meiosis and mitosis to produce eight ascospores. The asci form the inner layer of a fruiting body known as an ascocarp. (*Inset*) Ascocarps of the rainforest cup fungus (*Cookeina tricholoma*). (*James L. Castner*)

Sexual reproduction in the black bread mold occurs when the hyphae of two genetically different mating types, designated plus (+) and minus (−), grow into contact with one another. An individual hypha mates only with a hypha of a different mating type; that is, sexual reproduction can occur only between a hypha of a plus (+) strain and a hypha of a minus (−) strain. Because there is no obvious sexual differentiation between the two mating types, it is not appropriate to refer to them as "male" and "female."

When hyphae of opposite mating types meet, hormones are produced that cause the hyphae to come together. Plus and minus nuclei then fuse to form a diploid nucleus, the zygote. A zygospore develops, providing a thick protective covering around the zygote. The zygospore may lie dormant for several months and can survive desiccation and extreme temperatures. Meiosis probably occurs during or just before germination of the zygospore. When the zygospore germinates, an aerial hypha develops with a sporangium at the top. Haploid spores form within the sporangium by mitosis. When these spores are released, they germinate to form new hyphae with either + or − nuclei. Only the zygote of a black bread mold is diploid; all of the hyphae and asexual spores are haploid.

Ascomycetes (Sac Fungi) Reproduce Sexually by Forming Ascospores

Phylum Ascomycota is a large group of fungi, the **ascomycetes**, consisting of about 30,000 species. The ascomycetes are sometimes referred to as **sac fungi** because their sexual spores are produced in little sacs called **asci** (sing. *ascus*). Their hyphae usually have septa, but these cross walls are perforated so that cytoplasm can move from one cell to another.

Ascomycetes include most yeasts; the powdery mildews; most of the blue-green, pink, and brown molds that cause food to spoil; saprotrophic cup fungi; and the edible morels and truffles. Some ascomycetes cause serious plant diseases such as dutch elm disease, chestnut blight, ergot disease on rye, and powdery mildew on fruits and ornamental plants (see Focus On: Fighting the Chestnut Blight).

In most ascomycetes, asexual reproduction involves the production of spores called **conidia**, which are pinched off at the tips of certain specialized hyphae known as **conidiophores** (conidia-bearers) (Figure 21-3). Sometimes called "summer spores," conidia are a means of rapidly propagating new mycelia when environmental conditions are favorable; each conidium that lands on a suitable food source rapidly grows

Prior to the 20th century, the American chestnut (*Castanea dentata*) was a very important tree, both ecologically and economically, in eastern forests. Throughout its natural range, which extended from Maine to Georgia and west to Illinois, the chestnut tree provided food and shelter for forest animals, including squirrels, birds, and insects. The tree grew rapidly, forming a straight, tall trunk that provided valuable, decay-resistant wood for house foundations and interior trim, furniture, railroad ties, and fence posts. The wood was also an important source of tannin, a chemical used to convert animal hides to leather. The nuts of the American chestnut were harvested, roasted, and eaten directly or ground into flour.

Early in this century, the chestnut blight fungus (*Cryphonectria parasitica*), an ascomycete, was accidentally imported on diseased oriental chestnuts and quickly attacked native chestnuts, which had no resistance to it. First identified in New York in 1904, the blight killed or damaged several billion mature North American chestnut trees—

almost every specimen throughout the tree's entire natural range—by the late 1940s (see figure).

A spore of the fungus enters a chestnut tree through a wound or crack in the bark, and the hyphae grow rapidly, attacking the inner bark. When the trunk is girdled (completely encircled) by the fungus, the portion of the plant above the injury dies. The fungus does not kill the roots, however, and they periodically sprout branches, only to be killed back by the blight once again. The spores are carried by wind, water, and animals from tree to tree.

Although the American chestnut is on the brink of extinction, there are hopes for its eventual recovery. Some biologists have bred the American chestnut with the Chinese chestnut, a related species that is resistant to chestnut blight. With careful selective breeding over many years, it may be possible to develop a disease-resistant variety of American chestnut.

Other biologists are approaching the problem in a different way—by trying to isolate strains of the chestnut blight fungus that are less virulent (less deadly). Such a strain was first identified in Europe in the 1950s; some of the trees there have been able to recover from the disease. More recently, a less virulent strain of the chestnut blight fungus has appeared in the United States, and, even more encouraging, it appears that this strain is slowly spreading, replacing the more deadly strain. (It is not known *why* the less virulent strain is replacing the more virulent one.)

Biologists determined that a certain virus causes the reduced virulence of the chestnut blight fungus. They genetically engineered a synthetic form of the virus and introduced it into the virulent strain of the fungus, which subsequently lost its virulence. They plan to treat infected chestnut trees with the benign fungal strain, which they hope will spread rapidly and eventually replace the virulent strain.

The massive trunks of American chestnuts photographed in the late 1800s in the Great Smoky Mountains. These trees and almost all other American chestnuts were killed by the chestnut blight fungus during the first part of this century. (*Inset*) Chestnuts that sprout from the roots do not grow very large before they are killed back by the chestnut blight. (*Photo courtesy of the American Chestnut Foundation; inset, Marion Lobstein*)

into a new mycelium, which in turn produces conidia. Conidia occur in various shapes, sizes, and colors in different species; the color of the conidia is what gives many of these molds their characteristic brown, blue-green, pink, and other tints.

Some species of ascomycetes have genetically distinct mating strains; others are self-fertile and have the ability to mate with themselves. Sexual reproduction takes place after two hyphae grow together and their cytoplasm mingles. Within this fused structure the two nuclei come together, but they do not fuse. New hyphae develop from the fused structure, and the cells of these hyphae are dikaryotic. These $n + n$ hyphae form a fruiting body, known as an **ascocarp**, that is characteristic of the species (Figure 21-3). This is where the asci develop.

Within a cell that develops into an ascus, the two nuclei fuse and form a diploid nucleus, the zygote. Each zygote then undergoes meiosis to form four haploid nuclei. This process is usually followed by one mitotic division of each of the four nuclei, resulting in the formation of eight haploid nuclei. Each haploid nucleus (surrounded by cytoplasm) develops into an **ascospore**, so that there are usually eight haploid ascospores within the ascus. The ascospores are released when the tip of the ascus breaks open. Individual ascospores are carried by air currents, often for long distances. If an ascospore lands in a suitable location, it germinates and forms a new mycelium.

Yeasts are unicellular ascomycetes that reproduce asexually by budding and sexually by forming ascospores. During sexual reproduction, two haploid yeasts fuse, forming a diploid zygote. The zygote undergoes meiosis, and the resulting haploid spores remain enclosed for a time within the original diploid cell wall. This sac of spores corresponds to an ascus and ascospores. Yeasts are essential in making bread and in fermenting alcoholic beverages (discussed later in this chapter).

Basidiomycetes (Club Fungi) Reproduce Sexually by Forming Basidiospores

The 16,000 or so species that make up the *phylum Basidiomycota* include the most familiar of the fungi—mushrooms, bracket fungi, and puffballs—as well as some destructive plant parasites of important crops, such as wheat rust and corn smut (Figure 21-4).

Basidiomycetes, or **club fungi**, derive their name from the fact that they develop a **basidium** (pl. *basidia*), a structure comparable in function to the ascus of ascomycetes. Each basidium is an enlarged, club-shaped hyphal cell, at the tip of which four **basidiospores** develop (Figure 21-5). Note that basidiospores develop on the *outside* of the basidium, whereas ascospores develop *within* the ascus. The mature basidiospores are released, and when they come in contact with the proper environment, each develops into a new mycelium.

The mycelium of a basidiomycete such as the cultivated mushroom introduced at the beginning of the chapter consists of a mass of white, branching, threadlike hyphae that occur mostly underground (see Focus On: The Giant Fungus). The hyphae are divided into cells by septa, but, as in ascomycetes, the septa are perforated and allow cytoplasmic streaming between cells.

Compact masses of hyphae called buttons develop along the mycelium. Each button grows into the structure that we ordinarily call a mushroom. More formally, the mushroom, which consists of a stalk and cap, is referred to as a **basidiocarp**. The lower surface of the cap usually consists of many thin plates called **gills**, which extend radially from the stalk to the edge of the cap. The basidia develop on the surfaces of these gills.

Each individual fungus produces millions of basidiospores, and each basidiospore has the potential, should it be transported to an appropriate environment, to give rise to a new **primary mycelium**. Hyphae of a primary mycelium consist of monokaryotic cells, each of which contains a single haploid nucleus. When, in the course of its growth, such a hypha encounters another hypha of a genetically different mating type, the two fuse. As in the ascomycetes, the two haploid nuclei remain separate within each cell. In this way a **secondary mycelium** with dikaryotic hyphae is produced, in which each cell contains two separate haploid nuclei.

The $n + n$ hyphae of the secondary mycelium grow extensively and eventually form compact masses, which are the mushrooms, or basidiocarps. Each basidiocarp actually consists of a mat of intertwined hyphae. On the gills of the mushroom the dikaryotic nuclei fuse, forming diploid zygotes. These are the only diploid cells that form in the life history of a basidiomycete. Each zygote undergoes meiosis, resulting in four haploid nuclei. These nuclei move to the outer edge of the basidium. Finger-like extensions of the basidium develop, and the nuclei and some cytoplasm move into them; each of these extensions becomes a basidiospore, which is attached to the rest of the basidium by a delicate stalk. When the stalk breaks, the basidiospore is released.

Imperfect Fungi Have No Known Sexual Stage

About 17,000 species of fungi have been assigned to the *phylum Deuteromycota* and are commonly called **deuteromycetes**. They are also known as **imperfect fungi** because many have not been observed to have a sexual stage in their life cycles. Should further study reveal a sexual stage, these species will be reassigned to the appropriate phyla. Most deuteromycetes reproduce only by means of conidia and so are closely related to the ascomycetes; a few appear to be more closely related to the basidiomycetes.

(a)

(b)

(c)

(d)

(e)

Figure 21-4   Diversity in basidiomycete fruiting bodies. (a) Basidia line the gills of the jack-o'-lantern mushroom (*Omphalotus olearius*), a poisonous species whose gills glow green in the dark. **(b)** Rounded earthstars (*Geastrum saccatum*) release puffs of microscopic spores after the sac wall is hit by a raindrop or brushed by animals. This fungus is common in leaf litter under trees. **(c)** A giant puffball (*Calvatia gigantea*) in White Plains, New York. At maturity, a dried-out puffball often has a pore through which the basidiospores are discharged as a puff of dust. **(d)** The stinkhorn (*Phallus ravenelii*) has a foul smell that attracts flies, which help disperse the slimy mass of basidiospores. **(e)** Bracket fungi (*Fomes* sp) grow on both dead and living trees, producing shelflike fruiting bodies. Basidiospores are produced in pores located beneath each shelf. (a, *Dennis Drenner;* b, *James L. Castner;* c, *Ed Kanze/Dembinsky Photo Associates;* d, *Visuals Unlimited/Richard D. Poe;* e, *Richard H. Gross*)

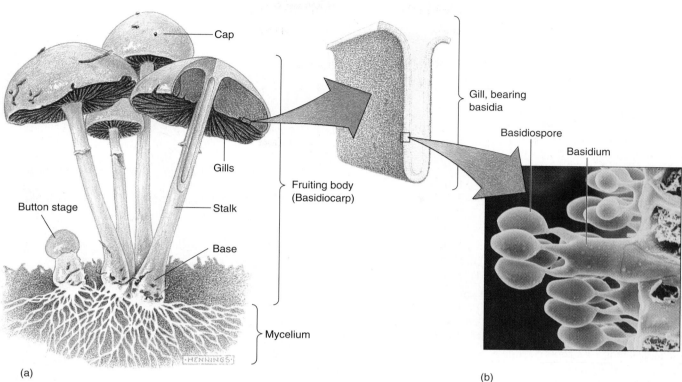

Figure 21-5   Sexual reproduction in the basidiomycetes. (a) Compacted hyphae form the basidiocarp commonly called a mushroom. Numerous basidia are borne on the gills. **(b)** Each basidium produces four basidiospores, which are attached to the basidium. (b, *Biophoto Associates*)

Well-known deuteromycetes include *Penicillium chrysogenum,* which is grown commercially to produce the antibiotic penicillin; *Epidermophyton floccosum,* which causes athlete's foot; and *Candida albicans,* which causes candidiasis ("yeast" infections).

FUNGI ARE PART OF DUAL ORGANISMS CALLED LICHENS

Although a **lichen** looks like a single organism, it is actually a symbiotic association between a photosynthetic organism and a fungus. The photosynthetic partner is usually a green alga or a cyanobacterium, and the fungus is most often an ascomycete. In some lichens in tropical regions, the fungal partner is a basidiomycete. The algae or cyanobacteria found in lichens are also found as free-living species in nature, but the fungal components of lichens are generally found only as part of lichens.

In the laboratory the fungal and photosynthetic partners of a lichen can be separated and grown separately in appropriate culture media. The alga or cyanobacterium grows more rapidly when separated, whereas the fungus grows slowly and requires a culture medium that provides many complex carbohydrates. Generally, the fungus does not produce fruiting bodies when separated from its photosynthetic partner; as part of a lichen, however, the fungus does produce fruiting bodies. The photosynthetic organism and the fungus can be reassembled as a lichen, but only if they are placed together in a culture medium under conditions that cannot support either of them independently.

What is the nature of this partnership? In the past the lichen was considered a definitive example of mutualism, a symbiotic relationship equally beneficial to both species. The photosynthetic partner carries on photosynthesis, producing carbohydrate molecules for itself and the fungus, and the fungus obtains water and minerals for the photosynthetic partner as well as protects it against desiccation. Some biologists, however, think that the lichen partnership is not a true case of mutualism but one of controlled parasitism of the photosynthetic partner by the fungus; microscopic examination reveals that some of the algal cells have been penetrated and destroyed by fungal hyphae.

There are approximately 20,000 species of lichens, which typically possess one of three different growth forms—crus-

Focus On
The Giant Fungus

Giants are known in the plant and animal kingdoms (for example, the giant sequoia and the blue whale), but until recently, the fungi were all thought to be relatively small organisms. In 1992 several biologists reported a saprotrophic fungus that is one of Earth's largest and oldest organisms, living in a forest in Michigan (see figure). This fungus, *Armillaria bulbosa,* commonly called the honey mushroom, has a vast underground network of hyphae and "tentacles," or rhizomorphs. The rhizomorphs, which snake through the upper layers of the soil in search of dead or dying tree roots to digest and absorb, are composed of compacted strands of hyphae and resemble shoestrings in thickness.

The biologists used a type of genetic "fingerprinting" to confirm that a single individual fungus occupies at least 15 hectares (37 acres) of the forest floor—an area equivalent to more than 33 football fields! Moreover, they found that the fungus is territorial; no other individual honey mushroom was found in the 15 hectares. Near the end of each summer, the giant fungus produces thousands of honey-colored mushrooms. But, according to the DNA data, none of the millions of spores produced by the mushrooms appears to have grown into a new individual within the area occupied by the giant fungus.

Because rhizomorphs grow at a rate of about 8 inches per year, the age of the fungus was calculated as 1500 years. At least one forest fire (in 1928) is known to have destroyed the trees in the area inhabited by this fungal giant, but it survived underground. It probably thrived immediately after the fire because of the many dead tree roots that became available for food.

The discovery of a giant fungus has prompted biologists to speculate that even larger fungi may exist elsewhere under the soil. For example, a single fungus of a related species of *Armillaria* in the state of Washington is claimed to occupy over 600 hectares (1500 acres), but so far this claim

The honey mushroom (*Armillaria bulbosa*) is among the Earth's largest and longest-lived organisms. (*Johann N. Bruhn*)

has not been substantiated with DNA evidence. Clearly, fungi are the giants of the forest, a fact that helps confirm their pivotal role as saprotrophs in the forest ecosystem.

---

tose, foliose, and fruticose (Figure 21-6). **Crustose** lichens are flat and grow tightly attached to the rock or whatever they are growing on; **foliose** lichens are also flat, but they have leaflike lobes and are not so tightly attached; **fruticose** lichens grow erect and are branched and shrublike.

Able to tolerate extremes of temperature and moisture, lichens grow everywhere that life can be supported at all, except in heavily polluted cities. For example, there are 350 lichen species that exist in the arctic region as compared with only 2 species of flowering plants. Many other lichen species are at home in steaming equatorial rain forests. They grow on tree trunks, mountain peaks, and bare rock. In fact, lichens are often the first organisms to colonize bare rocky areas and play a role in the formation of soil. Lichens gradually make tiny cracks in the rocks to which they cling, facilitating weathering of the rocks by wind and rain.

The "reindeer mosses" of arctic regions are not mosses but lichens that serve as the main source of food for the caribou (reindeer) herds of the region. Some lichens produce colored pigments; one pigment, orchil, is used to dye woolens, and another, litmus, is widely used in chemistry laboratories as an acid-base (pH) indicator.

Lichens vary greatly in size. Some are almost invisible, whereas others, such as the reindeer mosses, may cover kilometers of land with a growth that is ankle deep. Lichens grow slowly, and the radius of a crustose lichen may increase by less than a millimeter each year. Some mature crustose lichens are known to be thousands of years old.

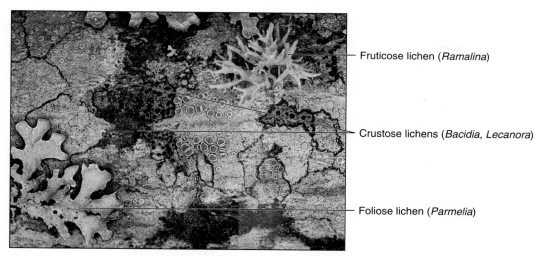

Fruticose lichen (*Ramalina*)

Crustose lichens (*Bacidia, Lecanora*)

Foliose lichen (*Parmelia*)

Figure 21-6   Lichens vary in color, shape, and overall appearance. Crustose lichens grow tightly attached to tree trunks, rocks, and some other surfaces. Foliose lichens are leaflike in appearance, whereas fruticose lichens are branching and shrublike. (*Fred M. Rhoades*)

Lichens absorb minerals mainly from the air and from rainwater but can also absorb them directly from the surface on which they grow. They cannot excrete the elements that they absorb, and perhaps for this reason they are very sensitive to toxic compounds in the environment. A reduction in lichen growth has been used as an indicator of air pollution, especially sulfur dioxide. Absorption of such toxic compounds results in damage to the chlorophyll of the photosynthetic partner. The return of lichens to an area indicates an improvement in air quality in that vicinity.

Lichens reproduce mainly by asexual means, usually fragmentation, in which bits of the lichen break off and, if they land in a suitable place, establish themselves as new lichens. Some lichens release special dispersal units called **soredia** that contain cells of both partners (Figure 21-7). In others, the alga reproduces asexually by mitosis, whereas the fungus produces ascospores. The ascospores are dispersed by wind and may find an appropriate photosynthetic partner only by chance.

FUNGI ARE ECONOMICALLY IMPORTANT

The same powerful digestive enzymes that enable fungi to decompose wastes and dead organisms also permit them to reduce wood, fiber, and food to their components with great efficiency. Various fungi cause incalculable damage to stored goods and building materials each year. Bracket fungi, for example, cause enormous losses by decaying wood—both stored lumber and the wood of living trees.

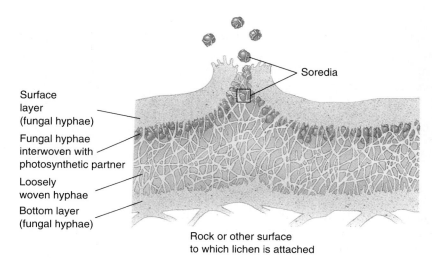

Soredia

Surface layer (fungal hyphae)

Fungal hyphae interwoven with photosynthetic partner

Loosely woven hyphae

Bottom layer (fungal hyphae)

Rock or other surface to which lichen is attached

Figure 21-7   A cross section of a typical lichen shows distinct layers. Soredia are asexual reproductive structures composed of clusters of algal or cyanobacterial cells enclosed by fungal hyphae.

Fungi Provide Beverages and Food

People exploit the ability of yeasts to produce ethyl alcohol and carbon dioxide from sugar by fermentation in order to make wine and beer and bake bread. Wine is produced when yeasts ferment fruit sugars; beer is made when yeasts ferment grain, usually barley. During the process of making bread, carbon dioxide produced by the yeast becomes trapped in the dough as bubbles, which cause the dough to rise and give leavened bread its light texture. Both the carbon dioxide and the alcohol produced by the yeast evaporate during baking (Figure 21-8).

The unique flavors and smells of cheeses such as Roquefort and Camembert are produced by the action of the deuteromycete *Penicillium* (Figure 21-8). *Penicillium roquefortii,* for example, is found in caves near the French village of Roquefort, and only cheeses produced in that area can be called Roquefort cheese.

Many cultures have used fungi to improve the nutrient quality of the diet. For example, in the Orient the deuteromycete *Aspergillus tamarii* and other microorganisms (bacteria and yeasts) are used to produce soy sauce by fermentation of soybeans. Soy sauce enhances other foods with more than its special flavor; to the low-protein rice diet it adds vital amino acids from both the soybeans and the fungi themselves.

Among the basidiomycetes there are some 200 species of edible mushrooms and about 70 species of poisonous ones, sometimes called toadstools. Some edible mushrooms are cultivated commercially (recall *Agaricus bisporus* in the chapter introduction). Morels, which superficially resemble mushrooms, and truffles, which produce underground fruiting bodies, are ascomycetes (Figure 21-9). These gourmet delights

Figure 21-8   Wine, beer, bread, and distinctive cheeses are produced in part by fungi. Yeasts ferment fruits (wine) or grains (beer), producing ethyl alcohol. That same process produces the carbon dioxide bubbles responsible for making bread rise. The bluish splotches in the cheese are patches of conidia. (*Raymond Tschoepe*)

(a)

(b)

Figure 21-9   Edible ascomycetes. (a) The common morel (*Morchella esculenta*) and **(b)** the truffle (*Tuber* sp) are expensive gourmet treats. Both are ascocarps that produce ascospores. Truffles are subterranean ascocarps that people find with the help of trained dogs or pigs. Here truffles are shown whole and sectioned. (a, *Richard Shiell/ Dembinsky Photo Associates;* b, *Visuals Unlimited/John D. Cunningham*)

Figure 21-10   The destroying angel (*Amanita virosa*). This extremely poisonous mushroom is recognizable, as are other amanitas, by the ring of tissue around its stalk and the underground cup from which the stalk protrudes. About 50 grams (2 ounces) of this mushroom could kill an adult man. Initial symptoms include vomiting, diarrhea, and cramps. If the poisoning goes untreated, liver and kidney failure occur, resulting in death in 5 to 10 days. (*James W. Richardson/CBR Images*)

are now cultivated from mycorrhizae on the roots of tree seedlings.

Edible and poisonous mushrooms can look very much alike and may even belong to the same genus. There is no simple way to distinguish edible mushrooms from poisonous ones; they must be identified by an expert. Some of the most poisonous mushrooms belong to the genus *Amanita*. Toxic species of this genus are appropriately called such names as "destroying angel" (*Amanita virosa*) (Figure 21-10) and "death cap" (*Amanita phalloides*). Eating part of a single cap of either of these mushrooms can be fatal.

Ingestion of certain species of mushrooms causes intoxication and hallucinations. Central American Indians still use the sacred mushrooms of the Aztecs, *Conocybe* and *Psilocybe*, in religious ceremonies, for their hallucinogenic properties. The chemical ingredient psilocybin, chemically related to lysergic acid diethylamide (LSD), is responsible for the trance-like state and colorful visions experienced by those who eat these mushrooms. Ingestion of psychoactive mushrooms is not recommended, because negative reactions vary consider-

ably—from mild indigestion to death. In addition, the possession and use of such mushrooms are illegal in most states.

Fungi Produce Useful Chemicals

In 1928 Alexander Fleming noticed that one of his petri dishes containing bacteria was contaminated by mold. The bacteria were not growing in the vicinity of the mold, which led Fleming to the conclusion that the mold was releasing some substance harmful to them. Within a decade of Fleming's discovery, penicillin produced by the deuteromycete *Penicillium notatum* was purified and used to treat bacterial infections. Penicillin, now produced by *Penicillium chrysogenum,* is still the most widely used and most effective antibiotic. Other drugs derived from fungi include the antibiotic griseofulvin, used clinically to inhibit the growth of athlete's foot and ringworm fungi, and cyclosporine, used to suppress immune responses in patients receiving organ transplants.

The ascomycete *Claviceps purpurea* infects the flowers of rye plants and other cereals. It produces a structure called an **ergot** where a seed would normally form in the grain head (Figure 21-11). When livestock eat ergot-contaminated grain or when humans eat bread made from ergot-contaminated rye flour, they may be poisoned by the toxic substances in the ergot. These substances may cause a condition called ergotism, which can involve nervous spasms, convulsions, psychotic delusions, and constriction of blood vessels that, if severe enough, can lead to gangrene. During the Middle Ages, ergotism occurred often and was known as St. Anthony's Fire. In the year 994, for example, an epidemic of St. Anthony's Fire caused more than 40,000 deaths. In 1722 the cavalry of Czar Peter the Great was felled by ergotism on the eve of the battle for the conquest of Turkey. This was one of several times in

Figure 21-11   The ascomycete *Claviceps purpurea* infecting rye flowers (*left*). This fungus produces a brownish-black structure called an ergot where a seed would normally form in the grain head. A healthy grain head is shown for comparison (*right*). (*Dennis Drenner*)

recorded history that a fungus changed the course of human affairs.

Lysergic acid, one of the constituents of an ergot, can be used to produce LSD. Some of the compounds produced by ergot are now used clinically in small quantities as drugs to induce labor, to stop uterine bleeding (recall that it constricts blood vessels), to treat high blood pressure, and to relieve one type of migraine headache.

Fungi can be used as biological control agents to prevent damage to plants from insect pests, without the use of chemical insecticides. For example, the citrus mealybug is not a serious pest on citrus trees in Florida because it is naturally controlled by the zygomycete *Entomophthora fumosa*. It is interesting that Florida's climate favors the growth of *E. fumosa*, whereas California's climate does not. As a result, *E. fumosa* cannot be used to inhibit the citrus mealybug in California.

Other fungi are grown commercially to produce citric acid and other industrial chemicals. Yeasts and other fungi are also increasingly being modified by recombinant DNA techniques to make them produce important biological molecules such as hormones.

Fungi Cause Many Important Diseases of Plants

Fungi are responsible for many serious plant diseases, including epidemic diseases that spread rapidly and often result in complete crop failures. Damage may be localized in certain tissues or structures of the plant, or the disease may be systemic and spread throughout the entire plant. Fungal infections may cause stunting of plant parts or of the entire plant, they may cause growths similar to warts, or they may kill the plant.

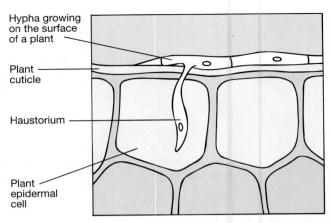

Figure 21-12   Haustoria produced by the powdery mildew fungus penetrate leaf epidermal cells. Powdery mildew disease causes yellowing and curling of leaves and a decline in photosynthesis, which results in stunted plants.

A plant may become infected after hyphae enter through stomata (pores) in the leaf or stem or through wounds in the plant body. Alternatively, the fungus may produce an enzyme that dissolves the plant's cuticle, after which it can easily invade the plant's tissues. As the fungal mycelium grows, it may remain mainly between the plant cells or it may penetrate the cells. Parasitic fungi often produce special hyphal branches called **haustoria** that penetrate host cells and obtain nourishment from the cytoplasm (Figure 21-12).

Some important plant diseases caused by ascomycetes are powdery mildews, chestnut blight, dutch elm disease, apple scab, and brown rot, which attacks cherries, peaches, plums, and apricots (Figure 21-13a).

Diseases caused by basidiomycetes include smuts and rusts that attack a variety of plants including the cereals—

Figure 21-13   Fungi are important plant pathogens. (a) Brown rot of peaches is caused by *Monilinia fruiticola,* an ascomycete. **(b)** Corn smut on an ear of sweet corn. *Ustilago maydis* is the basidiomycete that causes corn smut. (a, *Visuals Unlimited/A.L. Jones;* b, *James L. Castner*)

(a)

(b)

corn, wheat, oats, and other grains (Figure 21-13b). Some of these parasites, such as wheat rust, have complex life cycles that involve two or more different plants, during which several kinds of spores are produced. For example, the wheat rust must infect an American barberry plant to complete the sexual stage of its life cycle. Since this fact has been known, the eradication of American barberries in wheat-growing regions has reduced infection with wheat rust. Wheat rust has not been eliminated by eradication of the barberry, however, because the fungus overwinters on wheat at the southern end of the Grain Belt and forms asexual spores. During the spring, wind blows these spores for hundreds of miles, reinfecting northern areas of the United States and Canada.

Certain imperfect fungi also cause plant diseases. Examples include verticillium wilt on potatoes, which is caused by the deuteromycete *Verticillium* sp, and bean anthracnose, which is caused by the deuteromycete *Colletotrichum lindemuthianum.*

Fungi Cause Certain Diseases of Humans

Some fungi cause superficial infections in which only the skin, hair, or nails are infected. Ringworm and athlete's foot are examples of superficial fungal infections; both are caused by imperfect fungi. Candidiasis, an infection of mucous membranes of the mouth or vagina, is among the most common superficial fungal infections; it, too, is caused by an imperfect fungus.

Other fungi cause systemic infections; that is, they infect internal tissues and organs and may spread through many regions of the body. Histoplasmosis is an infection of the lungs caused by inhaling the spores of a fungus abundant in bird droppings. Most people in the eastern and midwestern parts of the United States have been exposed to this fungus at one time or another, and fortunately, if one is infected, the infection usually stays in the lungs and is of short duration. If the infection spreads into the bloodstream, however, it can be quite serious.

Most pathogenic fungi cause systemic infections only when the body's immunity is lowered. For example, the deuteromycete *Aspergillus fumigatus* does not usually cause disease in humans but does cause aspergillosis in people with defective immune systems, such as AIDS patients. During the course of this disease, the fungus can invade the lungs, heart, brain, kidneys, and other vital organs and can result in death.

STUDY OUTLINE

I. Fungi are eukaryotes with cell walls composed of chitin.
  A. Fungi lack chlorophyll and are heterotrophic, absorbing predigested food through the cell wall and plasma membrane.
  B. A fungus may be unicellular or multicellular.
    1. The body of a mold consists of a mass of long, branched hyphae, called a mycelium.
    2. In the zygomycetes, the hyphae are coenocytic (undivided by septa).
    3. In other fungi, perforated septa are present that divide the hyphae into individual cells.
  C. Fungi reproduce by means of spores, which may be produced sexually or asexually. When a fungal spore comes into contact with an appropriate food source, it germinates and begins to grow a mycelium.
II. Fungi are ecologically significant.
  A. Many fungi are saprotrophs and break down organic compounds.
  B. Mycorrhizae are mutualistic relationships between fungi and the roots of plants. The fungus supplies minerals to the plant, and the plant secretes organic compounds needed by the fungus.
III. Fungi are classified into phyla based on their modes of sexual reproduction.
  A. Zygomycetes produce sexual spores called zygospores. The black bread mold is a representative of this group.
  B. Ascomycetes produce asexual spores called conidia and sexual spores called ascospores; the latter are produced in asci. Ascomycetes include yeasts, cup fungi, morels, truffles, and pink and green molds.
  C. Basidiomycetes produce sexual spores called basidiospores on the outside of a basidium; in mushrooms, basidia develop on the surfaces of gills. Basidiomycetes include mushrooms, puffballs, rusts, and smuts.
  D. Deuteromycetes are imperfect fungi in which a sexual stage has not been observed. Most reproduce by forming conidia (asexual spores). Deuteromycetes include species of *Penicillium* and *Aspergillus.*
IV. A lichen is a symbiotic association between a fungus and an alga or cyanobacterium in which the fungus benefits from the association. Lichens have three main growth forms—crustose, foliose, and fruticose.
V. Fungi include species of both positive and negative economic importance.
  A. Mushrooms, morels, and truffles are used as food; yeasts are vital in the production of wine, beer, and bread; certain fungi are used to produce cheeses and soy sauce.
  B. Fungi are used to make penicillin and other antibiotics; ergot is used to produce certain drugs; other fungi make citric acid and many other industrial chemicals.
  C. Fungi cause many plant diseases, including wheat rust, dutch elm disease, and chestnut blight; they cause human diseases such as ringworm, athlete's foot, histoplasmosis, and candidiasis.

SELECTED KEY TERMS

ascomycete, p. 349
ascospore, p. 351
ascus, p. 349
basidiomycete, p. 351
basidiospore, p. 351
basidium, p. 351

chitin, p. 345
conidia, p. 349
deuteromycete, p. 351
form phylum, p. 347
fruiting body, p. 346

heterotroph, p. 345
hyphae, p. 345
lichen, p. 353
mold, p. 345
mycelium, p. 345

mycorrhizae, p. 347
saprotroph, p. 345
yeast, p. 345
zygomycete, p. 347
zygospore, p. 347

REVIEW QUESTIONS

1. What characteristics distinguish fungi from other organisms?
2. How does the body plan of a yeast differ from that of a mold?
3. What is the difference between a hypha and a mycelium? Between a fruiting body and a mycelium?
4. Describe the ecological significance of each of the following: saprotrophic fungi, lichens, mycorrhizae.
5. Distinguish between fungi that are saprotrophs and those that are parasites.

6. List the four phyla of fungi, and give several characteristic features of each.
7. Describe the life cycle of a typical mushroom.
8. Briefly describe three important fungal diseases of plants and three fungal diseases of humans.
9. Label the following:

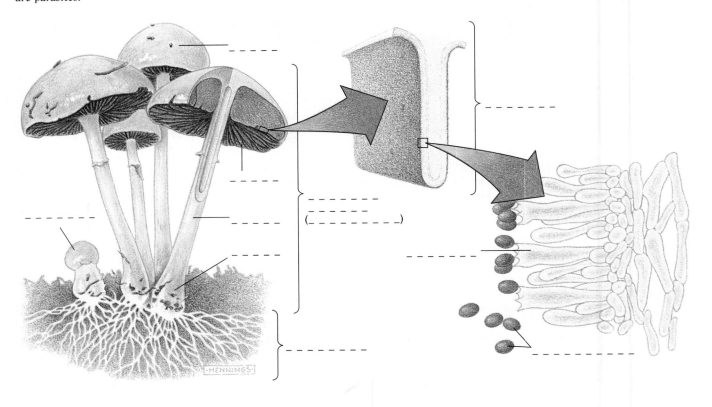

THOUGHT QUESTIONS

1. If you do not see any mushrooms in your lawn, can you conclude that there are no fungi living there? Why or why not?
2. Why are mushrooms and puffballs usually located aboveground, whereas the fungal mycelium is underground?
3. Under what kinds of environmental conditions might it be more advantageous for a fungus to reproduce asexually? Sexually?

4. What measures can you suggest to prevent bread from becoming moldy?
5. Biologists have determined that many mycorrhizal fungi are sensitive to a low pH. What human-caused environmental problem might prove catastrophic for these fungi? How might this problem affect their plant partners?

SUGGESTED READINGS

The Audubon Society Field Guide to North American Mushrooms, New York, Alfred A. Knopf, 1981 (8th printing, 1992). A beautifully illustrated guide to common fungi.

Herrington, J., "Institutions: The American Chestnut Foundation," *Environment,* Vol. 34, No. 10, December 1992. Describes the work done by the American Chestnut Foundation, a private nonprofit organization, to find a permanent solution to the chestnut blight.

Hoffman, M., "Yeast Biology Enters a Surprising New Phase," *Science,* Vol. 255, March 20, 1992. Baker's yeast can be induced to grow like a filamentous mold.

Newhouse, J.R., "Chestnut Blight," *Scientific American,* July 1990. An account of the chestnut blight fungus that has almost completely eradicated the American chestnut in the United States.

Scitil, K., "Fungus Among Us," in "Biology 1992," *Discover,* January 1993. Discusses the giant fungus discovered in Michigan in 1992.

Walsh, R., "Seeking the Truffle," *Natural History,* January 1996. A fascinating account of the history of truffles in France.

22

INTRODUCTION TO THE PLANT KINGDOM: BRYOPHYTES

Several hundred species of peat mosses (*Sphagnum* spp) grow in bogs, swamps, and other wet places throughout the world. Also known as sphagnum moss and bog moss, peat moss is a "giant" among mosses because it is capable of growing 30 centimeters (1 foot) or more in length; most mosses are never more than 5 centimeters (2 inches) long. *Sphagnum* often forms "quaking bogs," which are spongy mats that grow over the surfaces of bodies of water.

Commercially, peat mosses are the most important mosses. One of the distinctive features of *Sphagnum* is the presence of large, empty cells in the whitish-green "leaves," which apparently function to absorb and hold water. These water storage cells are also found in the "stems." Its capacity to absorb water makes peat moss a useful packing material for shipping live plants as well as a good soil conditioner. Peat moss in soil soaks up and retains moisture, and its acidic nature makes it a particularly good soil conditioner for acid-loving plants such as azaleas and rhododendrons.

Peat moss is resistant to decay, in part because it is acidic. The acid and the anaerobic conditions of the bog retard the growth of bacterial and fungal decomposers. As a result, dead peat mosses accumulate as thick deposits—some several me-

Peat mosses (*Sphagnum* spp) grow along the edges of ponds, slow-moving streams, and bogs.
(*Carlyn Iverson*)

ters in depth—under the growing mat of living peat mosses. Over time, the dead organic material compresses to form *peat*. In some countries—Ireland, for example—peat is taken from the bogs, cut into blocks, dried, and burned as fuel. Peat is also burned to produce smoke, which gives the barley malt used to make Scotch whiskeys its characteristic flavor.

In some bogs, the lowest layers of peat are thousands of years old. When the pollen and spores found in various layers are carefully sampled and accurately identified, they provide a history of the changing plant communities in a given area. This information provides indirect evidence of long-term temperature and precipitation changes for that locality.

Occasionally the remains of humans have been uncovered during excavations of old peat bogs in Ireland and other parts of Europe. In such cases the clothing and features are remarkably well-preserved, because the acidic conditions of bogs inhibit decay. In a peat bog in western Denmark, a man's body, estimated to be 2000 years old, was found with a noose about its neck. Archaeologists think that other bodies discovered in the peat bogs are the remains of people sacrificed during religious ceremonies.

PLANTS DESCENDED FROM ANCIENT GREEN ALGAE

About 440 million years ago, planet Earth would have seemed a most inhospitable place because, although life abounded in the oceans, it did not yet exist in abundance on land. The oceans were filled with vast numbers of fish, mollusks, and crustaceans as well as countless microscopic algae, and the water along rocky coastlines was home to large seaweeds. Occasionally, perhaps, an animal would crawl out of the water onto land, but it never stayed there permanently because there was little to eat—not a single blade of grass, a fruit, or a seed was to be found.

During the next 30 million years, an interval corresponding roughly to the Silurian period in geological time, plants appeared in abundance and colonized the land. Where did they come from? Although plants living today exhibit great diversity in size, form, and habitat, they are all thought to have evolved from a common ancestor, an ancient green alga. Biologists infer this relationship because modern green algae share a number of biochemical and metabolic traits with modern plants. Green algae and plants contain the same photosynthetic pigments: chlorophylls *a* and *b*, carotenes, and xanthophylls. Also, both store their excess carbohydrates as starch and possess cellulose as a major component of their cell walls. In addition, certain details of cell division, including the formation of a cell plate during cytokinesis, are shared by plants and some green algae.

COMPLEX PHOTOSYNTHETIC ORGANISMS ARE PLACED IN THE PLANT KINGDOM

The plant kingdom comprises thousands of species that live in every conceivable habitat, from the frozen arctic tundra to lush tropical rain forests. Plants, which range in size from minute, almost microscopic duckweeds and water-meal to massive giant sequoias, are complex multicellular organisms that typically obtain energy by photosynthesis (Figure 22-1). Plants use the green pigment chlorophyll to absorb radiant energy from the sun; this energy is then converted to the chemical energy found in carbohydrates. In addition to chlorophylls *a* and *b*, all plants have accessory pigments that assist in trapping sunlight for photosynthesis: **xanthophylls** (yellow pigments) and **carotenes** (orange pigments).

One important difference between plants and algae is that the aerial portion of a plant is covered by a waxy **cuticle**. A cuticle is essential for existence on land because it helps prevent the desiccation, or drying out, of plant tissues by evaporation. Plants are rooted in the ground and, unlike animals, cannot move to wetter areas during dry spells; therefore, a cuticle is critical to a land plant's survival.

Plants obtain the carbon they need for photosynthesis from the atmosphere as carbon dioxide (CO_2). In order for the CO_2 to be fixed into organic molecules such as sugar, it must first diffuse into the chloroplasts that are inside green plant cells. Because the external surfaces of stems and leaves are covered by a waxy cuticle, however, gas exchange through the cuticle between the atmosphere and the insides of cells is negligible. To facilitate gas exchange, tiny pores called **stomata** (sing. *stoma*) dot the surfaces of leaves and stems; algae lack stomata.

Plants possess multicellular sex organs called **gametangia** (sing. *gametangium*), while the gametangia of algae are single-celled (Figure 22-2). Each gametangium has a layer of sterile (nonreproductive) cells around the delicate gametes (eggs or sperm cells) it produces. This outer jacket of sterile cells protects the gametes from desiccation. In plants, after fertilization occurs, the fertilized egg develops into a multicellular **embryo** *within* the female gametangium; in this way, the embryo is protected during its development. In algae, the fertilized egg develops away from its gametangium. (In some algae the gametes are released into the water before fertilization, whereas in others the fertilized egg is released.)

(a)

(b)

Figure 22-1 Size variation in plants. (a) The giant sequoias (*Sequoiadendron giganteum*) of California are the world's largest plants, growing to 76 meters (250 feet) in height with a massive trunk many meters in diameter. **(b)** Duckweeds (*Spirodela* sp) are tiny floating aquatic plants about 1 centimeter (3/8 inch) long. If you look closely at the frog's body, you'll see minute green dots. Each is a water-meal (*Wolffia* sp) plant. These tiny floating herbs, about 1.5 millimeters (1/16 inch) long, are the smallest known flowering plants.
(a, *John Mielcarek/Dembinsky Photo Associates*; b, *Carlyn Iverson*)

There Are Four Major Groups of Plants

The plant kingdom consists of four main groups of living plants: bryophytes, ferns and their allies,[1] gymnosperms, and the flowering plants, which are also called angiosperms (Table 22-1 and Figure 22-3). The mosses and other **bryophytes**

---

[1]Fern allies are plants (whisk ferns, horsetails, and club mosses) that are similar to ferns in many respects.

(from the Greek words meaning "moss plant") are **nonvascular plants**—they lack a vascular, or conducting, system to transport dissolved nutrients, water, and essential minerals throughout the plant body. In the absence of such a system, bryophytes rely on diffusion and osmosis to obtain needed materials. This reliance means that bryophytes are restricted in size; if they were much larger, some of their cells could not obtain all the necessary materials in sufficient quantities.

The other three groups of plants—ferns, gymnosperms, and flowering plants—are **vascular plants**. They possess two

Figure 22-2 The reproductive structures of plants and algae differ. (a,b) In algae, gametangia are generally single-celled. When the gametes are released, the only thing remaining is the wall of the original cell. **(c,d)** In plants, gametangia are multicellular, but only the inner cells become gametes. The gametangium is surrounded by a protective layer of sterile cells.

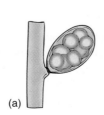

(a)

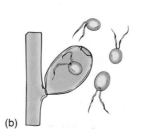

(b)

(c)

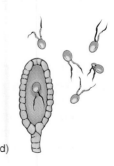

(d)

TABLE 22-1   The Plant Kingdom

Nonvascular Plants

I. Nonvascular plants with a dominant gametophyte generation (bryophytes)
   [handwritten: land plants]
   Phylum Bryophyta (mosses)
   Phylum Hepaticophyta (liverworts)
   Phylum Anthocerotophyta (hornworts) *[handwritten: - rare]*

Vascular Plants

II. Vascular plants with a dominant sporophyte generation
   A. Seedless plants
      Phylum Pterophyta (ferns)
      Phylum Psilotophyta (whisk ferns)
      Phylum Sphenophyta (horsetails)
      Phylum Lycophyta (club mosses)

   B. Seed plants
      1. Plants with naked seeds (gymnosperms)
         Phylum Coniferophyta (conifers)
         Phylum Cycadophyta (cycads)
         Phylum Ginkgophyta (ginkgoes)
         Phylum Gnetophyta (gnetophytes)

      2. Seeds enclosed within a fruit
         Phylum Anthophyta (angiosperms or flowering plants)
            Class Dicotyledones (dicots)
            Class Monocotyledones (monocots)

vascular tissues: **xylem** for water and mineral conduction and **phloem** for conduction of organic molecules such as sugar. A key step in the evolution of vascular plants was the development of the ability to produce **lignin**, a strengthening polymer found in the walls of cells that provide support and conduction. The stiffening property of lignin enabled plants to grow tall and dominate the landscape. In turn, the successful occupation of the land by plants made the evolution of terrestrial (land-dwelling) animals possible by providing them with both habitat and food (see Focus On: The Environmental Challenges of Living on Land).

Ferns and their allies are seedless vascular plants that reproduce and disperse primarily via haploid spores. Gym-nosperms and flowering plants are vascular plants whose primary means of reproduction is seeds. Gymnosperms bear seeds unprotected on a cone. Flowering plants, also known as angiosperms, produce seeds enclosed within a fruit.

Use Care When Comparing One Plant Group to Another

In comparing groups of plants, it is convenient to use terms such as *lower* and *higher, simple* and *complex*, and *primitive* and *advanced*. Do not take such terms to imply, however, that plants labeled higher, complex, or advanced are better or more nearly perfect than others. Rather, these labels are used in a

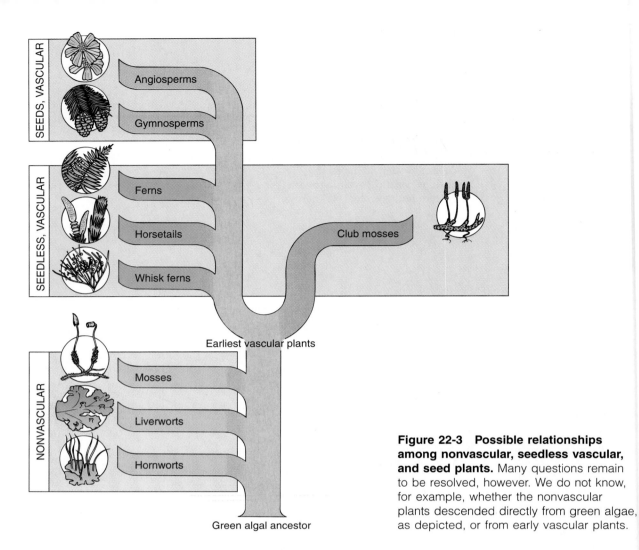

Figure 22-3 Possible relationships among nonvascular, seedless vascular, and seed plants. Many questions remain to be resolved, however. We do not know, for example, whether the nonvascular plants descended directly from green algae, as depicted, or from early vascular plants.

comparative sense to describe hypothesized evolutionary relationships. For example, the terms *higher* and *lower* usually refer to the level at which a particular group diverged from a main line of evolution. It is customary, for instance, to refer to mosses and liverworts as lower plants because they are thought to have originated near the base of the phylogenetic tree of the plant kingdom. However, neither mosses nor liverworts are primitive in all structural or physiological characteristics. Each has become highly specialized to its own lifestyle.

THE PLANT LIFE CYCLE EXHIBITS AN ALTERNATION OF GENERATIONS

Like certain algae, fungi, and other organisms, plants have clearly defined **alternation of generations**; that is, each plant spends part of its life in the haploid stage and part in the diploid stage (Figure 22-4). The haploid portion of the life cycle is called the **gametophyte generation** because it gives rise to haploid gametes by *mitosis*. When two gametes fuse, the diploid portion of the life cycle, called the **sporophyte generation**; begins. The sporophyte generation gives rise to haploid **spores** by *meiosis;* these spores represent the first stage in the gametophyte generation.

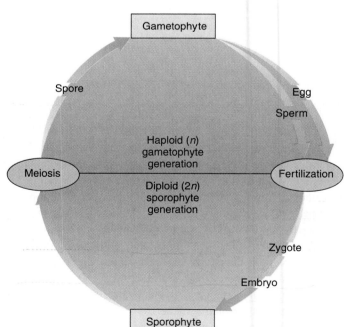

Figure 22-4 The basic plant life cycle, in which a plant alternates generations, spending part of its life in the haploid gametophyte stage and part in the diploid sporophyte stage. All plants have modifications of this cycle.

Focus On

The Environmental Challenges of Living on Land

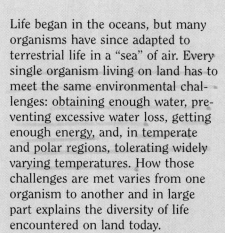

Life began in the oceans, but many organisms have since adapted to terrestrial life in a "sea" of air. Every single organism living on land has to meet the same environmental challenges: obtaining enough water, preventing excessive water loss, getting enough energy, and, in temperate and polar regions, tolerating widely varying temperatures. How those challenges are met varies from one organism to another and in large part explains the diversity of life encountered on land today.

Let's compare how plants and vertebrates (animals with backbones) meet several terrestrial challenges.

Obtaining Enough Water Animals are motile and walk, slither, fly, run, or crawl to water sources; this mobility requires not only the ability to move (with the use of skeletal and muscular systems) but the ability to sense the water's presence (with the nervous system).

Plants adapted very differently to the challenge of obtaining water: they have roots that not only anchor them in the soil but absorb water and dissolved minerals. Some plants have shallow, spreading root systems that obtain moisture from precipitation as it drains into the soil. Other plants have root systems that penetrate deeply into the soil to reach groundwater.

Preventing Excessive Water Loss The outer layers of terrestrial vertebrates and plants protect the moist inner tissues from drying out. Vertebrates that are adapted to living on land have accumulations of a water-insoluble protein called keratin in their epithelial cells. Keratin is particularly thick in reptiles, where it forms scales that greatly reduce water loss by evaporation.

Plants possess a water-insoluble waxy coating called a cuticle over their epidermal cells. Plants that are adapted to moister habitats (such as water lilies) may have a very thin cuticle, whereas those adapted to drier environments (such as cacti) often possess a thick, crusty cuticle.

Getting Enough Energy Animals are heterotrophs and eat plants or other animals that eat plants. This means that animals could not become permanent colonizers of land until plants were established.

Almost all plants are autotrophs and must absorb enough sunlight for effective photosynthesis. Plants that are relatively tall avoid the shade of competitors; the evolution of lignin-reinforced supporting cells such as fibers facilitated such growth, since plants lack skeletal systems for support. Other plants are adapted to lower light intensities and so are able to grow in the shade of larger plants, albeit more slowly.

Tolerating High Temperatures Air temperature on land varies greatly, particularly in temperate regions. Many animals avoid the heat by resting in the shade or by burrowing in the ground during the day; these animals become active at night when it is cooler. Mammalian skin has sweat glands that secrete a watery solution called sweat when the air temperature is high. The evaporation of sweat off the skin cools the body.

Plants also rely on evaporative cooling, although they do not produce sweat. Plants lose large quantities of water through their stomata in a process known as transpiration. As this water evaporates, it carries heat with it.

Tolerating Low Temperatures Vertebrates deal with the low temperatures of winter in several ways. Mammalian hair and bird feathers provide insulation and enable the body to conserve heat by trapping air next to the skin's surface. In addition, some animals avoid cold weather by migrating to warmer climates or by passing the winter in a dormant state called hibernation.

Many plants (biennials and perennials) also overwinter in a dormant state. The aboveground parts of some plants die during the winter, but the underground parts remain alive and dormant; the following spring, these underground parts resume metabolic activity and develop new aboveground shoots. Likewise, the seeds of certain plants remain dormant and do not germinate until they have been exposed to near-freezing temperatures for a certain period of time. Many trees are deciduous and shed their leaves, remaining bare for the duration of their dormancy. By shedding its leaves, a plant reduces water loss during the cold winter months, when obtaining water from the soil is difficult since roots cannot absorb water from ground that is cold or frozen. Thus, the shedding of leaves is actually an adaptation to the "dryness" of winter.

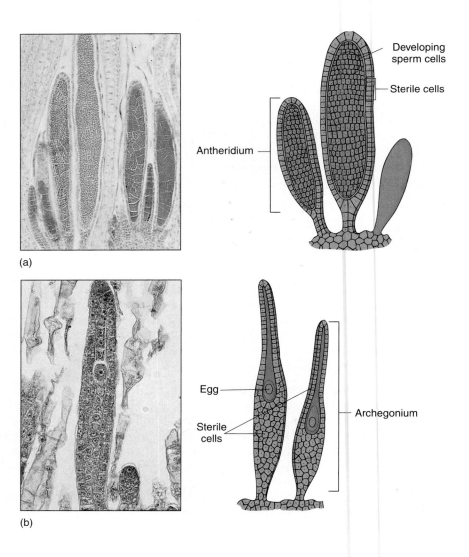

(a)

(b)

Figure 22-5   The gametangia (sex organs) of plants are multicellular structures. (a) Each moss antheridium, the male gametangium, produces numerous sperm cells. **(b)** Each moss archegonium, the female gametangium, produces a single egg. (a, *George S. Ellmore;* b, *Dennis Drenner*)

Let us examine alternation of generations more closely. The haploid gametophyte plant produces male gametangia called **antheridia** (sing. *antheridium*) in which numerous sperm cells form (Figure 22-5a). Female gametangia called **archegonia** (sing. *archegonium*), each bearing a single egg, are also formed by the gametophyte plant (Figure 22-5b). Sperm cells are transported to the archegonium in a variety of ways, such as by water, animals, and wind. They swim down the neck of the archegonium, and one sperm cell fuses with the egg. This process, known as **fertilization**, results in a fertilized egg, or **zygote**.

The diploid zygote is the first stage in the sporophyte generation. The zygote divides by mitosis and develops into a multicellular embryo, which is supported, nurtured, and protected within the archegonium of the gametophyte plant. Eventually, the embryo matures into a multicellular sporophyte plant containing special cells that are capable of dividing by meiosis. Each of these **spore mother cells** undergoes meiosis to form four haploid spores. All spores produced by plants are haploid and are the result of meiosis. This is in contrast with algae and fungi, which may produce haploid spores by either meiosis or mitosis.

The haploid spores represent the first stage in the gametophyte generation. Each spore divides by mitosis to form a multicellular gametophyte plant, and the cycle continues. Thus, plants have an alternation of generations—they alternate between a haploid gametophyte stage and a diploid sporophyte stage.

BRYOPHYTES ARE NONVASCULAR PLANTS

Bryophytes are considered the simplest plants, in part because they are the only nonvascular plants. As mentioned earlier, because they have no means of transporting water, essential minerals, and dissolved sugar extensive distances within their

(a)

(b)

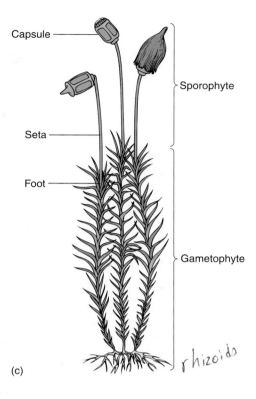

(c)

Figure 22-6 Mosses. (a) Mosses cover several rocks in the Great Smoky Mountains National Park. **(b)** A close-up of moss gametophytes. Mosses grow in dense clusters. **(c)** After sexual reproduction takes place in mosses, the sporophyte grows out of the gametophyte. The gray-green represents haploid gametophyte tissue, and the yellow-green is diploid sporophyte tissue. (a, *Willard Clay/Dembinsky Photo Associates;* b, *Carlyn Iverson*)

bodies, bryophytes are typically quite small. Although some bryophytes possess a cuticle, others do not and instead absorb water directly through the surfaces of their leafy shoots. They generally require a moist environment for active growth and reproduction, although a few bryophytes are tolerant of dry areas.

The bryophytes are divided into three distinctive phyla: mosses (phylum Bryophyta), liverworts (phylum Hepaticophyta), and hornworts (phylum Anthocerotophyta).[2] These three groups of plants differ in many ways and may or may not be closely related. They are usually studied together, however, because they lack vascular tissues and have similar life cycles.

Mosses Have a Dominant Gametophyte Generation

Mosses (*phylum Bryophyta*), with about 9000 species, usually live in dense colonies or beds on moist soil, rocks, or tree bark (Figure 22-6). Each individual plant has tiny rootlike structures called **rhizoids** that anchor it. Each plant also has a slender stemlike structure that bears leaflike blades. This structure may grow upright or along the ground (prostrate). Because they have no specialized vascular tissues, mosses do not possess true roots, stems, or leaves. (Some moss species have water-conducting cells and sugar-conducting cells, although these cells are not as specialized or as effective as the vascular tissues found in the vascular plants.)

In mosses the gametophyte generation is a leafy, green, often perennial plant that bears its gametangia at its tip (Figure 22-7). Many moss species have male plants that bear antheridia and separate female plants that bear archegonia; others produce antheridia and archegonia on the same plant.

[2]Plants were traditionally classified into divisions rather than phyla. The International Botanical Congress recently approved the use of the phylum designation for plants.

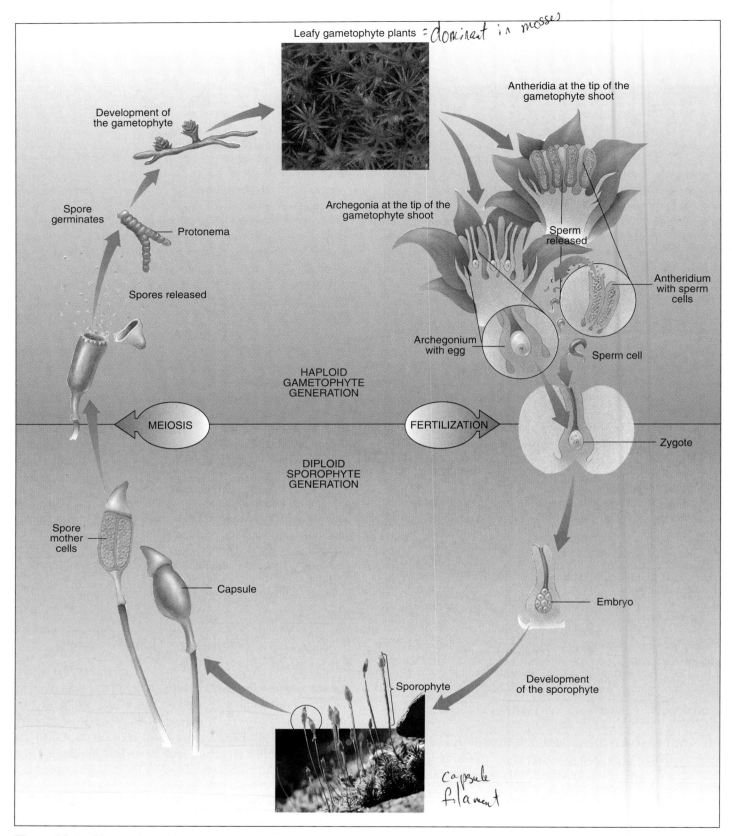

Leafy gametophyte plants = *dominant in mosses*

Development of the gametophyte

Antheridia at the tip of the gametophyte shoot

Spore germinates

Protonema

Archegonia at the tip of the gametophyte shoot

Sperm released

Antheridium with sperm cells

Spores released

Archegonium with egg

Sperm cell

HAPLOID GAMETOPHYTE GENERATION

MEIOSIS

FERTILIZATION

Zygote

DIPLOID SPOROPHYTE GENERATION

Spore mother cells

Capsule

Embryo

Sporophyte

Development of the sporophyte

capsule filament

Figure 22-7   Alternation of generations in mosses. The dominant generation is the gametophyte, represented by the leafy green plants, which are typically a few millimeters to several centimeters tall. The sporophyte generation grows out of the archegonium at the top of the gametophyte. Mosses require water as a transport medium for sperm cells during fertilization. (Top, *Rod Planck/Dembinsky Photo Associates;* bottom, *David Cavagnaro*)

Minute flagellated sperm cells are released from antheridia during rainy weather and are transported to archegonia by splashing raindrops. A raindrop lands on the top of a male gametophyte plant, and sperm cells are discharged into it from the antheridia. When another raindrop lands on the male plant, it may splash the first sperm-laden droplet into the air and onto the top of a nearby female plant. Or insects may touch the sperm-laden fluid and carry it a considerable distance. Once the sperm-filled water lands on a female moss plant, a haploid sperm cell swims down the neck of the archegonium and fuses with the haploid egg. The need for water as a transport medium for sperm cells is considered a primitive characteristic, passed down from algal ancestors, that mosses have retained.

The diploid zygote that formed as a result of fertilization grows by a series of mitotic divisions into a multicellular embryo, which develops into a moss sporophyte capable of producing spores. This sporophyte generation grows out of the top of the female gametophyte, to which it is permanently attached and on which it depends for nutrition. The sporophyte is initially green and photosynthetic but turns golden brown at maturity. A moss sporophyte plant is composed of three parts: a **foot**, which is embedded in the tip of the leafy gametophyte shoot and anchors the sporophyte to the gametophyte; a **seta**, or stalk; and a **capsule**, the sporangium (spore case) that contains spore mother cells.

The spore mother cells in each capsule undergo meiosis to form as many as 50 million haploid spores. When the spores are mature, the capsule bursts open from the pressure that has developed as it dried out, and it releases the spores. Wind or rain then carries the spores to other places. If a moss spore lands in a favorable spot, it germinates and grows into a branching, filamentous thread of green cells called a **protonema**. The protonema, which resembles a filamentous green alga, eventually forms buds, each of which grows into a leafy gametophyte plant; and the life cycle continues.

The haploid gametophyte generation is considered the dominant generation in mosses because it is larger, more persistent, and nutritionally independent. In contrast, the sporophyte generation in mosses, which is relatively short-lived, is attached to and nutritionally dependent on the gametophyte plant.

The name "moss" is commonly used for plants that have no biological relationship to the mosses. For example, reindeer moss is actually a lichen that is an important type of vegetation in the arctic tundra, while Spanish moss is a flowering plant and club moss is a relative of ferns.

Mosses are a significant part of their environment

Although mosses are typically inconspicuous in the environment, they are important ecologically. By frequently colonizing rock that was previously colonized by lichens, they play a role in forming soil. Moss plants, which form mats that cover the rock, eventually die, forming a thin soil in which grasses and other plants can grow. Because they grow packed together in dense colonies, mosses hold the soil in place and help prevent soil erosion. At the same time, they retain moisture that they and other organisms need. Waxwings and other birds use moss, along with twigs and grass, as nesting material.

Liverworts Are Either Thalloid or Leafy

Like other bryophytes, **liverworts** (*phylum Hepaticophyta*) lack vascular tissue and are small, generally inconspicuous plants restricted largely to damp environments. Liverworts grow on moist soil, rocks, old stumps, and tree bark. They are common, particularly in moist coniferous forests. About 6000 species of liverworts exist.

The gametophytes of some liverworts, such as those in the widely distributed genus *Marchantia*, are quite different from those of mosses. The gametophyte body form consists of a flattened, branching, ribbon-like **thallus** (pl. *thalli*), which is lobed (Figure 22-8). On the underside of the liverwort thallus are unicellular, threadlike rhizoids that anchor the plant to the soil. Although *Marchantia* and other thalloid liverworts are distinctive in appearance, they are not typical of the liverwort group as a whole. Most liverworts have a leafy appearance, rather than a lobed thallus, and are superficially very similar to mosses, with prostrate leafy shoots and rhizoids.

Figure 22-8 The gametophytes of some liverworts consist of flattened, ribbon-like lobes. The common liverwort (*Marchantia polymorpha*) thallus produces gemmae cups. Each gemmae cup contains tiny reproductive bodies, called gemmae, that are dispersed by splashing raindrops and can then grow into new liverwort thalli. (*Carlyn Iverson*)

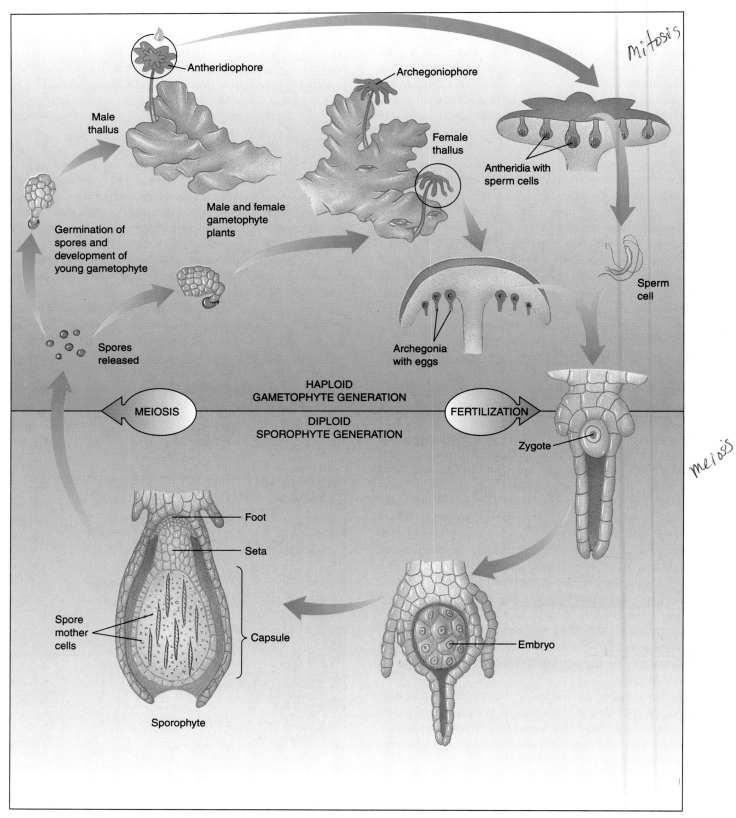

Figure 22-9   The life cycle of the common liverwort (*Marchantia polymorpha*), which has male and female structures on separate plants. The dominant generation is the gametophyte, represented by the ribbon-like male and female thalli.

The life cycle of liverworts is basically the same as that of mosses. The haploid gametophyte plant is considered the dominant generation, and the sporophyte is relatively short-lived. The common liverwort (*Marchantia polymorpha*) has male and female structures on separate plants (Figure 22-9). The female thallus produces stalked structures (*archegoniophores*) that bear archegonia. The male thallus produces stalked structures (*antheridiophores*) that bear antheridia. Like mosses, liverworts require water as a transport medium for flagellated sperm cells. Following fertilization, the sporophyte generation grows out of the archegonium of the gametophyte.

Sexual reproduction in liverworts, as in mosses, involves the production of archegonia and antheridia on the haploid gametophyte thallus. In *Marchantia*, archegonia and antheridia are usually produced on separate gametophyte plants. Archegonia are borne on the undersides of stalked, umbrella-like archegoniophores, and antheridia form on the upper surfaces of stalked disclike antheridiophores. The sporophyte, which develops from a zygote, is attached to and nutritionally dependent on the gametophyte plant (Figure 22-10). It is only a few millimeters long and, like moss sporophytes, consists of a foot, a seta, and a capsule.

Asexual reproduction in liverworts occurs in two ways. In one process, the gametophyte forms tiny reproductive bodies called **gemmae** (sing. *gemma*), which are borne in a saucer-shaped structure, the **gemmae cup**, directly on the liverwort thallus (see Figure 22-8). Splashing raindrops and small animals aid in the dispersal of gemmae. When a gemma lands in a suitable place, it grows into a new gametophyte thallus.

In the second asexual process, the thallus branches and grows. As the individual lobes elongate, they extend the thallus. When the older part of the thallus that originally attached the individual lobes dies, each extended lobe becomes a separate plant. Both of these mechanisms of reproduction, gemmae and thallus branching, are asexual because they do not involve the fusion of gametes or meiosis.

Liverworts are so named because their thalli superficially resemble the lobes of a liver

During the Middle Ages, many Europeans believed in the **doctrine of signatures**, which held that each disease or ailment could be treated by a specific plant and that each plant's "signature," or symbolic physical feature, provided a clue to its use. Because their form suggested the lobes of an animal liver,

(a)

Sporophytes

Archegoniophore

Gametophyte thallus

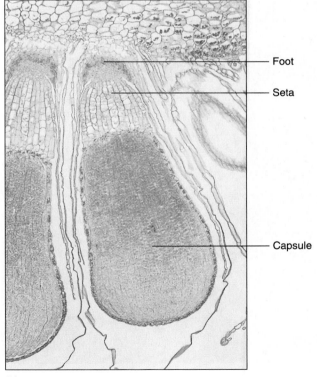

Foot

Seta

Capsule

(b)

Figure 22-10   The sporophyte generation of the liverwort *Marchantia polymorpha*. (a) After fertilization of the egg within an archegonium on the archegoniophore, the diploid sporophyte generation develops. It hangs upside down under the finger-like projections of the archegoniophore and resembles a miniature coconut hanging on a palm tree. **(b)** The liverwort sporophyte, which is always attached to and dependent on the gametophyte plant, has the same basic structure as the moss sporophyte, with its foot, seta, and capsule. Meiosis occurs in the capsule, producing haploid spores. (a, *Walter H. Hodge/Peter Arnold, Inc.*; b, *James Mauseth University of Texas*)

thalloid liverworts were thought to have medicinal value for the treatment of liver ailments. Modern medical research has not supported the doctrine of signatures, and liverworts do not help cure any diseases of the liver.

Hornworts Are Inconspicuous Thalloid Plants

Hornworts (*phylum Anthocerotophyta*), shown in Figure 22-11, are a small group of about 100 species of bryophytes whose gametophytes superficially resemble the liverworts. Hornworts are found in disturbed habitats such as fallow fields and roadsides.

At a phylogenetic level, it is not clear whether hornworts are closely related to other bryophytes. For example, their cell structure, which includes a single large chloroplast in each cell, is more like certain algal cells than it is like plant cells. In contrast, mosses, liverworts, and other plants have many disc-shaped chloroplasts per cell.

In hornworts, archegonia and antheridia are embedded in the gametophyte thallus so that they are not visible as in *Marchantia*. After fertilization and development, a needle-like sporophyte projects out of the gametophyte thallus, forming a spike, or "horn"—hence the name *hornwort*. A single gametophyte plant often produces multiple sporophytes. Meiosis occurs within each sporangium, and spores form. The sporangium splits open from the top to release the spores; each spore has the potential to give rise to a new gametophyte thallus. One unique feature of hornworts is that the sporophyte, unlike those of mosses and liverworts, continues to grow from its base for the remainder of the gametophyte plant's life.

(a)

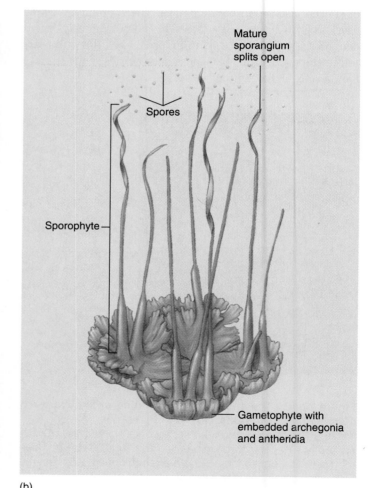

(b)

Figure 22-11   A typical hornwort. (a) The gametophyte with mature sporophytes of the common hornwort (*Anthoceros natans*). **(b)** The leafy green thallus, which is 1 to 2 centimeters (less than 1 inch) in diameter, is the gametophyte generation. After fertilization and development, the sporophytes project up out of the thallus, forming "horns." (*Robert A. Ross*)

THE EVOLUTION OF BRYOPHYTES IS OBSCURE

Although all plants are thought to have evolved from green algal ancestors, the mosses and other bryophytes are not in a direct path to the vascular plants; that is, vascular plants did not have bryophyte ancestors. Some fossil evidence indicates that mosses and other bryophytes are ancient plants, perhaps representing an evolutionary sideline that developed from ancestral green algae. Alternatively, bryophytes may have evolved from early vascular plants by becoming simpler and losing their vascular tissue. The fossil record of ancient bryophytes is incomplete, consisting mostly of spores and small tissue fragments, and can be interpreted in different ways; as a result, it does not provide a definitive answer on bryophyte evolution.

In 1995, in the journal *Nature*, an exciting fossil discovery was reported: a complete fossil plant about 400 million years old. This fossil, obtained from a coal deposit in the United Kingdom, resembles modern-day liverworts in many respects, but interestingly, its spores are virtually identical to those found in 460-million-year-old rocks. This discovery is significant because it indicates that bryophytes may have been some of the earliest plants to colonize land.

STUDY OUTLINE

I. Plants probably evolved from green algal ancestors.
   A. Plants and green algae have similar biochemical characteristics: the same pigments (chlorophylls *a* and *b*, carotenes, and xanthophylls), cell wall components (cellulose), and carbohydrate storage material (starch).
   B. Plants and green algae share similarities in certain fundamental processes such as cell division.

II. The colonization of land by plants required the evolution of structural, physiological, and reproductive adaptations.
   A. Plants possess a waxy cuticle for protection against water loss and stomata for the gas exchange that is necessary for photosynthesis.
   B. Plants produce gametes in multicellular gametangia that contain a protective layer of sterile cells; antheridia produce sperm cells, and archegonia produce eggs.

III. Plants have a complex life cycle that involves an alternation of generations between a haploid gametophyte and a diploid sporophyte.
   A. The gametophyte generation produces haploid gametes by mitosis.
   B. These gametes fuse to form a diploid zygote in a process known as fertilization.
   C. The first stage in the sporophyte generation is the zygote, which develops into a multicellular embryo protected and nourished by the gametophyte plant.

   D. The mature sporophyte plant has spore mother cells that undergo meiosis, producing haploid spores that are the first stage in the gametophyte generation.

IV. Bryophytes are small, fairly simple plants.
   A. Several advancements that bryophytes have in comparison with green algae include the possession of a cuticle, stomata, and multicellular gametangia.
   B. The gametophyte is the dominant generation in the bryophyte life cycle; that is, it grows independently of the sporophyte and is usually perennial. The sporophyte generation is always permanently attached to and nutritionally dependent on the gametophyte.
   C. Bryophytes, although adapted for a terrestrial existence, require water to complete the fertilization process.

V. There are three phyla of bryophytes: mosses, liverworts, and hornworts.
   A. Moss gametophytes are leafy plants that grow from a filamentous protonema; a moss sporophyte consists of a capsule, a seta, and a foot. Mosses are important ecologically, in part because they help to form soil and prevent soil erosion.
   B. Liverwort gametophytes are either leafy shoots or branched ribbon-like thalli; liverwort sporophytes are less complex than those of mosses.
   C. Hornwort gametophytes are thalloid, and their sporophytes form hornlike projections out of the gametophyte thallus.

SELECTED KEY TERMS

REVIEW QUESTIONS

1. What features distinguish plants from algae?
2. Which group of algae is thought to have been the ancestor of plants? Cite evidence for your answer.
3. What are the most important environmental challenges that plants experience on land, and what adaptations do plants possess to meet these challenges?
4. Define alternation of generations, and distinguish between gametophyte and sporophyte generations.
5. What limits the size of bryophytes? Why do most bryophytes live in moist environments?
6. List the three phyla of bryophytes.
7. How are mosses, liverworts, and hornworts similar? How is each group distinctive?
8. How are mosses ecologically important?

9. Label the following diagram.

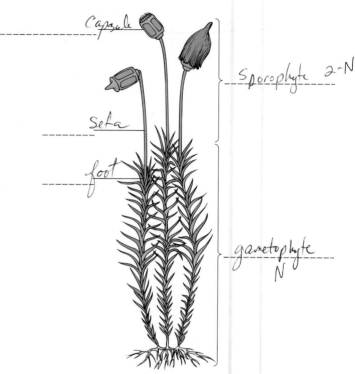

Capsule

Sporophyte 2-N

Seta

foot

gametophyte N

THOUGHT QUESTIONS

1. Fossils of trees are easier to find than fossils of bryophytes. Offer a possible explanation.
2. Suppose you are given an unknown plant. How could you determine whether or not it is a bryophyte?

3. Develop a hypothesis to explain why mosses grow in clusters or mats. How would you test your hypothesis?

RECOMMENDED READINGS

Crum, H., *A Focus on Peatlands and Peat Bogs*, Ann Arbor, University of Michigan Press, 1988. Among other topics, this book examines the economic importance of peat mosses.

Levanthes, L.E., "Mysteries of the Bog," *National Geographic* 171:3, March 1987. Discusses peat bogs and the mosses that form them.

CHAPTER 23

THE PLANT KINGDOM: SEEDLESS VASCULAR PLANTS

One of the best ways to learn about plants is to grow them. With a little care and attention, you can maintain a delicate maidenhair fern (*Adiantum* sp) indoors even if you live in a small apartment or a dormitory. Most of the approximately 200 species of maidenhair ferns are native to tropical American forests. Maidenhair ferns are popular as houseplants because they are attractive and easy to grow, and they add a special outdoor dimension to a room.

The leaves, or fronds, of maidenhair ferns typically grow 30 to 38 centimeters (12 to 15 inches) long and consist of shiny, dark black stalks that bear delicate blades divided into small, fanlike leaflets. (The maidenhair fern is so named because the slender leaf stalks resemble human hair.) The fronds of the maidenhair fern are borne on a horizontal, fast-spreading underground stem called a rhizome, which grows just beneath the surface of the soil.

Like all ferns, the maidenhair reproduces by forming spores. When mature, the outer edges of fertile leaves (leaves that bear spores) typically fold over the

Maidenhair ferns (*Adiantum* sp) are a good choice for a college student with little background in caring for plants, because they thrive indoors with a minimum of care. (*Marion Lobstein*)

spore cases, which are produced on the undersides of the leaf margins. Although it is possible to grow a new maidenhair fern from a spore, a lot of patience and special care are required. For that reason, most amateur plant growers purchase maidenhair ferns from a florist or plant nursery.

Like all houseplants, the maidenhair fern is a fascinating organism that reacts to the care (or lack of care) it receives. It prefers a bright filtered light, but do not place it in direct sunlight, or the delicate leaves will scorch. Normal room temperature is generally fine for maidenhair ferns, although it would be best to place them above (not in) a tray or saucer of wet pebbles to maintain a high humidity if the air temperature rises above 24°C (75°F). As with all plants, maidenhair ferns need to be watered more frequently when the air is warmer, but be careful to avoid overwatering. Add enough water so that the soil is slightly moist but not waterlogged. This can be done by moistening (but not soaking) the soil and then watering again only after the top inch of the soil has dried out.

377

SEEDLESS VASCULAR PLANTS INCLUDE FERNS AND THEIR ALLIES

In Chapter 22 we learned that the bryophytes are plants with dominant gametophytes whose sporophytes are attached to and dependent on the gametophyte plants. The bryophytes are also characterized by something they do *not* have: they are the only plants that lack vascular tissues. In contrast, virtually all plants other than bryophytes possess vascular tissues and have long-lived, dominant sporophytes.

This chapter introduces ferns and fern *allies,* seedless vascular plants that reproduce by forming spores rather than seeds. Ferns are an ancient group of plants that are still successful today, as evidenced by their considerable diversity. Three groups of plants—whisk ferns, club mosses, and horsetails—are considered allies of the ferns because they possess vascular tissues, reproduce by forming spores, and share similarities in their life cycles.

The most important advancement of the ferns and their allies as compared with algae and bryophytes is the presence of specialized vascular tissues (xylem and phloem) for support and conduction. Recall from Chapter 10 that **xylem** conducts water and dissolved minerals and **phloem** conducts carbohydrates, predominantly sucrose; both tissues also provide support.

Because this system of internal conduction can transport water, dissolved minerals, and carbohydrates over great distances to all parts of the plant, it enables vascular plants to attain greater sizes than mosses. Although ferns in temperate areas are relatively small plants with small underground stems, there are tree ferns in the tropics with erect stems that may grow to heights of 23 meters (75 feet) (Figure 23-1). The ferns and fern allies all have stems with vascular tissues, and most have vascularized roots and leaves as well.

The Leaf Evolved as the Main Organ of Photosynthesis

There are two basic types of leaves, microphylls and megaphylls, that represent two distinct lines of evolution (Figure 23-2). The **microphyll,** which is usually small and possesses a single vascular strand (that is, a single vein), is thought to have evolved from small outgrowths, or *enations,* of stem tissue. Club mosses, which are fern allies, are the only group of plants with microphylls.

Figure 23-1   The Tasmanian tree fern (*Dicksonia antarctica*). Tree ferns, most of which are native to tropical rain forests, are found in Australia, New Zealand, South Africa, and South America. (*Dennis Drenner*)

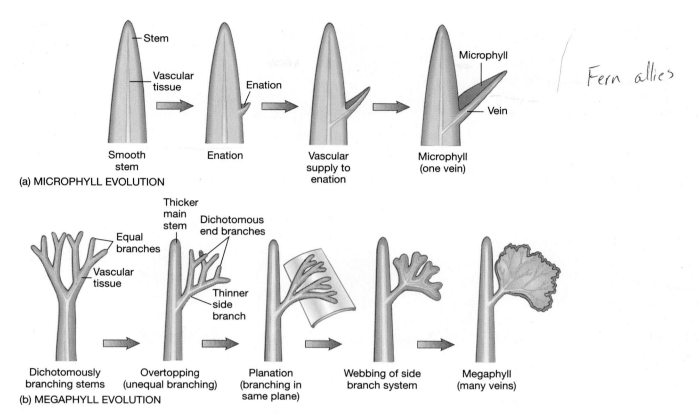

(a) MICROPHYLL EVOLUTION

Fern allies

(b) MEGAPHYLL EVOLUTION

Figure 23-2 There are two kinds of leaves. (a) Microphylls probably originated as outgrowths (enations) of stem tissue that developed a single vascular strand later. Club mosses have microphylls. **(b)** Megaphylls, which are more complex and have multiple veins, probably evolved from the evolutionary modification of side branches. Webbing is the evolutionary process in which the spaces between close branches are filled with chlorophyll-containing cells. Ferns, horsetails, gymnosperms, and flowering plants have megaphylls.

Megaphylls are thought to have evolved from side branches, which gradually filled in with additional tissue (that is, chlorophyll-containing cells), by a process known as *webbing*, to form most leaves as we know them today. Megaphylls possess a complex vein system consisting of more than one vascular strand, as would be expected if they evolved from branch systems. Ferns, horsetails (which are fern allies), and seed plants (gymnosperms and flowering plants) possess megaphylls.

FERNS ARE SEEDLESS VASCULAR PLANTS WITH LARGE MEGAPHYLLS

Most of the 11,000 species of **ferns** (*phylum Pterophyta*) are terrestrial, although a few have adapted to aquatic habitats (Figure 23-3). Although ferns range from the tropics to the Arctic Circle, the greatest number of species are found in tropical rain forests, where they perch high in the branches of trees. In temperate regions, ferns commonly inhabit swamps, marshes, moist woodlands, and stream banks, although some species can be found in fields, rocky crevices on cliffs or mountains, or even deserts. The most common fern species throughout the world is bracken (*Pteridium aquilinum*), a rugged, coarse, "weedy" plant that grows well on poor soil and is uncommon in the moist habitats favored by other ferns.

Ferns Have a Dominant Sporophyte Generation

The life cycle of ferns involves a clearly defined alternation of generations (Figure 23-4). The ferns that are grown as houseplants or seen in natural settings are the diploid, or sporophyte, generation. A fern sporophyte is typically composed of a horizontal underground stem, called a **rhizome,** that bears thin, wiry roots and relatively large, conspicuous leaves (megaphylls) called **fronds.** The fronds have two functions, photosynthesis and reproduction (discussed shortly). Fern sporophytes are **perennials;** that is, they live for a number of years. In geographical areas with pronounced winters, the leaves die each autumn, and the underground rhizome produces new leaves each spring.

(a)   (b)   (c)

Figure 23-3   Representative ferns. (a) The Christmas fern (*Polystichum acrosti-choides*) is green at Christmas, making it a popular holiday decoration. This fern was photographed in the Great Smoky Mountains. **(b)** The mosquito fern (*Azolla caroliniana*) is a free-floating aquatic fern that does not resemble "typical" ferns. It sometimes grows so densely across ponds that it reportedly smothers mosquito larvae. **(c)** The staghorn fern (*Platycerium bifurcatum*) is native to Australian rain forests and is widely cultivated elsewhere. In nature the staghorn fern is an epiphyte and grows attached to tree trunks but derives no nourishment from them (a, *Ed Reschke;* b, *Visuals Unlimited/W. Ormerod;* c, *Carlyn Iverson*)

When each young frond first emerges from the ground, it is tightly coiled. Because it resembles the end of a violin, this structure is called a **fiddlehead** (Figure 23-5). As a fiddlehead grows it unrolls from the tip and expands to form the frond. Fern fronds are usually compound (that is, the blade is divided into several leaflets), with the leaflets forming beautifully complex leaves. Fronds, roots, and rhizomes are considered true plant organs because they contain vascular tissues.

The conspicuous plant body of the fern—the sporophyte generation—forms spores by meiosis. Spore production usually occurs on the undersides of the fronds, where fertile areas on the leaves develop **sporangia** (sing. *sporangium*), or spore cases, in which spore mother cells form. The sporangia are frequently borne in clusters called **sori** (sing. *sorus*) on the fronds, although in some species sporangia are borne on separate stalks. Within the sporangia, spore mother cells undergo meiosis to form haploid spores.

The sporangium of many ferns is characterized by a row of cells with unevenly thickened cell walls. As the relative humidity changes, tensions develop that break the sporangium open along the thin-walled cells, thrusting the spores into the air. Fern spores are disseminated in air currents, often over great distances. When the spores land in suitable places, such as moist soil or cracks in rocks, they germinate and grow by mitosis into mature gametophyte plants.

The gametophyte generation of ferns, which bears no resemblance to the sporophyte generation, is a tiny (about the size of half of a fingernail), green, often heart-shaped structure that grows flat against the ground (Figure 23-6). The fern gametophyte, called a **prothallus**, lacks vascular tissues and has tiny rootlike rhizoids that anchor it in the ground. As the prothallus matures, it produces both female gametangia (archegonia) and male gametangia (antheridia) on its underside. The archegonia, which are flask-shaped and located in the central region of the prothallus near the "notch," each contain a single nonmotile egg. Numerous sperm cells are produced in the spherical antheridia, which are found scattered among the rhizoids. The sperm cells are shaped like tiny corkscrews, each with many flagella.

Like the mosses, ferns have retained the primitive requirement of water to accomplish fertilization. If a thin film of water is on the ground beneath the prothallus, it provides

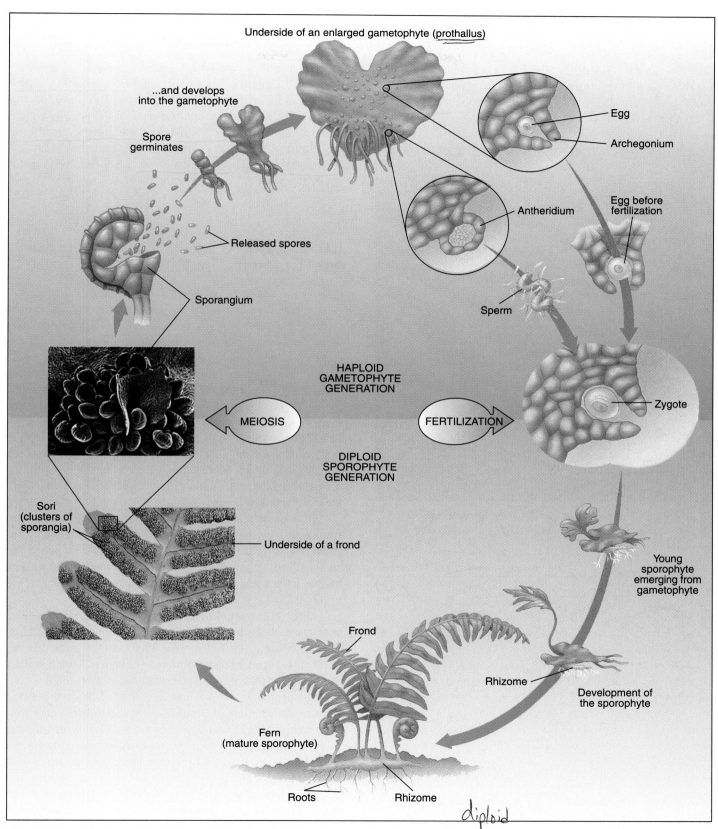

Figure 23-4   The fern life cycle. Note the clearly defined alternation of generations between the haploid gametophyte and diploid sporophyte stages. (Upper left, *Biophoto Associates;* lower left, *Ed Reschke*)

the transport medium in which the flagellated sperm cells swim to the archegonium. The neck of the archegonium provides a passageway from the archegonium surface to the egg. After one sperm cell fertilizes the egg in an archegonium, the resulting diploid zygote grows by mitosis into a multicellular embryo. At this stage in its life, the sporophyte embryo is attached to and dependent on the gametophyte, but as the embryo develops into a small sporophyte plant, the prothallus withers and dies.

The fern life cycle, which can be completed in anywhere from 4 to 18 months, has a clearly defined alternation of generations between the diploid sporophyte plant, with its rhizome, roots, and fronds, and the haploid gametophyte plant (the prothallus). The sporophyte generation is dominant not only because it is larger than the gametophyte but because it persists for an extended period of time, whereas the gametophyte dies soon after reproducing.

Figure 23-5 Fiddleheads exhibit the characteristic growth pattern of new fern leaves. (*Marion Lobstein*)

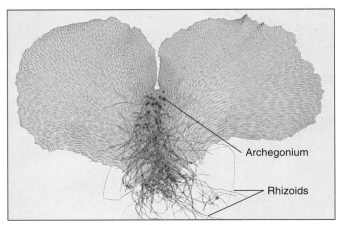

Archegonium

Rhizoids

Figure 23-6 The prothallus is the gametophyte generation of a fern. The dark spots near the notch of the "heart" are archegonia; no antheridia are visible. (*Carolina Biological Supply Company/Phototake-NYC*)

LACKING ROOTS AND LEAVES, WHISK FERNS ARE THE SIMPLEST VASCULAR PLANTS

Only about 12 species of **whisk ferns** (*phylum Psilotophyta*) exist today, and the fossil record contains several extinct species. *Psilotum nudum* (Figure 23-7a), a representative whisk fern, lacks true roots and leaves but has a vascularized stem. *Psilotum*, which lives in tropical and subtropical areas, has both a horizontal underground rhizome, bearing rootlike rhizoids, and erect aboveground stems that grow to 61 centimeters (2 feet) in height. The aerial stem branches extensively, resembling a whisk broom—hence the common name "whisk fern." The rhizome with its attached rhizoids performs the functions of a root—that is, it absorbs water and dissolved minerals—while the aerial stems are green and photosynthetic. Small, scalelike projections extend from the stem, but these are not considered leaves because they lack vascular tissues.

Whenever the stem forks or branches, it always divides into two essentially equal halves. This **dichotomous** branching is considered a primitive characteristic. In contrast, when the stems of most plants branch, one stem is more vigorous and becomes the main stem, whereas the other becomes a lateral branch.

Conspicuous sporangia, which are borne on the ends of short lateral branches, contain spore mother cells that undergo meiosis to form haploid spores. After being dispersed by air currents to a suitable environment, the spores, which require darkness to germinate, develop into haploid prothalli.

The prothalli of whisk ferns are difficult to study because they are minute (about 2 millimeters in diameter) and grow underground (Figure 23-7b). They are nonphotosynthetic due to their subterranean location, and they apparently have a symbiotic relationship with fungi, which provide them with nourishment.

Most species of whisk ferns are extinct, and the few surviving whisk ferns live mainly in the tropics. Although whisk ferns do not closely resemble ferns in appearance, they are considered fern allies because of similarities in their life cycles.

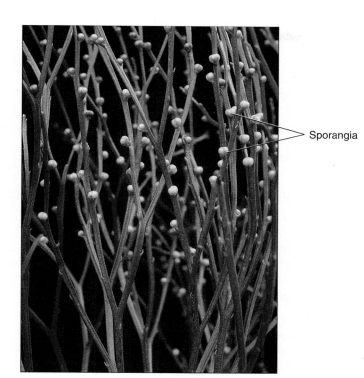

Sporangia

(a)

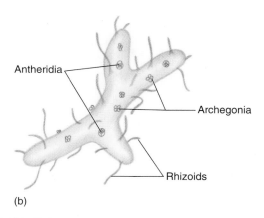

Antheridia

Archegonia

Rhizoids

(b)

Figure 23-7   The whisk fern (*Psilotum nudum*). (a) The sporophyte of *P. nudum*. The stem is the main photosynthetic organ of this rootless, leafless vascular plant. Sporangia, which are initially green but turn yellow as they mature, are borne on short lateral branches directly on the stems. *Psilotum nudum* has been cultivated for centuries in Japan, where numerous varieties are highly prized. **(b)** The gametophyte of *P. nudum* lives underground, nourished by soil fungi. (a, *John Arnaldi*)

Although the structure of whisk ferns has been carefully studied in recent years, botanists disagree about how to interpret these unique vascular plants that lack roots and leaves. Most botanists consider whisk ferns to be the surviving representatives of ancient vascular plants.

HORSETAILS HAVE HOLLOW, JOINTED STEMS AND SMALL MEGAPHYLLS

Millions of years ago, **horsetails** (*phylum Sphenophyta*), some of which were as large as trees, were among the Earth's dominant plants (Figure 23-8). Ancient horsetails are still significant to us today, because approximately 300 million years ago their dead remains, along with the remains of ancient ferns and club mosses, became the material from which our current vast coal deposits developed. (See Plants and People: Ancient Plants and Coal Formation, and Plants and the Environment: The Environmental Effects of Mining Coal.)

(*Text continues on page 386.*)

Figure 23-8   *Calamites* was an ancient horsetail the size of a small tree. Like modern-day horsetails, *Calamites* had an underground rhizome where roots and aerial shoots originated.

PLANTS AND People

Ancient Plants and Coal Formation

The industrial society in which we live depends on energy from fossil fuels, which formed from the remains of ancient organisms. One of our most important fossil fuels is coal, which is burned to produce electricity, to heat homes, and to manufacture items made of steel and iron. Although coal is mined from the ground as minerals are, it is not a mineral like gold or aluminum but is an organic material, formed from ancient plants.

Much of the coal we use today formed from the prehistoric remains of ancient vascular plants, particularly those of the Carboniferous period, which occurred approximately 300 million years ago (see figure). Of the five main groups of plants that contributed to coal formation, three were seedless vascular plants: the club mosses, horsetails, and ferns. The other two important groups of coal formers were seed plants: the seed ferns (now extinct) and early gymnosperms.

It is hard to imagine that the small, relatively inconspicuous club mosses, ferns, and horsetails of today could have been so significant in forming the vast beds of coal in the ground. However, many of the members of these groups that existed during the Carboniferous period were giants by comparison and formed vast forests. Some giant club mosses, for example, grew almost 40 meters (130 feet) tall.

The climate during the Carboniferous period was warm and mild, and plants grew year round because of the favorable weather conditions. Forests of these plants occurred in low-lying areas that were periodically flooded. When the water level receded, the plants would become established again.

When these large plants died or blew over during storms, they decomposed incompletely because they were covered by the swampy water. The anaerobic (oxygenless) conditions of the water prevented wood-rotting fungi from decomposing the plants,

and anaerobic bacteria do not decompose wood rapidly. Thus, over time the partially decomposed plant material accumulated and consolidated.

When the water level rose and flooded the low-lying swamps, layers of sediment formed over the plant material. Over time, heat and pressure built up in these accumulated layers and converted the plant material to coal and the sediment layers to sedimentary rock. Much later, geological upheavals raised the layers of coal and sedimentary rock.

Coal is usually found in underground layers, called seams, that vary from 2.5 centimeters (1 inch) to more than 30 meters (100 feet) in thickness. Coal of different grades (lignite, bituminous, and anthracite) formed as a result of different temperatures and pressures. The largest coal deposits are in North America, Russia, and China, but deposits are also found in the Arctic islands, western Europe, India, South Africa, Australia, and eastern South America.

The plants of the Carboniferous period included giant ferns, horsetails, and club mosses. (*No. GEO 85638c, Field Museum of Natural History, Chicago*)

KEY

1-13	club mosses
14-16	seed ferns
17-19	ferns
20-21	horsetalis
22	early gymnosperm
23	primitive insect
24	early dragonfly
25,26	early roaches

Strobilus —

(a)

(b)

Figure 23-9   Representative horsetails (*Equisetum* sp). (a) In some horsetail species, both fertile shoots, which bear conelike strobili, and vegetative (nonreproductive) shoots are unbranched. **(b)** Other horsetail species have unbranched fertile shoots and highly branched vegetative shoots with whorls of branches that resemble a bushy horse's tail. (a, *John Arnaldi;* b, *Ed Reschke*)

The few surviving horsetails, about 15 species in the single genus *Equisetum*, grow mostly in wet, marshy habitats and are small (less than 1.3 meters, or 4 feet, tall) but very distinctive plants (Figure 23-9). They are widely distributed on every continent except Australia.

Horsetails have true roots, stems, and leaves. Their hollow, jointed stems are impregnated with silica, which gives them a gritty feeling. The erect stems arise from perennial rhizomes that also bear wiry roots. The scalelike leaves, which are small megaphylls, are fused in whorls at each node (the area on the stem where leaves attach) and are quite small. Horsetail leaves are usually tinged brown, and the green stem is the main organ of photosynthesis.

The word *Equisetum* comes from the Latin words *equus*, "horse," and *saeta*, "bristle." These plants were named horsetails because certain vegetative (nonreproductive) stems have whorls of branches that give the appearance of a bushy horse's tail. In pioneering days, horsetails were called "scouring rushes" and, because of their gritty texture, were used to scrub out pots and pans along stream banks.

Each reproductive branch of a horsetail bears a terminal conelike **strobilus**. The strobilus is composed of several stalked structures, each of which bears 5 to 10 sporangia in a circle around a common axis.

The horsetail life cycle is similar in many respects to the fern life cycle. In horsetails, as in ferns, the sporophyte is the conspicuous plant, whereas the gametophyte is a minute lobed thallus ranging in width from the size of a pinhead to about 1 centimeter (less than 1/2 inch) across. The sporophyte and gametophyte plants are both photosynthetic and nutritionally independent at maturity. Like ferns, horsetails require water as a medium for flagellated sperm cells to swim to the egg.

CLUB MOSSES ARE SMALL PLANTS WITH MICROPHYLLS

Like horsetails, **club mosses** (*phylum Lycophyta*) were important plants millions of years ago, when species that are now extinct often attained great size (Figure 23-10). These large treelike plants, like the ancient horsetails, were major contributors to the formation of coal deposits.

The 1000 or so species of club mosses living today, represented by the genus *Lycopodium* (Figure 23-11), are small (less than 25 centimeters, or 10 inches, tall), attractive plants commonly found in temperate woodlands. Club mosses are

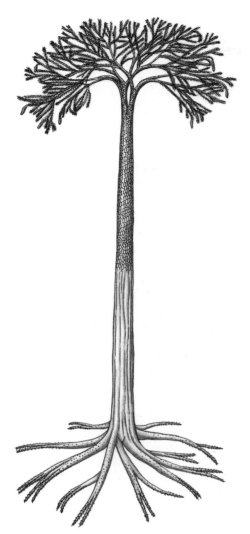

Figure 23-10 *Lepidodendron* **was an ancient club moss the size of a large tree.** Numerous fossils of *Lepidodendron* were preserved in coal deposits, particularly in Great Britain and the central United States.

Strobilus

(a)

(b)

Figure 23-11   Two species of *Lycopodium,* club mosses. **(a)** *Lycopodium annotinum.* **(b)** *Lycopodium complanatum.* Although club mosses superficially resemble mosses, they are fern allies. The sporophyte plant has small, scalelike leaves (microphylls) that are evergreen. Spores are produced in sporangia on fertile leaves clustered in a conelike strobilus (*as shown in a*) or scattered along the stem. (a, *Dwight R. Kuhn;* b, *Ed Reschke*)

evergreen and are often fashioned into Christmas wreaths and other holiday decorations. In some areas they are endangered from overharvesting.

The club mosses vividly exemplify one of the problems with the use of common names in biology. Their most common names are "club mosses" (because they are small-leaved like the mosses) and "ground pines" (because some that are common in pine forests superficially resemble miniature trees), yet these plants are neither mosses nor pines but are most closely allied to the ferns.

Club mosses possess true roots, stems (both rhizomes and erect or trailing aboveground stems), and small, scalelike, single-veined leaves (microphylls). Sporangia are borne on fertile leaves in conelike strobili at the tips of stems or scattered along the stems.

The life cycle of *Lycopodium* is similar in many respects to that of the ferns. As in ferns, both sporophyte and gametophyte are independent plants at maturity. In *Lycopodium,* the sporophyte is the conspicuous generation, whereas the gametophyte is a small thallus ranging from microscopic size to about 2.5 centimeters (1 inch) in length. The gametophytes

Environment

The Environmental Effects of Mining Coal

Coal mining, especially surface mining (also called strip mining), has substantial effects on the environment. In surface mining, vegetation and topsoil are completely removed by bulldozers, giant power shovels, and wheel excavators to expose the coal seam (see figure). The coal is then scraped out of the ground and loaded into railroad cars or trucks. Approximately 60 percent of the coal mined in the United States is obtained by surface mining, which destroys plant and animal habitat and increases soil erosion and water pollution.

Prior to the passage of the 1977 Surface Mining Control and Reclamation Act (SMCRA), abandoned surface mines were usually left as large open pits. Cliffs of excavated rock called highwalls, some more than 30 meters (100 feet) high, were left exposed. Acid and toxic mineral drainage from such mines and the burial of topsoil or its removal by erosion prevented most plants from recolonizing the land naturally. Streams were polluted with sediment and acid mine drainage. Dangerous landslides occurred on land that was unstable due to the lack of vegetation.

Surface-mined land can be restored to prevent such degradation and to make the land productive for other purposes, but restoration is extremely expensive. Before the existence of laws requiring coal companies to reclaim the land, few restorations of surface-mined land took place. Since 1977 the SMCRA has required coal companies to restore areas that have been surface-mined.

Sometimes this restored land is used for wildlife habitat—for example, rangelands for wild grazing animals in some western states, and wetlands in midwestern and eastern states. The Eastern Kentucky Regional Airport is situated on a reclaimed mining site, as is a tree farm in Garrett County, Maryland. Other post-mining land uses include camping and recreational areas, such as golf courses; farmlands; sanitary landfills; cemeteries; and housing developments.

Land that was strip-mined prior to 1977 is gradually being restored as well, with money from a tax that coal companies pay on currently mined coal. However, because so many abandoned coal mines exist, it is doubtful that they can all be restored, and only those that are most dangerous from a health and safety viewpoint are reclaimed.

Strip mining at a lignite coal mine in eastern Texas. (*Chuck Meyers, Office of Surface Mining, Department of the Interior*)

are either photosynthetic or subterranean (and therefore nonphotosynthetic). Like whisk fern gametophytes, the subterranean gametophytes of club mosses apparently form a symbiotic relationship with fungi that provides them with nourishment. Like ferns, *Lycopodium* requires water as a medium in which flagellated sperm cells can swim to the egg in the archegonium.

SOME FERNS AND CLUB MOSSES ARE HETEROSPOROUS

In the life cycles we have examined so far, plants produce only one type of spore as a result of meiosis. This condition, known as **homospory**, is found in bryophytes (see Chapter 22), horsetails, whisk ferns, and most ferns and club mosses. Certain ferns and club mosses, however, are **heterosporous** and produce two different types of spores. Figure 23-12 illustrates the generalized life cycle of a heterosporous plant.

Spike moss (*Selaginella* sp), a small, delicate club moss, is an example of a heterosporous plant (Figures 23-13 and 23-14). Its strobilus bears two kinds of sporangia, microsporangia and megasporangia. **Microsporangia** are sporangia that produce **microspore mother cells**, which undergo meiosis to form tiny haploid **microspores**. Each microspore develops into a male gametophyte that produces sperm cells within antheridia.

Figure 23-13 Spike moss (*Selaginella* sp) is a small, mosslike plant that is widely distributed in the tropics and subtropics, with a few species in temperate areas. *Selaginella* is heterosporous and produces both microspores (which develop into male gametophytes) and megaspores (which develop into female gametophytes). (*Dennis Drenner*)

Megasporangia in the *Selaginella* strobilus produce **megaspore mother cells**. When megaspore mother cells undergo meiosis, they form haploid **megaspores**, each of which develops into a female gametophyte that produces eggs within archegonia.

In *Selaginella*, the development of male gametophytes from microspores and the development of female gametophytes from megaspores occur within their respective spores. As a result, the male and female gametophytes are not truly free-living, unlike the gametophytes of other seedless vascular plants.

Heterospory was a significant development in plant evolution, because it was the forerunner of the evolution of seeds. Heterospory also occurs in the two most successful groups of plants existing today, the gymnosperms and the flowering plants, both of which produce seeds.

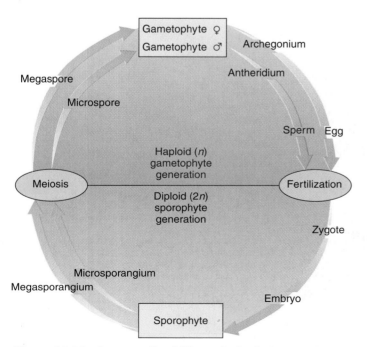

Figure 23-12 A generalized life cycle for heterosporous plants, which produce two types of spores, microspores and megaspores.

VASCULAR PLANTS AROSE OVER 400 MILLION YEARS AGO

According to the fossil record, the vascular plants arose about 420 million years ago. Ferns and fern allies, an ancient group of plants that extends back nearly 400 million years, were of considerable importance as Earth's dominant plants in past ages. Fossil evidence indicates that many species of these plants were the size of immense trees. Many ferns and most fern allies are extinct today; a few small representatives of the ancient groups survive.

precursors of seeds; stay inside megaspore

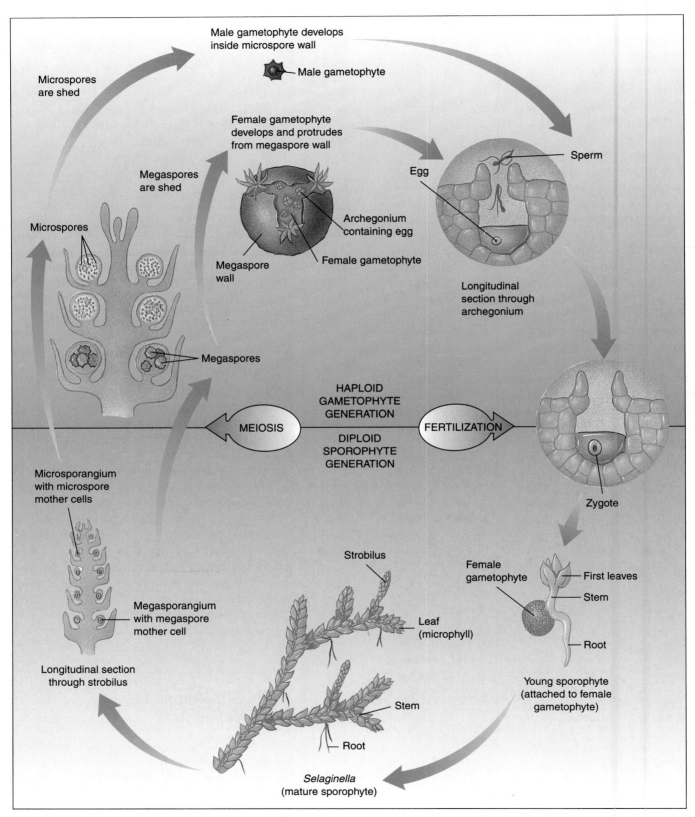

Figure 23-14   The life cycle of spike moss (*Selaginella* sp). Because spike moss is heterosporous, unisexual gametophytes form.

The oldest known vascular plants belong to a separate phylum (*phylum Rhyniophyta*), all of whose species became extinct about 380 million years ago. *Rhynia* sp, for example, was an early vascular plant that superficially resembled whisk ferns in that it consisted of leafless upright stems that branched dichotomously from an underground rhizome (Figure 23-15). *Rhynia* lacked roots, although it possessed absorptive rhizoids. Sporangia formed at the ends of some of its branches.

SEEDLESS VASCULAR PLANTS HAVE SOME ECOLOGICAL AND ECONOMIC IMPORTANCE

Ferns and fern allies are ecologically important, in part because—like other plants—they help to form soil. Their branching underground rhizomes and roots or rhizoids help hold the soil in place, thereby preventing erosion. Ferns are sometimes present in stages of *ecological succession*, the orderly sequence of changes in the kinds of plants living in a particular area. Some animals, such as muskrats, eat seedless vascular plants that grow along stream banks.

Ferns are of limited economic importance, although they are widely cultivated for their aesthetic appeal. Because of the beauty and variety of their fronds, ferns such as Boston fern, maidenhair fern, and staghorn fern are cultivated in homes, gardens, and conservatories, and florists often add fern leaves to floral arrangements.

The fiddleheads of some fern species, such as the ostrich fern (*Matteuccia struthiopteris*), are harvested in early spring, boiled or steamed, and eaten. Fiddleheads are sold fresh (as a seasonal item), frozen, and canned and are particularly popular in New England and the Canadian Maritime provinces.

The most economically important seedless vascular plants are the ones that died about 300 million years ago. The coal deposits that formed from the remains of these ancient ferns, club mosses, and horsetails are of great relevance to humans. Coal powered the steam engine and supplied the energy needed for the Industrial Revolution of the 19th century. Today coal is used primarily by utility companies to produce electricity and, to a lesser extent, by heavy industries such as steelmaking.

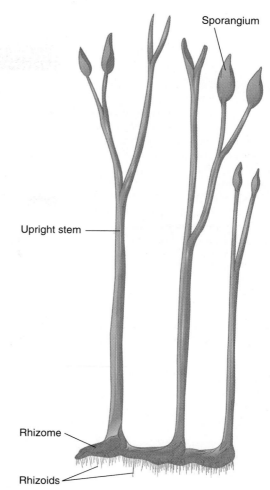

Figure 23-15 A reconstruction of *Rhynia* sp, a leafless plant that was one of Earth's earliest vascular plants and is now extinct. Fossils show that *Rhynia*, which probably lived in marshes, had rhizoids, rhizomes, upright stems, and sporangia.

STUDY OUTLINE

I. Seedless vascular plants have several advancements as compared to algae and bryophytes, including the possession of vascular tissues and alternation of generations with a dominant sporophyte generation.
 A. Two types of leaves evolved in the plant kingdom.
  1. Microphylls, small leaves that are thought to have evolved from lateral projections of stem tissue, are characteristic of club mosses.
  2. Megaphylls, leaves that probably evolved from branch systems, are characteristic of all vascular plants other than club mosses.
 B. Four phyla of seedless vascular plants exist today—ferns, whisk ferns, horsetails, and club mosses. All reproduce by forming spores and have essentially similar life cycles.

II. Ferns are the largest and most diverse group of seedless vascular plants.
 A. Fern sporophytes have rhizomes, roots, and megaphylls. Their leaves bear sporangia, which produce haploid spores.
 B. The fern gametophyte, called a prothallus, develops from a haploid spore and bears both archegonia and antheridia.
 C. Ferns have motile sperm cells, and their fertilization depends on water as a transport medium for these cells.

III. Sporophytes of whisk ferns consist of dichotomously branching rhizomes and erect stems; they lack true roots and leaves.

IV. Horsetail sporophytes have roots, rhizomes, aerial stems that are hollow and jointed, and leaves that are small megaphylls.

V. Sporophytes of club mosses consist of roots, rhizomes, erect branches, and leaves that are microphylls.

VI. Vascular plants are either homosporous or heterosporous.

    A. Homospory, the production of one kind of spore, is characteristic of whisk ferns, horsetails, some club mosses, and most ferns.

    B. Heterospory, the production of two kinds of spores (microspores and megaspores), is found in certain club mosses and ferns and in all seed plants.

       1. Microspores give rise to male gametophytes that produce sperm cells.

       2. Megaspores give rise to female gametophytes that produce eggs.

VII. Vascular plants evolved some 420 million years ago.

    A. The ferns and fern allies were Earth's dominant plants in past ages.

    B. *Rhynia* was one of the earliest vascular plants.

    C. Coal formed largely from the prehistoric remains of ancient ferns, club mosses, and horsetails. These plants existed during the Carboniferous period, approximately 300 million years ago.

S E L E C T E D K E Y T E R M S

club moss, p. 386
dichotomous, p. 382
fern, p. 379
frond, p. 379
heterospory, p. 389

homospory, p. 389
horsetail, p. 383
megaphyll, p. 379
megaspore, p. 389
microphyll, p. 378

microspore, p. 389
phloem, p. 378
prothallus, p. 380
rhizome, p. 379
sorus, p. 380

sporangium, p. 380
strobilus, p. 386
whisk fern, p. 382
xylem, p. 378

R E V I E W Q U E S T I O N S

1. Explain why bryophytes and seedless vascular plants are classified in different phyla.
2. Contrast the specific stages of alternation of generations in the mosses and in the ferns.
3. Distinguish between microphylls and megaphylls.
4. In terms of the number of species, which group of seedless vascular plants is most diverse?
5. Name and briefly describe the three groups of fern allies. What features do club mosses, horsetails, and whisk ferns share with ferns?

6. How are whisk ferns distinctive from other seedless vascular plants? How do whisk ferns survive without true roots or leaves?
7. Why are certain horsetails called "scouring rushes"?
8. How is the common name "club moss" misleading?
9. Define heterospory, and explain how heterospory modifies the plant life cycle.
10. Label the following diagram.

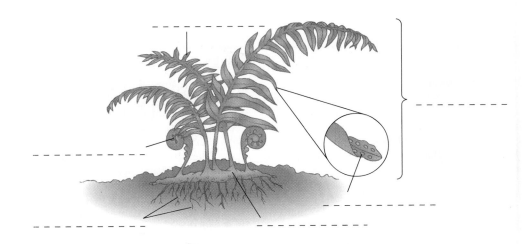

THOUGHT QUESTIONS

1. Which of the living plants studied in this chapter would you consider the most primitive? Why?
2. Suppose you are given an unknown seedless vascular plant. How could you determine whether it is a fern, a whisk fern, a horsetail, or a club moss?

SUGGESTED READINGS

Camus, J.M., A.C. Jermy, and B.A. Thomas, *A World of Ferns*, London, Natural History Museum Publications, 1991. An outstanding book on ferns.

Cobb, B., *A Field Guide to Ferns and Their Related Families*, The Peterson Field Guide Series, Boston, Houghton Mifflin Company, 1963. Describes various species of ferns and fern allies that are common in northeastern and central North America.

Foster, F.G., *Ferns To Know and Grow*, Portland, Timber Press, 1992. A guide to growing ferns.

THE PLANT KINGDOM: GYMNOSPERMS

The dawn redwood (*Metase-quoia glyptostroboides*) is a fast-growing tree closely related to the coastal redwood. The trunk of the dawn redwood flares out at the base, and the tree has a pyramidal shape. Although not as large as the coastal redwood, the dawn redwood grows to an impressive height of 30.5 meters (100 feet) or more. Its leaves, which are small (slightly more than 1 centimeter long), needle-like, and arranged horizontally in two rows on small lateral branchlets, resemble those of the coastal redwood. Unlike those of the coastal redwood, however, the leaves and lateral branchlets of the dawn redwood are deciduous and turn an attractive pinkish-brown color before falling off in autumn. The dawn redwood is a seed plant that produces its seeds in small, rounded, dark brown cones.

The scientific discovery of the dawn redwood, often called a "living fossil," is interesting. Dawn redwood fossils were discovered in the

A dawn redwood (*Metasequoia glyptostroboides*) with autumn coloration. (*Visuals Unlimited/Laurel Wallace*)

19th century but were not correctly identified until 1941. Because botanists had not observed living specimens, they presumed that the dawn redwood was extinct. A few years later, however, the dawn redwood was discovered growing in small temple groves in an isolated valley in Szechwan, China.

The dawn redwood descends from an ancient line of seed plants. From about 100 million to 25 million years ago, the *Metasequoia* genus consisted of multiple species and was the most common and most widely distributed genus in the redwood family. Today the dawn redwood is the sole surviving species of this genus.

The dawn redwood is commonly grown in North American and European gardens and on college campuses, thanks to botanists from the Arnold Arboretum in Massachusetts, who collected the seeds in China and distributed them. The dawn redwood can also be propagated from stem cuttings.

After reading this chapter, you should be able to:

1. *Compare the features of seeds with those of spores and discuss the adaptive advantages of plants that reproduce by seeds.*
2. *Summarize the features that distinguish gymnosperms from ferns and fern allies.*
3. *Name and briefly describe the four phyla of gymnosperms.*
4. *Label a diagram of the life cycle of pine, and compare its sporophyte and gametophyte generations.*
5. *Describe the ecological and economic significance of gymnosperms.*
6. *Trace the evolution of gymnosperms from seedless vascular plants.*

A Seed Contains a Young Plant and Nutritive Tissue

Until now, our discussion of plants has focused on bryophytes and on ferns and their allies, all of which reproduce by means of *spores*, which are haploid reproductive units that give rise to gametophyte plants. Although the most successful and widespread plants also produce spores, their primary means of reproduction and dispersal is by **seeds**, each of which consists of an embryonic sporophyte plant and nutritive tissue surrounded by a protective coat. Seeds develop from the reproductive structures of the female gametophyte and the tissues associated with it. The two groups of seed plants, gymnosperms and angiosperms (flowering plants), exhibit the greatest evolutionary complexity in the plant kingdom and are the dominant plants in most habitats today. Indeed, one could think of the present time as the "Age of Seed Plants."

Seeds are reproductively superior to spores in several ways. First a seed contains a well-developed multicellular young plant with embryonic root, stem, and leaves already formed, whereas a plant spore is a single cell. The parent plant protects and supports the young plant in the seed during its development; spores do not receive such protection.

Second, a seed contains a food supply. After *germination*, in which the young plant begins to grow and establish itself as an independent plant, it is nourished by food stored in the seed until it becomes self-sufficient. Because a spore is a single cell, minimal food reserves exist to sustain the plant that develops from a germinating spore.

Third, a seed is protected by a multicellular seed coat. Both seeds and spores can survive for extended periods of time at reduced rates of metabolism and then germinate when conditions become favorable.

Seeds and seed plants have been intimately connected with the development of human civilization. From prehistoric times, early humans collected and used seeds for food. The food supply stored in the seed contains proteins, oils, carbohydrates, and vitamins that are nourishing for humans as well as for germinating plants. Also, because it is easy to store seeds, provided they are kept dry, humans could collect seeds during times of plenty and save them for times of need. Few other foods can be stored as conveniently or for as long. Although most seeds that humans consume are produced by flowering plants, seeds of certain gymnosperms—the piñon pine (*Pinus edulis*), for example—are edible. They are usually sold as "pine nuts."

Gymnosperms and Flowering Plants Bear Seeds

The two groups of seed plants are the **gymnosperms** and the **angiosperms**, or flowering plants. The word *gymnosperm* is adapted from the Greek for "naked seed." Gymnosperms, which are woody trees, shrubs, or vines, produce seeds that are either totally exposed or are borne on the scales of female strobili (cones).

Pine, spruce, fir, hemlock, and *Ginkgo* trees are examples of gymnosperms; some of the most interesting plants in the plant kingdom, including a number of record holders, are gymnosperms. For example, one of the world's most massive organisms is the General Sherman, a giant sequoia in Sequoia National Park, California, that is over 81.6 meters (272 feet) tall and has a girth of over 23.7 meters (79 feet) at a distance 1.5 meters (5 feet) above ground level. The world's tallest tree is a coastal redwood that measures almost 117 meters (385 feet) in height. The oldest living trees are bristlecone pines. One has been dated by tree ring analysis as 4900 years old (Figure 24-1).

The term *angiosperm* derives from a Greek expression that translates loosely to "seed enclosed in a vessel or case." Flowering plants, which produce their seeds within a fruit, include such diverse plants as grasses, oaks, water lilies, maples, roses, cacti, and buttercups.

Figure 24-1 The bristlecone pine (*Pinus aristata*), found in mountainous parts of California, Nevada, and Utah, is the world's longest-lived tree. Bristlecone pines are conifers—gymnosperms that produce their seeds in cones. (*John Gerlach/Dembinsky Photo Associates*)

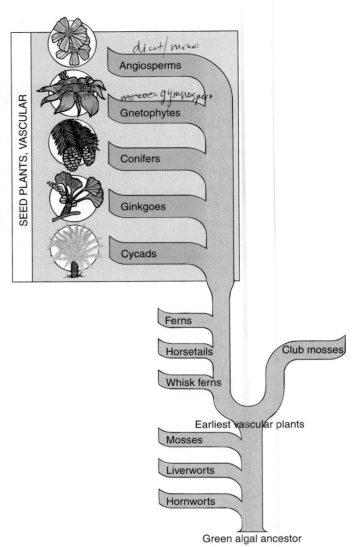

Figure 24-2 This phylogenetic tree illustrates possible relationships among plants. The five groups at the top of the tree are the seed plants, which are the focus of this chapter and Chapter 25. Four of these groups (cycads, ginkgoes, conifers, and gnetophytes) are gymnosperms. The angiosperms, or flowering plants, have more species than all other plant groups combined.

Gymnosperms and flowering plants both possess vascular tissues: xylem for the conduction of water and dissolved minerals, and phloem for the conduction of carbohydrates. Both exhibit alternation of generations in their life cycles; that is, they spend a portion of their lives in the diploid sporophyte stage and a portion in the haploid gametophyte stage. The sporophyte generation is the dominant stage in each group, and the gametophyte generation is significantly reduced. Unlike the plants we have considered so far (bryophytes and ferns and most of their allies), the gymnosperms and flowering plants do not have free-living gametophytes; instead, the gametophyte is attached to and nutritionally dependent on the sporophyte generation. Finally, both gymnosperms and flowering plants are heterosporous and produce two types of spores, microspores and megaspores.

GYMNOSPERMS ARE CLASSIFIED INTO FOUR PHYLA

The four phyla into which gymnosperms are classified represent four different evolutionary lines (Figure 24-2). Most gymnosperms belong to the phylum Coniferophyta and are commonly called conifers. Two phyla of gymnosperms are evolutionary remnants of gymnosperms that were more significant in the past: the Ginkgophyta and the palmlike or fernlike Cycadophyta. The fourth phylum of gymnosperms, the

Gnetophyta, is a collection of some very unusual plants that share certain advances not found in the other gymnosperms.

CONIFERS ARE WOODY PLANTS THAT BEAR SEEDS IN CONES

The **conifers** (*phylum Coniferophyta*), which include pines, spruces, hemlocks, and firs, are woody trees or shrubs that produce annual additions of secondary tissues (wood and bark); there are no herbaceous (nonwoody) conifers. The wood (secondary xylem) is composed of tracheids—long, tapering cells with pits through which the water and dissolved minerals move from one cell to another.

Many conifers produce **resin**, a viscous, clear or translucent substance consisting of several organic compounds that may protect the plant from attack by fungi or insects. The resin collects in **resin ducts**—tubelike cavities that extend throughout the plant's roots, stems, and leaves. Resin is produced and secreted by cells lining the resin ducts.

Most conifers have leaves (megaphylls) called **needles** that are commonly long and narrow, tough, and leathery (Figure 24-3). Pines bear clusters of two to five needles, depending on the species. In a few conifers such as American arborvi-

tae (*Thuja occidentalis*), however, the leaves are scalelike and cover the stem.

Most conifers are **evergreen** and bear their leaves throughout the year (Figure 24-4). Pine needles, which are evergreen, have a thick, waxy cuticle and stomata that are sunken in depressions (see Figure 8-12). These features are water-conserving adaptations that enable the pine tree to retain its needles throughout the winter, when low temperatures inhibit absorption of water by roots. Only a few conifers, such as the dawn redwood, larch, and bald cypress, are **deciduous** and shed their needles at the end of the growing season.

Most conifers are **monoecious**, which means that they have separate male and female reproductive parts in different locations on the same plant. These reproductive parts are generally borne in strobili that are commonly called **cones**—hence the name *conifer*, which means "bears cones."

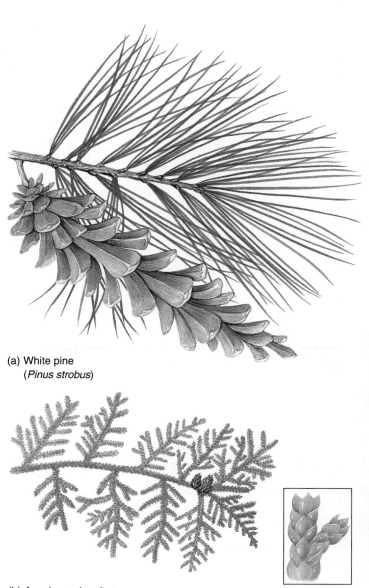

(a) White pine
(*Pinus strobus*)

(b) American arborvitae
(*Thuja occidentalis*)

Figure 24-3   Leaf variation in conifers. (a) In white pine (*Pinus strobus*), leaves are long, slender needles that occur in clusters of five. **(b)** In American arborvitae (*Thuja occidentalis*), leaves are small and scalelike (*see inset*).

Figure 24-4   The Colorado blue spruce (*Picea pungens*) is evergreen and bears its needles throughout the year.
Evergreen leaves have special adaptations for surviving challenging environmental conditions such as cold weather. (*Dennis Drenner*)

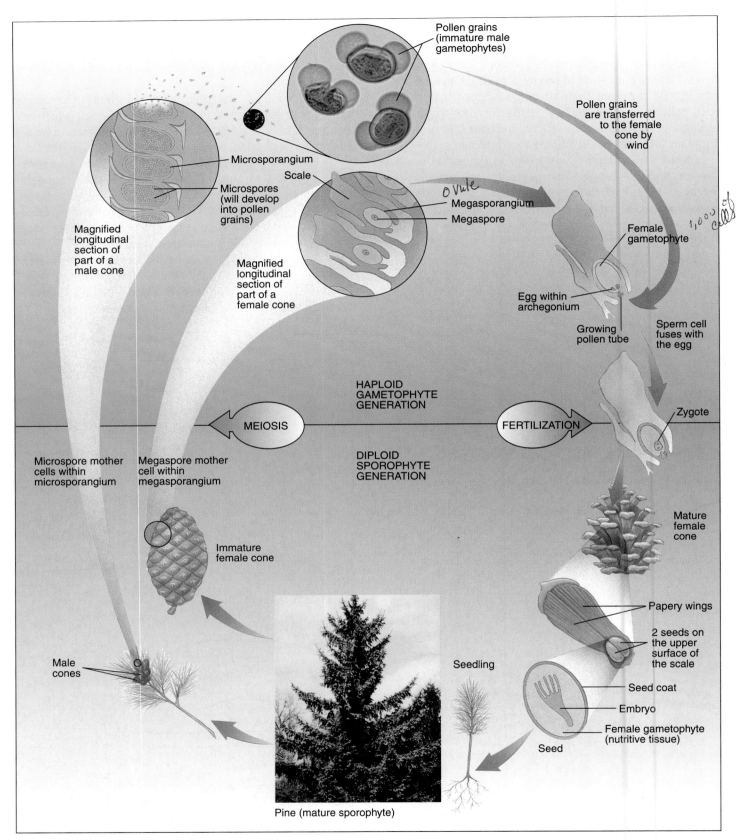

Pollen grains (immature male gametophytes)

Pollen grains are transferred to the female cone by wind

Microsporangium

Microspores (will develop into pollen grains)

Magnified longitudinal section of part of a male cone

Scale

Ovule

Megasporangium

Megaspore

Female gametophyte

1,000 cells

Magnified longitudinal section of part of a female cone

Egg within archegonium

Growing pollen tube

Sperm cell fuses with the egg

HAPLOID GAMETOPHYTE GENERATION

MEIOSIS

FERTILIZATION

Zygote

Microspore mother cells within microsporangium

Megaspore mother cell within megasporangium

DIPLOID SPOROPHYTE GENERATION

Immature female cone

Mature female cone

Papery wings

2 seeds on the upper surface of the scale

Male cones

Seedling

Seed coat

Embryo

Female gametophyte (nutritive tissue)

Seed

Pine (mature sporophyte)

Figure 24-5 The life cycle of pine. Unlike seedless vascular plants, gymnosperms produce windborne pollen and have nonmotile sperm cells. Pines and other gymnosperms are not dependent on water as a transport medium for male gametes. Their reproduction is totally adapted for life on land. (Top, *Manfred Kage/Peter Arnold, Inc.;* Bottom, *Dennis Drenner*)

398

The approximately 550 species of conifers occupy extensive areas of the Earth today and are the dominant vegetation in the taiga, the vast, forested northern regions of Alaska, Canada, Europe, and Siberia. In addition, conifers are important in the Southern Hemisphere, particularly in wet mountains of temperate and tropical areas in South America, Australia, New Zealand, and Malaysia. Southwestern China, with more than 60 species of conifers, has the greatest regional diversity of conifer species on Earth. California, New Caledonia (an island west of Australia), southeastern China, and Japan also have considerable diversity of conifer species.

Pines Represent a Typical Conifer Life Cycle

The genus *Pinus*, by far the largest genus in the conifers, consists of about 100 species. A pine tree is a mature sporophyte plant. Because pine is heterosporous, it produces two kinds of spores, microspores and megaspores, in separate cones (Figure 24-5). The male cones, usually 1 centimeter or less in length, are smaller than female cones and generally occur in dense clusters on the ends of lower branches each spring (Figure 24-6). The more familiar female cones, which are on the tree year-round, are usually located on the upper branches of the pine tree and bear seeds after sexual reproduction has occurred (Figure 24-6). Female cones vary considerably in size among pine species. The sugar pine (*Pinus lambertiana*) of California produces the world's longest female cones, which reach lengths of 60 centimeters (2 feet).

The woody scales of the female cones (also called seed cones) are considered by many botanists to be **sporophylls**, leaflike structures that bear megasporangia at their bases. Within each megasporangium is a *megaspore mother cell* that undergoes meiosis to produce four haploid *megaspores*. One of these develops into a *female gametophyte*, which produces one egg within each of several *archegonia*. The other three megaspores are nonfunctional and soon break down. When the megasporangium is ready to receive pollen, it produces a sticky droplet by the opening where the pollen grains land.

Each male cone (also called a pollen cone) is composed of sporophylls, leaflike scales that bear microsporangia. At the base of each sporophyll are two microsporangia, which contain numerous *microspore mother cells*. Each microspore mother cell undergoes meiosis to form four haploid *microspores*. Microspores then develop into *male gametophytes*, which are extremely reduced in size in gymnosperms. Each immature male gametophyte, also called a **pollen grain**, consists of four cells, two of which—a *generative cell* and a *tube cell*—are involved in reproduction. The other two cells soon degenerate. Two large air sacs occur on each pollen grain, providing buoyancy for wind dissemination. Pollen grains are shed from the male cones in great numbers, and some are carried by wind currents to the immature female cones. **Pollination**, the transfer of pollen to the female cone, occurs in

the spring during a period of a week or 10 days, after which the pollen cones wither and drop off the tree.

One of the many pollen grains that adhere to the sticky female cone grows a **pollen tube**, an outgrowth that digests its way through the female gametophyte tissue to the egg within the archegonium. The tube cell (which is involved in pollen tube growth) and the generative cell enter the pollen tube, and the generative cell divides to form two nonmotile (nonflagellated) sperm cells. When the pollen tube reaches the female gametophyte, it discharges the two sperm cells near the egg. One of these sperm cells fuses with the egg to form a zygote, or fertilized egg, which is the first stage in the sporophyte generation. The other sperm cell degenerates. The zygote undergoes a series of mitotic divisions, developing into a young multicellular embryo in the seed.

The developing embryo, which consists of an embryonic root and an embryonic shoot with several cotyledons (embryonic leaves), is embedded in haploid female gametophyte tissue that becomes the nutritive tissue in the mature pine seed. The embryo and nutritive tissue are surrounded by a

Figure 24-6 Male and female cones in lodgepole pine (*Pinus contorta*). (*Bottom*) Clusters of golden male cones produce copious amounts of pollen in the spring. (*Top*) Mature, dark brown female cones have opened to shed their seeds. The location of female cones above male cones facilitates cross pollination between different trees, because it is unlikely that wind will blow pollen grains directly upward to female cones on the same tree. (*Visuals Unlimited/Walt Anderson*)

Focus On

The Trend from Motile Cells in the Algae to Nonmotile Cells in Seed Plants

A stepwise progression is evident in reproductive adaptations to life on land. Many algae produce flagellated reproductive cells—both spores and sperm cells—that can swim through water. Although reproduction by flagellated spores and sperm cells is an advantage in aquatic environments where algae thrive, it may not be adaptive on land, particularly in locations where there are extended dry periods. In such terrestrial sites,

the production of nonmotile, airborne spores and sperm cells may be more

advantageous than motile reproductive cells. Thus, a general survey of algae and plants shows that (1) algae generally have *motile spores and sperm cells*; (2) the bryophytes and seedless vascular plants have *nonmotile spores and motile sperm cells*; and (3) most seed plants, including conifers, gnetophytes, and flowering plants, have *nonmotile spores and sperm cells*.

---

tough protective seed coat that extends out to form a thin papery wing at one end of the seed. This wing enables the seed to be dispersed by air currents.

A long period of time elapses between the appearance of female cones on a tree and the maturation of pine seeds in those cones. When pollination occurs in the spring, the female cone is still immature, and the megaspores have not yet formed. During the following year, the female reproductive tissue gradually matures, and eggs form within archegonia. Meanwhile, the pollen tube grows slowly through the female tissues to the archegonia. Fertilization, the union of the egg and the sperm cell, occurs during the spring of the following year, and the embryo begins to develop. Seed maturation takes several additional months. Some seeds are shed as soon as they mature, whereas others remain within the female cones for weeks, months, or even years before being shed.

In the pine life cycle, the sporophyte generation is dominant and the gametophyte generation is restricted in size to microscopic structures in the cones. Although the female gametophyte produces archegonia, the male gametophyte is so reduced that it does not produce antheridia (male gametangia). The gametophyte generation in pines, as in all seed plants, depends totally on the sporophyte generation for nourishment.

Some Conifers Produce Fleshy Fruitlike Structures

Some conifers do not produce woody female cones. As shown in Figure 24-7, each yew (*Taxus* sp) seed is almost completely surrounded by a fleshy cuplike covering that is an outgrowth from the base of the seed. The seed covering is red and attracts birds, which eat them and disperse the seeds. Although the

yew is an attractive ornamental and is planted extensively, it is not widely known that its bark, leaves, and seeds are very poisonous. Children should be warned not to eat the brightly colored seed coverings, so that they do not accidentally ingest the seeds.

Figure 24-7   The seeds of yews (*Taxus* sp) are borne in fleshy cuplike structures. Birds and other animals eat the fleshy structures and disperse the seeds. (*Edgar E. Webber*)

Juniper (*Juniperus* sp) bears its seeds in fleshy cones that resemble berries. The fleshy scales of the cones are fused together and completely envelop the seeds. At maturity, these cones are colored orange, red, brown, purple, or black, depending on the species. Birds eat the cones and swallow the seeds, which are dispersed in their droppings. Gin makers dry the fleshy cones of the common juniper (*Juniperus communis*) and use them to flavor the gin.

Conifers Do Not Require Water for Transport of Sperm Cells

Air currents transfer pine pollen to female cones, and non-motile sperm cells accomplish fertilization by moving through a pollen tube to the egg. Therefore, pine and other conifers are plants whose reproduction is totally adapted for life on land (see Focus On: The Trend from Motile Cells in the Algae to Nonmotile Cells in Seed Plants).

CYCADS HAVE COMPOUND LEAVES AND SIMPLE SEED CONES

The **cycads** (*phylum Cycadophyta*) were a very important plant group during the Triassic period, which occurred approximately 248 million years ago and is sometimes referred to as the "Age of Cycads." The few surviving cycads—about 140 species—are tropical and subtropical plants with stout, trunklike stems and compound leaves that resemble those of palms or tree ferns (Figure 24-8a). Many cycads are endangered species, including Sago cycas (*Zamia integrifolia*), which is native to Florida (Figure 24-8b,c).

Reproduction in cycads is similar to that in pines except that cycads are **dioecious** and therefore have seed cones on female plants and pollen cones on male plants. Their seed structure is most like that of the earliest seeds, and cycads have also retained the primitive feature of motile sperm cells, each of which possesses many short, hairlike cilia. These motile sperm cells are a vestige retained from the ancestors of cycads, in which sperm cells swam from antheridia to archegonia. Cycad pollen, however, is carried by air or possibly insects to the female plants; there the pollen germinates and grows a pollen tube down which the sperm cells pass to get to the egg. In other words, despite having motile sperm cells, cycads do not need water as a transport medium for fertilization.

Figure 24-8 Cycads. (a) This older cycad in South Africa resembles a palm. **(b,c)** Sago cycas (*Zamia integrifolia*), the only cycad native to the United States, is found in Florida. Like *Zamia*, most cycads are short plants (under 2 meters tall). The female strobilus of *Z. integrifolia* is depicted in part c. (a, *Walter H. Hodge/Peter Arnold, Inc.*; b, *John Arnaldi*)

(a)

(b)

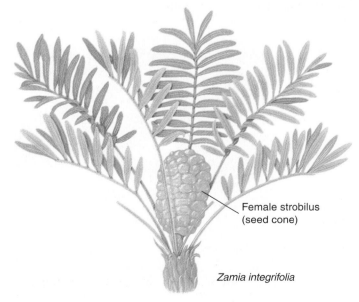

Female strobilus (seed cone)

Zamia integrifolia

(c)

GINKGO IS THE ONLY SURVIVING SPECIES IN ITS PHYLUM

There is only one living species in the *phylum Ginkgophyta*—the **ginkgo**, or maidenhair tree (*Ginkgo biloba*; Figure 24-9), so named because of the resemblance of its fan-shaped leaves to those of the maidenhair fern, discussed in the Chapter 23 introduction. Native to eastern China, ginkgo has been cultivated for centuries. It has been found growing wild in only two locations in China, and it is likely that it would have become extinct had it not been cultivated in Chinese monasteries. *Ginkgo* is the oldest known genus (and species) of living trees. Scientists have discovered fossil ginkgo leaves and wood that are 200 million years old and nearly identical to those of the modern-day ginkgo.

Ginkgo is widely cultivated in North America and Europe today, particularly in parks and along city streets, where it is planted frequently because it is somewhat resistant to air pollution and disease. Its leaves are deciduous and turn a beautiful golden color in the fall before being shed.

Like cycads, ginkgoes are dioecious, with separate male and female trees, and have ciliated sperm cells, an evolutionary vestige that is no longer required because the ginkgo produces airborne pollen. Its seeds are completely exposed rather than occurring within cones. Only female trees produce seeds, a fact you should remember if you ever wish to plant ginkgoes. As the seeds mature, their outer fleshy coverings give off a foul odor, which makes the female trees undesirable. Some cities and towns have ordinances prohibiting the planting of female ginkgoes. However, in China and Japan, where people eat the seeds, female trees are more common.

GNETOPHYTES INCLUDE THREE UNUSUAL GYMNOSPERMS

The *phylum Gnetophyta* is composed of about 70 species in three diverse genera: *Gnetum, Ephedra,* and *Welwitschia*. Although very diverse in appearance, **gnetophytes** share certain features that make them clearly more advanced than the rest of the gymnosperms. For example, gnetophytes have more efficient water-conducting cells, called vessel elements, in their xylem. Flowering plants also possess vessel elements in their xylem, but no gymnosperms except the gnetophytes do. Also, the cone clusters produced by some of the gnetophytes resemble flower clusters, and certain details in their life cycles resemble those of flowering plants.

The genus *Gnetum* contains tropical vines, shrubs, and trees with broad leaves similar to those of flowering plants

(a)

(b)

Figure 24-9   The ginkgo, or maidenhair, tree (*Ginkgo biloba*). (a) A young male ginkgo tree. **(b)** Close-up of a branch from a female ginkgo tree, showing the exposed seeds and the leaves, which resemble those of the maidenhair fern. (a, *Carlyn Iverson;* b, *Marion Lobstein*)

(a)

(b)

Figure 24-10   Gnetophytes have some advanced features that other gymnosperms lack. (a) The leaves of *Gnetum gnemon* resemble those of flowering plants. The exposed seeds are red to yellow when ripe. **(b)** A male joint fir (*Ephedra* sp) has pollen cones clustered at the nodes. Most species of *Ephedra* are dioecious and have separate male and female plants. Nineteenth-century pioneers used species native to desert areas in the American Southwest to make a beverage called Mormon tea. **(c)** *Welwitschia* is a gnetophyte that is native to deserts in southwestern Africa. The two leaves of this extremely unusual plant grow from the edges of a short, wide stem. (a, *Visuals Unlimited/John D. Cunningham;* b, *David Cavagnaro;* c, *Patti Murray*)

(c)

(Figure 24-10a). Species in the genus *Ephedra* include many shrubs and vines found in deserts and other dry temperate and tropical regions (Figure 24-10b). Some *Ephedra* species resemble horsetails in that they possess jointed green stems with tiny leaves. Commonly called joint fir, *Ephedra* has been used medicinally for centuries; today an Asiatic species is the source of ephedrine, which stimulates the heart and raises blood pressure. Ephedrine is sold over the counter in weight-control medications and herbal energy-boosters; in 1996 at least 15 deaths were reported from chronic use or overdose of products containing *Ephedra*.

The third gnetophyte genus, *Welwitschia*, contains a single species that is found in southwest African deserts (Figure 24-10c). The majority of *Welwitschia*'s body, a long taproot, grows underground; its aboveground stem forms a shallow disc, up to 0.9 meter (3 feet) in diameter, from which two ribbon-like leaves extend. These two leaves continue to grow from the stem throughout the plant's life, but their ends are usually broken and torn by the wind, making it appear that *Welwitschia* has numerous leaves. When *Welwitschia* reproduces, it forms cones around the edge of its disclike stem.

PLANTS AND People

Old-Growth Forests of the Pacific Northwest

The most commercially valuable public forest land in the United States is in northern California, western Oregon, and western Washington. A drama involving people's jobs and the environment unfolded there during the late 1980s and early 1990s. At stake were thousands of jobs and the future of 3 million acres of old-growth (virgin) coniferous forest, along with the existence of wildlife that depend on the forest. One of these forest animals, the northern spotted owl, came to symbolize the confrontation (see figure).

Biologists regard the few remaining old-growth forests of Douglas fir, western hemlock, and spruce in the Pacific Northwest as a living laboratory that demonstrates the complexity of one of the few natural ecosystems not extensively altered by humans. To environmentalists, the old-growth forests, with their 2000-year-old trees, are a national treasure to be protected and cherished. These stable forest ecosystems provide biological habitats for many species, including the northern spotted owl, an

The northern spotted owl, an endangered species found primarily in old-growth forests in the Pacific Northwest. (*Anthony Mercieca/Dembinsky Photo Associates*)

endangered species found primarily in old-growth forests. Provisions of the Endangered Species Act require the government to protect the habitat

of endangered species so that their numbers can increase. Enforcement of this law in the Pacific Northwest required the suspension of logging where the owl lives, in about 3 million acres of federal forest. Beginning in 1990, a court order blocked logging in these old-growth forests.

The timber industry bitterly opposed any such action, stating that thousands of jobs would be lost if the northern spotted owl habitat was set aside. Many rural communities in the Pacific Northwest do not have diversified economies; timber is their main source of revenue. Thus, a major confrontation over the future of the old-growth forest ensued between the timber industry and environmentalists.

The situation was more complex than simply jobs versus the environment, however. The timber industry had already been declining in terms of its ability to support people in the Pacific Northwest. For example, during the decade between 1977 and 1987, logging in Oregon's national forests increased by more than 15

FOSSILS PROVIDE CLUES ABOUT THE EVOLUTION OF SEED PLANTS

One of the groups of plants that evolved from ancestral seedless vascular plants was the **progymnosperms**, all now extinct (Figure 24-11). Progymnosperms had two features that their immediate ancestors lacked: (1) leaves that were megaphylls and (2) woody tissue—that is, secondary xylem—similar to that of modern conifers. Progymnosperms retained a primitive feature, however: they reproduced by spores.

Scientists have found fossils of several progymnosperms with reproductive structures intermediate between those of spore plants and those of seed plants. For example, the evolution of microspores into pollen grains and of megasporangia into ovules (seed-producing structures) can be traced in

fossil progymnosperms. Seed-producing plants appeared during the late Devonian period, more than 360 million years ago. The fossil record indicates that seed plants apparently evolved independently several times.

As mentioned previously, fossilized remains of ginkgoes are found in 200-million-year-old rocks, and other groups of gymnosperms were well established by 160 to 120 million years ago. Although the gymnosperms are an ancient group, some questions persist about the exact pathways of gymnosperm evolution. The fossil record indicates that progymnosperms probably gave rise to conifers and to another group of extinct plants called **seed ferns**. In turn, the seed ferns, which were seed-bearing woody plants with fernlike leaves, probably gave rise to cycads and possible ginkgoes as well as to several plant groups that are now extinct. The origin of gnetophytes remains unclear.

percent, whereas employment for loggers, sawmill operators, and the like dropped by 15 percent—an estimated 12,000 jobs—during the same period. The main cause of this decline was the automation of the timber industry. Thus, even if the timber industry were allowed to continue logging the old-growth forests, automation would probably continue to eat away at the number of logging jobs.

Another factor is that the timber industry in that region had not been operating *sustainably*; that is, trees were removed faster than the forest could regenerate. If the industry continued to log at its 1980s rates, most of the remaining old-growth forest would disappear within 20 years.

Fortunately, timber is not as important to the economy of the Pacific Northwest as it used to be. This change started long before the northern spotted owl controversy. Northwestern states have rapidly become more economically diversified, and by the late 1980s the timber industry's share of the economy in Oregon and Washington was less

than 4 percent. Interestingly, during the years in which logging in old-growth forests was blocked (1990–1994), Oregon reported its lowest unemployment rate in more than 20 years. Although it lost timber jobs, Oregon attracted several high-technology companies, thus creating more jobs than it lost. Many timber workers are being retrained for high-technology and other kinds of careers.

President Bill Clinton recognized that a broad new agenda was needed in the dwindling forests of the Pacific Northwest—an agenda that would take into account the multiple issues and interests involved. Early in his term of office, Clinton convened a Timber Summit in Portland, Oregon, with representatives of all sides of the issue attending. The Clinton Plan that resulted from this summit was a compromise that Clinton hoped would satisfy both environmental and timber interests. A federal judge approved it on December 22, 1994.

As a result of the Clinton Plan, limited logging has resumed in national federal forests of Washington,

Oregon, and northern California. However, only about one fifth of the logging that occurred during the 1980s is now allowed. The plan, along with previous congressional and administrative actions, reserves about 75 percent of the federal timberlands to protect the northern spotted owl and other species, especially salmon and other fish. Fish are protected by no-logging areas (known as riparian buffers) along streams, which preserve the aquatic ecosystems and reduce sedimentation caused by soil erosion.

As happens with many compromises, neither the ecologists and environmentalists nor the timber-cutting interests are happy with the Clinton Plan. Some environmentalists think the plan does not protect the forest adequately, while timber-cutting interests challenge the forestry laws and the Endangered Species Act on which the Clinton plan was based. In January 1996, the U.S. Supreme Court upheld the legislation that protects the northern spotted owl and other endangered species.

Gymnosperms Are Ecologically and Economically Significant

The conifers are the most important phylum of gymnosperms both ecologically and commercially, because they are far more common than the other, relatively obscure gymnosperms. Conifers are the predominant trees in about 35 percent of the world's forests. In North America, western and northern forests are primarily coniferous, as are those in the southeastern United States.

Coniferous forests play an essential role in the environment. Their roots hold vast tracts of soil in place, reducing soil erosion. Forests are important *watersheds* (areas of land drained by river systems) because they absorb, hold, and slowly release water. Watersheds provide a well-regulated flow

of water to the river system, even during dry periods, and help to control floods. In addition, forests provide food and shelter to animals, fungi, and many other organisms (see Plants and People: Old-Growth Forests of the Pacific Northwest).

The conifers are also of considerable economic importance for both recreation and timber. Recreational uses of forests include such activities as camping, backpacking, picnicking, and observing nature. These activities, which require equipment, supplies, and guides, support a large recreation industry.

In the United States, about 80 percent of our timber crop comes from conifers. Humans use conifers for lumber (especially for building materials and paper products) and various substances such as turpentine and resins, which are ingredients in varnishes and certain plastics. Because of their attractive appearance, conifers are grown commercially for land-

Figure 24-11   A reconstruction of *Archaeopteris*, a progymnosperm identified from fossils discovered in eastern North America. Although its wood resembled that of modern gymnosperms, its leaves were fernlike, and some of them bore sporangia.

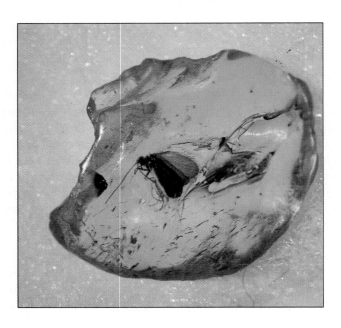

Figure 24-12   A prehistoric caddis fly embedded in Baltic amber has been preserved almost perfectly for 35 million years. (*Carlyn Iverson*)

scape design and for Christmas trees. Amber, the fossilized resin of prehistoric conifers, is often cut and polished to make jewelry. Some amber contains fossilized insects, spiders, small lizards, and plant fragments that appear perfectly preserved (Figure 24-12).

Three trees are highlighted in this discussion as representative conifers with commercial value: the Douglas fir from western forests, the red spruce from northern forests, and the loblolly pine from southeastern forests.

Douglas fir (*Pseudotsuga menziesii*), the state tree of Oregon, is a major timber tree in both the Rocky Mountain and Pacific forests of the West (Figure 24-13a). Its wood, which is strong, durable, and attractive when stained, is used for many products, including telephone poles, railroad ties, fences, paper, and plywood (more plywood is made from Douglas fir than from any other tree).

Red spruce (*Picea rubens*), the provincial tree of Nova Scotia, is an important timber tree in northern coniferous forests (Figure 24-13b). Its primary range extends from southeastern Canada to New England, and it is found in the Appalachian Mountains.[1] The wood of red spruce is relatively lightweight but strong and is used to make such products as oars and ladders as well as musical instruments such as violins and pianos. Much of the red spruce that is cut today is used to make paper, paperboard (for containers), and fiberboard. Small red spruces are harvested and sold as Christmas trees.

Loblolly pine (*Pinus taeda*), one of the southern yellow pines, is the most abundant timber tree in the southern coniferous forest (Figure 24-13c). It is also planted extensively on tree plantations. Loblolly pine is used primarily to make paper, paperboard, and fiberboard. It is also valued as a construction material and is used to make turpentine.

---

[1]Red spruce is declining and dying throughout much of its range in the Appalachians. Environmental pollution, including acid rain, is thought to be a contributing factor.

(a) Douglas fir (*Pseudotsuga menziesii*)   (b) Red spruce (*Picea rubens*)   (c) Loblolly pine (*Pinus taeda*)

Figure 24-13   Three commercially important conifers. (a) Douglas fir (*Pseudotsuga menziesii*) grows along the Pacific coast and in the Rocky Mountains. **(b)** Red spruce (*Picea rubens*), found in eastern Canada and the northeastern United States, also extends southward to the Great Smoky Mountains. **(c)** Loblolly pine (*Pinus taeda*) is widely distributed through the southeastern United States.

STUDY OUTLINE

I. Seeds are reproductively superior to spores.
   A. Each seed contains a well-developed plant embryo and a food supply.
   B. Gymnosperms and flowering plants reproduce by seeds.
II. Gymnosperms are vascular plants with "naked seeds," that is, seeds that are either totally exposed or are borne on the scales of cones.
   A. Gymnosperms are advanced in relation to ferns, in several ways.
      1. Gymnosperms produce windborne pollen.
      2. The gametophyte generation is not free-living in gymnosperms but instead is attached to and nutritionally dependent on the sporophyte.
      3. Gymnosperms are heterosporous.
   B. Pine produces two kinds of spores in separate cones.
      1. Microspores are produced in male cones, which occur in clusters at the ends of lower branches each spring. Microspores give rise to pollen grains (male gametophytes).
      2. Megaspores are produced in female cones, which are on the tree year-round, usually on the upper branches. Each megaspore develops into a female gametophyte.
      3. Female cones bear seeds after sexual reproduction (pollination and fertilization) has occurred.

III. Gymnosperms are classified into four phyla.
   A. Conifers, by far the largest group of gymnosperms, are woody plants that bear needle-like or scalelike leaves and produce their seeds in cones.
   B. Cycads are palmlike or fernlike in appearance but reproduce in a manner similar to pines. There are relatively few surviving members of this once-large phylum.
   C. *Ginkgo*, the only surviving species in its phylum, is a deciduous, dioecious tree. Female ginkgoes produce seeds with fleshy coverings directly on branches.
   D. Gnetophytes share a number of advances in relation to the rest of the gymnosperms, including the presence of more efficient water-conducting cells, called vessel elements, in their xylem.
IV. Seed plants evolved from seedless vascular plants.
   A. Progymnosperms were seedless vascular plants that possessed megaphylls and woody tissue similar to those of modern gymnosperms.
      1. Progymnosperms probably gave rise to conifers.
      2. Progymnosperms probably gave rise to seed ferns, which in turn may have given rise to cycads and possibly ginkgoes.
   B. The evolution of gnetophytes is unclear.

SELECTED KEY TERMS

cone, p. 397
conifer, p. 396
cycad, p. 401
deciduous, p. 397
dioecious, p. 401

evergreen, p. 397
ginkgo, p. 402
gnetophyte, p. 402
gymnosperm, p. 395
monoecious, p. 397

needle, p. 397
pollen grain, p. 399
pollen tube, p. 399
pollination, p. 399
progymnosperm, p. 404

resin, p. 397
resin duct, p. 397
seed, p. 395
seed fern, p. 404
sporophyll, p. 399

REVIEW QUESTIONS

1. Why are seeds such a significant evolutionary development?
2. List the four phyla of gymnosperms, and briefly describe each.
3. Are the spores produced by pine homosporous or heterosporous? Explain.
4. How does pollination occur in the gymnosperms? Fertilization?
5. List several ways in which conifers are advanced in relation to ferns.

6. What features do cycads, ginkgoes, and gnetophytes share with conifers?
7. How are conifers ecologically important?
8. What are some of the commercial uses of conifers?

THOUGHT QUESTIONS

1. Which of the following gymnosperms could be successfully grown on your campus grounds? Explain your answers. (a) Douglas fir (b) Cycad (c) *Ginkgo* (d) Red spruce
2. If you were given the wood of an unknown plant and asked to identify it as either a conifer or a gnetophyte, what would you look for? How could you identify the unknown plant if you were given a few of its leaves?

3. Describe the evolutionary changes that had to take place as ancient seedless vascular plants evolved into gymnosperms.
4. Explain how a bristlecone pine can live for thousands of years yet remain "physiologically young." (*Hint:* The answer to this question is related to secondary growth.)

SUGGESTED READINGS

Cushman, J.H., Jr., "Judge Approves Plan for Logging in Forests Where Rare Owls Live," *The New York Times*, December 22, 1994. The Clinton Plan for logging in the Pacific Northwest has been given the seal of approval by a federal judge.

Knickerbocker, B., "Beauty Fuels Growth Spurt in Northwest," *The Christian Science Monitor*, January 11, 1996. Although pockets of high unemployment still exist in the Pacific Northwest, a diversifying economy in the region is making the "jobs versus the environment" issue less of a controversy.

Poinar, G.O., "Still Life in Amber," *The Sciences*, Vol. 34, No. 2, 1993. The fossilized remains of many insects and other small animals have been preserved in amber.

THE PLANT KINGDOM: FLOWERING PLANTS

The common foxglove (*Digitalis purpurea*), a member of the snapdragon family, is an herbaceous biennial native to Europe and parts of Asia. Having been introduced widely to other parts of the world, including North America, the common foxglove is a popular garden plant today. It consists of a basal rosette of long leaves from which a flower stalk up to 152 centimeters (5 feet) tall grows. The tube-shaped flowers, which occur in an *inflorescence*, a cluster of flowers on a common axis, are about 5 centimeters (2 inches) long and droop slightly on the flower stalk. Although some varieties of common foxglove are pink or white, the flowers are usually purple and are spotted inside.

Besides being a lovely ornamental in flower gardens, the common foxglove is also the source of digitalis, an important medicine for the treatment of heart disease. Drugs produced from digitalis make the heart muscle contract strongly,

The nodding flowers of common foxglove (*Digitalis purpurea*) make it a popular garden plant. The leaves of common foxglove have medicinal value. (*Ed Reschke*)

thereby increasing the flow of blood from the heart and thus improving the body's circulation. Digitalis also helps restore regular heart beats in people suffering from congestive heart failure. The active ingredients in digitalis are cardiac glyco-

sides, lipids similar in structure to cholesterol.

The scientific investigation that led to the widespread use of digitalis to treat heart problems occurred during the 18th century in England. A physician, Dr. William Withering, noted that a woman dying from *dropsy*, a buildup of fluid in the tissues,[1] miraculously recovered after drinking a tea made from the leaves and other parts of the common foxglove plant. Over a period of many years of experimentation, Withering determined that the plant's leaves contained the most active ingredients. He then carefully tested increasing strengths of the dried, powdered leaves to determine the safest, most effective dosage. The painstaking care with which Withering carried out his research enabled physicians to administer digitalis safely and thereby save or prolong thousands of lives during the past 200 years. Had Withering been less precise in his studies, he almost certainly would have caused many deaths, because at higher doses digitalis is a powerful poison.

---

[1] *Edema*, the modern term for dropsy, is a symptom of heart disease.

The common foxglove is an appropriate introduction to the flowering plants because it provides a good example of how humans rely on plants; indeed, our survival as a species depends on them. All of our major food crops are flowering plants, including important grain crops such as rice, wheat, and corn. Woody flowering plants, such as oak, cherry, and walnut trees, provide us with valuable lumber. Other flowering plants supply us with fibers such as cotton, medicines such as digitalis, and products as diverse as cork, rubber, tobacco, coffee, and aromatic oils for perfumes. **Economic botany** is the subdiscipline of botany that deals with plants of economic importance, most of which are flowering plants.

Many earlier chapters of this book were devoted exclusively or primarily to the flowering plants. Chapters 5 (tissues and the multicellular plant body), 6 (roots), 7 (stems), 8 (leaves), and 9 (flowers, fruits, and seeds) examined the structures of flowering plants. This chapter emphasizes their life cycle and their economic significance.

LEARNING OBJECTIVES

After reading this chapter, you should be able to:

1. *Summarize the features that distinguish flowering plants from gymnosperms.*
2. *Distinguish between dicots and monocots and give specific examples of each class.*
3. *Label a diagram of the life cycle of flowering plants, compare its sporophyte and gametophyte generations, and describe double fertilization.*
4. *Trace the evolution of flowering plants from gymnosperms.*
5. *Describe the ecological and economic significance of the flowering plants.*
6. *Identify whether each of the following flowering plant families is a dicot or a monocot, and describe its distinguishing characteristics: magnolia, walnut, cactus, mustard, rose, pea, potato, grass, orchid, and agave families.*

FLOWERING PLANTS PRODUCE FLOWERS, FRUITS, AND SEEDS

Flowering plants, or **angiosperms**, are classified in the *phylum Anthophyta* (derived from the Greek *anthus*, "flower," and *phyt*, "plant").[2] Flowering plants are vascular plants that reproduce sexually by forming flowers and, after a unique double fertilization, seeds within fruits. The main way in which flowering plants differ from gymnosperms, the other group of seed plants, is that the **ovules**, the structures that contain egg cells and that develop into seeds following fertilization, are *enclosed* within an **ovary**, which later becomes a fruit (Figure 25-1 and Table 25-1). The ovary is the enlarged base of a **carpel** or the enlarged bases of a group of fused carpels. As discussed in Chapter 9, the female part of the flower is also referred to as a **pistil** and may be simple, consisting of a single carpel, or compound, consisting of several to many fused carpels.

The flowering plants are the most successful and abundant plants today. They are remarkably diverse and have adapted to almost every habitat. With at least 235,000 species, they are Earth's dominant plants. Flowering plants come in a wide variety of sizes and forms, from herbaceous violets to ivy vines to massive eucalyptus trees. Some flowering plants—tulips and roses, for example—have large, conspicuous flowers, whereas others, such as grasses and oaks, produce flowers that are small and inconspicuous.

The Two Classes of Flowering Plants Are the Monocots and Dicots

Phylum Anthophyta is divided into two classes, *Monocotyledones*, commonly called **monocots**, and *Dicotyledones*, commonly called **dicots**.[3] Examples of monocots include grasses, orchids, irises, onions, lilies, and coconut palms. The dicot class includes oaks, roses, mustards, cacti, blueberries, and sunflowers. Dicots are more diverse and include many more

---

[2]Some botanists prefer to classify flowering plants in phylum Magnoliophyta instead of phylum Anthophyta. The International Botanical Congress has not yet standardized phylum names, although the matter is currently under study.

[3]Botanists who prefer to classify flowering plants in phylum Magnoliophyta usually prefer class Magnoliopsida instead of class Dicotyledones, and class Liliopsida instead of class Monocotyledones. Like phylum names, class names have not yet been standardized by the International Botanical Congress.

TABLE 25-1   A Comparison of Gymnosperms and Flowering Plants   *Angiosperms*

Characteristic	Gymnosperms	Flowering Plants
Growth habit	Woody trees and shrubs	Woody or herbaceous
Conducting cells in xylem	Tracheids	Vessel elements and tracheids
Reproductive structures	Cones (usually) *strobili*	Flowers
Pollen grain transfer	Wind	Animals, wind, or water
Fertilization	Egg and sperm cell → zygote	Double fertilization: Egg and sperm cell → zygote 2 Polar nuclei and sperm cell → endosperm
Seeds	Exposed or borne on scales of cones	Enclosed within fruit

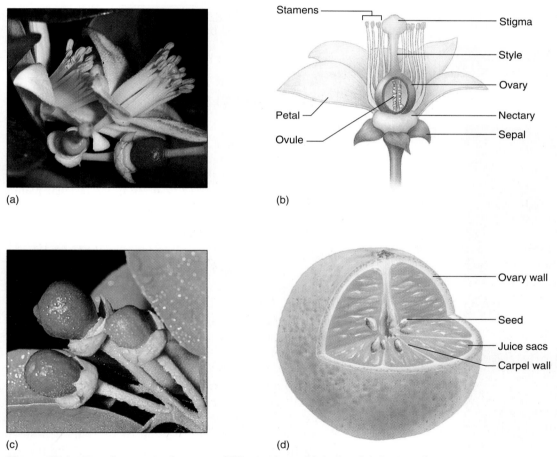

(a)

Stamens — Stigma
Style
Ovary
Petal —
Ovule —
Nectary
Sepal

(b)

(c)

Ovary wall
Seed
Juice sacs
Carpel wall

(d)

Figure 25-1   Development of orange (*Citrus sinensis*) fruits. (a) Orange flowers. Note that two of the flowers have lost their petals and stamens, revealing the stigma, style, and ovary of the pistil. **(b)** Diagram of an orange flower. Note the organ that secretes nectar, known as a nectary, in the diagram and in photo *a*. **(c)** Developing ovaries. The stigma and style have dropped off. **(d)** A cross section through an orange. Each section of an orange is a carpel; the ovary that develops into the fruit consists of several fused carpels. (a,c, *Visuals Unlimited/William J. Weber*)

(a)

(b)

Figure 25-2   There are two classes of flowering plants.
(a) Monocots such as this nodding trillium (*Trillium cernuum*) have their floral parts in threes. Note the three green sepals, three white petals, six stamens, and three stigmas, which are part of a compound pistil composed of three fused carpels.
(b) Dicots such as this *Tacitus bellus* have their floral parts in fours or fives. Note the five petals, ten stamens, and five pistils. (a, *Ed Reschke;* b, *Richard H. Gross*)

species (at least 170,000) than the monocots (at least 65,000). Figures 25-2 and 25-3 provide a comparison of some of the features of the two groups.

Most monocots are herbaceous plants with long, narrow leaves that have parallel venation (the main leaf veins run parallel to one another). The flower parts of monocots usually occur in threes or multiples of three. For example, a flower might have three sepals, three petals, six stamens, and a compound pistil consisting of three fused carpels. Monocot seeds have a single **cotyledon**, or embryonic seed leaf, and **endosperm**, nutritive tissue, is usually present in the mature seed.

Dicots may be herbaceous (like a tomato plant) or woody (like a hickory tree). Their leaves vary in shape, but they are usually broader than monocot leaves and have netted venation. Flower parts in dicots usually occur in fours or fives or multiples of four or five. Two cotyledons are present in seeds of dicots, and endosperm is often absent in the mature seed, having been absorbed by the two cotyledons.

THE LIFE CYCLE OF FLOWERING PLANTS INCLUDES DOUBLE FERTILIZATION

Flowering plants have an alternation of generations in which the sporophyte generation is larger and nutritionally independent and the gametophyte generation is microscopic in size and nutritionally dependent on the sporophyte (Figure 25-4). Like gymnosperms and certain other vascular plants, flowering plants are heterosporous and produce two kinds of spores, microspores and megaspores.

Sexual reproduction takes place in the flower, which usually consists of sepals, petals, stamens, and pistil (see Figure 9-1). Recall from Chapter 9 that the stamen, the male reproductive organ, consists of a thin stalk (the filament) and an anther, where pollen forms. The pistil consists of a stigma (where the pollen lands), a style (the neck through which the pollen tube must grow), and an ovary, which contains one or more ovules.

Each ovule within an ovary contains a megaspore mother cell that undergoes meiosis to produce four haploid megaspores. Usually, three of these megaspores disintegrate, and one develops by mitotic divisions into a female gametophyte, also called an **embryo sac**. The embryo sac consists of seven cells with eight nuclei. Six of these cells, including the egg cell, contain a single nucleus each, and a central cell has two nuclei, called **polar nuclei**. The egg and the central cell with two polar nuclei are directly involved in fertilization; the other five cells in the embryo sac apparently have no direct role in the fertilization process and later disintegrate.

Each pollen sac, or microsporangium, of the anther contains numerous microspore mother cells, each of which undergoes meiosis to form four haploid microspores. Every

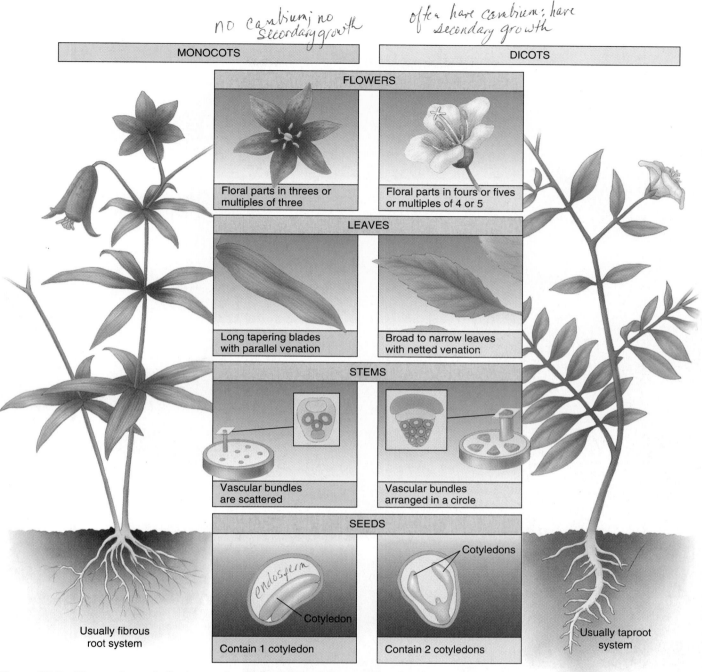

no cambium; no secondary growth

often have cambium; have secondary growth

MONOCOTS

DICOTS

FLOWERS

Floral parts in threes or multiples of three

Floral parts in fours or fives or multiples of 4 or 5

LEAVES

Long tapering blades with parallel venation

Broad to narrow leaves with netted venation

STEMS

Vascular bundles are scattered

Vascular bundles arranged in a circle

SEEDS

endosperm

Cotyledon

Cotyledons

Contain 1 cotyledon

Contain 2 cotyledons

Usually fibrous root system

Usually taproot system

Figure 25-3   Monocots and dicots can be distinguished from each other in several ways.

microspore develops into an immature male gametophyte, also called a pollen grain (Figure 25-5). Pollen grains are extremely small; each consists of two nuclei, the tube nucleus and the generative nucleus.

The anther sacs split open and begin to shed pollen grains. Pollination, the transfer of pollen from the anther to the stigma, is accomplished by a variety of agents, including wind, water, insects, and other animal pollinators. If a pollen grain lands on the stigma of a plant of the same species, it germinates; that is, the *tube nucleus* forms a pollen tube that

grows down the style and into the ovary. Next, the *generative nucleus* divides to form two nonflagellated male gametes called sperm cells. The sperm cells move down the pollen tube and are discharged into the embryo sac.

A phenomenon occurs during sexual reproduction in flowering plants that does not happen anywhere else in the living world. When the two sperm cells enter the embryo sac, *both* fuse with other cells. One sperm cell fuses with the egg, forming a diploid zygote that will grow by mitosis and develop into a multicellular embryo in the seed. The second sperm

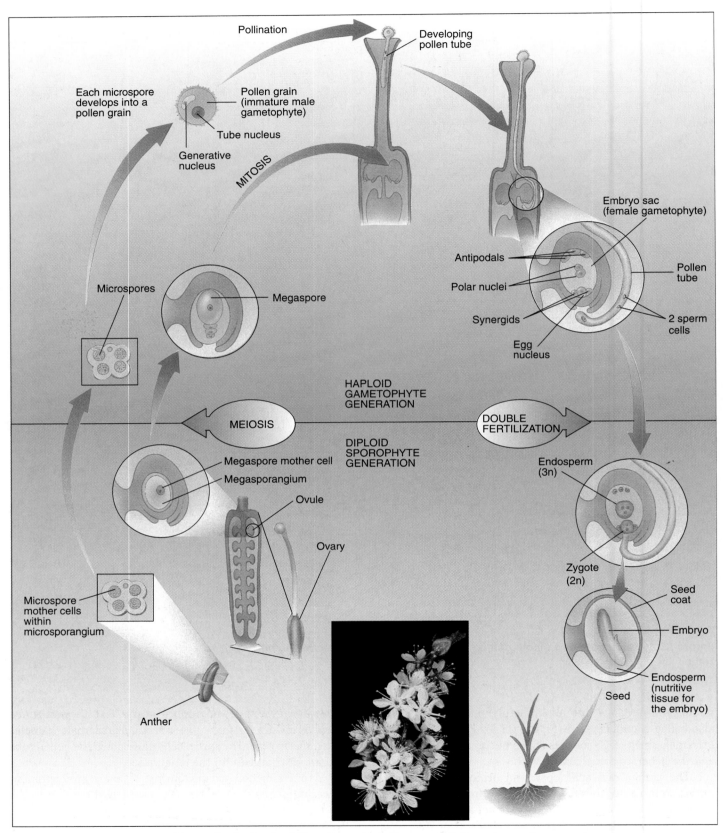

Figure 25-4   A generalized life cycle of a typical flowering plant. The most distinctive feature of the flowering plant life cycle is double fertilization. (*Visuals Unlimited/Marc Epstein*)

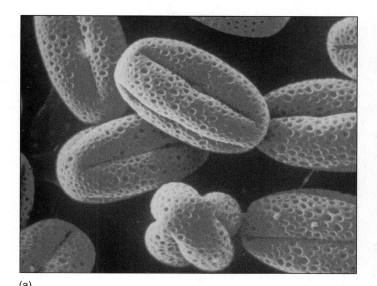

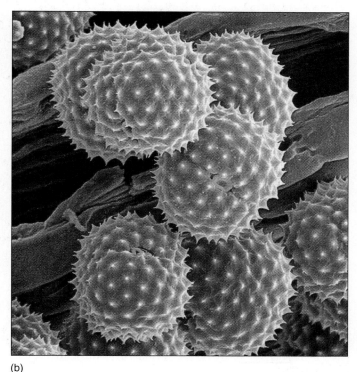

(a)

(b)

Figure 25-5   Scanning electron micrographs of representative pollen grains, many of which are so distinctive that they can be identified at the species level. (a) Lemon (*Citrus limon*) pollen. **(b)** Ragweed (*Ambrosia bidentata*) pollen. (a, *Visuals Unlimited/G. Shih and R. Kessel;* b, *David Scharf/Peter Arnold, Inc.*)

cell fuses with the two haploid polar nuclei of the central cell, forming a **triploid** ($n + n + n = 3n$) cell that grows by mitosis and develops into the *endosperm*, a nutrient tissue that nourishes the embryo. This process, which involves two separate cell fusions, is called **double fertilization** and, with one exception, is found only in flowering plants. In 1990 double fertilization in the gymnosperm *Ephedra nevadensis* was reported. It differs from double fertilization in flowering plants, however, in that an additional zygote, rather than endosperm, is produced.

SEEDS AND FRUITS DEVELOP AFTER FERTILIZATION

As a result of double fertilization and subsequent growth and development, each seed contains a young plant embryo and nutritive tissue (the endosperm), both of which are surrounded by a protective seed coat. In monocots the endosperm persists and is the main source of food in the mature seed. In most dicots the endosperm nourishes the developing embryo, which subsequently stores food in its cotyledons.

As a seed develops from an ovule following fertilization, the ovary wall surrounding it enlarges dramatically and develops into a fruit. In some instances, other tissues associated with the ovary also enlarge to form part of the fruit. Fruits serve two purposes: to protect the developing seeds from desic-

cation as they grow and mature and to aid in the dispersal of seeds. For example, animals often swallow edible fruits whole, including the seeds, which are dispersed in their droppings. Once a seed lands in a suitable place, it may germinate and develop into a mature sporophyte plant that produces flowers, and the life cycle continues as described.

Apomixis Is the Production of Seeds Without the Sexual Process

Sometimes flowering plants produce embryos in seeds without the fusion of gametes. This process is known as **apomixis**. Seed production by apomixis is a form of asexual reproduction, and—because there is no fusion of gametes—the embryo is genetically similar to the original parent. (The embryo may not be *identical* to the parent, because mutations may occur.) Apomixis has an advantage over other methods of asexual reproduction in that the seeds produced by apomixis can be dispersed by methods associated with sexual reproduction. Examples of plants that reproduce by apomixis include dandelions, citrus trees, blackberries, garlic, and certain grasses.

Mangos

MANY FEATURES OF FLOWERING PLANTS ACCOUNT FOR THEIR SUCCESS

Why are the flowering plants so successful both in terms of their ecological dominance and in terms of their great number of species? Seed production as the primary means of reproduction and dispersal, an adaptation shared with the gymnosperms, is clearly significant and provides a definite advantage over seedless vascular plants. The presence of closed carpels, which allow seeds to develop enclosed within a fruit, and the process of double fertilization increase the likelihood of reproductive success in the flowering plants.

In addition to their highly successful reproductive structures and processes, flowering plants have other advanced features that have helped them survive and flourish. For example, most flowering plants possess very efficient water-conducting cells, called vessel elements, in their xylem in addition to tracheids. In contrast, the xylem of almost all seedless vascular plants and gymnosperms consists exclusively of tracheids. Most flowering plants also have efficient carbohydrate-conducting cells, called sieve-tube members, in their phloem. With the exception of flowering plants and gnetophytes, a phylum of gymnosperms, vascular plants lack vessel elements and sieve-tube members.

The leaves of flowering plants, with their broad, expanded blades, are structured for maximum efficiency in photosynthesis. Abscission of these leaves during cold or dry spells reduces water loss and thus has enabled some flowering plants to expand into habitats that would otherwise be too harsh for survival. The stems and roots of flowering plants are often modified for food or water storage, another feature that helps flowering plants to survive in severe environments.

Probably the most crucial factor in the success of flowering plants is the overall adaptability of the sporophyte generation. This adaptability is evident in the great diversity exhibited by species of flowering plants. The cactus, for instance, with its stem that stores water, leaves (as spines) that protect against thirsty herbivorous animals, and thick waxy cuticle that reduces water loss, is remarkably well adapted for desert environments. On the other hand, the water lily is well adapted for wet environments, in part because it has air channels that provide adequate oxygen to stems and roots living in oxygen-deficient water and mud. Thus, flowering plants are remarkably adapted to their environments.

FOSSILS PROVIDE CLUES ABOUT THE EVOLUTION OF FLOWERING PLANTS

The flowering plants are the most recent group of plants to evolve. The fossil record, although incomplete, indicates that flowering plants probably descended from gymnosperms. By the middle of the Jurassic period, approximately 180 million years ago, several gymnosperm lines existed, with features reminiscent of flowering plants. Among other traits, these derived gymnosperms possessed leaves with broad, expanded blades and the first closed carpels. It is evident that beetles

were visiting these plants, and biologists have suggested that perhaps this relationship was the beginning of coevolution (mutual adaptation) between plants and their animal pollinators.

One important task facing **paleobotanists** (biologists who study fossil plants) is to determine which of the ancient gymnosperms are in the direct line of evolution leading to the flowering plants. Based on structural data, most botanists think that flowering plants arose only once—that is, that there is only one line of evolution from the gymnosperms to the flowering plants. The gnetophytes are the gymnosperm group that many botanists consider the closest living relatives of flowering plants; both structural and molecular data support this conclusion.

The oldest trace of flowering plants in the fossil record is of fossil pollen grains in Cretaceous rocks some 127 million years old. The oldest fossilized flowers are about 120 million years old. By 90 million years ago, during the late Cretaceous period, flowering plants had diversified and had begun to replace gymnosperms as Earth's dominant plants. Fossils of flowering plants outnumber those of gymnosperms and ferns in late Cretaceous deposits, indicating the rapid success of flowering plants once they appeared (Figure 25-6).

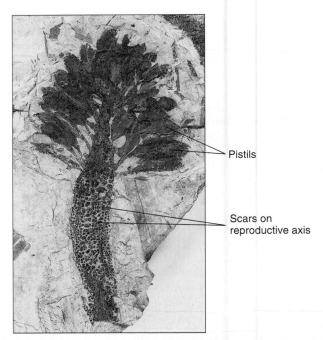

Pistils

Scars on reproductive axis

Figure 25-6   The fossil of a part of a flower of *Archaeanthus linnenbergeri*, an extinct plant that lived during the Cretaceous period, about 95 to 98 million years ago. The scars on the receptacle may have been where stamens, petals, and sepals were originally attached but abscised (fell off). Many spirally arranged pistils were still attached at the time that this flower was fossilized. (*Courtesy of D. Dilcher*)

Based on the fossil record as well as both structural and molecular data of living angiosperms, the first flowering plants are thought to have been dicots that were weedy shrubs or small herbaceous plants rather than trees. Monocots, which are not as well represented in the Cretaceous fossil record as dicots, apparently originated from dicots.

A SURVEY OF TEN FLOWERING PLANT FAMILIES SHOWS THEIR DIVERSITY AND IMPORTANCE

The flowering plants, as the principal photosynthetic organisms on land, provide other organisms with a continual supply of energy that they must have to survive. Animals require flowering plants directly or indirectly. A deer browsing on magnolia twigs during the winter is using plants directly. A garter snake that consumes a toad that ate grasshoppers that consumed leaves is using plants indirectly.

Early humans depended on flowering plants for their survival—for the food they ate, the shelters they built, the clothing they wore, and the fires they burned. Even today, humans who live in complex, highly industrialized societies still depend on flowering plants and use them in a variety of ways. Ten of the more than 300 families of flowering plants are highlighted here to show their importance to humans and to the biosphere. Seven of these—the magnolia, walnut, cac-

tus, mustard, rose, pea, and potato families—are dicots, and three—the grass, orchid, and agave families—are monocots.

The Magnolia Family Includes Ornamentals and Timber Trees

About 220 species of trees and shrubs native to temperate and tropical areas of Asia and the Americas are classified in the magnolia family (*Magnoliaceae*). Members of the magnolia family are easy to recognize because they possess simple alternate leaves and large, conspicuous flowers that contain numerous stamens and pistils. The fruit is conelike in appearance. Based on its floral structure, many biologists regard the magnolia family as the most primitive family of flowering plants alive today.

Two representative members of the magnolia family are the southern magnolia and the tuliptree (Figure 25-7). The southern magnolia, with its shiny evergreen leaves and fragrant creamy-white flowers, is a popular ornamental that is planted widely in parks and gardens. Many other species of the magnolia family are also cultivated extensively as ornamentals.

The tuliptree, also known as the yellow poplar, is the tallest of the eastern hardwoods. It is a fast-growing tree of some commercial importance, particularly in the eastern part of the United States. The soft, fine-grained wood of this forest tree is used for a variety of products from furniture to toys. The tuliptree is also planted as an ornamental. Songbirds and squirrels eat its seeds.

(a)

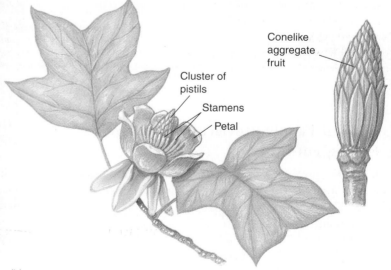

Cluster of pistils

Stamens

Petal

Conelike aggregate fruit

(b)

Figure 25-7   The magnolia family. (a) Southern magnolia (*Magnolia grandiflora*) is highly prized for its beauty. **(b)** The tuliptree (*Liriodendron tulipifera*) is a valuable ornamental and is also harvested for its wood. (*Marion Lobstein*)

Plants AND People

The Medicinal Value of Flowering Plants

From extracts of cherry and horehound for cough medicines to chemical compounds in periwinkle and autumn crocus for cancer therapy, derivatives of roots, stems, leaves, and flowers play important roles in the treatment of illness and disease. About 25 percent of all prescription medicines contain one or more active ingredients extracted from plants.

Many of the plant-produced chemicals with medicinal properties are *alkaloids*, bitter-tasting organic compounds that contain nitrogen. For example, the rosy periwinkle produces two alkaloids, vinblastine and vincristine, which are used to treat two kinds of cancer—Hodgkin's disease, which generally afflicts young adults, and childhood leukemia. Other important alkaloids with medicinal value include quinine, morphine, and reserpine. Quinine, an antimalarial drug, comes from the bark of yellow cinchona. Morphine, used medically to relieve pain, is extracted from the opium poppy.

An ethnobotanist consults with a Tirio Indian in Suriname about the uses of a rainforest plant. (*Courtesy of Mark Plotkin, Conservation International*)

The Walnut Family Provides Nuts and Wood for Making Furniture

About 50 to 60 species of walnuts, hickories, and pecans belong to the walnut family *(Juglandaceae)*. All are deciduous trees native to the temperate and subtropical areas of Asia and North and South America (Figure 25-8). The leaves are pinnately compound and arranged alternately on the stem. Members of the walnut family bear separate male and female flowers on the same plant. The male flowers, which occur in **catkins**, clusters of tiny flowers borne on a pendulous stalk, produce pollen that is carried by the wind to female flowers. The female flowers are small, lack petals, and usually occur individually on small, erect stalks.

The walnut family is ecologically and economically important for its edible nuts and its wood. In forests, wildlife ranging from opossums to wild turkeys feed on the nuts. English walnuts—native to Asia, not England as their common name implies—and pecans, native to North America, are widely cultivated for their nuts. The edible portion of the "nut" is the greatly enlarged cotyledons of the embryo within the seed. In addition to being eaten directly, walnut and pecan seeds are pressed to yield oil, which is an important ingredient in many soaps and cosmetics.

Walnut and pecan trees produce a strong, rich, fine-grained wood that is used to make furniture, interior woodwork, and even the interiors of Rolls Royces. The wood of black walnut, native to eastern North America, is among the

The Indian snakeroot is the source of reserpine, an alkaloid used to treat hypertension and some psychiatric disorders.

Various medicinal plants that we employ today have been used for centuries in folk medicine. The modern medical establishment has learned to respect reports about medicinal plants in the traditions of indigenous peoples, because they provide clues about which plants to test. Studies show that plants identified as useful by shamans (native healers) and traditional plant users are up to 60 percent more likely to have medicinal value than plants that are randomly collected. **Ethnobotany**, the study of the traditional uses of plants by indigenous people, helps pharmaceutical companies identify medicinal plants. Unfortunately, much of this knowledge is disappearing as tribes are exposed to modern ways and their traditions are rapidly forgotten.

Only about 5000 of the 235,000 species of flowering plants have been investigated for their medicinal value. It is likely that the remaining 230,000 species contain other valuable alkaloids with medicinal properties. Unfortunately, we may never have the opportunity to find out, because many of the wild plant species that have medicinal potential are themselves threatened as human activities destroy wildlife habitats.

The rosy periwinkle, for example, is native to Madagascar, an extremely poor island nation with a human population that is rapidly increasing. The 13.3 million people living in Madagascar, with a per capita gross national product of $250 (U.S. dollars), have not benefited from the $180 million earned annually by the sale of vinblastine and vincristine from the rosy periwinkle. It is estimated that 80 percent of the 10,000 plant species on Madagascar are found nowhere else in the world.

Yet during the past 45 years, most of the forests of Madagascar have been destroyed by impoverished people, who clear the forests for agriculture.

New programs seek to distribute proceeds from sales and patents directly to indigenous peoples. The Suriname Biodiversity Prospecting Initiative, announced in 1993 by Conservation International, is the first such agreement of its kind. This 5-year enterprise pays money to the people of Suriname, a small country in South America, for new drugs derived from the plants that they identify in their rain forests (see figure). With support from the U.S. government, universities, and non-profit organizations, this initiative is the first-ever partnership between indigenous people and a private pharmaceutical company.

most valuable hardwood trees in North America; as a result of its commercial value, black walnut has been cut extensively.

The Cactus Family Is Important for Its Ornamentals

The cactus family (*Cactaceae*) consists of over 2000 species of perennial herbs, vines, shrubs, and small trees that are native to North and South America, although some were introduced from the Americas and subsequently became naturalized in China, India, and the Mediterranean. The greatest number and variety of cacti occur in Mexico; in the United States cacti are most abundant in Texas, Arizona, and New Mexico. Most cacti live in deserts and are succulent plants—

that is, they store water in their stems or leaves. However, a few cacti, such as the Christmas cactus of Brazil, live as epiphytes on trees in tropical rain forests (Figure 25-9a).

Most cacti have **spines**, modified leaves borne on a more or less succulent stem.[4] The stem functions not only for water storage—to enable the plant to survive periods of drought—but as the main organ of photosynthesis (Figure 25-9b). In certain cacti the succulent stems are prominently ribbed during periods of drought, but during rainy periods

---

[4]Other flowering plant families also have succulent members. Some of these, like the euphorbias, closely resemble cacti in appearance and are often confused with them.

the stems swell and the ribs fill in, much like an extended accordion. Cacti roots vary depending on the growth habits of the particular species; many have an extensive fibrous root system that covers a large area horizontally but may extend vertically only a few inches. Each flower contains numerous sepals, petals, and stamens and a compound pistil composed of two to many fused carpels (Figure 25-9c); after fertilization, berries containing many tiny black or brown seeds develop.

Because of the fascinating adaptations of cacti that enable them to survive in deserts, botanists are interested in how cacti evolved. It is thought that the ancestors of cacti may have been vinelike plants with large leaves; such plants would have required a moist environment. As the climate changed and became much drier, many species of plants may not have been able to adapt and undoubtedly became extinct. The cacti probably survived this drying trend through a gradual reduction in leaf size and the evolution of succulent stems.

Cacti are of economic importance primarily as ornamentals, although some (for example, prickly pear) are cultivated for their fleshy edible fruits. Cacti have wide appeal as houseplants, not only for their attractive flowers and varied spine and rib formations, but also because they can tolerate a lot of neglect (such as forgetting to water them). Some cactus species are in danger of extinction because they have been overcollected for "desert landscaping."

The Mustard Family Contains Many Important Food Crops

The mustard family (*Brassicaceae* or, alternatively, *Cruciferae*) consists of about 3000 species, many of which are important food crops, ornamentals, and weeds. Members of the mustard family are found worldwide, although the greatest number of species are native to temperate regions of Asia and the Mediterranean.

(a)

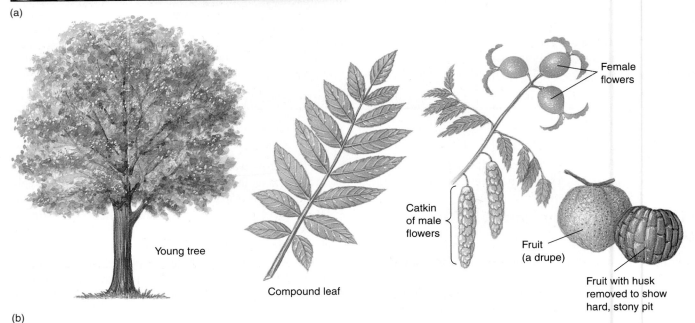

(b)

Figure 25-8   The walnut family. (a) A pecan tree (*Carya illinoinensis*) bearing many fruits. The pecan "nut" is actually a seed. **(b)** The pinnately compound leaves and green fruits of black walnut (*Juglans nigra*) make it an easy tree to recognize. (*Horticultural Photography, Corvallis, Oregon*)

(a)

(b)

Stamens

Stigma

Petals

(c)

Figure 25-9   The cactus family. (a) The Christmas cactus (*Schlumbergera bridgesii*) is popular as a houseplant because of its magenta flowers, which appear around Christmas. Native to Brazil where it grows as an epiphyte in rain forests, the Christmas cactus has numerous jointed, spineless stems. The individual stem segments resemble leaves. **(b)** The barrel cactus (*Echinocactus grusonii*) is a short cylindrical cactus with prominent ribs. The "barrel" can grow to 0.9 meters (3 feet) thick. Native to Mexico and the southwestern United States, where it is found on rocky hillsides, the barrel cactus forms yellow flowers at the top of the stem. **(c)** Cactus flowers contain numerous sepals, petals, and stamens. The compound pistil consists of two to many fused carpels. In this photo of a prickly pear cactus (*Opuntia humifusa*), the petals, stamens, and stigma are evident. (*a, Visuals Unlimited/M. Long; b, Carlyn Iverson; c, Marion Lobstein*)

Most species are annual or perennial herbs, although a few are shrublike. The leaves are simple and arranged alternately on the stem. The flowers are usually arranged in an inflorescence. Each flower almost always contains four sepals, four petals, six stamens, and a compound pistil composed of two fused carpels; the petals are arranged in the shape of a cross, or cruciform—hence the family name *Cruciferae* (Figure 25-10a).

The mustard family, particularly its species that are food crops, is of considerable economic importance. One of the most important members of the mustard family is *Brassica oleracea*, which includes cabbage, cauliflower, broccoli, kale, kohlrabi, collards, and brussels sprouts (see Chapter 16). Turnips, radishes, watercress, and Chinese cabbage are also in the family, and the condiments mustard and horseradish are made from members of the mustard family. Oil is extracted

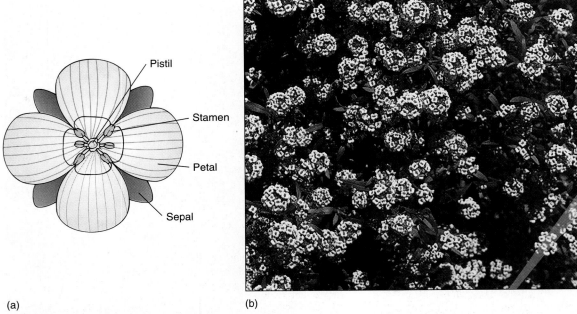

(a)   (b)

Figure 25-10   The mustard family. (a) A face view of a typical mustard flower, showing the four petals arranged in the shape of a cross. **(b)** Sweet alyssum (*Lobularia maritima*) is a popular garden plant used as a ground cover. Note the similarity in appearance between sweet alyssum flowers and mustard. (*Visuals Unlimited/John D. Cunningham*)

from the seeds of other species, such as oilseed rape. The mustard family also includes many ornamentals, such as sweet alyssum (Figure 25-10b) and honesty. Shepherd's purse and weedy mustards are bothersome weeds.

The Rose Family Is Important for Its Fruits and Ornamentals

The rose family (*Rosaceae*), which includes apples, cherries, strawberries, and other fruits as well as roses and many other ornamentals, is a large and important family of about 3370 species. It includes woody trees, shrubs, and herbaceous plants distributed worldwide, although most members of the rose family are concentrated in northern temperate regions.

Leaves of members of the rose family are alternate and can be either simple or compound. The flowers, which are pollinated by insects, are large and showy and typically have five sepals, five petals, numerous stamens, and one to many pistils (Figure 25-11a,b). Wild roses always have five petals, and the numerous petals of cultivated roses adhere to the five-fold rule discussed earlier in this chapter; that is, the petal number is some multiple of five.

Many important fruiting plants in the temperate region are members of the rose family, including apples, pears, peaches, nectarines, cherries, apricots, plums, almonds, strawberries, raspberries, and blackberries (Figure 25-11c). The apple, one of the most commercially important fruits in

the world, is eaten raw, baked, roasted, sauteed, and candied. Apples are also used to make applesauce, apple butter, jelly, vinegar, and beverages such as apple cider and apple juice.

Many members of the rose family are cultivated as ornamentals because of the beauty and fragrance of their flowers; flowering apples, cherries, and plums are important ornamentals, as is the world's most widely cultivated flower, the rose. Thought to have been the first flower that humans cultivated, roses may have grown in ancient Greek gardens as long ago as the fifth century BC. Roses were also important to the Romans, the early Chinese, and the Moors. The rose was a symbol of the early Christian church and the emblem of a number of European kings. It was cultivated for its medicinal value during the Middle Ages and is still grown in parts of Europe and Asia for its fragrant oil, from which perfumes are made. Throughout history, humans have glorified the rose in literature, music, and art.

The Pea Family Provides Important Food Crops

The pea family (*Fabaceae* or, alternatively, *Leguminosae*), which includes as many as 17,000 species throughout the world, contains food and forage crops of major importance. The pea family members are herbs, vines, shrubs, or trees whose leaves are usually pinnately compound with an alter-

(a)

(b)

(c)

Figure 25-11   The rose family. Flowers of **(a)** Mexican plum (*Prunus mexicana*) and **(b)** strawberry (*Fragaria* sp) are characterized by five sepals, five petals, and numerous stamens. Mexican plum has a single pistil, whereas strawberry has numerous pistils. **(c)** Apples, apricots, blackberries, cherries, nectarines, peaches, plums, raspberries, and strawberries are all members of the rose family. (*a,b, James Mauseth, University of Texas; c, Carlyn Iverson*)

nate leaf arrangement. The roots of most members of the pea family contain nodules in which nitrogen-fixing bacteria convert atmospheric nitrogen to nitrogenous compounds that the plant can use. The presence of these nodules enables members of the pea family to thrive in relatively poor soil.

Flowers occur in erect or pendulous clusters. The five sepals and five petals in each flower sometimes vary in size and degree of fusion with one another, making the flower distinctly irregular in shape (Figure 25-12a,b). Each flower usually contains ten stamens and a simple pistil. The fruit that forms after reproduction occurs is a legume, a dry fruit that splits open along two seams (Figure 25-12c). (The term *legume* refers both to a fruit type and to any of the members of the pea family, all of which produce this fruit type.)

The most important members of the pea family are food and forage crops. The seeds and pods of many species—such as garden peas, chick peas, green beans and kidney beans,

broad beans, lima beans, soybeans, peanuts, and lentils—are particularly nutritious because they are rich in proteins and oil. The pea family is second in importance as a source of human food after the grass family, which contains the cereal grains. Every major civilization since the dawn of agriculture has incorporated a grain-legume combination into its cuisine for balanced nutrition. Examples of such pairings are rice and soybeans, corn and beans, and barley and lentils. Each of these combinations provides the proper complement of essential amino acids that human cells cannot synthesize.

Important forage crops—that is, crops on which livestock such as beef and dairy cattle feed—include red clover and alfalfa (Figure 25-12d). The importance of forage crops cannot be overestimated; in the United States, more land is used to grow forage crops than to grow all other crops combined. Members of the pea family that are forage crops also produce nectar, which bees use to make honey.

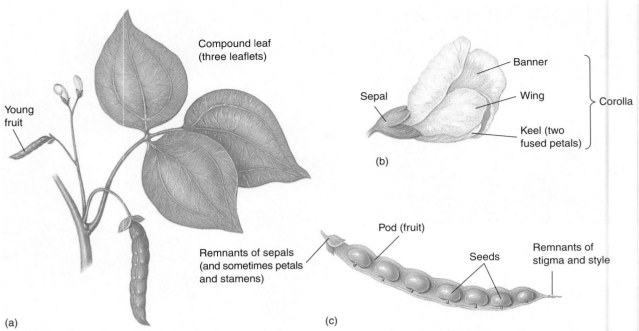

Compound leaf
(three leaflets)

Young
fruit

Banner

Sepal

Wing

Corolla

Keel (two
fused petals)

(b)

Pod (fruit)

Remnants of sepals
(and sometimes petals
and stamens)

Seeds

Remnants of
stigma and style

(a)   (c)

Figure 25-12   The pea family. (a,b,c) The common bean
(*Phaseolus vulgaris*). **(a)** Part of a bean plant, showing flowers
and developing fruits. **(b)** Close-up of a bean flower, showing
its irregular shape. **(c)** The mature fruit, a legume, opened to
reveal the seeds. **(d)** Alfalfa (*Medicago sativa*), one of the
most valuable forage crops, is grown in many parts of the
world. Because alfalfa produces more protein per acre than
any other forage crop, it is a nutritious food for livestock.
(*Carlyn Iverson*)

(d)

Important ornamentals in the pea family include sweet
peas, lupines, and wisteria. Other species within the pea family are sources of timber, fibers, dyes, tannins, and many other
products.

The Potato Family Is Important for Its Food Crops and Drugs

The 2000 to 3000 species in the potato family (*Solanaceae*)
are found worldwide, with many species centered in Central
and South America and in Australia. The potato family consists primarily of herbs, although a few members are shrubs
or small trees. The potato family is of great importance to
humans because it contains many food crops, ornamentals,
and medicinal and poisonous plants.

Leaves of members of the potato family are simple although
highly varied, with alternate leaf attachment. The flowers are
usually regular and typically contain five sepals, five petals,
five stamens, and a compound pistil composed of two fused
carpels (Figure 25-13a). The fruit that develops from the ovary

is either a berry or a capsule and contains a large number of
seeds.

Some of the most important fruit and vegetable crops—
such as white potatoes, tomatoes, green and red peppers, chili
peppers, pimentos, eggplants, and paprika—occur in the
potato family. Notable ornamentals include the petunia,
grown for its attractive flowers, and the Chinese lantern plant,
grown for its ornamental fruits (Figure 25-13b).

Many members of the potato family contain toxic *alkaloids* that have medicinal value at low concentrations. Deadly

(a)   (b)

Figure 25-13   The potato family. (a) Potato (*Solanum tuberosum*) flowers. Note the five fused petals. The yellow stamens and stigma/style are also visible. **(b)** The Chinese lantern plant (*Physalis alkekengi*) is an ornamental in the potato family that is cultivated for its fruits, each of which is surrounded by an orange-red calyx of fused sepals. (a, *Edgar E. Webber;* b, *David Cavagnaro*)

nightshade, also called belladonna, is the source of the alkaloid atropine, a substance used to treat poisoning from nerve gas and certain pesticides. In addition, drugs known as tropane alkaloids are extracted from belladonna and used to treat heart irregularities. Jimsonweed, which contains the hallucinogenic alkaloid scopolamine, was ingested by native Americans to produce visions during important rituals such as puberty rites, but its overall toxicity prevented its widespread use; chewing the seeds can be fatal. In the introduction to Chapter 8, we discussed the toxicity of tobacco, a member of the potato family that is widely cultivated for its leaves, which are smoked, chewed, or snuffed.

The Grass Family Is the Most Important Family of Flowering Plants

The grass family (*Poaceae* or, alternatively, *Gramineae*), which contains about 9000 species distributed throughout the world, is the most important family of flowering plants because it includes the cereal grains such as rice (see Chapter 4 introduction), wheat, oats, and corn. Grasses are monocots and are most often annual or perennial herbs; bamboos,

which are somewhat "woody," are a notable exception.[5] Each grass leaf has parallel venation and consists of a long, narrow blade and a sheath that wraps around the stem above the nodes (Figure 25-14a). The stem may be erect or creeping and is usually hollow except at the nodes, where it is always solid. Grasses have fibrous root systems.

Flowers of grasses occur in inflorescences and are highly modified. Petals are quite reduced or absent (because grasses are wind-pollinated, they have little need for petals), and each flower typically contains three stamens and a compound pistil consisting of two or three fused carpels (Figure 25-14a). The fruit is usually a seedlike grain.

Grasses, which grow in virtually all of Earth's major terrestrial ecosystems, are of major ecological significance. In the great grasslands of the world, such as prairies and savannas, grasses are the principal vegetation. Many wild animals, from large hoofed mammals such as bison to tiny rodents, depend on grasses for food.

---

[5]The hardness of bamboo is due to an abundance of fibers, not to the production of secondary tissues as in woody dicots and gymnosperms.

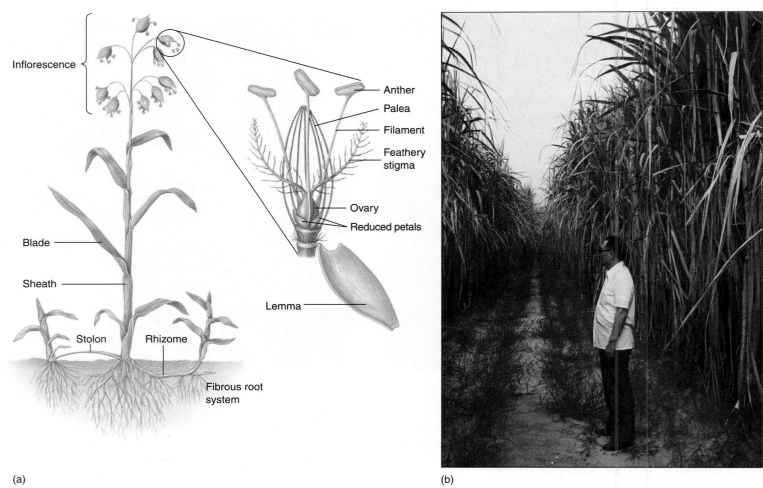

Figure 25-14   The grass family. (a) The growth habit of a representative grass plant. Grass flowers, which are wind-pollinated, are highly modified. Note, for example, the long, feathery stigmas, adapted to efficiently catch windborne pollen. **(b)** Sugarcane (*Saccharum officinarum*) supplies most of the world's sugar. Sugarcane plants photosynthesize very efficiently and produce more calories (of carbohydrates) per acre than any other crop. (*Visuals Unlimited/Sylvan Wittwer*)

The grass family supplies humans with many important foods. In addition to corn, wheat, and rice, grasses provide other important grains, such as barley, rye, and oats. Sugarcane, also a member of the grass family, supplies much of the world's table sugar (Figure 25-14b). Grasses also provide forage and fodder for domesticated animals, thereby feeding humans indirectly. In many parts of the world, grasses provide shelter, notably bamboo for buildings and thatch (grass leaves) for roofing. Ryegrasses, Bermuda grass, bluegrasses, and fescues are some of the grasses that are widely cultivated as turf for lawns and golf courses; other grasses are grown as ornamentals in flower gardens. Some grasses are significant weeds.

The Orchid Family Exhibits Great Variation in Flower Structure

The orchid family (*Orchidaceae*), which occurs worldwide, contains at least 18,000 species and is one of the largest families of flowering plants. All orchids are perennial herbs. About half of these species are terrestrial and grow in soil, and the other half are epiphytic and grow attached to other plants or rocky surfaces. One fourth to one third of all orchid species are in danger of extinction as a result of the destruction of their natural habitats and overcollection by orchid hobbyists.

The roots of terrestrial orchids are like those of other plants, whereas aerial roots of epiphytic orchids are thick and

(a)

(b)

(c)

Figure 25-15   The orchid family. (a) The moth orchid (*Phalaenopsis* hybrid) has photosynthetic aerial roots. **(b)** A *Cymbidium* hybrid has pseudobulbs that function as storage organs. **(c)** Orchid flowers, such as *Dendrobium* sp, are distinctive in that the third petal forms a lip. (a, *John Arnaldi;* b,c, *Carlyn Iverson*)

covered with **velamen**, a layer of dead cells that absorbs and holds water. The aerial roots of many orchids are unusual in that they are green and photosynthesize (Figure 25-15a).

Some orchids have a single upright stem that bears leaves and flowers. Other orchids have **pseudobulbs** (false bulbs), which are thickened stems at or above the soil level that arise from a horizontal rhizome and serve as a storage organ for food and water (Figure 25-15b). Each pseudobulb bears one or more leaves and eventually a flower stalk; after it flowers, the pseudobulb slowly dies. Orchid leaves, which are alternate and often occur in two rows, are usually fleshy. Each leaf is simple and encircles the stem as a sheath.

Orchid flowers vary extensively, occurring in many different colors, shapes, and sizes, although their overall pattern is unmistakable. Each flower consists of three petal-like sepals and three petals, the third petal forming a lip that is different in color and shape from the rest of the sepals and petals (Figure 25-15c). The reproductive structures are borne in the center of the flower, with both male and female parts fused into a single structure called a *column*.

In orchids the pollen is aggregated into masses called **pollinia** (sing. *pollinium*) that are carried intact to other flowers by insects, hummingbirds, bats, and sometimes even frogs. Some orchids produce a quick-drying glue that adheres the pollinium to the body of the pollinator. After fertilization occurs, numerous very tiny seeds develop within the fruit, a podlike *capsule*.

Orchids, which are monocots, are widely cultivated by florists and orchid fanciers for their fascinating flowers, and breeders have developed many thousands of hybrids by crossing different species. Vanilla flavoring is an extract from the immature capsules of an orchid native to tropical America.

The Agave Family Is Best Known for Its Ornamentals

The agave family (*Agavaceae*) contains about 600 species and is found throughout tropical and semitropical areas, particularly in arid regions. Plants in this family are perennial herbs or woody shrubs and trees. The leaves of plants in the agave family are crowded into a rosette at the base of the stem and are stiff, fleshy, narrow, and sharp-pointed (Figure 25-16). Many agaves are succulents.

Agaves bear their flowers in inflorescences. Each flower consists of three petal-like sepals and three petals, joined to form a tube; six stamens; and a compound pistil composed of three fused carpels. After fertilization, numerous seeds develop in the fruit (a capsule or berry).

Many agaves are cultivated as ornamentals, both outdoors in semiarid regions and indoors as houseplants. The century plant, quite popular as an outdoor ornamental, is so named because it was mistakenly thought to flower only once in a century. Actually, the century plant can flower once it is 10 years old or so, after which it dies. At maturity some century plants, such as *Agave franzosinii*, are quite large; each leaf may reach 2.4 meters (8 feet) in length.

In addition to serving as important ornamentals, some species in the agave family—for example, bowstring hemp and sisal hemp—produce leaf fibers from which twine, ropes, mats, and fishing nets are made. Several species are used to make alcoholic beverages; in Mexico and Central America, the sweet sap of plants in the *Agave* genus is collected and fermented to make pulque, which can be drunk directly or distilled to make tequila.

Figure 25-16   The agave family. The century plant (*Agave* sp) is a succulent with sword-shaped leaves arranged as a rosette around a short stem. Note the floral stalk. Shown is *Agave shawii*. (*Visuals Unlimited/R. Gustafson*)

S T U D Y   O U T L I N E

I. Flowering plants (angiosperms) are Earth's dominant plants and form flowers that function in sexual reproduction.
   A. Flowering plants possess vascular tissues, as do ferns and gymnosperms, and produce seeds, as do gymnosperms.
   B. Unlike those of gymnosperms, the ovules of flowering plants are enclosed within an ovary. After fertilization, the ovules become seeds and the ovary develops into a fruit.
II. Monocots and dicots are the two classes of flowering plants.
   A. Monocots have floral parts in threes or multiples of three, and their seeds contain one cotyledon; the nutritive tissue in their mature seeds is endosperm.
   B. Dicots have floral parts in fours or fives or multiples of four or five, and their seeds contain two cotyledons; the nutritive tissue in their mature seeds is usually in the cotyledons.
III. Flowering plants have an alternation of generations in which the sporophyte generation is larger and nutritionally independent, and the gametophyte generation is reduced to only a few microscopic cells.
   A. Animals transfer the pollen of many flowering plants.
   B. Double fertilization results in the formation of a diploid zygote and triploid endosperm tissue.

IV. It is thought that flowering plants arose from ancient gymnosperms that possessed advanced features, such as closed carpels and leaves with broad, expanded blades.
   A. Flowering plants probably arose only once.
   B. The first flowering plants were probably weedy shrubs or small herbaceous plants.
   C. Monocots are thought to have evolved from dicots.
V. Flowering plants have been classified into more than 300 families.
   A. Magnolias, walnuts, cacti, mustards, roses, peas, and potatoes are representative dicot families.
      1. The magnolia family is important as ornamentals and as a source of timber. Examples include the southern magnolia and the tuliptree.
      2. The walnut family provides nuts for food and wood for making furniture. Examples include English walnut, black walnut, and pecan.
      3. The cactus family is important as ornamentals. Examples include prickly pear and Christmas cactus.
      4. The mustard family contains many important food crops. Examples include cabbage, broccoli, cauliflower, turnip, and mustard.
   5. The rose family is commercially important for its fruits and ornamentals. Examples are apple, pear, plum, cherry, apricot, peach, strawberry, raspberry, and rose.
   6. The pea family includes important food crops. Examples are garden pea, chick pea, green bean, soybean, lima bean, peanut, red clover, and alfalfa.
   7. The potato family is important for its food crops and chemicals used as drugs. Examples include potato, tomato, green pepper, eggplant, petunia, and deadly nightshade (belladonna).
   B. Grasses, orchids, and agaves are representative monocot families.
      1. The grass family is the most important family of flowering plants. Examples include rice, wheat, corn, oats, barley, rye, sugarcane, and bamboo.
      2. The orchid family is one of the largest families of flowering plants and contains a greater variety of flowers than any other family. An example is the vanilla orchid.
      3. The agave family is best known for its ornamentals. Examples include the century plant, sisal hemp, and bowstring hemp.

SELECTED KEY TERMS

angiosperm, p. 410
apomixis, p. 415
carpel, p. 410
catkin, p. 418
cotyledon, p. 412
dicot, p. 410

double fertilization, p. 415
economic botany, p. 410
embryo sac, p. 412
endosperm, p. 412
ethnobotany, p. 419
flowering plant, p. 410

monocot, p. 410
ovary, p. 410
ovule, p. 410
paleobotany, p. 416
pistil, p. 410

polar nucleus, p. 412
pollinium, p. 425
pseudobulb, p. 425
spine, p. 419
triploid, p. 415

REVIEW QUESTIONS

1. Name at least three features that distinguish flowering plants from other plants.
2. Explain how reproduction in flowering plants differs from that in gymnosperms, including the fertilization process, whether the seeds are exposed or enclosed, and whether endosperm is produced.
3. How does pollination occur in flowering plants? Fertilization?
4. What special advantage does asexual reproduction by apomixis have over other kinds of asexual reproduction (such as corms and bulbs, discussed in Chapter 7)?
5. What are the two classes of flowering plants, and how can one distinguish between them?
6. Describe the evolutionary changes that had to take place as ancient gymnosperms evolved into flowering plants.
7. How are flowering plants ecologically important?
8. How are flowering plants important to humans?
9. Specify the family in which each of the following flowering plants is classified: raspberry, pecan, barrel cactus, red clover, jimsonweed, century plant, wheat, tuliptree, turnip, garden pea, tomato, sisal hemp.

THOUGHT QUESTIONS

1. Why are conifers and flowering plants placed in different phyla?
2. In contrast to the cones of gymnosperms, which are either male or female, most flowers of flowering plants contain both male and female reproductive structures. Explain how bisexual flowers might be advantageous for flowering plants.
3. Water lilies are adapted to wet environments, whereas cacti and agaves are adapted to dry environments. For each plant, specify at least two adaptations that might have been selected for (or against) during the course of its evolution.

4. You are given a plant that you have never seen before (see figure). Determine whether it is a dicot or a monocot. What are the features that helped you make this determination?

(*Carlyn Iverson*)

SUGGESTED READINGS

Hughes, M.S., "Potato, Potahto—Either Way You Say It, They're a Peel," *Smithsonian*, October 1991. A history of the potato and its importance to humanity.

Plotkin, M., *Tales of a Shaman's Apprentice*, New York, Viking, 1993. Mark Plotkin, America's most famous ethnobotanist, presents a fascinating account of his experiences with indigenous people of the rain forests, who shared their uses of forest plants with him.

Roach, M., "Secrets of the Shamans," *Discover*, November 1993. Pharmaceutical companies are scouring the world, looking for plants that may provide new medicines.

Stix, G., "Village Pharmacy," *Scientific American*, May 1992. A short article about the neem tree, a flowering plant that contains compounds that show promise as insecticides, medicines, and birth control agents.

Thompson, J., "Cotton, King of Fibers," *National Geographic*, Vol. 185, No. 6, June 1994. Examines the importance of cotton, both today and in the past.

Weissmann, G., "Aspirin," *Scientific American*, January 1991. The story of aspirin, which is produced from derivatives of the bark of the willow tree, and how it works.

PART

PLANT ECOLOGY

CHAPTER 26

ECOSYSTEMS

The Chesapeake Bay on the east coast of the United States is the world's richest *estuary,* a semi-enclosed body of water where fresh water drains into the ocean. Biological diversity and productivity abound in the Chesapeake Bay, as they do wherever fresh water and saltwater form a *salinity gradient,* a gradual change from unsalty fresh water to salty ocean water. The salinity gradient in the Chesapeake Bay results in three distinct marsh communities: freshwater marshes at the head of the bay, brackish (moderately salty) marshes in the middle bay region, and salt marshes at the mouth of the bay. Each community has its own characteristic plants and other organisms.

Sail to one of the salt-marsh islands in the Chesapeake Bay, such as South Marsh Island, and you can explore a salt-marsh community

Cordgrass (*Spartina alterniflora*) growing in a salt marsh. (*Connie Toops*)

that is relatively unaffected by humans. A salt marsh presents a monotonous view—miles and miles of flooded meadows of cordgrass (*Spartina* spp). Cordgrass is a perennial grass with a horizontal rhizome and a stout, hollow stem. Its tough, flat leaves, which grow up to 41 centimeters (16 inches) long, taper to a narrow point. Cordgrass has short, intermediate, and tall forms; the tall form grows in deeper water, whereas the short form is found in shallow water.

High salinities (although not as high as that of ocean water) and twice-daily tidal inundations create a challenging environment to which only a few plants, such as cordgrass, have adapted. Those that have adapted to the salt-marsh environment

thrive—in part because essential minerals such as nitrates and phosphates, which drain into the marsh from the land, promote their rapid growth. Cordgrass and the microscopic algae and cyanobacteria that live on and around it capture energy from the sun and use it to manufacture carbohydrates by photosynthesis.

Many small aquatic animals depend on the cordgrass for food, shelter, and hiding places (to avoid being eaten). Cordgrass is eaten directly by some animals, and when the cordgrass dies, its remains (called detritus) provide nutrients for many other inhabitants of the salt marsh and the bay. The marsh crab, for example, is a common inhabitant that eats cordgrass as well as small animals and detritus. The meadow vole is a small rodent that constructs its nest of cordgrass on the ground above the high-tide zone and eats cordgrass and insects. The marsh periwinkle, a snail that moves along the cordgrass skimming off attached algae for its food, climbs up the cordgrass to avoid becoming prey for larger marsh animals such as northern diamondback terrapins, a species of turtle found in brackish environments.

Cordgrass is an appropriate plant with which to begin our discussion of ecology because, like other plants, it is essential to the survival of many organisms. The marsh crab, meadow vole, and marsh periwinkle all interact with and depend on cordgrass. Cordgrass plays a pivotal role in the salt-marsh community because it contributes to the flow of energy and the cycling of materials in this ecosystem.

After reading this chapter, you should be able to:

1. *Distinguish among population, community, ecosystem, biosphere, and ecosphere.*
2. *Describe what is meant by an organism's ecological niche.*
3. *Characterize producers, consumers, and decomposers and describe the primary role of each.*
4. *Define competition, predation, and symbiosis, and distinguish among mutualism, commensalism, and parasitism.*
5. *Diagram the carbon, nitrogen, and hydrologic cycles.*
6. *Summarize the concept of energy flow through ecosystems.*
7. *Define ecological succession, and distinguish between primary succession and secondary succession.*

ECOLOGISTS STUDY THE HIGHEST LEVELS OF BIOLOGICAL ORGANIZATION

The concept of ecology was first developed in the 19th century by Ernst Haeckel, who also created its name—*eco* from the Greek word for "house" and *logy* from the Greek word for "study." Thus, ecology literally means the study of one's house. In the study of ecology, nature is something like a great estate in which the physical environment (Earth's climate, soils, rocks, water, and atmosphere) and organisms interact in an immense and complicated web of relationships. **Ecology**, then, is the study of the interactions among different organisms and between organisms and their physical environments.

The focus of ecology can be local and very specific or global and quite generalized, depending on the viewpoint of the scientist. Thus, one ecologist might determine the temperature or light requirements of a single species of oak, another might study all the organisms that live in a forest where the oak is found, and another might examine how matter cycles between the forest and surrounding communities.

How does ecology fit into the organization of the biological world? In Chapter 1 we noted that one of the characteristics of life is its high degree of organization. Of all the levels of organization in the biological world, ecologists are most interested in the levels above that of the individual organism. Individual organisms are arranged into **populations**, members of the same species that live together in the same area at the same time. A population ecologist, for example, might study a population of pines or a population of marsh grass.

Populations are organized into **communities**, which are all the populations of organisms that live and interact within an area. A community ecologist might study how organisms interact with one another—including who eats whom—in a deciduous-forest community or in an alpine-meadow community. **Ecosystem**, a more inclusive term than community, encompasses a community and its environment. Thus, an ecosystem includes not only all the interactions among the organisms of a community but the interactions between organisms and their physical environment. An ecosystem ecologist, for example, might examine how temperature, light, precipitation, and soil affect the organisms living in a salt-marsh community or a rainforest community (Figure 26-1).

Figure 26-1   Many ecosystem studies require elaborate equipment. Here, a blimp tows in parts of an observation station at the top of the rain forest to study the bird, insect, and small plant communities in the forest canopy. (*Raphael Gaillarde/Gamma Liaison*)

All the communities on the Earth are organized into the **biosphere**, which includes all of the Earth's organisms. In addition to the biosphere, there are divisions of the Earth's physical environment on which organisms depend: the atmosphere, hydrosphere, and lithosphere. The *atmosphere* is the gaseous envelope surrounding the Earth, the *hydrosphere* is the Earth's supply of water (liquid and frozen, fresh and salty), and the *lithosphere* is the soil and rock of the Earth's crust. The term **ecosphere** encompasses the biosphere and its interactions with the atmosphere, hydrosphere, and lithosphere. Ecologists who study the biosphere or ecosphere examine the complex interrelationships among the Earth's atmosphere, land, water, and organisms.

Ecology, the broadest field within the biological sciences, has links to every other biological discipline and to other disciplines—geology and earth science, for example—that are not traditionally part of biology. Since humans are biological organisms, all human activities, including economics and politics, have profound ecological implications.

An Organism's Role in the Ecosystem Is Its Ecological Niche

Every organism has its own ecological **niche**, its role within the structure and function of an ecosystem. A complete description of an organism's ecological niche involves all dimensions and aspects of the physical, chemical, and biological factors that the organism needs to survive, to remain healthy, and to reproduce. Among other things, the niche includes the physical surroundings in which an organism lives (its **habitat**), how it obtains nourishment, which organisms consume it, which organisms it competes with, and how it interacts with and is influenced by the abiotic (nonliving) components of its environment, such as light, temperature, and moisture. The niche, then, represents the totality of an organism's adaptations, its use of resources, and the lifestyle to which it is fitted.

Ecosystems Contain Producers, Consumers, and Decomposers

On the basis of how they get nourishment, the organisms of an ecosystem can be divided into three categories: producers, consumers, and decomposers. Most ecosystems contain representatives of all three groups, which interact extensively with one another.

Sunlight is the source of energy that powers almost all life processes on Earth. **Producers**, also called **autotrophs**, perform photosynthesis, the biological process in which light energy is captured and transformed into the chemical energy of organic molecules, such as carbohydrates, which are man-

ufactured from carbon dioxide and water. By incorporating the organic molecules that they manufacture into their own cells, producers make their bodies or body parts potential food sources for other organisms. Plants are the most significant producers on land; in aquatic environments algae and certain types of bacteria are important producers. In a salt-marsh community, cordgrass, algae, and photosynthetic bacteria are all important producers.

Animals are **consumers**; that is, they are **heterotrophs** that use the tissues of other plant and animal organisms as a source of food energy and body-building materials. Consumers that eat producers are called **primary consumers**, which usually means that they are exclusively *herbivores* (plant eaters). Cattle and deer are examples of primary consumers, as is the marsh periwinkle in the salt-marsh community. **Secondary consumers** eat primary consumers and include flesh-eating *carnivores,* which consume other animals exclusively. Lions and tigers are examples of carnivores, as is the northern diamondback terrapin in the salt-marsh community. Other consumers, called *omnivores,* eat a variety of plant and animal organisms. Bears, pigs, and humans are examples of omnivores; the meadow vole, which eats cordgrass and insects in the salt-marsh community, is also an omnivore.

Some consumers, called **detritus feeders**, or *detritivores,* ingest **detritus**, which is dead organic matter that includes animal carcasses, leaf litter, and feces. Detritivores, such as snails, crabs, clams, and worms, are especially abundant in aquatic habitats, where they burrow in the bottom muck and consume the organic matter that collects there. Earthworms are terrestrial (land-dwelling) detritus feeders, as are termites, beetles, snails, and millipedes.

Many consumers do not fit readily into a single category of herbivore, carnivore, omnivore, or detritivore. To some degree, these organisms modify their food preferences as the need arises. For example, in the salt-marsh community, marsh crabs are both predators and detritus feeders. Furthermore, some organisms change their food preferences over their lifetimes. Tadpoles are primary consumers, but adult frogs are secondary consumers.

Decomposers, also called **saprotrophs**, are microbial heterotrophs that break down organic material and use the decomposition products as a source of energy. Decomposers, which complete the work of detritus feeders in consuming dead organisms and waste products, typically release simple inorganic molecules, such as carbon dioxide and mineral salts, that can then be reused by producers. Bacteria and fungi are important examples of decomposers. For example, sugar-metabolizing fungi are first to invade dead wood; they consume the wood's simple carbohydrates. When they exhaust those, other fungi, often aided by termites and bacteria, complete the decomposition of the wood by digesting cellulose, a complex carbohydrate that is the main constituent of wood.

Environment

Agricultural Ecosystems and Pests

Agricultural ecosystems differ from natural ecosystems in that a farmer's fields are managed to maximize the growth of one species—corn, for example—and minimize the presence of other species. Under cultivation, the crop is watered, fertilized, and protected as much as possible from pests, including weeds, insects, and disease organisms. These low-diversity fields are **monocultures**—areas planted in one crop.

One of the disadvantages of monocultures is that they are more prone to damage by pests than are natural ecosystems. A cultivated field is a very simple ecosystem. In contrast, natural ecosystems are extremely complex and contain many species, including predators and parasites of pest species, and plant species that are not suitable as food for pests.

A monoculture reduces the dangers and accidents that might befall a pest as it searches for food. In a forest a Colorado potato beetle would have a hard time finding anything to eat, but a 500-hectare potato field is like a big banquet table set just for this pest. It eats, prospers, and reproduces. With few natural predators (such as the ground beetle) and plenty of food present, the pest population thrives and grows. Increasing numbers of Colorado potato beetles voraciously consume the potato leaves, reducing the plants' ability to produce large tubers for harvest.

Toxic chemicals called **pesticides** are the most common means of controlling pests in agriculture. Pesticides reduce the amount of a crop lost through competition with weeds, consumption by insects, and diseases caused by plant pathogens. Agriculture is the sector that uses the most pesticides in the United States—approximately 74 percent of the estimated 0.43 million metric tons (0.48 million tons) used each year. Farmers in the United States spend $4.1 billion annually on pesticide treatment for crops. Pesticide use is usually justified economically, in that farmers save an estimated $3 to $5 in crops for every $1 they invest in pesticides.

Although pesticides have their economic benefits, they are accompanied by a number of environmental and human health problems. For one thing, many pest species develop a genetic resistance to pesticides after repeated exposure to them. Also, pesticides rarely affect the pest species alone, and the balance of nature, including predator-prey relationships, is upset. Certain pesticides concentrate at higher levels of the food web, where large predators are vulnerable.

Humans who apply and work with pesticides may be at risk of pesticide poisoning, usually from a short-term exposure to a high concentration of pesticide, and cancer, which may result from long-term exposure to low concentrations of pesticide. In addition, people who eat traces of pesticide on food are concerned about its long-term effects.

Fortunately, pesticides are not the only weapon in the arsenal, and we have many alternative ways to control pests. Sometimes routine agricultural practices can be altered in such a way that a pest is adversely affected. Mulching the ground with a thick layer of clean straw is a cultivation method that helps control the Colorado potato beetle, for example. Crop rotation (for instance, planting a crop other than potatoes the year after potatoes are grown) also helps control Colorado potato beetles, because the adult beetles spend the winter in the soil and emerge to find a less desirable food source. A combination of varied pest control methods in agriculture, often including a limited use of pesticides, is known as **integrated pest management.**

Natural communities contain a balanced representation of producers, consumers, and decomposers (see Plants and the Environment: Agricultural Ecosystems and Pests to learn about an *unbalanced* ecosystem). Producers and decomposers are necessary for the long-term survival of any ecosystem. Through photosynthesis, producers provide both food and oxygen for the rest of the community. Decomposers are also indispensable, because without them, dead organisms and waste products would accumulate indefinitely. Without decomposers, essential elements such as potassium, nitrogen, and phosphorus would remain permanently in dead organisms and organic compounds and therefore would be unavailable for use by new generations of organisms. Consumers also play an important role in communities by maintaining a balance between producers and decomposers.

ORGANISMS INTERACT IN A VARIETY OF WAYS

No species exists independent of other organisms. The producers, consumers, and decomposers of a community interact with one another in a variety of complex ways, and each forms associations with other organisms. Competition, predation, and symbiosis are important interactions that occur among species in an ecosystem.

Competition for the Same Resource Occurs Between Organisms

In **competition**, two or more organisms simultaneously require a single resource, which is usually in limited supply. The resources for which plants commonly compete include water, light, soil minerals, and growing space. For example, consider two plant species growing in a desert. In such an environment, competition is largely for water, and the species whose root growth results in the greatest acquisition of water has an advantage. The species that cannot obtain as much water must be able to survive with limited growth or it dies. Different combinations of structural and physiological traits interact to determine how well one plant can compete against another in a given environment.

Competition occurs both within a species and between species. Competition between two or more individuals belonging to the *same* species, known as **intraspecific competition**, is a major factor that limits population size. Competition between two or more individuals of *different* species, known as **interspecific competition**, also affects population size and sometimes results in the local extinction of one or more competing species.

In Predation, One Organism Feeds on Another Organism

Predation is the consumption of one species, the *prey,* by another, the *predator.* It includes both herbivores eating plants and carnivores eating other animals. Predation has resulted in an "arms race," with the evolution of predator strategies (more efficient ways to catch prey) and prey strategies (better ways to escape the predator). A predator that is more efficient at catching prey exerts a strong selective force on that prey. Over time, the prey may evolve some sort of countermeasure. In turn, the countermeasure acquired by the prey acts as a strong selective force on the predator. This interdependent

evolution between two interacting species is known as **coevolution** (see Chapter 9).

Plants, of course, cannot escape predators by fleeing, but they possess a number of adaptations that protect them from being eaten. The presence of spines, thorns, tough leathery leaves, or even thick wax on leaves discourages foraging herbivores from grazing. Other plant adaptations involve an array of protective chemicals that are unpalatable or even toxic to herbivores. The active ingredients in such plants as marijuana, opium poppy, tobacco, and peyote cactus, for example, are thought to discourage the foraging of herbivores.

Milkweeds are an excellent example of the biochemical coevolution between plants and herbivores. Milkweeds produce alkaloids and cardiac glycosides, chemicals that are poisonous to all animals except a small group of insects (Figure 26-2). During the course of evolution, these insects acquired the ability to either tolerate or metabolize the milkweed toxins. As a result, they can eat milkweeds without being poisoned. They avoid competition from other herbivorous insects, since few other insects can eat milkweed leaves. Predators also learn to avoid milkweed-tolerant insects, which accumulate the toxins in their tissues and are usually brightly colored to

Figure 26-2   The common milkweed (*Asclepias syriaca*) is protected by its toxic chemicals. Its leaves are poisonous to most herbivores except monarch caterpillars (*shown*) and a few other insects. (*Patti Murray*)

announce that fact. The black, white, and yellow banded cater-pillar of the monarch butterfly is an example of a milkweed feeder.

Symbiosis Is a Close Biological Association Between Two Species

Symbiosis is an intimate relationship or association between individuals of two or more species. The partners of a symbiotic relationship, called **symbionts**, may either benefit from, be unaffected by, or be harmed by the relationship. The thousands, or even millions, of symbiotic associations in nature are all products of coevolution, and they may be classified into several different types: mutualism, commensalism, and parasitism.

In mutualism, benefits are shared

Mutualism is a symbiotic relationship in which both partners benefit. For instance, mycorrhizae are mutualistic associations between fungi and the roots of almost all plants. The fungus absorbs phosphorus and other essential minerals from the soil and provides them to the plant, and the plant provides the fungus with carbohydrates produced by photosynthesis. Plants not only grow more vigorously in the presence of mycorrhizae but are better able to tolerate environmental stresses such as drought and high soil temperatures (see Figure 6-14).

A remarkable example of mutualism occurs in the rain forests of South and Central America between ants or wasps and flowering vines. Vines that produce bright red flowers are usually pollinated by hummingbirds, which obtain a reward of nectar from the blossoms. Some insects, however, try to rob the flower of its nectar without performing the important task of pollination. That is where the ants or wasps come into the picture. These insects patrol the plant and attack any would-be nectar robber (Figure 26-3). In exchange for their vigilance, the insect guards are rewarded with nectar, produced in the leaves.

Commensalism is taking without harming

Commensalism is a type of symbiosis in which one organism benefits and the other is neither harmed nor helped. One example of commensalism is the relationship between a rain-forest tree and its **epiphytes**, smaller plants that grow on other plants (Figure 26-4). Spanish moss, orchids, and bromeliads are examples of epiphytes. Most epiphytes anchor themselves onto tree branches but do not obtain nutrients or water directly from the trees. Their location enables them to obtain adequate light, rainwater dripping down the branches, and required minerals, which are washed out of the trees' leaves by rainfall. Thus, the epiphyte benefits from the association while the tree remains largely unaffected.

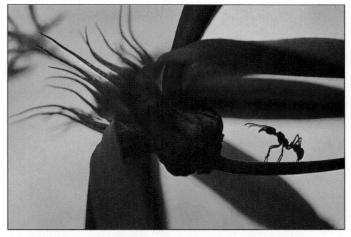

Figure 26-3   Mutualism. An ant guards the nectar of a passionflower (*Passiflora* sp) blossom against nectar-robbing thieves. The nectar is thus reserved for its intended recipients, pollen-transporting hummingbirds. The ant benefits from this association because it is rewarded with nectar produced by special glands in the leaves. (*Michael and Patricia Fogden*)

Parasitism is taking at another's expense

Parasitism is a symbiotic relationship in which one member, the **parasite**, benefits and the other, the **host**, is adversely affected. The parasite obtains nourishment from its host, and although a parasite may weaken its host, it rarely kills it. Many parasites do not cause disease, but some do. When a parasite causes disease and sometimes the death of a host, it is called a **pathogen**.

The mistletoes are some of the best-known parasitic seed plants, mainly because of their popularity at Christmastime.

Figure 26-4   Commensalism. Epiphytes are small plants that grow on larger plants. They are particularly common in the humid tropics, although a few epiphytes even occur in deserts. (*Carlyn Iverson*)

Figure 26-5 Parasitism. In this fascinating example of parasitism, a fungus (*Puccinia monoica*) parasitizes a North American plant called rock cress (*Arabis holboellii*), causing it to change its growth and produce, with the fungus's help, fake flowers. The "flowers" are so realistic that they attract insect pollinators, which help the fungus (instead of the plant) reproduce. (*Photograph by B.A. Roy, reprinted by permission from* Nature, *Vol. 362, cover and page 57*)

About 1300 species of mistletoes are known to parasitize conifer and hardwood trees, particularly in the tropics. Mistletoes anchor themselves to their hosts by means of adventitious roots that penetrate and branch through the hosts' tissues (see Figure 6-11). Mistletoes are "water robbers"; that is, they take water and minerals from their hosts, but not carbohydrates produced by photosynthesis.

One of the most fascinating examples of parasitism involves the fungus *Puccinia monoica,* which parasitizes rock cress (*Arabis holboellii*), a plant widely distributed across the northern part of North America. When *Puccinia* infects rock cress, it doesn't kill the plant, but it changes the plant's growth pattern. An infected plant grows much taller than normal and produces a cluster of leaves at its top. The fungal mycelium, which is bright yellow, grows on and eventually covers these leaves, giving them the appearance of buttercup flowers (Figure 26-5). The fungus also secretes a sugary solution and produces a strong scent, imitating the nectar and aroma of flowers. This elaborate mimicry increases the chances of successful fungal reproduction by attracting insects to the fungal "flowers." Sexual reproduction in *Puccinia* requires the union of nuclei from two different mating types. Insects are attracted to the "flowers," where they eat the sugary syrup. As the insects feed, pieces of the fungus cling to their bodies. Flying from one *Puccinia* "flower" to another, the insects distribute complementary mating types from fungus to fungus over a broad area. Rock cress pays for this parasitic relationship with altered growth and an early death.

MATTER CYCLES THROUGH ECOSYSTEMS

Matter, the material of which organisms are composed, cycles from the living world to the nonliving physical environment and back again in what are called **biogeochemical cycles**.

The Earth is essentially a *closed system,* a system from which matter cannot escape. The materials used by organisms cannot be "lost," although they can end up in locations, such as the ocean floor, that are outside the reach of most organisms. However, materials are usually reused and are often recycled both within and among ecosystems. Three different biogeochemical cycles of matter—carbon, nitrogen, and water—are particularly important to organisms.

Carbon Dioxide Is the Pivotal Molecule of the Carbon Cycle

Carbon must be available to organisms because proteins, carbohydrates, and other molecules that are essential to life contain carbon. Carbon is present in the atmosphere as a gas, carbon dioxide (CO_2), which makes up approximately 0.03 percent of the atmosphere. It is present in the ocean in a dissolved form and is also found in rocks such as limestone. Carbon cycles between the abiotic environment, including the atmosphere, and organisms (Figure 26-6).

During photosynthesis, plants remove CO_2 from the air and fix, or incorporate, it into complex organic molecules such as sugar. Thus, photosynthesis incorporates carbon from the abiotic environment into the biological compounds that make up an ecosystem's producers. These compounds are usually used as fuel for cellular respiration (the process that releases energy from organic compounds) by the producer that made them, by a consumer that eats the producer, or by a decomposer that breaks down the remains of the producer or consumer. Because the end products of aerobic respiration are CO_2, water, and energy for biological work, CO_2 is returned to the atmosphere by the process of respiration.

Sometimes the carbon in biological molecules isn't cycled back to the abiotic environment for some time. For example, a lot of carbon is stored in the wood of trees, where it may stay for several hundred years. In addition, millions of years ago, vast coal beds formed from the bodies of ancient trees that did not decay fully before they were buried. Similarly, the carbon-containing remains of unicellular marine organisms

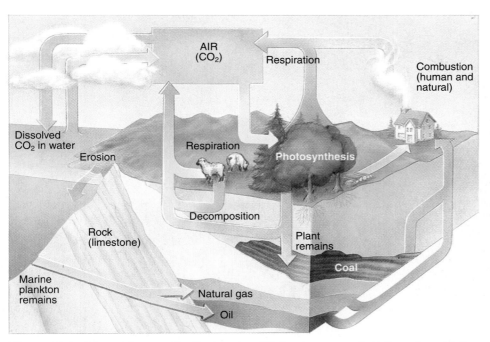

Figure 26-6   The carbon cycle. Carbon (as CO_2) enters organisms from the abiotic environment when plants and other producers photosynthesize. Carbon returns to the environment when organisms, including decomposers, respire. Combustion and erosion also return carbon to the abiotic environment. Fossil fuels, which are carbon-containing compounds formed from the remains of ancient organisms, and the carbon in limestone rock and marine animal shells may take millions of years to cycle back to living organisms.

that accumulated in the geological past probably gave rise to the underground deposits of oil and natural gas.

Coal, oil, and natural gas, called **fossil fuels** because they formed from the remains of ancient organisms, are vast depositories of carbon compounds that were the end products of photosynthesis millions of years ago. The carbon in coal, oil, natural gas, and wood may be returned to the atmosphere by the process of burning, or *combustion*. In combustion, organic molecules are rapidly oxidized (combined with oxygen) and converted to CO_2 and water with an accompanying release of heat and light.

Scientists think that most of the carbon that leaves the carbon cycle for millions of years is incorporated into the shells of marine organisms. When those organisms die, their shells sink to the ocean floor and are covered by sediments. Such shells accumulate to form seabed deposits hundreds of meters thick, which eventually are cemented together to become a sedimentary rock called limestone. The Earth's crust is dynamically active, and over millions of years, sedimentary rock on the bottom of the sea floor may be lifted and become land surfaces. For example, some cliffs in the Canadian Rockies are composed of limestone, indicating that that region was once beneath the ocean. When limestone is exposed by the process of geological uplift, it slowly erodes (wears away) through chemical and physical weathering processes, and its

carbon returns to the water and atmosphere, where it is available to participate in the carbon cycle again.

In the carbon cycle, then, (1) photosynthesis removes carbon from the abiotic environment and incorporates it into biological molecules, and (2) respiration, combustion, and erosion return carbon to the water and atmosphere of the abiotic environment.

Bacteria Are Essential to the Nitrogen Cycle

Nitrogen is crucial to life because it is an essential part of proteins, which are important structural components of cells and enzymes, and of nucleic acids. At first glance it would appear that a shortage of nitrogen for organisms is impossible: the Earth's atmosphere is about 80 percent nitrogen gas (N_2), a two-atom molecule. However, molecular nitrogen is so stable that it does not readily combine with other elements. Therefore, organisms cannot use nitrogen gas from the atmosphere to manufacture their proteins and nucleic acids. Nitrogen gas must be broken apart into individual nitrogen atoms before it can combine with other elements to form molecules.

Five steps occur in the nitrogen cycle (Figure 26-7): (1) nitrogen fixation, (2) nitrification, (3) assimilation, (4) ammonification, and (5) denitrification. Bacteria perform all of these steps except assimilation.

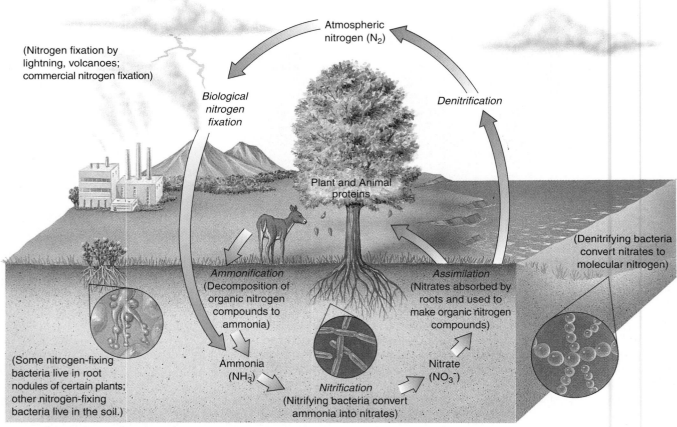

(Nitrogen fixation by lightning, volcanoes; commercial nitrogen fixation)

Atmospheric nitrogen (N₂)

Biological nitrogen fixation

Denitrification

Plant and Animal proteins

(Denitrifying bacteria convert nitrates to molecular nitrogen)

Ammonification (Decomposition of organic nitrogen compounds to ammonia)

Assimilation (Nitrates absorbed by roots and used to make organic nitrogen compounds)

(Some nitrogen-fixing bacteria live in root nodules of certain plants; other nitrogen-fixing bacteria live in the soil.)

Ammonia (NH₃)

Nitrification (Nitrifying bacteria convert ammonia into nitrates)

Nitrate (NO₃⁻)

Figure 26-7   The nitrogen cycle consists of five steps. (1) Nitrogen-fixing bacteria, including cyanobacteria, convert atmospheric nitrogen (N_2) to ammonia (NH_3). (2) Bacteria in the soil, known as nitrifying bacteria, convert ammonia to nitrate (NO_3^-). Nitrate is the main form of nitrogen absorbed by plants. (3) Plants assimilate nitrate when they produce proteins and nucleic acids; then animals eat plant proteins and produce animal proteins as a part of assimilation. (4) When plants and animals die, the nitrogen compounds in their remains are broken down by ammonifying bacteria. One of the products of this decomposition is ammonia. (5) Nitrogen is returned to the atmosphere by denitrifying bacteria, which convert nitrate to molecular nitrogen.

The first step in the nitrogen cycle, **nitrogen fixation**, involves the conversion of gaseous nitrogen (N_2) to ammonia (NH_3). This process is called nitrogen fixation because nitrogen is fixed (incorporated) into a form that organisms can use. Although considerable nitrogen is also fixed by combustion, volcanic action, and lightning discharges as well as by industrial processes, most nitrogen fixation is biological. Biological nitrogen fixation is carried out by nitrogen-fixing bacteria, including cyanobacteria, in soil and aquatic environments.

Nitrogen-fixing bacteria employ an enzyme called **nitrogenase** to break up molecular nitrogen and combine it with hydrogen. Because nitrogenase functions only in the absence of oxygen, the bacteria that use nitrogenase must insulate the enzyme from oxygen by some means. Some nitrogen-fixing

bacteria live beneath layers of oxygen-excluding slime on the roots of certain plants. But the most important nitrogen-fixing bacteria, *Rhizobium* spp, live in special swellings, or **nodules**, on the roots of legumes, such as beans or peas, and some woody plants (Figure 26-8a). The relationship between *Rhizobium* and its host plants is mutualistic: the bacteria receive carbohydrates from the plant, and the plant receives nitrogen in a form that it can use to make proteins and nucleic acids.

In aquatic habitats, most nitrogen fixation is performed by cyanobacteria. Some filamentous cyanobacteria have special oxygen-excluding cells called **heterocysts** that fix nitrogen (Figure 26-8b). Some water ferns have cavities in which cyanobacteria live, somewhat as *Rhizobium* lives in the root

(a)

Heterocysts

(b)

Figure 26-8   Organisms and nitrogen fixation. (a) Root nodules of a garden pea (*Pisum sativum*), which is a legume. Mutualistic *Rhizobium* bacteria live in these nodules on energy derived from sugars provided by their legume host. The bacteria fix nitrogen, some of which is used by the host plant. The ultimate death and decay of both partners enriches the soil with the nitrogen they have brought into chemical combination. **(b)** Many cyanobacteria fix nitrogen. Shown is *Anabaena,* a cyanobacterium with specialized cells called heterocysts where nitrogen fixation occurs. (a, *Dwight R. Kuhn;* b, *Dennis Drenner*)

nodules of legumes. Other cyanobacteria fix nitrogen in symbiotic association with certain plants or as the photosynthetic partners of certain lichens.

The second step of the nitrogen cycle is **nitrification**, the conversion of ammonia (NH_3) to nitrate (NO_3^-). Nitrification is a two-step process accomplished by aerobic soil bacteria. First the soil bacteria *Nitrosomonas* and *Nitrococcus* convert ammonia to nitrite (NO_2^-). Then the soil bacterium *Nitrobacter* oxidizes nitrite to nitrate. The process of nitrification furnishes these bacteria, called nitrifying bacteria, with energy.

In the third step, called **assimilation**, roots absorb either nitrate (NO_3^-) or ammonia (NH_3) that was formed by nitrogen fixation and nitrification, and incorporate the nitrogen in these molecules into plant proteins and nucleic acids. When animals consume plant tissues, they assimilate nitrogen as well, by ingesting the nitrogen compounds of plants and converting them to animal compounds. As a matter of fact, *all* organisms assimilate nitrogen.

The fourth step is **ammonification,** which is the conversion of biological nitrogen compounds to ammonia. Ammonification begins when organisms produce nitrogen-containing waste products. Land animals such as mammals and amphibians produce urea in urine. Other land animals such as reptiles, birds, and insects produce uric acid in their wastes. These substances, along with the nitrogen compounds in dead organisms, are decomposed, releasing the nitrogen into the abiotic environment as ammonia (NH_3). The bacteria that perform ammonification in both the soil and aquatic environments are called ammonifying bacteria. The ammonia produced by ammonification enters the nitrogen cycle and is available for the processes of nitrification and assimilation. As a matter of fact, most available nitrogen in the soil derives from the recycling of organic nitrogen by ammonification.

The fifth and final step of the nitrogen cycle is **denitrification**, which is the reduction of nitrate (NO_3^-) to gaseous nitrogen (N_2). Denitrifying bacteria reverse the actions of nitrogen-fixing and nitrifying bacteria; that is, denitrifying bacteria return nitrogen to the atmosphere as nitrogen gas. Denitrifying bacteria are anaerobic: they prefer to live and grow where there is little or no free oxygen. For example, some denitrifying bacteria are found deep in the soil near the water table, an environment that is nearly oxygen-free.

Water Circulates in the Hydrologic Cycle

Water continuously circulates from the ocean to the atmosphere to the land and back to the ocean, providing us with a renewable supply of purified water on land. This complex cycle, known as the **hydrologic cycle**, results in a balance between water in the ocean, on the land, and in the atmosphere (Figure 26-9).

Water moves from the atmosphere to the land and the ocean in the form of precipitation, such as rain, snow, sleet, or hail. Water then either evaporates from land and reenters the atmosphere directly or flows in rivers and streams to

coastal estuaries, where fresh water meets the ocean. The movement of water from land to the ocean is called **runoff**. Water also seeps downward in the soil to become **groundwater**. Groundwater supplies water to the soil, to streams and rivers, and to plants. Ultimately, however, the water that falls on land from the atmosphere makes its way back to the ocean.

Regardless of its physical form (solid, liquid, or vapor) or location, every molecule of water eventually moves through the hydrologic cycle. Tremendous quantities of water are cycled annually between the Earth and its atmosphere. The amount of water entering the atmosphere each year is estimated at 400,000 cubic kilometers (95,000 cubic miles). Ap-

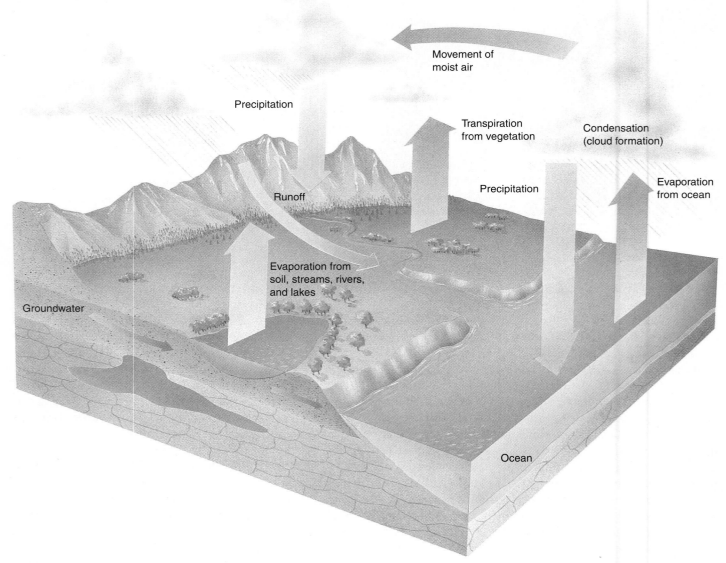

Figure 26-9   The hydrologic cycle. Water cycles from the ocean to the atmosphere to the land and back to the ocean. Although some water molecules are unavailable for thousands of years (those locked up in polar ice, for example), all water molecules eventually cycle through the hydrologic cycle.

proximately three fourths of this water reenters the ocean directly as precipitation over water; the remaining amount falls on land.

How do living organisms fit into the hydrologic cycle? Water makes up a large part of the mass of most organisms and is the solvent for most biological reactions. Water is also important because it supplies, through photosynthesis, the oxygen in the atmosphere. This oxygen, which is critical to the survival of aerobic organisms, is converted back to water during the biological process of cellular respiration. Another biological process, **transpiration**, is involved in the hydrologic cycle. Transpiration, the loss of water vapor from the aerial parts of plants, returns water to the atmosphere. Roughly 99 percent of the water absorbed from the soil by a plant is transported to the leaves, where it is lost by transpiration.

ENERGY FLOWS THROUGH ECOSYSTEMS IN A ONE-WAY DIRECTION

The passage of energy in a linear, or one-way, direction through an ecosystem is known as **energy flow** (Figure 26-10). Energy enters an ecosystem as the radiant energy of sunlight, some of which is trapped by plants during the process of photosynthesis. This energy, now in chemical form, is stored in the bonds of organic molecules such as glucose. When the molecules are broken apart by cellular respiration, the energy becomes available to do work such as tissue repair, production of body heat, or reproduction. As the work is accomplished, the energy escapes the organism and disperses into

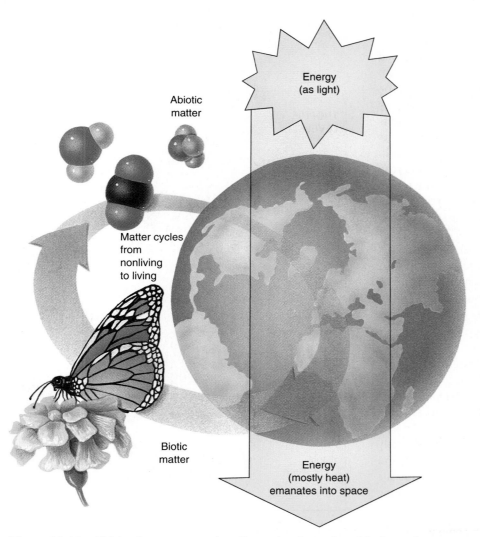

Figure 26-10   Although matter continually cycles from the abiotic environment to the organisms of ecosystems, energy flows one way through an ecosystem.

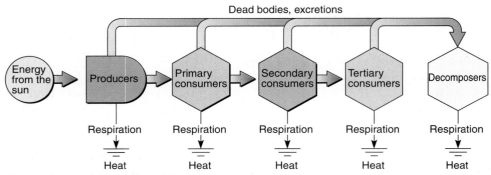

Figure 26-11   Energy flows in one direction through a food chain. Energy from an external source (the sun) enters a food chain and exits as a heat loss. Much of the energy acquired by a given trophic level (by primary consumers, for example) is used for metabolic purposes and is therefore unavailable to the next trophic level (in this example, secondary consumers).

the environment as low-quality heat. Ultimately, this heat energy radiates into space. Thus, once energy has been used by organisms, it becomes unavailable for reuse (see also the discussion of the second law of thermodynamics in Chapter 2).

In an ecosystem, energy flow can be traced through **food chains**, in which chemical energy passes from one organism to the next in a sequence (Figure 26-11). Producers form the beginning of the food chain by capturing the sun's energy through photosynthesis. Herbivores and omnivores eat plants to obtain the chemical energy of the producers' molecules as well as building materials from which to construct tissues. In turn, carnivores and omnivores consume herbivores and reap the energy stored in the herbivores' molecules. At the end of a food chain are decomposers; through cellular respiration, they break down organic molecules in the carcasses and body wastes of all other members of the food chain.

The most important thing to remember about energy flow in food chains is that it is linear, or unidirectional. That is, stored chemical energy can move from one organism to the next *as long as it is not used.* After an individual uses that energy to do work, however, it is unavailable for any other organism because it has changed into heat energy that disperses in the environment.

Each level in a food chain is called a **trophic level** (Figure 26-12). The first trophic level is made up of producers (organisms that photosynthesize), the second trophic level is primary consumers (herbivores), the third trophic level is secondary consumers (carnivores), and so on.

Food chains as simple as the one just described rarely occur in nature, since few organisms eat just one other kind of organism. More typically, the flow of energy and materials through ecosystems takes place in accordance with a range of choices of food on the part of each organism involved. In an ecosystem of average complexity, numerous alternative

pathways are possible. Thus, a **food web**, a complex of interconnected food chains in an ecosystem, is a more realistic model of the flow of energy through ecosystems (Figure 26-13).

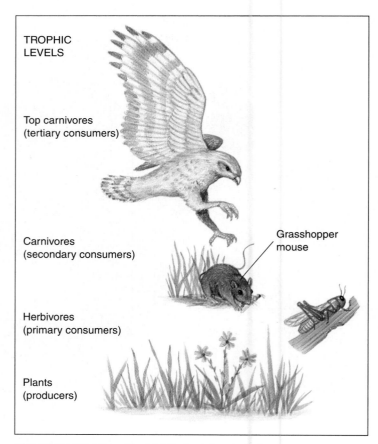

Figure 26-12   Trophic levels in a food chain.

Figure 26-13   A food web for a deciduous forest. This food web is greatly simplified compared to what actually happens in nature. Groups of species are lumped into single categories such as spiders, many other species are not included, and numerous links in the web are not shown.

Ecological Pyramids Illustrate How Ecosystems Work

An important feature of energy flow is that most of it dissipates into the environment when going from one trophic level to another. The relative energy value at each trophic level is often graphically represented by an **ecological pyramid**. There are three main types of pyramids: pyramids of numbers, pyramids of biomass, and pyramids of energy.

A **pyramid of numbers** shows the number of organisms at each trophic level in a given ecosystem (Figure 26-14). In most pyramids of numbers, each successive trophic level is occupied by fewer organisms. Thus, in a typical African grassland, the number of producers such as grasses is greater than the number of herbivores such as zebras and wildebeests, which is greater than the number of carnivores such as lions. Reverse pyramids of numbers, in which higher trophic levels have *more* organisms than lower trophic levels, are often observed among decomposers, parasites, tree-dwelling herbivorous insects, and similar organisms. One tree can provide food for thousands of leaf-eating insects, for example.

A **pyramid of biomass** illustrates the total biomass at each successive trophic level. **Biomass** is a quantitative estimate of the total mass, or amount, of living material. Its units of measure vary: biomass may be represented as total volume, as dry weight, or as live weight. Typically, a pyramid of biomass illustrates a progressive reduction of biomass in succeeding trophic levels (Figure 26-15). On the assumption that, on the average, about a 90 percent reduction of biomass occurs

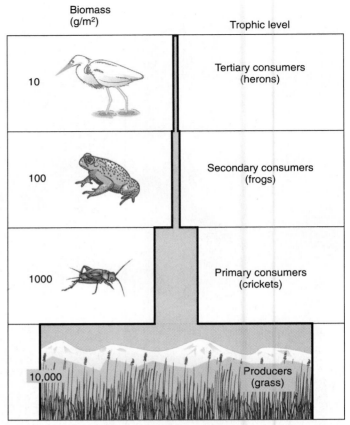

Figure 26-15   A pyramid of biomass for a hypothetical area of a temperate grassland. Pyramids of biomass are based on the biomass at each trophic level and typically resemble pyramids of numbers.

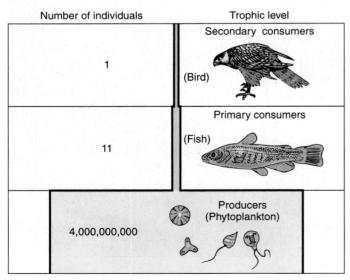

Figure 26-14   A pyramid of numbers is based on the number of organisms at each trophic level. Typically, there are more producers than primary consumers, more primary consumers than secondary consumers, and so on.

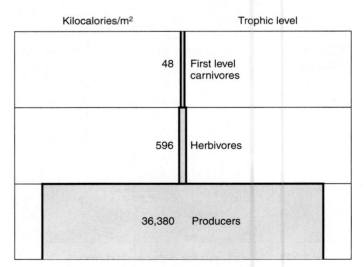

Figure 26-16   Energy flow—the functional basis of ecosystem structure—is represented by a pyramid of energy. The energy involved at each level is given in kcal/m².

for each trophic level,[1] 10,000 kilograms of grass should be able to support 1000 kilograms of crickets, which in turn support 100 kilograms of frogs. By this logic, the biomass of frog eaters (such as herons) could weigh, at the most, only about 10 kg. From this brief exercise, you can see that although carnivores may eat no vegetation, a great deal of vegetation is still required to support them.

A **pyramid of energy** shows the energy relationships of an ecosystem (Figure 26-16). It indicates the energy contents (usually expressed in kilocalories) of the biomasses of various trophic levels. On the whole, pyramids of energy resemble biomass pyramids in shape, but help to make another consequence of trophic levels clearer: *most food chains are short* because of the dramatic reduction in energy content that occurs at each trophic level.

ECOLOGICAL SUCCESSION IS COMMUNITY CHANGE OVER TIME

As discussed earlier in the chapter, a community consists of the populations of various species that live together in the same place. When a natural area has been disturbed—for example, by a volcanic eruption, an earthquake, or a forest fire—a new community does not spring into existence overnight but develops gradually through a series of stages. The process of community development over time, which involves species in one stage being replaced by other species, is called **ecological succession** or simply **succession**.

Ecological succession is usually described in terms of the changes in the species composition of the vegetation of an area, although each successional stage also has its own characteristic kinds of animals. The time intervals involved in ecological succession are on the order of tens, hundreds, or thousands of years, not millions of years as in the evolutionary time scale.

In Primary Succession, a Community Develops on Land Not Previously Inhabited by Plants

Primary succession is the change in species composition over time in an area that has not previously been inhabited by plants or other organisms; no soil exists when primary succession begins. Bare rock surfaces such as recently formed volcanic lava and rock scraped clean by glacial action are examples of sites where primary succession might occur (Figure 26-17).

---

[1]The 90 percent reduction in biomass is an approximation; actual biomass reduction varies widely in nature

(a)

(b)

(c)

Figure 26-17 Primary succession occurs after the retreat of glaciers. Although these photos were not taken in the same area, they show some of the stages of primary succession on glacial moraine (rocks, gravel, and sand deposited by a glacier). **(a)** The barren landscape exposed after the retreat of the glacier is initially colonized by lichens, then mosses. **(b)** At a later time, dwarf trees and shrubs colonize the area. **(c)** Still later, spruces dominate the community. (a,c, *Wolfgang Kaehler;* b, *Visuals Unlimited/Glenn N. Oliver*)

Exposed rock is an inhospitable environment that few organisms can tolerate. The temperature of bare rock may be quite high in the sunlight, and unless it is actually raining, the rock may be totally devoid of moisture. Whatever minerals are present in the rock are locked up in its hard crystalline structure, unavailable to organisms.

In primary succession on bare rock, the details vary from one site to another, but usually the first community that is observed consists of lichens. Because lichens are the first organisms to colonize bare rock, they are called **pioneers**. Lichens are able to live on the surfaces of many rocks and beneath the surfaces of porous rocks, in a sheltered and somewhat moister habitat. Lichens are very resistant to desiccation (drying out). They cease to grow when water is unavailable, then quickly resume active growth when moisture returns; they can absorb their own weight in water within moments of moistening.

With the passage of time, several important cumulative changes occur. First, the biomass of the lichen community increases. Second, lichens secrete acids that help break the rock apart. As a result, fine particles of rock become detached from the rock's surface or even within the rock itself. Third, as lichens die, their decomposing remains mix with the rock particles to form a rudimentary soil. Fourth, water is absorbed and retained in the tissues of the lichens and in the new, thin soil layer for longer periods of time than before. As all of these changes occur, an increasing number of tiny organisms move into the area and make their homes in the lichens and soil.

All of these changes—increased biomass, soil development, water retention, and an increased number of organisms—work together to moderate the harsh conditions under which the pioneer community must live, making it possible for mosses to grow there. In fact, since mosses can grow faster than lichens, they tend to replace the lichens. Their greater productivity leads to a greater accumulation of biomass and ultimately of soil. This leads to further habitat change. Over time the moss community may be replaced by drought-resistant ferns, followed in turn by tough grasses and herbs. Once sufficient soil has accumulated, grasses and herbs may be replaced by low shrubs, which in turn can be replaced by forest trees in several distinct stages. Primary succession from a pioneer community on bare rock to a forest community may take hundreds or thousands of years.

In Secondary Succession, a Community Develops After Removal of an Earlier Community

Secondary succession is the change in species composition over time in an area already substantially modified by a preexisting community; soil is already present at these sites. An abandoned field and an area cleared by a forest fire are common examples of sites where secondary succession occurs.

Secondary succession on abandoned farmland has been studied extensively by many ecologists since the late 1930s. Although it takes more than 100 years for all the stages of secondary succession to occur at a given site, it is possible for a single researcher to study old field succession in its entirety by observing different sites in the same area. Often the ecologist can accurately determine when each field was abandoned by examining court records.

Abandoned farmland in North Carolina is colonized by a predictable sequence of plant communities. The first year after cultivation ceases, annual and perennial herbs grow in abundance, but crabgrass (*Digitaria sanguinalis*) is by far the most common species. In the second year, horseweed (*Conyza canadensis*), a tall, coarse annual, is the main species. Horseweed does not remain dominant more than 1 year, however, because decaying horseweed roots inhibit the growth of young horseweed seedlings. In addition, horseweed does not compete well with plants that become established in the third year, including broomsedge (*Andropogon virginicus*). Typically, broomsedge outcompetes other plants because it is more tolerant of droughts. Broomsedge continues to be dominant for the next few years.

In years 5 to 15, the dominant plants in the abandoned farmland are pines, such as shortleaf pine (*Pinus echinata*) on drier sites or loblolly pine (*Pinus taeda*) on moister sites. Through shading and the buildup of litter (pine needles and branches) on the soil, pines produce conditions that cause the earlier dominant plants to decline in importance.

Over time, pines give up their dominance to hardwoods such as oaks and hickories. This stage of secondary succession depends primarily on the environmental changes produced by the pines. The pine litter causes soil changes, such as an increase in water-holding capacity, that are necessary for young hardwood seedlings to become established. In addition, young pine seedlings do not thrive in the shade of older trees, whereas hardwood seedlings do.

The Concept of a Climax Community Has Fallen Out of Favor

Ecologists initially thought that ecological succession inevitably led to a relatively stable community, known as a *climax community,* that was capable of maintaining itself as long as the climate did not change appreciably. More recently, however, this traditional view has been questioned. The apparent stability of a forest community, for example, is probably the result of the life spans of trees relative to the human life span. It is now recognized that mature climax communities are not in a state of permanent equilibrium but rather are in continual flux. A mature community changes in species composition and in the relative abundance of each species despite the fact that it retains a uniform appearance.

STUDY OUTLINE

I. The study of the interactions among different organisms and between organisms and their physical environment is called ecology. Ecologists study populations, communities/ecosystems, and the biosphere/ecosphere.
   A. A population is all the members of the same species that live together.
   B. A community is all the populations of organisms living in the same area; an ecosystem is a community and its abiotic environment.
   C. The biosphere is all the communities on Earth—in other words, all of the Earth's organisms. The ecosphere comprises the interactions between the Earth's biosphere, atmosphere, lithosphere, and hydrosphere.

II. An organism's ecological niche is its distinctive role in a community and encompasses the physical, chemical, and biological factors that it needs to survive, remain healthy, and reproduce.

III. Each organism in an ecosystem assumes the role of producer, consumer, or decomposer.
   A. Producers are the photosynthetic organisms at the bases of most food webs; they include plants, algae, and some bacteria.
   B. Consumers are heterotrophs that feed on other organisms.
   C. Decomposers are microbial heterotrophs (bacteria and fungi) that recycle the components of dead organisms and organic wastes.

IV. The organisms of a community interact with one another in many ways.
   A. During competition, two or more organisms simultaneously require the same resource, which is usually in limited supply.
      1. The resources for which plants commonly compete include water, light, soil minerals, and growing space.
      2. Intraspecific competition occurs among two or more individuals belonging to the same species, whereas interspecific competition occurs among two or more individuals of different species.
   B. Predation is the consumption of one species (the prey) by another (the predator).
      1. Predation includes both herbivores eating plants and carnivores eating other animals.
      2. During coevolution between predator and prey, the predator evolves more efficient ways to catch prey, and the prey evolves better ways to escape the predator.
      3. Plants possess spines, thorns, tough leathery leaves, thick wax, and an array of unpalatable or toxic chemicals to discourage herbivores.
   C. Symbiosis is any intimate association between two or more species.

1. Both partners in a mutualistic association benefit.
2. In commensalism, one organism benefits and the other is unaffected.
3. In parasitism, one organism (the parasite) benefits and the other (the host) is harmed.

V. All materials vital to life have biogeochemical cycles in which they continually cycle from the environment to organisms and back to the environment.
   A. Carbon enters plants and other producers as CO_2, which is incorporated into carbohydrate molecules by photosynthesis. Cellular respiration, combustion, and erosion return CO_2 to the atmosphere, making it available for producers again.
   B. The nitrogen cycle has five steps, four of which involve bacteria.
      1. Nitrogen fixation is the conversion of nitrogen gas to ammonia.
      2. Nitrification is the conversion of ammonia to nitrate, one of the main forms of nitrogen used by plants.
      3. Assimilation is the biological conversion, by plants, of nitrates or ammonia to proteins and other nitrogen-containing compounds, as well as the conversion of plant proteins to animal proteins.
      4. Ammonification is the conversion of organic nitrogen to ammonia.
      5. Denitrification converts nitrate to nitrogen gas.
   C. The hydrologic cycle, which continually renews the supply of water that is so essential to life, involves exchanges of water among land, atmosphere, and organisms.

VI. Energy flows through an ecosystem in one direction, from the sun to producer to consumer to decomposer. Much of the energy acquired by a given trophic level is used for metabolic purposes at that level and is therefore unavailable to the next trophic level.
   A. Food webs show the many alternative pathways that energy may take among the producers, consumers, and decomposers of an ecosystem.
   B. Ecological pyramids express the progressive reductions in numbers of organisms, biomass, and energy found in successively higher trophic levels.

VII. Ecological succession is the orderly replacement of one community by another.
   A. Primary succession begins in an area that has not previously been inhabited (for example, bare rock).
   B. Secondary succession begins in an area where there was a preexisting community and a well-formed soil (for example, abandoned farmland).

SELECTED KEY TERMS

REVIEW QUESTIONS

1. How is a community different from an ecosystem?
2. Distinguish between an organism's habitat and its ecological niche.
3. How might one distinguish between a producer and a decomposer? A consumer and a decomposer?
4. Distinguish between intraspecific competition and interspecific competition.
5. How is mutualism different from commensalism? How do parasitism and predation differ?
6. Why is the cycling of matter essential to the continuance of life on the Earth?
7. Diagram the carbon cycle, including the following processes: photosynthesis, respiration, and combustion.
8. List and describe the five steps in the nitrogen cycle.
9. Why is the concept of a food web generally preferable to that of a food chain?
10. Define ecological succession, and distinguish between primary and secondary succession.
11. Why does an agricultural ecosystem have a greater likelihood of pest problems than a natural ecosystem?

THOUGHT QUESTIONS

1. John Muir once said, "When one tugs at a single thing in nature, he finds it attached to the rest of the world." What did he mean?
2. Could a balanced ecosystem be constructed that contained only producers and consumers? Only consumers and decomposers? Only producers and decomposers? In each case, explain the reason for your answer.
3. Describe the ecological niche of humans. Do you think that our niche has changed during the past thousand years? Why or why not?
4. Is it possible to have an inverted pyramid of energy? Why or why not?
5. What kinds of environmental conditions trigger primary succession? Secondary succession?
6. Sometimes epiphytes are so abundant in tropical rain forests that they cover the entire surfaces of leaves and interfere with a larger plant's ability to photosynthesize. In this situation, would the epiphytes be an example of commensalism? If not, what type of interaction would they exemplify?

SUGGESTED READINGS

Beardsley, T., "Recovery Drill," *Scientific American,* November 1990. The recovery of Mount St. Helens is challenging some conventional ideas about how communities respond to environmental catastrophes.

Capon, B., *Plant Survival: Adapting to a Hostile World,* Portland, Ore., Timber Press, 1994. A fascinating account of how various plants are adapted to survive such environmental extremes as the heat and drought of a desert, the freezing cold of the Arctic, and the salt in a coastal marsh.

Gardner, G., "IPM and the War on Pests," *World Watch,* Vol. 9, No. 2, March/April 1996. Integrated pest management (IPM) offers a radical departure from the exclusive use of chemical pesticides in conventional agriculture.

Morrison, M., *Fire in Paradise,* New York, Harper Collins, 1993. Details of the September 1988 fire in Yellowstone National Park and its effects on the park and surrounding area.

Nelson et al., "Using a Closed Ecological System to Study the Earth's Biosphere: Initial Results from Biosphere 2," *BioScience,* Vol. 43, No. 4, April 1993. Biosphere 2 is a new type of research tool for ecologists studying biogeochemical cycles and other ecological processes.

Thornton, I., *Krakatau: The Destruction and Reassembly of an Island Ecosystem,* Cambridge, Mass., Harvard University Press, 1996. Summarizes the studies of primary succession on the Indonesian island Krakatau since an 1883 volcanic eruption destroyed all life and buried everything under a thick layer of hot ash.

CHAPTER 27

BIOMES OF THE WORLD

The common dandelion (*Tarax-acum officinale*) is a flowering plant that is native to Europe and Asia but is found worldwide in temperate climates. Dandelions are common in meadows and fields and on roadsides and lawns, where bees use the nectar to make honey and birds eat the seeds. Although this yellow-flowered plant is sometimes grown for salad greens, particularly in the Northeast, dandelions are generally considered bothersome weeds.

The dandelion body consists of a long taproot topped by a stem that bears leaves in a basal rosette. The flower head, borne on a hollow stem, consists of a cluster of small yellow flowers. Although the flowers contain both male and female structures, the seeds commonly develop asexually—without pollination or fertilization—by *apomixis*. The seed-balls, which are white, fluffy, and spherical, consist of many parachute-like fruits covered with silky hairs; the wind easily carries the fruits to new locations.

Dandelions are somewhat unusual in that their geographical range is so extensive. Only a few plants and animals—

The common dandelion (*Taraxacum officinale*), found in North America from Alaska to Florida, thrives in disturbed sites such as fields, lawns, and roadsides. (*Marion Lobstein*)

such as junipers, daisies, black bears, song sparrows, and honey-bees—have ranges as extensive as that of dandelions. Most organisms have more limited distributions. For instance, polar bears are found on ice floes and rocky shores of northern Alaska, Canada, and other polar regions, and coconut palms grow on the sandy beaches of warm Pacific islands and in other tropical places.

It is clear that species are not distributed uniformly throughout the Earth. But what governs their distribution? Basically, organisms live only in the areas to which they are adapted. The greater the physical differences among habitats, the greater the differences among the organisms that inhabit them: cacti could not survive in a salt marsh, and marsh grasses could not survive in the desert.

In this chapter we examine the Earth's major biomes, including their climates and soils, which largely determine which plants and animals are characteristic of each. We also consider two global environmental issues that affect biomes: global climate change and depletion of the ozone layer in the stratosphere.

After reading this chapter, you should be able to:

1. *Define* biome *and describe the nine major terrestrial biomes, including the climate, soil, and characteristic plants and animals of each.*
2. *Identify the biomes that make the best agricultural land, and explain why they are superior.*
3. *Describe a specific effect that humans have on each biome.*
4. *Name at least three greenhouse gases, and explain how greenhouse gases contribute to global warming.*
5. *Describe how global warming may affect sea level, precipitation patterns, plants and other organisms, and food production.*
6. *Cite the potential effects of ozone destruction in the stratosphere.*

BIOMES ARE DETERMINED LARGELY BY CLIMATE

A **biome** is a large, relatively distinct terrestrial region characterized by a particular combination of climate, soil, plants, and animals that is approximately the same regardless of where it occurs in the world (Figure 27-1). Because it is so large in area, a biome encompasses a number of interacting ecosystems. In terrestrial ecology, a biome is considered the next level of ecological organization above community and ecosystem.

A biome's boundaries are determined largely by invisible climate barriers, with temperature and precipitation being most important. Temperature is generally the overriding climate factor near the poles, whereas precipitation becomes more significant in temperate and tropical regions. Our discussion covers nine biomes: tundra, taiga, temperate rain forest, temperate deciduous forest, temperate grassland, chaparral, desert, savanna, and tropical rain forest.

Tundra Is the Northernmost Biome

Cold, boggy plains called **tundra** exist in the extreme northern latitudes wherever the snow melts seasonally. (The Southern Hemisphere has no equivalent of the arctic tundra, because no land exists in its equivalent latitudes.) In North America, tundra is found in northern Canada and Alaska.

Tundra is exposed to long, harsh winters and very short summers. Although the warm growing season is short (from 50 to 100 days, depending on location), the days are long. In many places the sun does not set for a considerable number of days in summer, although the amount of light at midnight is one-tenth that at noon. Much of the tundra receives little precipitation (10 to 25 centimeters, or 4 to 10 inches, per year), and most of it falls as rain during the summer and autumn.

Tundra soils, which for the most part are geologically young because they formed after the last ice age, are usually mineral-poor and have little organic litter. Although the soil's surface melts during the summer, tundra has a layer of permanently frozen ground called **permafrost**, which varies in depth and thickness. The permafrost interferes with drainage and prevents the roots of larger plants from becoming established. The limited precipitation of the tundra, in combination with its low temperatures, flat topography (surface features), and permafrost layer, produces a landscape of broad, shallow lakes, sluggish streams, and bogs (Figure 27-2a).

Few plant species are found in the tundra, but the individual species present often exist in great numbers. The dominant producers are mosses, lichens such as reindeer moss, grasses, and sedges, which are grasslike plants with triangular stems. Dwarf willows and other dwarf shrubs are common (Figure 27-2b); shrubs in the tundra seldom grow taller than 30 centimeters (12 inches).

The year-round animal life of the tundra includes brown lemmings, weasels, arctic foxes, snowshoe hares, musk oxen, ptarmigans, and snowy owls (which sometimes migrate south when food supplies are low). In the summer, large herds of barren-ground caribou migrate north into the tundra to graze on sedges, grasses, and dwarf willow. Mosquitos, blackflies, deerflies, and other insects survive the winter as eggs or pupae and occur in great numbers during summer.

Tundra regenerates very slowly after it has been disturbed, and even casual use by hikers can damage it. On portions of the arctic tundra, oil exploration and military use have inflicted damage that is likely to persist for hundreds of years.

Taiga Is Dominated by Conifers

Just south of the tundra is the **taiga**, or **boreal forest**, a huge evergreen forest biome that stretches across northern portions of North America as well as northern Europe (Scandinavia) and Asia (northern Russia) (Figure 27-3). In North America, taiga occurs throughout most of Canada and Alaska as well as in parts of New England and states around the Great Lakes. No biome comparable to the taiga occurs in the Southern Hemisphere.

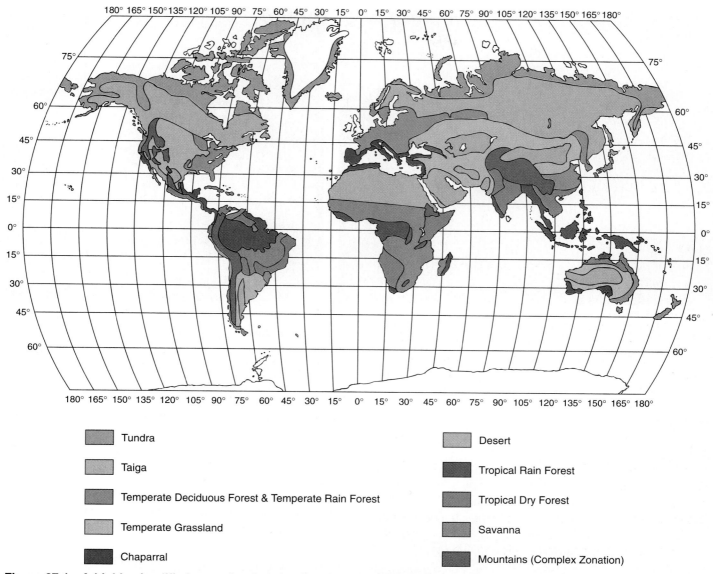

	Tundra			Desert
	Taiga			Tropical Rain Forest
	Temperate Deciduous Forest & Temperate Rain Forest			Tropical Dry Forest
	Temperate Grassland			Savanna
	Chaparral			Mountains (Complex Zonation)

Figure 27-1 A highly simplified map showing the distribution of the world's biomes. Although sharp boundaries are shown, biomes actually grade together at their boundaries. (*Adapted from Odum 1971*)

Winters are extremely cold and severe in the taiga, although not as harsh as in the tundra, and the growing season is somewhat longer than that of the tundra. Taiga receives as little as 50 centimeters (20 inches) of precipitation per year, and its soil is acidic, mineral-poor, and characterized by a deep layer of partly decomposed fir and spruce needles at the surface. Permafrost is either deep beneath the soil's surface (in the northern taiga) or absent (in the southern taiga). Taiga has numerous ponds and lakes where depressions were carved into the Earth's surface by the grinding ice sheets that covered that area during the last ice age.

Although deciduous flowering trees, such as aspen or birch, may form striking stands in the taiga, spruce, fir, and pine clearly dominate. Conifers possess drought-resistant adaptations such as <u>needle-like leaves, which have minimal surface areas to limit water loss.</u> Such adaptations enable the trees to withstand the "drought" of the northern winter, when roots cannot absorb water because the ground is frozen.

Some of the animals of the taiga are large species, such as wolf, bear, moose, and caribou, which migrate from the tundra to the taiga in autumn. However, most of the animal life is medium-sized to small, including rodents, rabbits, and

(a)

(b)

Figure 27-2   Arctic tundra. (a) The flat plains of Alaskan tundra are marshy and wet during the short summer months. This photo also shows the Susitna Glacier, which is flanked by mountains, in the background. **(b)** Arctic willow (*Salix arctica*) is a creeping shrub that rarely exceeds 10 centimeters (4 inches) in height. (a,b, *Visuals Unlimited/Steve*

Figure 27-3   The taiga, or boreal forest. These coniferous forests occur in cold regions of the Northern Hemisphere adjacent to the tundra. (*Charlie Ott/Photo Researchers, Inc.*)

fur-bearing predators such as lynx, sable, and mink. Most species of birds are abundant in the taiga during the summer but migrate to warmer climates in the winter. Insects are also abundant.

Most of the taiga is not well suited to agriculture because of its short growing season and mineral-poor soil. However, the taiga yields vast quantities of lumber and pulpwood for making paper products.

Temperate Rain Forest Occurs in the Mid-latitudes Where Annual Precipitation Is High

A coniferous **temperate rain forest** occurs on the northwest coast of North America (from northern California to Alaska), in southeastern Australia, and in southern South America (Figure 27-4). It is characterized by cool weather, dense fog, and high precipitation. Annual precipitation in this biome ranges from about 200 to 380 centimeters (80 to 152 inches) and is augmented by condensation of water from dense coastal fogs. The proximity of temperate rain forest to the coastline moderates the temperature so that there is a narrow seasonal fluctuation; winters are mild and summers are cool. The soil in temperate rain forest is relatively mineral-poor, although its organic content may be high.

The dominant vegetation in the North American temperate rain forest is large coniferous evergreen trees, such as western hemlock, Douglas fir, Sitka spruce, and western red cedar. The temperate rain forest is also rich in epiphytic

Figure 27-4   Trees of the temperate rain forest in Washington include Douglas fir, western hemlock, and western red cedar. Moisture-loving ferns, mosses, and lichens grow on the trees as well as the ground. (*Terry Donnelly/Dembinsky Photo Associates*)

Figure 27-5   Temperate deciduous forests exhibit dramatic seasonal changes. The onset of autumn turns the leaves into a variety of golds, oranges, reds, and browns. A variety of hardwoods grow in temperate deciduous forest, including oak, hickory, beech, tuliptree, sweetgum, and dogwood. (*Dennis Drenner*)

vegetation—smaller plants such as mosses, club mosses, lichens, and ferns that grow nonparasitically on the large trees. Squirrels, deer, and numerous bird species are among the animals to be found in the temperate rain forest.

Although temperate rain forest is one of the richest wood producers in the world and supplies us with lumber and pulpwood, it is also one of the most complex ecosystems. Care must be taken to avoid overharvesting the original old-growth forest, because it takes hundreds of years for such an ecosystem to develop. The logging industry typically clearcuts old-growth forest and replants the area with a monoculture (a single species) of tree seedlings that it can harvest in 40- to 60-year cycles. Thus, once harvested, the old-growth forest ecosystem never has a chance to redevelop.

Temperate Deciduous Forest Occurs in the Midlatitudes Where Annual Precipitation Is Relatively High

Temperate deciduous forest occupies most of the eastern United States, southeastern Canada, Europe (except for the northernmost and southernmost regions), and eastern Asia.

It consists of a dense canopy of broad-leaved trees that overlie saplings and shrubs.

Summers are hot, winters are pronounced, and annual precipitation ranges from about 75 to 125 centimeters (30 to 50 inches). The soil of a temperate deciduous forest typically consists of a topsoil rich in organic material and a deep, clay-rich lower layer. As organic materials decay, mineral ions are released. If they are not immediately absorbed by the roots of the living trees, these ions leach into the clay, which may retain them.

The temperate deciduous forests of the northeastern and middle-eastern United States are dominated by broad-leaved hardwood trees that lose their foliage annually, such as oak, hickory, beech, ash, and maple (Figure 27-5). In the southern reaches of the temperate deciduous forest, the number of broad-leaved evergreen trees, such as magnolia and some oak species, increases.

Temperate deciduous forest originally contained a variety of large mammals such as puma, wolf, deer, bison, and bear, plus many small mammals and birds. Reptiles and amphibians abounded, together with a denser and more varied insect life than exists today. Much of the original temperate deciduous forest was removed by logging and land clearing during the 18th and 19th centuries, although large areas have regenerated. For example, 35 percent of Vermont was forest in 1850, as compared to 80 percent today. Where it has been allowed to regenerate, however, temperate deciduous forest is often in a seminatural state; that is, it is highly modified by humans.

Worldwide, temperate deciduous forests were among the first biomes to be converted to agricultural use. In Europe and Asia, for example, many soils that originally supported temperate deciduous forests have been cultivated for thousands of years without a substantial loss in fertility. During the 20th century, however, intensive agricultural practices have been widely adopted, resulting in the degradation of perhaps 15 to 20 percent of Earth's total agricultural land. Most damage to farmland has been done since the end of World War II.

Temperate Grasslands Occur in the Mid-latitudes Where Precipitation Is Moderate

Grasses are the dominant vegetation of **temperate grasslands**, which typically occur in the drier continental interiors of North America (*prairie*) and Eurasia (*steppe*). In the Southern Hemisphere, temperate grasslands are found in East and South Africa (*veld*) and Argentina (*pampas*).

Summers are hot, winters are cold to mild, and rainfall is often uncertain in temperate grasslands. Annual precipitation averages 25 to 75 centimeters (10 to 30 inches). Grassland soils are some of the best agricultural soils in the world. As a result of low precipitation, minerals in grassland soils tend to accumulate just below the topsoil instead of leaching to deeper ground. Grassland soil also has considerable organic material, because the aboveground parts of many grasses die off each winter and contribute to the organic content of the soil. The roots and rhizomes survive the winter underground and send up new aboveground growth the following spring.

The North American Midwest is an excellent example of a temperate grassland. Few trees grow there except near rivers and streams, and perennial grasses grow in great profusion in the thick, fertile soil. Prairie violets, purple prairie clover, tall goldenrod, scarlet paintbrush, and many other herbaceous flowering plants grow among the grasses. Before humans had a major impact on temperate grasslands, certain species of grasses grew as tall as a person on horseback, and the land was covered with large herds of grazing animals, particularly bison. The principal predators were wolves, although in sparser, drier grasslands coyotes took their place. Smaller animals included prairie dogs and their predators (foxes, black-footed ferrets, and various birds of prey), grouse, reptiles, and great numbers of insects.

Shortgrass prairies are temperate grasslands, with less precipitation than the moister grasslands just described but with greater precipitation than deserts (Figure 27-6). They range from Alberta and the Dakotas south to Texas. Shortgrass prairies have sparser vegetation than the moister grasslands,

Figure 27-6   Temperate grasslands are mostly treeless but contain a profusion of grasses and other herbaceous flowering plants. This photo shows a shortgrass prairie at the Theodore Roosevelt National Historic Park in North Dakota. (*Willard Clay/Dembinsky Photo Associates*)

and some bare soil may be exposed in dryer areas. Native grasses and other herbaceous flowering plants of the shortgrass prairies are drought-resistant.

The temperate grasslands were so well suited to agriculture that few remnants remain. The American Midwest, the Ukraine, and other moist temperate grasslands became the breadbaskets of the world because they provide ideal growing conditions for crops such as corn and wheat, both of which are grasses.

Chaparral Is a Thicket of Evergreen Shrubs and Small Trees

Some temperate environments have mild winters with abundant rainfall combined with extremely dry summers. Such **Mediterranean climates**, as they are called, occur not only around the Mediterranean Sea but also in California, western Australia, portions of Chile, and South Africa. In the North American Southwest, this Mediterranean-type community is known as **chaparral**. Chaparral soils are thin and not very

Figure 27-7   Chaparral in the Santa Monica Mountains, California. Drought-resistant evergreen shrubs and small trees are the main vegetation in the chaparral biome. Chaparral develops where hot, dry summers alternate with mild, rainy winters. (*Visuals Unlimited/John Cunningham*)

fertile. Frequently fires occur naturally in this habitat, particularly in late summer and autumn, and native plants are adapted to the fires (see Plants and the Environment: Fire and Terrestrial Ecosystems).

Chaparral vegetation looks strikingly similar in different areas of the world, even though the individual species are quite different. Chaparral is usually dominated by a dense growth of evergreen shrubs but may contain drought-resistant pine and scrub oak trees (Figure 27-7). During the rainy season the habitat may be lush and green, but many plants are dormant during the hot, dry summer. Trees and shrubs often have **sclerophyllous leaves**—hard, small, leathery leaves that resist water loss. Some chaparral plants, such as sagebrush, are noted for **allelopathy**, in which their roots or leaves secrete toxic substances that inhibit the establishment of competing plants nearby. Many plants are also specifically fire-adapted and grow best in the rainy months following a fire. Such growth is possible in part because fire releases the minerals that were tied up in the aboveground parts of plants. The underground parts are not destroyed by fire, and with the new availability of essential minerals, the plants sprout vigorously during the winter rains. Mule deer, wood rats, chipmunks, lizards, and many species of birds are the common animals of the chaparral.

The fires that occur at irregular intervals in California are often quite costly to humans because they consume expensive homes built in the hilly chaparral vegetation. Efforts to control the naturally occurring fires often backfire. Denser, thicker vegetation tends to accumulate when periodic fires are prevented; consequently, when a fire does occur, it is much more severe.

Desert Occurs Where Little Precipitation Falls

Deserts are very dry areas in temperate and tropical regions. Although deserts occur on every continent, most are found along the Tropic of Cancer (between 15 and 30 degrees north of the Equator) or the Tropic of Capricorn (between 15 and 30 degrees south of the Equator). The Sahara Desert in North Africa, approximately the size of the United States, is the world's largest desert. In the world's driest desert, the Atacama Desert in Chile, precipitation is barely measurable.

The dryness of the desert atmosphere leads to wide daily temperature extremes—daytime heat and nighttime cold. Deserts vary greatly according to the amounts of precipitation they receive, which are generally less than 25 centimeters (10 inches) per year. Desert soils are low in organic material but often high in mineral content; in some regions the concentrations of certain minerals in the soil reach toxic levels.

Plant cover, both perennial and annual, is sparse in deserts, and much of the desert soil is exposed. Plants in North American deserts include creosote bush, cacti, yuccas, and widely scattered bunchgrass (Figure 27-8). Perennial desert plants such as the creosote bush may have reduced leaves or no leaves, an adaptation that enables them to conserve water. Other desert plants, such as mesquite, shed their leaves for most of the year and grow only during the brief moist season. Wide spacing of plants allows each one to obtain enough of the available soil moisture around it to survive. Many desert plants possess defensive spines, thorns, or toxins to prevent desert animals from feeding on them in this food- and water-deficient environment.

Desert animals tend to be small. In the heat of the day they remain under cover or periodically return to shelter, and at night they come out to forage or hunt. In addition to desert-adapted insects, there are many specialized desert reptiles, including lizards, tortoises, and snakes. Mammals include rodents such as the American kangaroo rat, which does not have to drink water but can subsist solely on the water content of its food plus metabolically generated water. American deserts are habitat for jackrabbits, and Australian deserts, for kangaroos. Desert carnivores such as the African fennec fox and some birds of prey, especially owls, live on the rodents and rabbits.

Savanna Is a Tropical Grassland with Scattered Trees

A **savanna** is a tropical grassland with widely scattered clumps of low trees. The world's largest savanna occupies about one third of the African continent. Savanna also occurs in parts of South America (northern Argentina), northern Australia, and India. The Serengeti Plain of East Africa is the best-known savanna.

The savanna biome occurs in warm, tropical areas of low rainfall or highly seasonal rainfall with prolonged dry periods

(a)

Figure 27-8   Desert plants are strikingly adapted to the demands of their environment. (a) The moister deserts of North America frequently contain large cacti, such as this organ-pipe cactus (*Lemaireocereus* sp). Photosynthesis is carried out in the stem, which also serves to store water. Leaves are modified into spines, which discourage herbivores from eating cactus tissue. **(b)** The Joshua tree (*Yucca brevifolia*) is one of the characteristic flowering plants of the Mojave Desert in the southwestern United States. Its evergreen leaves, which occur in dense clusters at the ends of branches, are stiff and narrow and have sharply pointed tips. (a, *Stan Osolinski/Dembinsky Photo Associates;* b, *Terry Donnelly/Dembinsky Photo Associates*)

(b)

(Figure 27-9). Because the temperatures in savannas vary little throughout the year, seasons are regulated by precipitation, rather than by temperature as in temperate grasslands. Annual precipitation is 85 to 150 centimeters (34 to 60 inches). Savanna soil is relatively infertile because it is low in essential minerals.

Savanna is characterized by wide stretches of grasses interrupted by occasional trees such as *Acacia* sp, which bristles with thorns that protect it against herbivores. Both trees and grasses have fire-adapted features, such as extensive underground root systems, which enable them to survive the fires that periodically sweep across dry savanna.

The African savanna contains the greatest assemblage of hoofed mammals in the world—great herds of wildebeest, antelope, giraffe, zebra, and the like. Large predators, such as lions and hyenas, kill and scavenge the herds. In areas of seasonally varying rainfall, the herds and their predators follow annual migration routes.

Humans are rapidly converting savannas to rangeland for cattle and other livestock, which are replacing the big herds of game animals. Severe overgrazing in places has converted some savannas to deserts (see Plants and the Environment: Desertification in Chapter 6).

Figure 27-9   Savanna. Grasslands such as this one, with widely scattered acacia trees, are found in East Africa. Parts of South America, Australia, and India also have areas of savanna. (*Mike Barlow/Dembinsky Photo Associates*)

Environment

Fire and Terrestrial Ecosystems

Fires started by lightning are an important environmental force in many geographical areas. The areas most prone to fires have wet seasons followed by dry seasons: vegetation that grows and accumulates during the wet season dries out enough during the dry season to burn easily. When lightning hits the ground, it ignites the dry organic material, and a fire spreads through the area.

Fires have several effects on the environment. First, combustion frees the minerals that were locked in dry organic matter. The ashes remaining after a fire are rich in potassium, phosphorus, calcium, and other minerals essential for plant growth. Thus, vegetation flourishes after a fire. Second, fire removes plant cover and exposes the soil, which stimulates the germination of seeds that require bare soil and encourages the growth of shade-intolerant plants. Third, fire can cause increased soil erosion because it removes plant cover, leaving the soil more vulnerable to wind and water.

The evolutionary effects of fire on plants have been studied extensively. Grasses adapted to fire have underground stems and buds that are unaffected by the blaze sweeping over them; after the fire kills the aerial parts, these underground parts send up new shoots. Fire-adapted trees such as bur oak and ponderosa pine have thick bark that is resistant to fire. (In contrast, fire-sensitive trees such as many hardwoods have thin bark.) Certain pines, such as jack pine and lodgepole pine, depend on fire for successful reproduction, because the heat of the fire opens the cones and releases the seeds.

Fires were part of the natural environment long before humans appeared, and many terrestrial ecosystems have adapted to fire. African savanna, California chaparral, North American grasslands, and pine forests of the southern United States are some of the fire-adapted biomes. Fire helps maintain grasses as the dominant vegetation in grasslands—for example, by removing fire-sensitive hardwood trees.

The influence of fire on plants became even more pronounced once humans appeared, because humans set fires both deliberately and accidentally, making fire a more common occurrence. Humans purposely set fires for several reasons: to facilitate the growth of grasses and shrubs that many game animals require; to clear the land for agriculture and human development; and, in times of war, to reduce enemy cover.

Humans also try to prevent fires, and sometimes that effort can have its own disastrous consequences. When fire is excluded from a fire-adapted ecosystem, organic litter accumulates. As a result, when a fire does occur, it is much more destructive. To avoid this situation, humans sometimes conduct *controlled burns*, a tool of ecological management in which the organic litter is deliberately burned before it has accumulated to levels that can fuel dangerous, destructive fires. Prevention of fire also converts grassland to woody vegetation and facilitates the invasion of fire-adapted conifer forests by fire-sensitive trees. In this instance, controlled burns can be used to suppress oaks and other hardwoods, thereby maintaining the natural fire-adapted ecosystem (see figure).

Fire is a necessary tool of ecological management. Here, a controlled burn helps maintain a ponderosa pine (*Pinus ponderosa*) stand in Oregon. (*Joan Landsberg, USDA Forest Service*)

Tropical Rain Forests Occur Where Temperature and Precipitation Are Uniformly High

Tropical rain forests, which are lush equatorial forests, are found near the Equator in Central and South America, Africa, and Southeast Asia. The world's largest tropical rain forest is in the Amazon River basin of northern South America.

Tropical rain forests occur where the temperature is warm throughout the year and precipitation occurs almost daily (Figure 27-10). The annual precipitation of tropical rain forests is 200 to 450 centimeters (80 to 180 inches). Much of this precipitation comes from locally recycled water that enters the atmosphere by transpiration from the forest's own trees.

Tropical rain forests are often found in areas with ancient soil that has been extensively leached by heavy precipitation, producing a soil that is poor in both minerals and organic content. Since the climate is warm and moist, decay organisms and detritus-feeding ants and termites decompose organic litter quite rapidly. The roots of plants quickly absorb minerals from the decomposing material. Thus, most of the minerals of tropical rain forests are tied up in the vegetation, not the soil.

Of all the biomes, the tropical rain forest is unexcelled in species diversity and variety. No single species dominates the tropical rain forest; one can travel for 0.4 kilometer (0.25 mile) without encountering two individuals of the same species of tree.

The trees of tropical rain forests are usually evergreen flowering plants. Their roots are often shallow and form a mat almost 1 meter (about 3 feet) thick in the upper layer of the soil, which absorbs minerals released from decomposition of organic material. Buttress roots are a feature of many tropical rainforest trees; they hold the tree trunks upright and aid in the extensive distribution of the shallow roots (Figure 27-11).

The vegetation of tropical rain forests is not dense at ground level except near stream banks or where a fallen tree has opened the canopy. The continuous canopy of leaves overhead produces a dark habitat with an extremely moist microclimate. A fully developed rain forest has at least three distinct stories of trees (Figure 27-12). The *upper story* consists of the crowns of occasional very tall trees, some 50 meters

Figure 27-10   A broad view of tropical rainforest vegetation along a riverbank in Southeast Asia. Except at riverbanks, tropical rain forest has a closed canopy that admits little light to the forest floor. (*Frans Lanting/Minden Pictures*)

Figure 27-11   Tropical rainforest trees typically possess elaborate buttresses that support them in the shallow, often wet soil. Shown are buttress roots on Australian banyan (*Ficus macrophylla*). (*John Arnaldi*)

Figure 27-12   A profile of the layers of trees in a tropical rain forest. There are at least three stories, with the middle story being so continuous that very little light penetrates to the forest floor.

(164 feet) or more in height, and is entirely exposed to direct sunlight. The dense *middle story*, which reaches a height of 30 to 40 meters (100 to 130 feet), forms a continuous canopy of leaves that lets in very little sunlight for the support of the sparse understory. The *understory* consists of shrubs and herbs specialized for life in the shade, as well as seedlings of taller trees.

The trees in a tropical rain forest support extensive communities of smaller epiphytic plants, such as orchids and bromeliads. Although epiphytes grow in the crotches of branches, on bark, or even on the leaves of their hosts, they use their host trees only for physical support, not for nourishment.

Since little light penetrates to the understory, some of the plants living there are adapted to climb on already-established host trees. Tropical vines, some as thick as a human thigh, twist up through the branches of tropical rainforest trees (see the section on vines in Chapter 7).

In the rain forest, animals are abundant and varied; there are huge numbers of insect, reptile, and amphibian species as well as birds. Most rainforest mammals, such as sloths and monkeys, live only in the trees, although some large ground-

dwelling mammals, including elephants, are also at home in rain forests.

Human population growth and industrialization in tropical countries may spell the end of most or all tropical rain forests in your lifetime. At current rates of deforestation, scientists estimate that almost all tropical rain forests will be destroyed by 2030. It is likely that many rainforest organisms will become extinct before they have even been scientifically described.

ALTITUDE AFFECTS ECOSYSTEMS

Hiking up a mountain is similar to traveling toward the North Pole with respect to the major vegetation types encountered. This is because it gets colder as one climbs a mountain, just as it does when one travels north. The types of plants growing on the mountain change along with the temperature.

The base of a mountain in Colorado, for example, may have open stands of oak trees, which shed their leaves every autumn. Above that altitude, where the climate is colder and more severe, is a coniferous forest of pine and Douglas fir; above that is a *subalpine forest* that resembles the taiga. Higher still, where the climate consists of cool summers and long, snowy winters, *alpine tundra* occurs, with vegetation composed of grasses, sedges, and small tufted plants. The very top of a tall mountain is often covered with a permanent ice or snow cap, similar to the nearly lifeless polar land areas.

Important environmental differences exist between high altitudes and high latitudes, however, which affect the types of organisms found in each place. Alpine tundra typically lacks permafrost and receives more snow and more high winds than does arctic tundra. Also, the high parts of temperate mountains do not experience the great extremes of daylength associated with the changing seasons in biomes at high latitudes. Furthermore, the intensity of solar radiation, including ultraviolet radiation, is greater at high elevations than at high latitudes, because at high elevations the sun's rays pass through less of the Earth's atmosphere.

Plant life of the alpine tundra varies with different amounts of wind and precipitation at the higher elevations. The most exposed areas support only a few lichens and short plants such as mountain heather and mountain douglasia, which grow as cushions or mats in rocky hollows that afford some protection from the wind. Hardy flowering plants, such as western anemone, alpine gold, and mountain primrose, are found in the more protected areas of alpine tundra. Mountain goats, sheep, marmots, meadow voles, birds, and insects live in alpine tundra.

Some Mountains Have Rain Shadows

Mountains tend to remove moisture from air by causing air masses to rise, which cools them so that water vapor condenses into water droplets, leading to precipitation. If prevailing winds blow onto a mountain range, precipitation occurs primarily on the windward slopes (the side from which the wind blows) of mountains. This situation exists on the North American west coast, where precipitation falls on the western slopes of the Cascade Range and the Sierra Nevada. Downwind, or leeward (in this case, east of the mountain range), a low-precipitation **rain shadow** develops, often generating deserts (Figure 27-13).

HUMANS ARE ALTERING THE ECOSPHERE

We have seen that humans have affected each biome, and although the biomes have been presented as discrete entities, none exists in isolation. Increasingly, the cumulative effects of all people are causing global changes. Two of these are changes in Earth's climate and destruction of the ozone layer, both of which are affected by the production of atmospheric pollutants.

The Greenhouse Effect Results from Air Pollution

The Earth may become warmer during the next century than it has been for the past 1 million years. The source of this change is five chemical compounds that are accumulating in the atmosphere as a result of human activities: carbon dioxide (CO_2), methane (CH_4), ozone (O_3), nitrous oxide (N_2O), and chlorofluorocarbons (CFCs).[1] Collectively these gases,

---

[1]CFCs are a group of chemicals that are familiar to most of us as aerosol propellants in spray cans (now banned in the United States and many other countries) and as the Freon cooling agent in refrigerators and air conditioners.

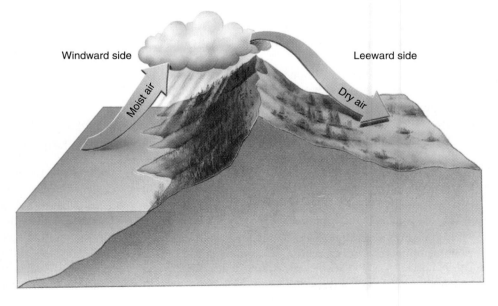

Figure 27-13   A rain shadow. When wind blows moist air over a mountain range, precipitation occurs on the windward side of the mountain, causing a dry rain shadow on the leeward side. Such a rain shadow occurs east of the Cascade Range in Washington state.

Windward side

Leeward side

Moist air

Dry air

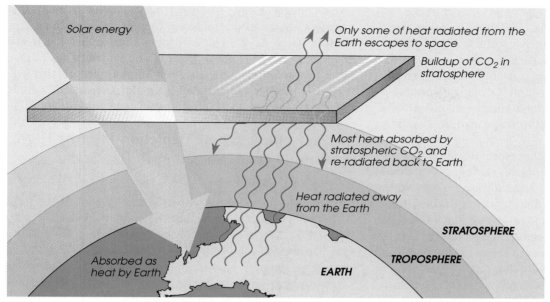

Solar energy

Only some of heat radiated from the Earth escapes to space

Buildup of CO_2 in stratosphere

Most heat absorbed by stratospheric CO_2 and re-radiated back to Earth

Heat radiated away from the Earth

STRATOSPHERE

TROPOSPHERE

Absorbed as heat by Earth

EARTH

Figure 27-14   The buildup of carbon dioxide and other greenhouse gases causes the greenhouse effect, which results in global warming. The increased level of carbon dioxide in the stratosphere is depicted in this diagram as a sheet of glass to emphasize that in this situation CO_2 functions much like the glass walls and roof of a greenhouse.

known as **greenhouse gases**, contribute to climate warming. The concentration of atmospheric CO_2, for example, has increased from about 280 parts per million approximately 200 years ago (before the Industrial Revolution began) to 360 parts per million today. The level of atmospheric CO_2 is still increasing, as are the levels of the other trace gases associated with global warming.

For example, every time you drive your car, the combustion of gasoline in the car's engine releases CO_2 and N_2O and triggers the production of O_3. Every day as tracts of rain forest are burned in the Amazon, CO_2 is released. One way in which CFCs get into the atmosphere is by escaping from old, leaking refrigerators and air conditioners. Decomposition in landfills is a major source of CH_4.

Climate warming occurs because heat that normally would dissipate into space is retained in the atmosphere by these gases (Figure 27-14). Some of the heat from the atmosphere is transferred to the oceans and raises their temperature, as well. As the atmosphere and oceans warm, the overall temperature of the Earth rises. Because CO_2 and other gases trap the sun's radiation in much the same way that the glass of a greenhouse does, global warming produced in this manner is known as the **greenhouse effect**.

Although virtually all scientists studying climate change agree on the existence of the greenhouse effect, the question of how much and how fast the Earth will warm is unclear. The global interactions among the atmosphere, oceans, and land are too complex and too large to study in the laboratory.

As a result, climate researchers develop models by using computer simulations. Most current models predict that a doubling of the concentration of CO_2 in the atmosphere will cause the average temperature of the Earth to increase by 2° to 5°C before the end of the 21st century, although the warming will probably not be uniform from region to region. At current rates of fossil fuel combustion and deforestation, scientists foresee that CO_2 will double within the next 50 years. However, the warming trend will be slower than the increasing CO_2 might indicate, because the oceans take longer than the atmosphere to absorb heat. It is generally thought that the second half of the 21st century will experience greater warming than the first half.

If the overall temperature of the Earth increases by just a few degrees, a major thawing of polar ice caps and glaciers may occur. In addition, the oceans may undergo thermal expansion, because water, like other substances, expands when it warms. These two changes may cause the sea to rise and flood low-lying coastal areas. These coastal areas may suffer additional damage from powerful hurricanes and typhoons.

Global warming is expected to change precipitation patterns, causing some areas to have more frequent droughts. At the same time, other areas may be more likely to have flooding. These factors could affect the availability and quality of fresh water in many locations.

Biologists have started to examine some of the potential consequences of global warming for organisms. Some species will undoubtedly become extinct, particularly those that can survive only within narrow temperature ranges, those confined to small reserves or parks, and those living in fragile ecosystems. Other species may survive, although their numbers and ranges will be greatly reduced. Some species may be able to migrate to new environments or adapt to the changing conditions in their present habitat. Some species may be unaffected by global warming, and others may come out of global warming as winners, with greatly expanded numbers and ranges.

Biologists generally agree that global warming will have an unusually severe impact on plants, because they cannot move about when conditions change. Although seeds may be dispersed over large areas, seed dispersal has definite limits in terms of how *fast* a plant species can migrate. Moreover, soil characteristics, water availability, competition with other species, and human alterations of natural habitats all affect the rate at which plants can move into a new area.

Global warming will increase problems for agriculture, which already has the challenge of providing enough food for a hungry world without doing irreparable damage to the environment. The rise in sea level will cause the inundation of river deltas, which are some of the world's best agricultural lands. Certain agricultural pests and disease-causing organisms may proliferate.

How should we deal with climate change?

Even if we immediately stopped polluting the atmosphere with greenhouse gases, the Earth would probably still have some climate change because of the greenhouse gases that have accumulated during the past 100 years. The amount and severity of global warming depend on the quantity of greenhouse gases we add to the atmosphere.

The development of alternatives to fossil fuels, such as solar energy, would reduce global warming by decreasing CO_2 emissions. The invention of technological innovations that trap the CO_2 emitted from smokestacks would help prevent global warming and yet allow us to use fossil fuels for energy.

One of the most effective ways to mitigate global warming involves forests. As you know, growing plants remove atmospheric CO_2 from the air and, during photosynthesis, incorporate the carbon into organic molecules that make up leaves, stems, and roots. We can lessen global warming by planting and maintaining new forests, thereby tying up the carbon dioxide in plant tissues for many years, and by limiting deforestation and thus preventing the release of CO_2 into the atmosphere from the burning or decomposition of trees. In addition, increasing the energy efficiency of automobiles and appliances would lessen global warming by reducing the output of CO_2.

In June 1992, representatives from around the world met in Rio de Janeiro, Brazil, for a summit conference officially called the United Nations Conference on Environment and Development. One of the issues examined at the conference was climate change. The participants drafted a treaty that curbs carbon dioxide emissions, thereby reducing the greenhouse effect. As of 1994, 166 nations had signed the treaty, which is now considered binding. The United States ratified the treaty in 1992, and President Clinton pledged that U.S. emissions of greenhouse gases would be reduced to 1990 levels by the year 2000, a goal mandated by the treaty.

Stratospheric Ozone Depletion Permits High Amounts of UV Radiation to Reach the Earth's Surface

Ozone (O_3) is a blue gas with a distinctive odor. It is a form of oxygen that is a human-made pollutant in the lower atmosphere, where it is one of the greenhouse gases and is a component of photochemical smog. Naturally produced ozone is, however, an essential part of the stratosphere. The **stratosphere**—the part of the atmosphere that encircles our planet some 10 to 45 kilometers (6 to 28 miles) above the Earth's surface—contains a layer of ozone that shields the Earth from much of the ultraviolet radiation coming from the sun (Figure 27-15).

Although it was first predicted in 1974, stratospheric ozone depletion was not reported until 1985, when a large "hole," or thin spot, was discovered in the ozone layer over Antarctica. The thinning is not a permanent feature but a seasonal phenomenon, evident for only a few months at the onset of the Antarctic spring in September. Every September the ozone hole has reappeared, and each year the layer of ozone has been thinner. In 1990 the minimum ozone concentration in the hole was 50 percent lower than the minimum 10 years earlier. And by 1992 there was clear evidence that stratospheric ozone was also being depleted over the Arctic. Probably the most disquieting news is that worldwide levels of stratospheric ozone have been decreasing for several decades.

The primary culprits responsible for ozone loss in the stratosphere are chlorofluorocarbons (CFCs), which are also implicated in the greenhouse effect. Other compounds that attack ozone include halons (found in many fire extinguishers), methyl bromide (used as a pesticide in agriculture), methyl chloroform (used to degrease metals), and carbon tetrachloride (used in many industrial processes, including the manufacture of pesticides and dyes).

These compounds drift up to the stratosphere, where ultraviolet radiation breaks them down into chlorine, fluorine, bromine, and carbon. Under certain conditions that exist in the stratosphere, chlorine is capable of reacting with ozone, converting it to molecular oxygen. The chlorine is not altered by this process; as a result, a single chlorine atom can break down thousands and thousands of ozone molecules.

Stratospheric ozone absorbs harmful ultraviolet radiation from sunlight. With the depletion of the ozone layer, greater levels of ultraviolet radiation reach the surface of the Earth.

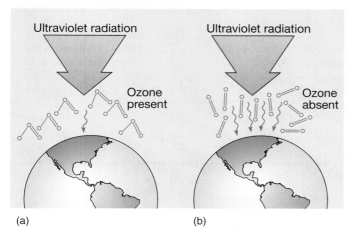

Figure 27-15   Ultraviolet radiation and the ozone layer.
(a) Ozone absorbs ultraviolet radiation, effectively shielding the Earth so that little UV reaches the surface (*red arrow*).
(b) When ozone is absent, more high-energy ultraviolet radiation penetrates the atmosphere and reaches the Earth's surface, where it harms organisms.

Excessive exposure to ultraviolet radiation is linked to a variety of health problems in humans, including cataracts, skin cancer, and weakened immune systems. Scientific evidence also documents crop damage from exposure to high levels of ultraviolet radiation.

Biologists are concerned that the ozone hole over Antarctica may damage the plankton that makes up the base of the food web for the southern ocean. A 1992 study confirmed that increased ultraviolet radiation is penetrating the surface waters around Antarctica and that the productivity of antarctic phytoplankton has declined by at least 6 to 12 percent as a result. If the productivity of phytoplankton continues to decline, the complex food web of Antarctica, which includes fishes, seals, penguins, whales, and vast populations of birds, will be at risk.

In 1987 representatives from 24 countries, including the United States, met in Montreal to sign the Montreal Protocol, an agreement to significantly reduce CFC production by 50 percent by 1998. Then, in 1990, more than 90 countries signed an agreement to phase out all use of CFCs by 2000. In 1992 the phase-out date for CFCs and similar compounds was moved to 1996. Because CFCs are very stable, they can survive in the atmosphere 120 years or more, so the adverse changes initiated by their release may not be quickly reversible. In the final analysis, however, the handling of ozone depletion may become one of the 20th century's most significant environmental success stories. When faced with a serious environmental crisis, governments, scientists, industrialists, and citizens of many nations worked together to help solve the problem.

STUDY OUTLINE

I. A biome is a large, relatively distinct terrestrial region characterized by a certain combination of climate, soil, plants, and animals that is approximately the same regardless of where it occurs in the world.
   A. Tundra, the northernmost biome, is characterized by a permanently frozen layer of subsoil called the permafrost and by low-growing vegetation adapted to extreme cold and a very short growing season.
   B. The taiga, or boreal forest, lies south of the tundra and is dominated by coniferous trees.
   C. Temperate rain forests, such as occur on the northwest coast of North America, receive high precipitation and are dominated by conifers.
   D. Temperate deciduous forests, which occur where precipitation is relatively high, are dominated by broad-leaved trees that lose their leaves seasonally.
   E. Temperate grasslands typically possess deep, mineral-rich soil and have moderate but uncertain precipitation.
   F. Deserts occur in temperate and tropical areas where precipitation rates are low. Desert organisms have specialized water-conserving adaptations.
   G. Tropical grasslands, called savannas, have widely scattered trees interspersed with grassy areas.
   H. The chaparral biome is characterized by thickets of small-leaved evergreen shrubs and small trees and by a climate of wet, mild winters and very dry summers.

I. Tropical rain forests are characterized by mineral-poor soil and very high rainfall that is evenly distributed throughout the year. Tropical rain forests have high species diversity, with at least three stories of forest foliage and many epiphytes.
II. The changes in vegetation that occur with increasing altitude on mountains resemble the changes in vegetation observed in a transition from warmer climates to colder climates.
III. Carbon dioxide and other greenhouse gases cause the atmosphere to retain heat and may contribute to global warming.
   A. Increases in CO_2 and other greenhouse gases (methane, nitrous oxide, CFCs, and ozone) in the atmosphere are arousing concerns about major climate changes that may occur during the next century.
   B. Global warming may cause a rise in sea level, changes in precipitation patterns, extinction of animals and plants, and problems for agriculture.
IV. The ozone layer in the stratosphere helps to shield the Earth from damaging solar ultraviolet radiation.
   A. The total amount of stratospheric ozone is declining, and large ozone holes develop over Antarctica and the Arctic each year.
   B. The attack on the ozone layer is caused by chlorofluorocarbons (CFCs) and similar chlorine-containing compounds.

SELECTED KEY TERMS

allelopathy, p. 457
biome, p. 452
chaparral, p. 456
desert, p. 457
greenhouse effect, p. 463

ozone, p. 464
permafrost, p. 452
rain shadow, p. 462
savanna, p. 457

sclerophyllous leaf, p. 457
taiga, p. 452
temperate deciduous forest,
    p. 455

temperate grassland, p. 456
temperate rain forest, p. 454
tropical rain forest, p. 460
tundra, p. 452

REVIEW QUESTIONS

1. What climate and soil factors produce each of the major biomes?
2. Name several representative plants and animals of each major terrestrial biome.
3. In which biome do you live? If your biome does not match the description given in this chapter, explain the discrepancy.
4. Which biomes are best suited for agriculture? Explain why each of the biomes you did not mention is unsuitable for agriculture.
5. Compare and contrast:
   a. Temperate rain forest and tropical rain forest
   b. Temperate grassland and savanna
   c. Chaparral and desert
   d. Tundra and taiga
   e. Temperate deciduous forest and temperate grassland
   f. Temperate grassland and desert
6. Describe the greenhouse effect, and name the most important greenhouse gas.
7. What environmental service does the stratospheric ozone layer provide? Describe what is causing the chemical destruction of ozone in the stratosphere.

THOUGHT QUESTIONS

1. Why do most animals of the tropical rain forest live in trees?
2. Why are fires not common in desert areas with hot, dry climates?
3. It has been suggested that the wisest way to "use" fossil fuels would be to leave them in the ground. How would this affect global warming? Energy supplies?
4. This statement was overheard in an elevator: "CFCs cannot cause the stratospheric ozone depletion over Antarctica because there are no refrigerators in Antarctica." Criticize the reasoning behind this statement.

SUGGESTED READINGS

Capon, B., *Plant Survival: Adapting to a Hostile World*, Portland, Ore., Timber Press, 1994. A fascinating account of how various plants are adapted to survive such environmental extremes as the heat and drought of a desert, the freezing cold of the tundra, and the salt in a coastal marsh.

Chadwick, D.H., "Roots of the Sky," *National Geographic*, Vol. 184, No. 4, October 1993. Wildlife flourishes in the patchy remnants of the North American prairie.

Goulding, M., "Flooded Forests of the Amazon," *Scientific American*, March 1993. During the rainy season, sections of the Amazonian rain forest are inundated, making them a uniquely aquatic ecosystem as well as a terrestrial one.

Kappel-Smith, D., "Fickle Desert Blooms: Opulent One Year, No-Shows the Next," *Smithsonian*, March 1995. Highlights flowering plants that have adapted to the arid conditions of the desert.

Moffett, M., "These Plants Claw and Strangle Their Way to the Top," *Smithsonian*, September 1993. Many unusual vines are found in tropical forests.

Rutzler, K., and I.C. Feller, "Caribbean Mangrove Swamps," *Scientific American*, March 1996. All about mangrove forests, which cover perhaps 70 percent of tropical coastlines and perform many environmental services.

APPENDICES

APPENDIX A
Metric Equivalents and Temperature Conversion

Some Common Prefixes

		Examples
kilo	1,000	a kilogram is 1,000 grams
centi	0.01	a centimeter is 0.01 meter
milli	0.001	a milliliter is 0.001 liter
micro (μ)	one millionth	a micrometer is 0.000001 (one millionth) of a meter
nano (n)	one billionth	a nanogram is 10^{-9} (one billionth) of a gram

The relationship between mass and volume of water (at 20°C)

$$1g = 1cm^3 = 1mL$$

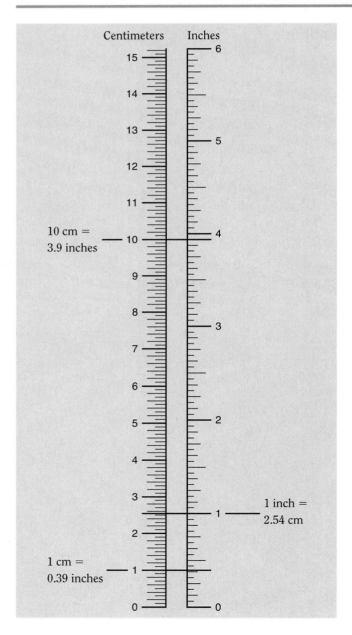

Centimeters Inches

10 cm = 3.9 inches

1 inch = 2.54 cm

1 cm = 0.39 inches

Some Common Units of Length

Unit	Abbreviation	Equivalent
meter	m	approximately 39 in
centimeter	cm	10^{-2} m
millimeter	mm	10^{-3} m
micrometer	μm	10^{-6} m
nanometer	nm	10^{-9} m

Length Conversions

1 in = 2.5 cm	1 mm = 0.039 in
1 ft = 30 cm	1 cm = 0.39 in
1 yd = 0.9 m	1 m = 39 in
1 mi = 1.6 km	1 m = 1.094 yd
	1 km = 0.6 mi

To convert	Multiply by	To obtain
inches	2.54	centimeters
feet	30	centimeters
centimeters	0.39	inches
millimeters	0.039	inches

Standard Metric Units

		Abbreviation
Standard unit of mass	gram	g
Standard unit of length	meter	m
Standard unit of volume	liter	L

Some Common Units of Volume

Unit	Abbreviation	Equivalent
liter	L	approximately 1.06 qt
milliliter	mL	10^{-3}L (1 mL = 1 cm^3 = 1 cc)
microliter	μL	10^{-6} L

Volume Conversions

1 tsp = 5 mL	1 mL = 0.03 fl oz
1 tbsp = 15 mL	1 L = 2.1 pt
1 fl oz = 30 mL	1 L = 1.06 qt
1 cup = 0.24 L	1 L = 0.26 gal
1 pt = 0.47 L	
1 qt = 0.95 L	
1 gal = 3.79 L	

To convert	Multiply by	To obtain
fluid ounces	30	milliliters
quart	0.95	liters
milliliters	0.03	fluid ounces
liters	1.06	quarts

Some Common Units of Mass

Unit	Abbreviation	Equivalent
kilogram	kg	10^3 g (approximately 2.2 lb)
gram	g	approximately 0.035 oz
milligram	mg	10^{-3} g
microgram	μg	10^{-6} g
nanogram	ng	10^{-9} g

Mass Conversions

1 oz = 28.3 g	1 g = 0.035 oz
1 lb = 453.6 g	1 kg = 2.2 lb
1 lb = 0.45 kg	

To convert	Multiply by	To obtain
ounces	28.3	grams
pounds	453.6	grams
pounds	0.45	kilograms
grams	0.035	ounces
kilograms	2.2	pounds

Energy Conversions

calorie (cal) = energy required to raise the temperature of 1 g of water (at 16°C) by 1°C

1 calorie = 4.184 joules

1 Kilocalorie (Kcal) = 1000 cal

Temperature Scales

Celsius (Centigrade) = °C	
Fahrenheit = °F	

Temperature Conversions

$$°C = \frac{(°F - 32) \times 5}{9}$$

$$°F = \frac{°C \times 9}{5} + 32$$

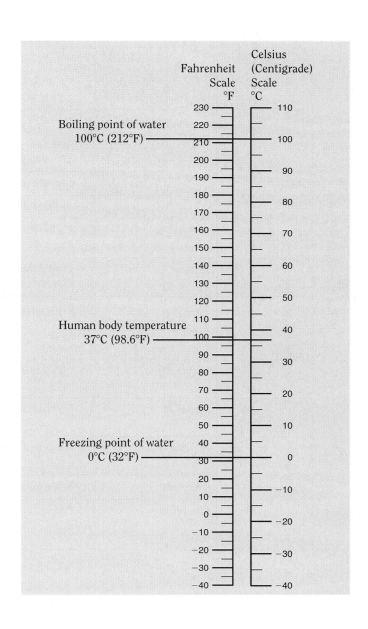

APPENDIX B
Major Periods in Earth's History

Time	Era	Period	Epoch	Geological/Climatic Conditions
10,000 years ago to present	Cenozoic	Quaternary	Holocene	End of last Ice Age; warmer climate; higher sea levels as glaciers melt
2	Cenozoic	Quaternary	Pleistocene	Four Ice Ages; glaciers in Northern Hemisphere
5	Cenozoic	Tertiary	Pliocene	Uplift and mountain-building; volcanoes; climate much cooler; North and South America join at Isthmus of Panama
25	Cenozoic	Tertiary	Miocene	Mountains form; climate drier and cooler
38	Cenozoic	Tertiary	Oligocene	Rise of Alps and Himalayas; most land low; volcanic activity in Rockies; climate cool and dry
55	Cenozoic	Tertiary	Eocene	Climate warmer
65	Cenozoic	Tertiary	Paleocene	Continental seas disappear; climate mild to cool and wet
144	Mesozoic	Cretaceous		Continents separate; Rockies form; other continents low; large inland seas and swamps; climate warm
213	Mesozoic	Jurassic		Continents low; inland seas; mountains form; continental drift begins; climate mild
248	Mesozoic	Triassic		Many mountains form; widespread deserts; climate warm and dry
286	Paleozoic	Permian		Glaciers; Appalachians form; continents rise and merge as Pangaea; climate variable
360	Paleozoic	Carboniferous		Lands low and swampy; climate warm and humid, becoming cooler later
408	Paleozoic	Devonian		Glaciers; inland seas
438	Paleozoic	Silurian		Continents mainly flat; flooding; climate warm
505	Paleozoic	Ordovician		Sea covers continents; climate warm
570	Paleozoic	Cambrian		Oldest rocks with abundant fossils; lands low; climate mild and wet

Million Years Before Present

Plants and Microorganisms	*Animals*
Decline of some woody plants; rise of herbaceous plants	Age of *Homo sapiens*
Extinction of many plant species	Extinction of many large mammals; humans evolve
Development of grasslands; decline of forests	Many grazing mammals; large carnivorous mammals; first known human-like primates
Flowering plants continue to diversify	Many new mammal species appear
Spread of forests; flowering plants	Apes appear; all present mammalian families are represented
Gymnosperms and flowering plants dominant	Mammals diversify; modern birds diverge
Many now-extinct woody flowering plants	Primitive mammals diversify
Rise of flowering plants	Dinosaurs reach peak, then become extinct; toothed birds become extinct; first modern birds; primitive mammals
Gymnosperms common	Large, specialized dinosaurs; first toothed birds; insectivorous marsupials
Gymnosperms dominant; ferns common	First dinosaurs; egg-laying mammals
Conifers diversify; cycads appear	Modern insects appear; mammal-like reptiles; extinction of many Paleozoic invertebrates
Forests of ferns, club mosses, horsetails, and gymnosperms; mosses and liverworts	First reptiles; spread of ancient amphibians; many insect forms; ancient sharks abundant
Plants diversify and become well-established; first forests; gymnosperms appear; bryophytes appear	Fishes diversify; amphibians appear; wingless insects appear; many trilobites
Vascular plants appear; algae dominant in aquatic environments	Fishes diversify; terrestrial arthropods; coral reefs common
Marine algae dominant	Invertebrates dominant; first fishes appear
Algae; bacteria and cyanobacteria; fungi	Age of marine invertebrates; most modern animal phyla represented

APPENDIX C
Answers to Review Questions

Chapter 1

1. People in highly developed countries consume too large a share of resources (that is, more than is needed to survive), resulting in pollution, resource depletion, and environmental degradation.
2. Plants exhibit a high degree of organization, take in and use energy, respond to stimuli, grow and develop, reproduce, possess DNA as genetic material, and adapt to their environment over time.
3. The hierarchical levels of biological organization are atoms → molecules → macromolecules → organelles → cells → tissues → organs → organisms → populations → communities → biosphere.
4. The cactus is adapted to an arid environment. Its thick, succulent stem stores water, and its spines discourage herbivores from eating the succulent stem tissue.
5. The six kingdoms of life are Archaebacteria, Eubacteria, Protista, Fungi, Animalia, and Plantae. Plants are complex multicellular organisms that obtain energy by photosynthesis and possess a cuticle, stomata, and multicellular gametangia.
6. The hierarchical levels of classification are species → genus → family → order → class → phylum → kingdom.
7. Two plants classified in the same species are more similar to one another than two plants classified in the same family. Classification is hierarchical, and each taxonomic level is more inclusive than the one below. (A family contains all of the species of all of the genera that are classified in that family.)
8. In the binomial system of nomenclature, each species is given a two-part name consisting of its genus and its specific epithet. The white oak, for example, is *Quercus alba*.
9. Scientific conclusions are inferred from available data, which are based on observation and experimentation. Conclusions based on faith, emotion, and intuition are not a part of science.
10. Inductive reasoning uses specific examples to draw a general conclusion, whereas deductive reasoning operates from generalities to specifics. A hypothesis is an educated guess that is testable by observation and experimentation; a theory is a hypothesis that is supported by a large body of observations and experiments; a principle is a theory that has withstood repeated testing and is the strongest statement we can make about the natural world.
11. A control is a group in which the experimental variable is kept constant. The control provides a standard of comparison that is used to verify the results of the experiment.

Chapter 2

1. An element is a substance that cannot be broken down into simpler substances by chemical changes, whereas a compound is a substance composed of elements that are united in fixed ratios; a compound can be broken down into its component elements. An atom is the smallest unit of an element, whereas a molecule is the smallest unit of a covalent compound. An atom has no electrical charge; an ion, because it has gained or lost electrons, has a negative or positive charge.
2. The atomic number is the number of protons in an atom, whereas the mass number is the total number of protons and neutrons.
3. The element phosphorus, $^{31}_{15}P$, has (a) 15 protons, 16 neutrons, and 15 electrons; (b) atomic number 15; and (c) mass number 31.
4. In a covalent bond between two atoms, pairs of electrons are shared; an ionic bond is an attraction between oppositely-charged ions.
5. Water is a polar molecule with a high freezing/melting point, high boiling point, and high heat of vaporization. Many substances dissolve in water.
6. A monosaccharide is a simple sugar, whereas a disaccharide consists of two monosaccharides that are covalently linked, and a polysaccharide consists of many monosaccharides that are covalently linked.
7. Neutral fats and oils consist of a glycerol covalently linked to one, two, or three fatty acids.
8. See Figure 2–10.

9. An enzyme is an organic catalyst that accelerates a specific chemical reaction by lowering the activation energy required for that reaction. They are essential because they permit chemical reactions to occur at a rate that supports life.

Chapter 3

1. A microscope's magnification is its ability to enlarge a small object; resolving power is the ability to show fine detail in that object.

2. Mitochondria and chloroplasts are alike in that both are cellular organelles involved in energy conversions. They differ in the kinds of energy conversions that take place in them: mitochondria are sites of aerobic respiration, and chloroplasts are sites of photosynthesis.

3. Proteins are made at the ribosomes along the rough ER using instructions from the nucleus. The protein is packaged into a membrane-bounded sac by the ER and sent to a Golgi body where the proteins are modified. The finished products are then packaged into a membrane-bounded vesicle that migrates to the plasma membrane, fuses with it, and releases the contents outside.

4. The strength in a woody plant is largely the result of the lignin in its secondary cell walls. A nonwoody plant supports itself largely as a result of the turgidity of its vacuoles, which press against the cell walls.

5. A membrane consists of a phospholipid bilayer in which various proteins are embedded. The nonpolar, hydrophobic fatty acid chains of the phospholipids project into the interior of the membrane, and the polar, hydrophilic heads of the phospholipids are on the two surfaces of the membrane. The fluid mosaic model is so-called because the phospholipids are *fluid* and move about laterally, while the proteins appear to form a *mosaic* pattern in the membrane.

6. A plant that is placed in a hypertonic solution will wilt because water always moves across a selectively permeable membrane from a hypotonic solution (in this case, the interiors of the plant cells) to a hypertonic solution.

7. In facilitated diffusion, a substance moves from an area of higher concentration to an area of lower concentration, and no metabolic energy is required. In active transport, which requires metabolic energy, a substance moves from an area of lower concentration to an area of higher concentration.

8. See Figure 3–4.

9. See Figure 3–13.

Chapter 4

1. In photosynthesis, light energy is captured by chlorophyll and certain other pigments and converted to the chemical energy in carbohydrate molecules, which are synthesized from water and carbon dioxide.

2. All organisms require an energy source to survive, and, with few exceptions, photosynthesis supplies that energy—either directly (for example, plants) or indirectly (for example, animals that eat plants).

3. The equation for photosynthesis is

$$6CO_2 + 12H_2O \xrightarrow{\text{Light}} C_6H_{12}O_6 + 6O_2 + 6H_2O$$

4. In the light-dependent reactions, light energy is absorbed and converted to chemical energy, which is temporarily stored in ATP and NADPH; water is split in the process. In the Calvin cycle, the ATP and NADPH that were produced during the light-dependent reactions supply the energy to fix CO_2, forming carbohydrates.

5. As energized electrons move along an electron transport chain, some of their energy is used to pump protons (H^+) across a membrane. This establishes a proton gradient (a high concentration of protons on one side of the membrane) that contains potential energy. These protons can cross the membrane only through special channels found in the enzyme ATP synthase, and as they pass through the membrane, energy is released that is used to synthesize ATP.

6. In a city's electric power plant, the chemical energy stored in fuel molecules, such as coal, is converted to another form, electrical energy, which provides the energy to heat water, produce light, and so on. In a cell's mitochondria, the chemical energy stored in fuel molecules, such as glucose, is converted to ATP, which provides the energy for growth, movement, reproduction, and so on.

7. The equation for aerobic respiration is

$$C_6H_{12}O_6 + 6O_2 + 6H_2O \longrightarrow$$
$$6CO_2 + 12H_2O + \text{Energy (as ATP)}$$

8. Hydrogens removed from the fuel molecule during glycolysis are temporarily accepted by NAD, forming NADH. When oxygen is present, the NADHs go to the mitochondria, where their hydrogens (or their electrons) are removed and transferred along an electron transport chain, releasing energy that is used to produce ATP.

9. Molecular oxygen (O_2) is the final acceptor at the end of the electron transport chain in the mitochondria. It combines with hydrogen to form water. When aerobic cells are deprived of oxygen, the electron transport system stops, and because the cells are unable to generate sufficient ATP, they die.

10. See Figure 4–12.

Chapter 5

1. Roots anchor the plant and absorb water and dissolved minerals from the soil. Shoots consist of stems, which bear leaves, the main organs of photosynthesis, and reproductive structures.

2. The dermal tissue system covers the plant's outer surface and provides protection. The vascular tissue system conducts water, dissolved minerals, and dissolved food (carbohydrates) throughout the plant body; it also provides structural support. The ground tissue system has a variety of functions, including photosynthesis, storage, secretion, and structural support.
3. Parenchyma cells, which have thin primary cell walls, function to photosynthesize, store materials, and secrete substances. Collenchyma cells, which have unevenly thickened primary cell walls, provide support, particularly in soft, nonwoody plant organs. Sclerenchyma cells, which have primary walls and thick secondary walls, provide support, particularly in woody plant parts.
4. Xylem conducts water and dissolved minerals, whereas phloem conducts food materials (carbohydrates) in solution. Both tissues also provide strength and support to the plant body.
5. Both epidermis and periderm function as the outer protective covering of the plant body. Epidermis covers the herbaceous plant body, and periderm, which comprises the outer bark, covers the woody plant body.
6. Meristems are sites of cell division and growth in plants.
7. Primary growth is an increase in stem and root length, is due to the activity of apical meristems, and occurs in all plants. Secondary growth is an increase in stem and root girth, is due to the activity of lateral meristems, and occurs in woody plants.

Chapter 6

1. Each plant has either a taproot system or a fibrous root system.
2. Roots function in absorption, conduction, storage, and anchorage. Epidermal root hairs aid in absorption; xylem and phloem are the conducting tissues; and cortex is the main storage tissue. Pericycle could be considered the tissue responsible for anchorage because it gives rise to lateral roots.
3. A root hair is a single-celled extension of the epidermis. Lateral roots are multicellular and originate from the pericycle.
4. If the central tissue of the root is xylem, it is probably a dicot, whereas if the central tissue is pith, it is probably a monocot.
5. The pathway of water from the soil through the root tissues in a herbaceous dicot root is root hair and epidermis → cortex → endodermis → pericycle → xylem.
6. Storage roots are enlarged, usually to store starch. Prop roots develop from branches or a vertical stem and provide support. Aerial "breathing" roots are found on plants that live in wet areas and may help get oxygen to the submerged roots. Some epiphytes have photosynthetic roots. Corms and bulbs may have contractile roots, which pull them deeper into the soil.
7. See Figure 6–4.
8. See Figure 6–8.

Chapter 7

1. Stems support the plant body, conduct, and produce new stem tissues. In herbaceous stems, support is provided by collenchyma and sclerenchyma tissues found in the cortex, xylem, and phloem; xylem and phloem are the conducting tissues; and meristems produce new tissues.
2. If the vascular bundles are arranged in a circle, it is probably a dicot, whereas if the vascular bundles are scattered throughout the stem, it is probably a monocot.
3. As the stem increases in girth, the epidermis, cortex, and primary phloem are gradually split apart and sloughed off. Cells of the pith and primary xylem die but remain in the center of the stem.
4. Periderm, secondary phloem, and remnants of certain primary tissues (epidermis, cortex, and primary phloem) are removed when the bark of a branch is peeled off.
5. Both vascular cambium and cork cambium are lateral meristems. Vascular cambium, which is located between the primary xylem and primary phloem, produces secondary xylem and secondary phloem, whereas cork cambium, which arises near the stem's surface, produces cork cells and cork parenchyma.
6. An annual ring is a concentric circle of wood, as seen in cross section, and usually represents one growing season's growth. An annual ring appears as a ring because of differences in cell size and cell wall thickness between secondary xylem formed at the end of one year's growth and that formed at the beginning of the next year's growth.
7. A rhizome is a horizontal underground stem; a tuber is an underground stem that is greatly enlarged for storing food; a bulb is an underground bud with a shortened stem and fleshy storage leaves; a corm is an underground stem that superficially resembles a bulb; a stolon is an aboveground horizontal stem with long internodes.
8. See Figure 7–3.
9. See Figure 7–4.

Chapter 8

1. See Figures 8–2 (a and d), 8–3, and 8–4.
2. The general equation for photosynthesis is

$$6CO_2 + 12H_2O \xrightarrow{\text{Light}} C_6H_{12}O_6 + 6O_2 + 6H_2O$$

Carbon dioxide enters the leaf through stomata and diffuses into all mesophyll cells because of their loose arrangement; water is delivered to the leaf by the xylem; light penetrates into the leaf's interior because the blade is thin and flat and because the epidermal cells are transparent. Sugar produced by photosynthesis is converted to sucrose and transported to other parts of the plant in the phloem; the oxygen produced as a by-product of photosynthesis diffuses into the atmosphere through the stomata.

3. The stomata are related to both photosynthesis and transpiration. The carbon dioxide needed for photosynthesis diffuses into the leaf's interior through its stomata, and the oxygen produced during photosynthesis diffuses out of the leaf through its stomata; most water that a plant transpires passes out of the plant through its stomata.

4. When water moves into the guard cells, making them turgid, they change shape and a pore forms, that is, the stoma opens. When water moves out of the guard cells, making them flaccid, the pore collapses, that is, the stoma closes.

5. Leaves adapted to arid conditions often have (1) a thick shape that gives them less surface area exposed to the dry air and (2) a very thick cuticle that reduces water loss. (*Note:* other answers are possible, such as the presence of sunken stomata and leaves that are reduced in size.)

6. Sunlight, warmer temperatures, lower humidity, and wind all increase the rate of transpiration. Stomata open in the presence of light and/or a low concentration of CO_2 in the leaves; during a drought, however, stomata remain closed, even when it is day and the concentration of CO_2 in the leaves is low.

7. Leaf abscission in autumn allows a plant to conserve water (transpiration becomes negligible after the leaves are shed) and therefore to survive the low temperatures of winter when little or no water can be absorbed from the soil.

8. Two plants with leaves adapted for trapping insects are the Venus flytrap and the pitcher plant. Leaves of the Venus flytrap are shaped like tiny bear traps that snap shut when an insect alights. Leaves of the pitcher plant are shaped so that rainwater collects in them; insects drown in the water before being digested.

9. The reasons for tree decline are many, including air pollution, insect damage, low winter temperatures, prolonged droughts, and bacterial, viral, and fungal diseases.

10. See Figure 8–5.

Chapter 9

1. Sexual reproduction involves the fusion of gametes and results in offspring that are genetically variable. Asexual reproduction does not involve the fusion of gametes and results in offspring that are virtually identical to one another and to the parent.

2. An imperfect flower is always incomplete, but an incomplete flower is not necessarily imperfect. An imperfect flower lacks either stamens or carpels, whereas an incomplete flower lacks one or more of these parts: sepals, petals, stamens, and carpels.

3. Pollination is the transfer of pollen from the anther to the stigma, and fertilization is the fusion of gametes. In flowering plants, fertilization can take place only if pollination has first occurred.

4. A seed is a structure containing a plant embryo and nutrients, whereas a fruit is a structure that surrounds and protects the seeds. Seeds develop from ovules, and fruits develop from ovaries.

5. A simple fruit develops from a single flower that has one pistil; tomatoes, cherries, pea pods, and corn kernels are examples. An aggregate fruit develops from a single flower that contains many separate pistils; raspberries are an example. A multiple fruit develops from many pistils of many flowers that grow in proximity to one another; pineapples are an example. An accessory fruit is one in which the major part of the fruit is tissue other than ovary tissue; strawberries and apples are examples.

6. Some animal-dispersed seeds have spines or barbs that get caught in the animal's feathers or fur, or in human clothing; these seeds are carried from one place to another. Other animal-dispersed seeds are swallowed when an animal eats the surrounding fruit tissue; these seeds are deposited with the animal's feces, usually in a different location from where they were eaten.

7. See Figure 9–1.

Chapter 10

1. Soil consists of inorganic minerals, organic matter, soil organisms, soil atmosphere, and soil water. Plants obtain essential minerals from both the inorganic and organic portions of the soil. (The organic material must first decompose before its minerals are available to the plant.) Soil atmosphere supplies oxygen to the roots, and soil water provides both water and a dilute solution of dissolved mineral ions. Soil organisms help plants by improving the soil; they do this by decomposing organic matter, cycling nutrients, aerating the soil, and adding organic matter to the soil.

2. Chemical, physical, and biological weathering processes gradually break rock into smaller and smaller pieces that eventually form the inorganic mineral portion of the soil.

3. If an element is essential for plant growth, plants grown in the absence of that element cannot develop normally or complete their life cycle. Essential elements must be shown to affect a plant's metabolism directly.

Essential elements are required by many kinds of plants.

4. Macronutrients are required in fairly large quantities (greater than 0.5 percent dry weight), whereas micronutrients are required in trace amounts.

5. Organic fertilizers are derived from organisms, have complex but variable compositions, and release their minerals slowly as the organic material decomposes. Inorganic fertilizers do not come from organisms, have relatively simple but precise chemical compositions, and release their minerals rapidly.

6. According to the tension-cohesion theory, the evaporative pull of transpiration causes a tension at the top of the plant that pulls water up the plant from the soil. Water can be pulled only if it forms an unbroken column in the plant, and this is possible because water is both cohesive with other water molecules and adhesive with xylem cell walls.

7. Root pressure is a mechanism for water transport in which water moves by osmosis from the soil into roots; this movement produces a pressure that pushes water up the xylem. Root pressure is not as important in larger plants as it is in smaller plants; also, root pressure is not as strong during dry periods of the year as it is in the spring when the soil is very wet.

8. Sugar is actively loaded into phloem at the source, and water moves into the phloem by osmosis, increasing the pressure at the source. Sugar is actively unloaded from phloem at the sink, and water moves out of the phloem by osmosis, decreasing the pressure at the sink. Materials move in the phloem from the source to the sink as a result of the difference in pressure between the two locations.

9. Saline soils are physiologically dry even when they are physically wet because their concentration of water is lower than that in plant cells (because they contain such a high concentration of dissolved salts). Water therefore moves by osmosis out of plant cells into the surrounding soil.

Chapter 11

1. Factors that may affect germination include oxygen, water, temperature, light, embryo maturity, and chemical inhibitors. Oxygen is required by most germinating seeds for aerobic respiration, and the germinating plant must respire because it has a high energy requirement. Water is required by germinating seeds because a watery environment allows the plant's metabolism to be turned on. The seeds of each species germinate at an optimal temperature that assures their likely survival in the environment to which they are adapted. Some seeds require exposure to a prolonged period of cold

(which corresponds to winter) before germinating; without a cold requirement, these seeds would perish because they would germinate immediately before or during winter. Many small seeds require exposure to light for germination. These seeds germinate if they are planted in shallow soil (which light can penetrate) but remain dormant if planted in deep soil where light cannot penetrate; if these seeds germinated in deep soil, their small amount of storage materials would be depleted before they were able to grow to the soil's surface. Some seeds have immature embryos that must develop further before germinating, whereas other seeds contain chemical inhibitors that prevent germination. These traits are adaptive because they delay germination until environmental conditions are favorable, thereby increasing the likelihood that the plant will survive to maturity.

2. Environmental cues are important to plant growth and development because they trigger sexual reproduction at a time of year that increases the likelihood that reproduction will be successful.

3. Phytochrome is a pigment that is involved in many developmental processes in plants. Phytochrome detects the time of year by determining the relative lengths of daylight and darkness. This information cues plants about when to initiate reproduction—that is, when to flower.

4. Vernalization is the promotion of flowering by exposure to low temperature for a period of time.

5. Photoperiodism is the response of plants to the duration and timing of light and dark; in other words, photoperiodism helps plants recognize the time of year. Circadian rhythms are regular rhythms in growth or activities that approximate a 24-hour day; in other words, circadian rhythms help plants recognize the time of day.

6. Turgor movements are movements caused by changes in turgor of certain cells; turgor movements are temporary. Tropisms are directional growth responses and are permanent.

7. A hormone is a chemical messenger that regulates one or more aspects of plant growth and development.

8. Auxin is transported from the shoot tip down the shaded side of a stem, causing the cells it comes into contact with to elongate. As a result of the differential growth of cells on the sunny and shaded sides, the shoot tip bends toward light.

9. (a) Gibberellin promotes seed germination, and abscisic acid inhibits it; cytokinin may also be involved in seed germination. (b) Auxin, gibberellin, and cytokinin are involved in stem elongation because they promote cell division and/or cell elongation. (c) Ethylene promotes the ripening of fruit, and both auxin and gibberellin are involved in fruit development, of which ripening is one aspect. (d) Ethylene promotes leaf abscission, whereas

auxin and cytokinin inhibit it. (e) Abscisic acid causes seed dormancy, whereas gibberellin breaks seed dormancy; cytokinin may also be involved in breaking dormancy.

10. Auxin promotes cell elongation, the formation of plant organs in tissue culture, apical dominance, and fruit development, but inhibits leaf abscission. Gibberellin promotes cell division, cell elongation, seed germination, flowering, and fruit development; it is also involved in breaking winter dormancy. Cytokinin promotes cell division and formation of plant organs in tissue culture, but inhibits apical dominance, senescence, and leaf abscission; cytokinins may also be involved in seed germination and winter dormancy. Ethylene promotes fruit ripening and leaf abscission, and it may also have a role in apical dominance. Abscisic acid promotes winter dormancy of plants and seed dormancy.

Chapter 12

1. See Figure 12–2. In DNA, adenine and thymine are complementary, as are cytosine and guanine.

2. Because DNA is a double helix consisting of two strands bound to one another by complementary base pairing, it is able to form exact copies of itself by using each of the strands as a template on which a new complementary strand is synthesized.

3. As a result of semiconservative replication, each double helix molecule formed during the replication process contains an old (i.e., conserved) strand and a newly synthesized strand.

4. DNA is synthesized during DNA replication, whereas RNA is synthesized during transcription. In DNA replication, DNA synthesis occurs on both strands of DNA at the same time, whereas in transcription, RNA synthesis occurs on only one of the DNA strands.

5. Both DNA and RNA are composed of nucleotides. DNA nucleotides contain deoxyribose and the bases adenine, guanine, cytosine, and thymine; RNA nucleotides contain ribose and the bases adenine, guanine, cytosine, and uracil. DNA is a double helix whose two strands are joined by hydrogen bonds between complementary bases; RNA is a single-stranded molecule that is sometimes folded elaborately and held in position by hydrogen bonds between complementary bases.

6. Messenger RNA (mRNA) carries, from the nucleus to the ribosome, the coded message that specifies what kind of polypeptide is to be synthesized. Ribosomal RNA (rRNA) is a structural part of ribosomes that must be present if translation is to occur. Transfer RNA (tRNA) molecules carry, to the ribosomes, specific amino acids, which are to be assembled into a polypeptide.

7. A triplet is a 3-base sequence in DNA that is complementary to a codon, a 3-base sequence in mRNA. A codon is, in turn, complementary to an anticodon, a 3-base sequence in tRNA. A triplet and its corresponding codon and anticodon specify a specific amino acid that is to be added to a polypeptide during protein synthesis at a ribosome.

8. The genetic code is based on triplets rather than pairs or single nucleotides because triplets—that is, trios of nucleotides—permit enough unique combinations to specify all 20 amino acids found in proteins.

9. The universality of the genetic code in organisms indicates that they are all related—that is, all organisms living today evolved from the same early life forms.

10. (a) Replication; (b) transcription; (c) translation.

Chapter 13

1. Chromatin is the complex of DNA and protein that makes up eukaryotic chromosomes. A chromosome is one of several rod-shaped bodies in the nucleus that contain hereditary units (genes). A chromatid is one of the two halves of a duplicated (doubled) chromosome.

2. Chromosomes contain genes arranged in a linear sequence.

3. During G_1, the cell grows and synthesizes enzymes used in DNA replication. During the S phase, DNA is replicated, along with other materials such as the proteins that are components of chromosomes. During G_2, protein synthesis occurs in preparation for cell division.

4. A cell that is diploid has two sets of chromosomes, whereas a cell that is haploid has one set of chromosomes. Homologous chromosomes, which occur in pairs, with one originating from the sperm cell and the other from the egg cell, are similar in structure and genetic constitution.

5. See Figure 13–2.

6. Cytokinesis is that part of cell division in which the cytoplasm is divided to form two daughter cells. Mitosis is the division of the cell's nucleus, whereas cytokinesis is division of the cell's cytoplasm.

7. During prophase I, synapsis and crossing over occur. In metaphase I, homologous chromosomes line up in pairs along the equator of the spindle. Homologous chromosomes separate during anaphase I. In telophase I, the nucleus may reorganize, the chromosomes may lengthen, and cytokinesis may occur. During prophase II, new spindle fibers form and the nuclear envelope breaks down (if it reformed during telophase I). In metaphase II, chromosomes line up along the equator, and in anaphase II, the centromeres that join sister chromatids split, and the chromatids, each now considered a chromosome, move to opposite poles. In

telophase II, nuclear envelopes form around each set of chromosomes, the chromosomes lengthen into chromatin, and cytokinesis occurs.

8. Crossing over results in an exchange of genetic material between homologous chromosomes and therefore increases genetic variability in offspring formed by sexual reproduction.

9. Each daughter cell produced by mitosis has the same number of chromosomes as the original cell, whereas each cell produced by meiosis has half the number of chromosomes of the original cell. Mitosis is a single division that produces two daughter cells, whereas meiosis consists of two successive cell divisions that produce four cells. Cells produced by mitosis are identical to each other and to the original cell, whereas cells produced by meiosis are genetically different from one another and the original cell.

10. The immediate product of meiosis in plants is spores; in animals, it is gametes.

Chapter 14

1. Genes are discrete units of hereditary information, and alleles are the alternate forms of a gene. Two genes do not occupy corresponding positions on homologous chromosomes, whereas two alleles do.

2. According to Mendel's principle of dominance, when two different alleles are present in an individual, only one—the dominant allele—is expressed.

3. Only one chromosome member of each homologous pair is received from each parent because meiosis, which reduces the chromosome number in gametes, separates homologous chromosomes from one another.

4. According to Mendel's principle of segregation, the two alleles of each gene separate into different gametes; that is, each gamete receives only one member of each pair.

5. A homozygous individual possesses a pair of identical alleles for a particular trait, and a heterozygous individual possesses a pair of unlike alleles for a particular trait.

6. Dominant alleles are always expressed when present, regardless of whether an individual is homozygous or heterozygous for the trait. Recessive alleles are expressed only when an individual is homozygous.

7. An individual's genotype is its genetic makeup, whereas an individual's phenotype is its appearance—that is, the physical expression of its genes.

8. According to Mendel's principle of independent assortment, the distribution to the gametes of alleles governing one trait are not affected by the distribution to the gametes of alleles governing another trait.

9. A test cross would not have to be performed to determine the genotype of pink four o'clocks. Because flower color in four o'clocks is incompletely dominant, pink four o'clocks can only be R^1R^2.

10. Incomplete dominance is a condition in which a single gene affects a trait, but neither member of a pair of alleles is completely expressed when the other is present. In polygenic inheritance, two or more genes affect the same trait in an additive fashion.

Chapter 15

1. Genetic engineering is the ability to take a specific gene from one cell and place it into another cell, where it is expressed.

2. Restriction enzymes are bacterial enzymes that recognize specific DNA base sequences and cleave the DNA molecule at that site. Ligase is an enzyme that joins pieces of DNA that were cleaved with a restriction enzyme. These enzymes are used in recombinant DNA research to isolate and join DNA segments from different organisms.

3. Restriction enzymes cleave DNA into more manageable fragments. Segments of DNA from different organisms are then joined by ligase to form recombinant DNA molecules.

4. A foreign gene may be implanted in a bacterial cell by using a plasmid as a vector. Plasmids are small circular DNA molecules that can have specific genes spliced into them, thereby forming recombinant plasmids. When bacterial cells take in recombinant plasmids, the gene of interest may be transcribed and translated, resulting in the production of novel proteins by the bacteria.

5. One way to insert genes into plant cells is by using the bacterium *Agrobacterium tumefaciens*. Recombinant plasmids are inserted into *Agrobacterium*, which subsequently infects a plant cell and inserts the plasmid, including the gene of interest, into the plant's chromosome.

6. A genetic probe is a radioactively labeled segment of RNA or single-stranded DNA that can be used to identify cells that contain a specific gene. A few cells from each bacterial colony are transferred to a filter, to which the genetic probe is then added. Because the probe is complementary to the specific gene, it binds to it, allowing its detection by x-ray film.

7. PCR is used to amplify DNA by subjecting the DNA to alternating heat and enzymatic treatments. When heated, a single DNA molecule separates into two single strands (recall that DNA is a double helix). The enzyme DNA polymerase then assembles new nucleotides along the single strands to form two double helices. Thus, one DNA molecule is amplified to two molecules. Repeating the procedure doubles the amount of DNA each time.

8. Short-term benefits of plant genetic engineering include developing plants that are resistant to insect pests, plants that are resistant to viral diseases, and plants that have longer shelf lives.
9. Strict safety guidelines for recombinant DNA research exist because there are unanswered concerns about releasing genetically engineered organisms into the environment.

Chapter 16

1. The four premises of Darwin's theory of evolution by natural selection are: one, each species produces more offspring than will survive to maturity; two, the individuals in a population exhibit heritable variation in their traits; three, organisms compete with one another for the resources needed for life; and four, the offspring with the most favorable combination of characteristics are most likely to survive and reproduce, passing those genetic characteristics to the next generation.
2. Lamarck thought that organisms possessed a vital force that drove them to evolve into organisms of greater complexity; scientific evidence of this vital force has not been demonstrated. Lamarck also thought that organisms could pass on traits they had acquired during their lifetimes to their offspring; this idea was discredited when the mechanisms of heredity were discovered.
3. According to Lamarck, short-necked ancestral giraffes stretched and elongated their necks when reaching up into trees for food; their offspring inherited slightly longer necks, which they stretched still further; over many generations, this process resulted in the giraffe's long neck. According to Darwin, ancestral giraffes had slightly varying neck lengths, which were determined by heredity; the giraffes with longer necks were more likely to obtain food and therefore to survive and reproduce; thus, giraffes with longer necks became more common in the population, and over many generations, the average neck length increased to that of the present-day giraffe.
4. Only inherited variations are important in evolution because variations that are not inherited cannot be passed on to the next generation.
5. Darwin thought that individuals passed traits on to the next generation, but he did not know how traits were passed from one generation to the next or why individuals vary in a population. The synthetic theory of evolution combines Darwin's theory of evolution by natural selection with today's knowledge of genetics.
6. Comparative anatomy of closely related organisms indicates they have underlying structural similarities. Biogeography shows that areas that have been isolated

from the rest of the world for a long time contain organisms unique to those areas. Molecular biology indicates that closely related organisms have longer DNA sequences in common. Fossils of ancient organisms often reveal evolutionary relationships among living species. (*Note:* other answers are possible.)
7. Two examples of mimicry are the yellow bee orchid, a plant whose flowers mimic female bees, and geometrid larvae, caterpillars that resemble twigs.
8. Cacti and spurges are not closely related, yet they have a remarkable resemblance to one another, presumably because they evolved similarly in response to similar environmental conditions in two different places.
9. According to the gradualism model of evolution, evolutionary change is due to a slow, steady accumulation of changes over time. In contrast, in the punctuated equilibrium model, evolution proceeds with long periods of inactivity followed by very active phases of rapid change.
10. Many island species are found nowhere else because they evolved in isolation on an island but could not extend their ranges because the surrounding ocean acted as a barrier to expansion.

Chapter 17

1. A population is a group of organisms of the same species that live in the same geographical area at the same time. A species consists of all the individuals with similar characteristics that in nature breed only with one another and have a close common ancestry. A gene pool consists of all the genes that are present in the individuals of a freely interbreeding population or species. Population genetics helped clarify the definition of species because the members of a species share a common gene pool.
2. Natural selection decreases genetic variation within a population by eliminating alleles that are less favorable for survival. Mutation increases the genetic variation in a population because it produces new alleles. Gene flow increases the genetic variation in a population, because it adds alleles that may not have been present in that population's gene pool before migration took place. Genetic drift decreases the genetic variation in a population as a result of random events.
3. If a mutation occurs in a somatic cell, the new allele cannot become established in the population because it will not be passed on to the next generation. Only mutations in reproductive cells become established in a population.
4. Over many generations, natural selection has resulted in organisms that are well-adapted to their environments. Because an organism's genetic material reflects how well-adapted the organism is, any changes in that

genetic material are more likely to be neutral or harmful than beneficial.

5. Three examples of reproductive isolation are: (1) two species flower at different times of the day, season, or year; (2) the floral structures of two species are different and, as a result, they have different pollinators; and (3) individuals of two species reproduce but their offspring die at an early stage of embryonic development. (*Note:* other answers are possible.)

6. Normal embryonic development is a complex process that requires the interaction and coordination of many genes. Development of interspecific hybrids usually fails because the genes of two different species do not interact properly.

7. Five examples of geographical barriers are oceans, glaciers, mountains, deserts, and large lakes. (*Note:* other answers are possible.)

8. Allopatric speciation is more likely to occur if the isolated population is small rather than large because genetic drift, which causes changes in allele frequencies, is a factor.

9. Allopolyploidy occurs when individuals of two species successfully reproduce and the chromosomes of the interspecific hybrid offspring spontaneously double some time before it undergoes meiosis.

10. Extinction of species provides empty ecological niches that are available for new species.

Chapter 18

1. An organism's common names come from different people in different regions where the organism is found; one organism can have many common names, some of which may also be applied to other organisms. An organism's scientific name consists of two words that are usually derived from Greek or Latin; each organism has a single scientific name that is shared with no other organism. Scientific names are used in scientific work to be accurate and avoid confusion.

2. Linnaeus used the binomial system of nomenclature, in which he shortened the scientific names of organisms to two words, the genus and the specific epithet.

3. A species is a group of organisms that have a common gene pool and can breed with one another under natural conditions but are reproductively isolated from other organisms.

4. The taxonomic levels in the hierarchical system, beginning with species: are species, genus, family, order, class, phylum, and kingdom.

5. Cladistics is an approach to classification that is based on recency of common ancestry.

6. Many biologists classify the bacteria in two kingdoms instead of one, based on molecular evidence—DNA

studies done in the late 1980s—that indicate the archaebacteria and eubacteria are very different from one another.

7. In contrast to the other kingdoms, the kingdom Protista consists of organisms that did not have a common ancestor.

8. Viruses are not cellular and therefore cannot be assigned to one of the six kingdoms.

Chapter 19

1. Archaebacteria and eubacteria differ from one another biochemically. For example, peptidoglycan occurs in the cell walls of eubacteria but not in those of archaebacteria. Plants and other eukaryotes appear to be more closely related to archaebacteria.

2. The cell walls of gram-positive bacteria are very thick and consist primarily of peptidoglycan, whereas the cell walls of gram-negative bacteria consist of a thin wall of peptidoglycan surrounded by an outer membrane.

3. Autotrophs are organisms that synthesize organic molecules from inorganic materials, whereas heterotrophs are organisms that cannot and therefore must obtain them from other organisms. A chemosynthetic autotroph is one that obtains its energy from inorganic chemicals (as compared to most autotrophs, which obtain their energy from light).

4. Some bacteria exchange genetic material by conjugation, in which genetic material is transferred through pili from one cell to another. (*Note:* other possible answers are transformation and transduction.)

5. See Figure 19–8.

6. An endospore is a thick-walled, highly resistant structure formed within certain bacterial cells to withstand adverse environmental conditions.

7. Saprotrophic bacteria are decomposers that break down dead organic materials into their simpler components, which are subsequently reused by other organisms (as raw materials for the synthesis of new cellular components).

8. See Figure 19–1.

Chapter 20

1. Protists are "simple" eukaryotic organisms. Most are aquatic. Protists are difficult to characterize because they are so varied in their sizes, modes of nutrition, methods of reproduction, and kinds of locomotion.

2. Unicellular protists are eukaryotic, whereas bacteria are prokaryotic.

3. Protists provide many important services and products for humans. Photosynthetic protists (the algae) are the base of food webs in aquatic ecosystems and therefore

support a wide variety of organisms, such as fish, on which humans depend for food. Humans eat certain protists (the brown and red algae, for example) and obtain a variety of products from them, such as diatomaceous earth and agar. Certain protists, such as the dinoflagellates that cause red tide and the water mold that causes late blight of potato, are harmful from a human viewpoint. (*Note:* other answers are possible.)

4. Algae are important producers in aquatic ecosystems. Zooxanthellae are protists that live symbiotically in the bodies of corals and other marine invertebrates; their contribution to the productivity of coral reefs is substantial. Larger algae such as kelps provide underwater habitats as well as food for many animals. Water molds are decomposers that break down dead organic material into simpler components. (*Note:* other answers are possible.)

5. Algae could be considered "simple" plants because, like plants, they are photosynthetic. Most biologists classify algae as protists rather than plants because they lack many structural features of plants. Algae, for example, do not possess a cuticle or produce gametes in multicellular gametangia.

6. Classification of algae into phyla is largely based on their pigments and energy-storage molecules.

7. The brown algal body typically consists of leaflike blades, stemlike stipes, and anchoring holdfasts.

8. Water molds could be considered fungi because, like fungi, their bodies are masses of threadlike hyphae and their mode of nutrition is by absorption of predigested food. Most biologists classify water molds as protists rather than fungi because they produce flagellated cells at some point in their life cycles (fungi have no flagellated cells), and their cell wall composition is often different from that of fungi.

Chapter 21

1. Fungi differ from bacteria because fungi are eukaryotic. They differ from other eukaryotes—the protists, plants, and animals—in their mode of nutrition, which is by absorbing predigested organic molecules.

2. The body of a mold consists of a mass of threadlike, branched hyphae. In contrast, yeast are single-celled fungi.

3. A hypha is one of the threadlike filaments that form the mycelium of a fungus; in other words, a mycelium consists of a mass of hyphae. A mycelium forms the vegetative (nonreproductive) body of a fungus, whereas a fruiting body is a reproductive structure in which sexual spores are produced; both mycelium and fruiting body consist of hyphae.

4. Saprotrophic fungi are ecologically important because they degrade organic wastes and dead organisms,

releasing simple raw materials such as carbon dioxide and water into the environment to be reused by other organisms. Lichens are ecologically important because they provide food for grazing animals in the Arctic; other lichens play a role in the formation of soil from rock. Mycorrhizae, mutualistic relationships between plant roots and fungi, are ecologically important because they provide plants with water and essential elements such as phosphorus.

5. Fungi that are saprotrophs obtain their nourishment from dead organisms, whereas fungi that are parasites obtain their nourishment from living organisms.

6. Zygomycetes, ascomycetes, basidiomycetes, and imperfect fungi are the four phyla of fungi. Zygomycetes produce asexual spores in a sporangium and thick-walled, resistant sexual spores (zygospores). Ascomycetes produce asexual spores called conidia, which pinch off of conidiophores; their sexual spores (ascospores) are formed in a saclike ascus. Basidiomycetes rarely produce asexual spores, but they form sexual spores (basidiospores) on a clublike basidium. Imperfect fungi reproduce asexually by forming conidia but do not appear to reproduce sexually.

7. The vegetative (nonreproductive) mycelium of a typical mushroom grows underground. Initially, it is a primary mycelium (that is, it is monokaryotic), but after it fuses with a hypha of a genetically different mating type, a secondary mycelium develops that is dikaryotic. When it is ready to reproduce, compact masses of hyphae called buttons develop along the secondary mycelium. Each button grows into a mushroom. On the gills of the mushroom the dikaryotic nuclei fuse and then undergo meiosis to form haploid basidiospores. Each basidiospore can germinate to produce a new primary mycelium.

8. Three important fungal diseases of plants are: (1) chestnut blight, which is caused by an ascomycete; (2) wheat rust, which is caused by a basidiomycete; and (3) bean anthracnose, which is caused by an imperfect fungus. Three fungal diseases of humans are ringworm and candidiasis, both of which are superficial infections, and histoplasmosis, which is a systemic infection that can be quite serious if it spreads into the bloodstream. (*Note:* other answers are possible.)

9. See Figure 21–5.

Chapter 22

1. Unlike algae, plants have a waxy covering, called a cuticle, over their aerial parts. The external leaf and stem surfaces of plants are dotted with tiny pores, called stomata, to facilitate gas exchange. Plants also produce gametes in multicellular gametangia.

2. It is thought that plants descended from an ancient green alga. Plants and green algae share several features, such as the same photosynthetic pigments and energy storage compound. Also, certain details of cell division are shared by plants and some green algae.

3. Plants have root systems that enable them to obtain enough water to survive. Plants have cuticles that help conserve water. Plants either grow tall so that they can obtain enough sunlight or are adapted to survive in lower light intensities—that is, in the shade of other plants. Plants transpire water, which cools them when the temperature is high. Many plants survive the low temperatures of winter by remaining dormant. (*Note:* other answers are possible.)

4. Alternation of generations is a life cycle characteristic of plants and certain other organisms in which they spend part of their life in a multicellular gametophyte stage and part in a multicellular sporophyte stage. The gametophyte generation is haploid, develops from a spore, and produces gametes, whereas the sporophyte generation is diploid, develops from a zygote, and produces spores.

5. Bryophytes lack vascular tissues and therefore are restricted in size because there is no efficient means of conducting materials throughout a large plant body. Lack of vascular tissues also explains why bryophytes generally require a moist environment; that is, they get their water and dissolved minerals from water at the soil's surface, which is absorbed by their bodies, rather than from roots anchored in the soil. In addition, some bryophytes must live in moist environments because they lack a water-conserving cuticle.

6. The three phyla of bryophytes are mosses (phylum Bryophyta), liverworts (phylum Hepaticophyta), and hornworts (phylum Anthocerotophyta).

7. Mosses, liverworts, and hornworts are plants that lack vascular tissues and have a dominant gametophyte generation. Moss gametophytes are leafy plants that grow from protonemata; their sporophytes project from the tops of the leafy gametophytes. Liverwort gametophytes are either leafy or thalloid; their sporophytes are either embedded within the thallus or are attached to tiny stalked structures. Hornworts are thalloid; their hornlike sporophytes project out of the thalli.

8. Mosses hold the soil in place and help prevent soil erosion. They provide habitat for small animals, and nesting material for certain birds. Mosses also play a role in forming soil.

9. See Figure 22–6c.

Chapter 23

1. Bryophytes and seedless vascular plants are classified in different phyla because they have significant differences: seedless vascular plants have vascular tissues and alternation of generations with a dominant sporophyte, and bryophytes do not.

2. In mosses the gametophyte generation is dominant, and the sporophyte is attached to and nutritionally dependent on the gametophyte. In ferns the gametophyte and sporophyte generations exist independently of one another, and the larger, longer-living sporophyte generation is dominant.

3. Microphylls are small leaves that are thought to have evolved from lateral projections of stem tissue; club mosses are the only plants with microphylls. Megaphylls are leaves that are thought to have evolved from branch systems; all vascular plants other than club mosses and whisk ferns have megaphylls.

4. Ferns are the most diverse group of seedless vascular plants.

5. Whisk ferns, horsetails, and club mosses are the three groups of fern allies. Whisk fern sporophytes have dichotomously branching rhizomes and erect stems; they lack roots and leaves. Horsetails have roots, rhizomes, aerial stems that are hollow and jointed, and leaves that are small megaphylls. Club mosses have roots, rhizomes, erect branches, and leaves that are microphylls. Ferns and fern allies are similar in that they possess vascular tissues, have a dominant sporophyte generation, and reproduce by forming spores.

6. Unlike other seedless vascular plants, whisk ferns lack true roots and leaves. They survive without these organs because their rhizomes absorb water and dissolved minerals, and their aerial stems are green and photosynthetic.

7. Certain horsetails are known as scouring rushes because pioneers used them to scrub out pots and pans.

8. The common name "club moss" is misleading because these plants are fern allies and are not close relatives of mosses.

9. Heterospory is the production of two types of spores, microspores and megaspores. Heterospory modifies the plant life cycle because each type of spore gives rise to a unisexual gametophyte: microspores produce male gametophytes, and megaspores produce female gametophytes.

10. See Figure 23–4.

Chapter 24

1. The production of seeds was a significant event in evolution because seeds, with their well-developed young plant embryos, large food reserves, and protective seed coats, are reproductively superior to spores.

2. The four phyla of gymnosperms are the conifers (phylum Coniferophyta), cycads (phylum Cycadophyta),

Ginkgo (phylum Ginkgophyta), and gnetophytes (phylum Gnetophyta). Conifers are woody, monoecious plants that bear needle-like or scalelike leaves and produce their seeds in cones. Cycads are palmlike or fernlike in appearance, dioecious, and produce their seeds in cones. Members of *Ginkgo,* the only species in its phylum, are deciduous, dioecious trees whose females produce seeds with fleshy coverings directly on their branches. Gnetophytes are a diverse group of gymnosperms that share a number of advances, including vessel elements.

3. Pine is heterosporous and produces microspores and megaspores.

4. Pollen is transferred from male cones to female cones by the wind. Once a pollen grain adheres to the sticky female pine cone, it grows a pollen tube through the female gametophyte tissue to the egg. Two sperm cells are discharged from the pollen tubes, and one of these fertilizes the egg.

5. Conifers are advanced in relation to ferns in that conifers produce windborne pollen, have a dependent gametophyte that is attached to and nutritionally dependent on the sporophyte, and are heterosporous.

6. Cycads, ginkgoes, gnetophytes, and conifers are vascular plants that reproduce by seeds that are either exposed or borne on the scales of cones.

7. Conifers are the predominant trees in about 35 percent of the world's forests. These forests are ecologically important because they hold the soil in place, serve as watersheds, and provide food and shelter to many organisms.

8. Conifers are commercially important in the recreation and timber industries. Conifer wood is harvested for building materials, paper products, turpentine, and resin. Conifers are also grown for Christmas trees and for landscape design.

Chapter 25

1. Flowering plants produce flowers and seeds enclosed within fruits. Flowering plants have a unique double fertilization process that results in the formation of a diploid zygote and triploid endosperm tissue.

2. In flowering plants, unlike gymnosperms, the ovule is enclosed by an ovary. Within the ovule, an embryo sac develops that contains an egg cell, two polar nuclei, and several other cells. Each pollen grain produces two sperm cells. Following pollination, double fertilization occurs in which one of the sperm cells unites with the egg cell to form a zygote, and the other sperm cell unites with the two polar nuclei to form endosperm. After fertilization, the ovule develops into a seed, and the surrounding ovary into a fruit.

3. Flowering plants transfer pollen in a variety of ways, such as wind, insects, birds, and bats. After pollination has occurred, flowering plants have a double fertilization process that results in the formation of both a zygote and endosperm.

4. Because plants that reproduce asexually by apomixis produce seeds and fruits, they are able to take advantage of seed- and fruit-dispersal methods associated with sexual reproduction.

5. Monocots and dicots are the two classes of flowering plants. Monocots have floral parts in threes and multiples of three, whereas dicots have floral parts in fours or fives or multiples of four or five. Monocot seeds contain one cotyledon, whereas dicot seeds contain two cotyledons. The nutritive tissue in mature monocot seeds is usually the endosperm, whereas the nutritive tissue in mature dicot seeds is usually the cotyledons.

6. The line of ancient gymnosperms that evolved into flowering plants developed, during the course of evolution, closed carpels and leaves with broad, expanded blades.

7. Flowering plants, as the principal photosynthetic organisms on land, provide animals and other organisms with food and shelter.

8. Flowering plants are important to humans as ornamentals (e.g., cacti) and for their timber (e.g., walnut); foods, including seeds (e.g., peas), fruits (e.g., cherries), stems (e.g., potatoes), roots (e.g., turnip), and leaves (e.g., cabbage); drugs (e.g., atropine); and spices (e.g., vanilla).

9. The raspberry belongs to the rose family; pecan to the walnut family; barrel cactus to the cactus family; red clover to the pea family; jimsonweed to the potato family; century plant to the agave family; wheat to the grass family; tuliptree to the magnolia family; turnip to the mustard family; garden pea to the pea family; tomato to the potato family; and sisal hemp to the agave family.

Chapter 26

1. A community is all the populations of organisms living in an area, whereas an ecosystem encompasses a community and its abiotic environment.

2. An organism's habitat is the environment where it lives; its ecological niche encompasses both its habitat and role in a community, including the physical, chemical, and biological factors that it needs to survive, remain healthy, and reproduce.

3. Producers are plants, algae, and certain bacteria that contain chlorophyll and photosynthesize. Decomposers are bacteria and fungi that break down dead organisms and organic wastes. Although consumers and decomposers are both heterotrophs, consumers are usually animals, whereas decomposers are bacteria and fungi.

4. Intraspecific competition is competition between two or more members of the same species, whereas interspecific competition is between two or more members of different species.

5. In mutualism, both organisms benefit from their association. In commensalism, one organism benefits and the other one is neither harmed nor helped. Predation is an interaction between two organisms in which one consumes the other, whereas parasitism is an association between two organisms in which one lives on or in the other at its expense; predators kill their prey, whereas parasites harm their hosts but do not necessarily kill them.

6. If matter were not returned to the abiotic environment from the biotic world, organisms would eventually run out of many of the essential raw materials, such as potassium, nitrogen, and phosphorus, that are required for life.

7. See Figure 26–6.

8. The five steps in the nitrogen cycle are: (1) nitrogen fixation, in which nitrogen gas is converted to ammonia; (2) nitrification, in which ammonia is converted to nitrate; (3) assimilation, in which nitrates and ammonia are converted to nitrogen-containing organic compounds; (4) ammonification, in which organic nitrogen is converted to ammonia; and (5) denitrification, in which nitrate is converted to nitrogen gas.

9. A food web more realistically describes the flow of energy and materials through ecosystems than does a food chain because few organisms eat just one other kind of organism.

10. Ecological succession is the sequence of vegetation changes in a community over time. Primary succession occurs on land that has not previously been inhabited by plants, whereas secondary succession takes place after some disturbance destroys the existing vegetation; initially, no soil is present on land undergoing primary succession, whereas soil is already present on land undergoing secondary succession.

11. Agricultural ecosystems are more likely to have pest problems than are natural ecosystems because agricultural ecosystems, being essentially monocultures, contain plenty of food and generally do not have many species that prey on or parasitize the pest species.

Chapter 27

1. Tundra is characterized by a long, extremely cold winter, an extremely short growing season, and little precipitation; it has a permanently frozen layer of subsoil called permafrost. Taiga has long, cold winters (although less severe than in the tundra) and little precipitation (although more than in the tundra); soil is generally mineral-poor, often with some permafrost deep below the surface. Temperate rain forest is characterized by cool weather, dense fog, and high precipitation; the soil is relatively mineral-poor but often has a high organic content. Temperate deciduous forest has pronounced winters, hot summers, and a relatively high amount of precipitation; the soil is typically rich in organic matter and contains minerals in the clay-rich lower layer. Temperate grassland has cold to mild winters, hot summers, and uncertain rainfall; its soil is typically rich in organic material and contains a mineral-rich layer just below the topsoil. Chaparral has mild winters with abundant rainfall and extremely dry summers; chaparral soils are thin and not very fertile. Savanna occurs in warm, tropical areas with low rainfall or seasonal rainfall; savanna soil is relatively low in minerals. Tropical rain forest occurs where the temperature is warm and precipitation occurs almost daily; its soil is low in minerals and organic material. Deserts occur in temperate and tropical regions and are characterized by little precipitation; desert soils are low in organic matter but often high in mineral content.

2. Tundra contains mosses, lichens, grasses, and sedges, along with caribou, lemmings, weasels, and snowshoe hares. Plants of the taiga include spruce, fir, and pine; animals include wolves, bears, rodents, and rabbits. Large, coniferous trees such as western hemlock, Douglas fir, Sitka spruce, and western red cedar are found in temperate rain forests; animals include squirrels, deer, and numerous bird species. Broad-leaved deciduous hardwood trees are representative plants, and deer, small mammals, birds, and reptiles are representative animals of the temperate deciduous forest. Temperate grassland contains perennial grasses and other herbaceous flowering plants, along with bison, coyotes, foxes, and various birds of prey. Plants of the chaparral include fire-adapted evergreen shrubs and scrub oak; animals include mule deer, wood rats, lizards, and many species of birds. Creosote bush, cacti, yuccas, and bunchgrass are found in North American deserts; animals include lizards, snakes, tortoises, and the kangaroo rat. Grasses and widely scattered trees such as *Acacia* are representative plants, and wildebeests, giraffes, lions, and hyenas are representative animals of the African savanna. Tropical rain forests contain numerous species of evergreen trees, epiphytes, and vines, along with many kinds of insects, amphibians, reptiles, and birds.

3. Answers will vary depending on where you live.

4. Those biomes best suited for agriculture are the temperate grassland and temperate deciduous forest. Desert, savanna, and chaparral are not suitable for agriculture because they lack enough precipitation during

the growing season. The growing season in tundra and taiga is too short. The soil is too infertile in tropical rain forest, desert, and savanna; the soil is marginally infertile in temperate rain forest.

5. (a) Both temperate rain forest and tropical rain forest receive a high annual precipitation. Tropical rain forest differs in that it does not have winter and summer seasons; also, species diversity is greater in tropical rain forest than in temperate rain forest. (b) Both temperate grassland and savanna occur in areas with uncertain rainfall. Savanna differs in that it does not have winter and summer seasons; also, savanna soil is infertile compared to that of temperate grassland. (c) Summer in chaparral has a climate similar to that of desert in that there is little or no precipitation. Chaparral has a rainy winter season that is not typical of temperate desert, however. (d) Tundra and taiga are characterized by long, cold winters and little precipitation; winters are colder and longer, and precipitation is less in tundra as compared to taiga. (e) Temperate deciduous forest and temperate grassland have distinct seasons (winter, spring, summer, and fall), but temperate grassland receives less precipitation than temperate deciduous forest. (f) Temperate grassland and temperate desert have distinct seasons, but temperate desert receives less precipitation than temperate grassland.

6. Warming produced by the buildup of CO_2 and other greenhouse gases, which trap the sun's radiation in much the same way that glass does in a greenhouse, is known as the greenhouse effect. The main greenhouse gases are carbon dioxide, methane, nitrous oxide, CFCs, and ozone.

7. Stratospheric ozone blocks much of the ultraviolet radiation that comes from the sun. The attack on the ozone layer is caused by chlorofluorocarbons and similar chlorine-containing compounds.

GLOSSARY

abscisic acid   A plant hormone involved in dormancy and in responses to stress.

abscission   The normal (usually seasonal) falling off of leaves or other plant parts, such as fruits or flowers.

accessory fruit   A fruit composed primarily of tissue other than ovary tissue. Apples and pears are accessory fruits.

achene   A simple dry fruit with one seed in which the fruit wall is separate from the seed coat. Sunflower fruits are achenes.

activation energy   The energy required to initiate a chemical reaction.

active transport   Energy-requiring transport of a molecule across a membrane from a region of low concentration to a region of high concentration.

adaptation   Any feature of an organism that improves its chances of surviving and producing offspring. Natural selection favors adaptations.

adaptive radiation   The evolution of several to many species from an unspecialized ancestor.

adenosine triphosphate (ATP)   An organic compound containing adenine, ribose, and three phosphate groups; of prime importance for energy transfers in biological systems.

adventitious   Arising in an unusual position; applies to an organ such as a root or bud.

aerobe   *See* aerobic.

aerobic   Growing or metabolizing only in the presence of molecular oxygen. Aerobic organisms are called aerobes. *Compare with* anaerobic.

aerobic respiration   The process by which cells use oxygen to break down organic molecules into waste products (such as carbon dioxide and water), with the release of energy that can be used for biological work.

aggregate fruit   A fruit that develops from a single flower with several to many pistils, such as a raspberry.

air pollution   Any of a variety of chemicals (gases, liquids, or solids) present at high enough levels in the atmosphere to harm plants and other organisms.

allele   One of two or more alternative forms of a gene. Alleles occupy corresponding positions (loci) on homologous chromosomes.

allelopathy   An adaptation in which toxic substances secreted by roots or shed leaves inhibit the establishment of competing plants nearby.

allopatric speciation   Evolution of a new species that occurs when one population becomes geographically separated from the rest of the species and subsequently evolves.

allopolyploid   A polyploid formed from an interspecific hybrid; contains two or more sets of chromosomes from each of two parent species.

alternation of generations   A life cycle, characteristic of plants and certain other organisms, in which the organism spends part of its life in a multicellular *n* gametophyte stage and part in a multicellular *2n* sporophyte stage. The gametophyte develops from a spore and produces gametes; the sporophyte develops from a zygote and produces spores.

amino acid   An organic compound containing an amino group (—NH$_2$) and a carboxyl group (—COOH). Amino acids may be linked together to form protein molecules.

anabolic reaction   *See* anabolism.

anabolism   The phase of metabolism in which simpler substances combine to form more complex substances; requires energy. *Compare with* catabolism.

anaerobe   *See* anaerobic.

anaerobic   Growing or metabolizing in the absence of molecular oxygen. Anaerobic organisms are called anaerobes. *Compare with* aerobic.

analogous organs   Organs that are similar in function or appearance but not because of a common evolutionary origin. *Compare with* homologous organs. *Also see* convergent evolution.

angiosperm   The traditional name for flowering plants. *See* flowering plant.

annual   A plant that completes its entire life cycle in one growing season. *Compare with* biennial and perennial.

annual ring   A layer of wood (secondary xylem) that forms in the stems and roots of woody plants growing in temperate areas or tropical areas with rainy and dry seasons. Usually one ring forms per year.

antheridium   The multicellular gametangium that produces sperm cells.

anticodon   A sequence of three nucleotides in transfer RNA that is complementary to, and combines with, the three-nucleotide codon on messenger RNA, thus helping to specify the addition of a particular amino acid to the end of a growing polypeptide chain.

apical dominance   The inhibition of axillary buds by the apical meristem.

apical meristem   An area of cell division at the tip of a stem or root; produces primary tissue.

apomixis   A type of reproduction in flowering plants in which seeds and fruits are formed asexually.

archaebacteria   *See* kingdom Archaebacteria.

archegonium   The multicellular gametangium that produces an egg.

artificial selection   *See* selective breeding.

ascomycete   A member of a phylum of fungi characterized by the production of nonmotile asexual conidia and sexual ascospores.

ascospore   One of a set of (usually eight) sexual spores contained in a special spore case (the ascus) of an ascomycete.

ascus   A saclike spore case in an ascomycete that contains sexual spores called ascospores.

asexual reproduction   Reproduction in which only one parent participates, no fusion of gametes occurs, and the genetic makeups of parent and offspring are virtually identical. *Compare with* sexual reproduction.

ATP synthase   An enzyme complex that synthesizes ATP from ADP, using the energy of a proton gradient; located in the inner mitochondrial membrane and in thylakoid membranes of chloroplasts.

autotroph   An organism that synthesizes organic molecules from inorganic materials. Plants and other photosynthetic organisms are autotrophs. *Compare with* heterotroph.

auxin   A plant hormone involved in various aspects of plant growth and development.

bacillus (pl. *bacilli*)   A rod-shaped bacterium.

basidiomycete   A member of a phylum of fungi characterized by the production of sexual basidiospores.

basidiospore   One of a set of sexual spores (usually four) borne on a club-shaped structure (a basidium) of a basidiomycete.

basidium   The clublike spore-producing organ of a basidiomycete that bears sexual spores called basidiospores.

berry   A simple fleshy fruit in which the fruit wall is soft throughout. Tomatoes, bananas, and grapes are berries.

biennial   A plant that takes 2 years (that is, two growing seasons) to complete its life cycle. *Compare with* annual and perennial.

binary fission   The equal division of one cell into two cells. Bacteria reproduce by binary fission.

binomial system of nomenclature   A system of naming organisms by combining the genus name and the species epithet.

biogeochemical cycle   The process by which matter cycles from the living world to the nonliving physical environment and back again; examples include the carbon, nitrogen, and hydrologic cycles.

biogeography   The study of the distribution of organisms.

biome   A large, relatively distinct terrestrial region characterized by a particular combination of climate, soil, plants, and animals that is approximately the same regardless of where it occurs in the world.

bioremediation   A method of cleaning up a hazardous waste site that uses organisms (usually bacteria or fungi) to break down the toxic pollutants.

biosphere   All of the Earth's living organisms; includes all the communities on Earth.

biotechnology   Broadly, the use of organisms to produce products that benefit humans. Today the term is usually interpreted more narrowly; *see* genetic engineering.

biotic pollution   The introduction of a foreign, or exotic, species into an area where it is not native. Exotic species often harm native species.

blade   The broad, flat part of a leaf.

bolting   The rapid elongation of a flower stalk that occurs when a plant initiates flowering; also caused by the application of gibberellin.

boreal forest   *See* taiga.

botany   The scientific study of plants. Also called plant biology.

brown alga   One of a phylum of predominantly marine algae that are multicellular and contain the pigments chlorophylls *a* and *c* and carotenoids, including the yellow-brown fucoxanthin.

bryophytes   Nonvascular, spore-producing plants that include mosses, liverworts, and hornworts.

bud   An undeveloped shoot that contains an embryonic meristem. A bud may be terminal (at the tip of the stem) or axillary (on the side of the stem).

bud scale   A modified leaf that covers and protects winter buds.

bulb   A rounded, fleshy underground bud that consists of a short stem with fleshy leaves. An onion bulb is an example.

bundle sheath   A ring of parenchyma or sclerenchyma cells surrounding the vascular bundle in a dicot or monocot leaf.

Calvin cycle A cyclic series of reactions, occurring in the light-independent phase of photosynthesis, that fixes carbon dioxide and produces glucose.

capsule (1) The portion of the moss sporophyte that contains spores. (2) A simple dry fruit that opens along many seams or pores to release seeds, such as cotton fruits. (3) A thick, sticky layer surrounding certain bacteria, external to the cell wall.

carbohydrate An organic compound containing carbon, hydrogen, and oxygen in the approximate ratio 1C:2H:1O. Examples are sugars, starches, and cellulose.

carpel The female reproductive unit of a flower, which bears ovules. *See* pistil.

caryopsis (pl. *caryopses*) *See* grain.

Casparian strip A band of waterproof material around the radial and transverse cell walls of the endodermis.

catabolic reaction *See* catabolism

catabolism The phase of metabolism in which complex organic molecules are broken down, with a release of energy that can be used to do biological work. *Compare with* anabolism.

catkin A cluster of tiny (much reduced) unisexual flowers borne on a pendulous stalk.

cell The basic structural and functional unit of life, which consists of living material bounded by a membrane. Some simple organisms are composed of single cells, whereas complex organisms are composed of many cells.

cell cycle The cyclic series of events in the life of a dividing eukaryotic cell; consists of interphase, mitosis, and cytokinesis.

cell plate A structure that develops across the cell's equator between newly formed nuclei in dividing cells; becomes the cell walls and plasma membranes of adjacent cells.

cellular respiration The cellular process in which energy of organic molecules is released for biological work.

cellular slime mold One of a phylum of fungus-like protists whose feeding stage consists of unicellular, amoeboid organisms that aggregate to form a pseudoplasmodium during reproduction.

cell wall A comparatively rigid supporting wall exterior to the plasma membrane in plants, fungi, bacteria, and certain protists.

center of origin The area where a given species evolved.

centromere A specialized region of a chromosome. At prophase, sister chromatids are joined in the vicinity of their centromeres.

chaparral A biome with a Mediterranean climate—that is, with mild, moist winters and hot, dry summers. Chaparral vegetation consists primarily of small-leaved evergreen shrubs.

chemiosmosis Synthesis of ATP, using the energy of a proton gradient established across a membrane; occurs during electron transport in both photosynthesis and aerobic respiration.

chitin A nitrogen-containing polysaccharide that forms the cell walls of many fungi.

chloroplast A chlorophyll-bearing organelle of certain plant cells; the site of photosynthesis.

chromatid One of the two halves of a duplicated (doubled) chromosome.

chromatin The complex of DNA and protein that makes up eukaryotic chromosomes.

chromosome One of several rod-shaped bodies in the cell nucleus that contain the hereditary units (genes).

circadian rhythm An internal rhythm that approximates the 24-hour day. Circadian rhythms are found in plants, animals, and other organisms.

citric acid cycle A series of aerobic chemical reactions in which fuel molecules are completely degraded to carbon dioxide and water with the release of metabolic energy; also known as the Krebs cycle.

cladistics An approach to classification based on recency of common ancestry.

cladogram A branching diagram that illustrates taxonomic relationships based on the principles of cladistics.

class A taxon composed of related, similar orders.

clay The smallest inorganic soil particles. *Compare with* silt and sand.

club moss One of a phylum of seedless vascular plants with a life cycle similar to that of ferns.

coccus (pl. *cocci*) A bacterium with a spherical shape.

codon A sequence of three nucleotides in mRNA that specifies individual amino acids or translation start and stop signals.

coenocytic Having a body that consists of a multinucleated cell; that is, the nuclei are not separated from one another by septa. Applies to a protist or fungus.

coevolution The interdependent evolution of two or more species that occurs as a result of their interactions over a long period of time. Flowering plants and their animal pollinators are an example of coevolution, because each has profoundly affected the other's characteristics.

collenchyma Living cells with moderately but unevenly thickened walls. Collenchyma cells provide flexible support for the primary plant body.

colony (1) In bacteria, a population of millions of genetically identical cells derived from a single cell, grown in culture. (2) In protists, an aggregation of loosely organized cells. (3) In plants, a population of individuals that live close to one another.

commensalism A type of symbiosis in which one organism benefits and the other one is neither harmed nor helped. *Compare with* mutualism and parasitism.

community An association of different species living together in a defined habitat with some degree of mutual interdependence. *Compare with* ecosystem.

competition An interaction between two or more organisms that simultaneously require the same resource, usually in a limited supply; includes intraspecific competition (within a species) and interspecific competition (between different species).

compost A natural soil-and-humus mixture that improves soil structure and fertility.

compound leaf A leaf with a blade that consists of two or more leaflets. *Compare with* simple leaf.

concentration gradient The variance of concentration of specific atoms or molecules from one area of a biological system (where the concentration is high) to another (where it is low).

condensation reaction A chemical reaction in which water is removed as two or more smaller molecules join to form a larger molecule.

cone A reproductive structure that produces either microspores or megaspores in many gymnosperms.

conidium (pl. *conidia*) An asexual spore that usually forms at the tip of a specialized hypha called a conidiophore.

conifer Any of a large phylum of gymnosperms that are woody trees and shrubs with needle-like, mostly evergreen leaves and with seeds in cones.

conjugation (1) A sexual phenomenon in certain protists that involves exchange or fusion of a cell with another cell. (2) A mechanism for DNA exchange in bacteria that involves cell-to-cell contact.

consumer An organism that cannot synthesize its own food from inorganic materials and therefore must use the bodies of other organisms as a source of energy and body-building materials; includes primary consumers (herbivores) and secondary consumers (carnivores). *Compare with* producer.

contractile root A specialized root, often found on bulbs and corms, that contracts and thereby pulls the plant deeper into the soil.

control That part of a scientific experiment in which the experimental variable is kept constant. The control provides a standard of comparison that enables the experimenter to verify the results of the experiment.

convergent evolution The independent evolution of structural or functional similarity in two or more unrelated organisms as a result of adaptation to similar environmental conditions.

cork Cells produced by the cork cambium. Cork cells are dead at maturity and function for protection.

cork cambium A lateral meristem that produces cork cells and cork parenchyma (phelloderm). Cork cambium and the tissues it produces make up the outer bark on a woody plant.

cork parenchyma One or more layers of parenchyma cells produced by the cork cambium. *Also called* phelloderm.

corm A short, thickened underground stem specialized for food storage and asexual reproduction—for example, the gladiolus corm.

cortex The ground tissue between the epidermis and the phloem in nonwoody roots and stems.

cotyledon The seed leaf of a plant embryo that often contains food stored for germination. Some cotyledons have functions other than food reserves, such as digestion of food stored in the endosperm and photosynthesis in the newly germinated seedling.

covalent bond A chemical bond involving one or more shared pairs of electrons.

crossing over The breaking and rejoining of homologous (nonsister) chromatids during meiotic prophase I, resulting in an exchange of genetic material.

cuticle A waxy covering over the epidermis of the aerial parts of a plant; reduces water loss.

cyanobacteria Prokaryotic photosynthetic microorganisms that possess chlorophyll and produce oxygen during photosynthesis. Formerly known as blue-green algae.

cycad Any of a phylum of gymnosperms that live mainly in tropical and subtropical regions and have stout stems (to 20 meters in height) and fernlike leaves.

cytokinesis The part of cell division in which the cytoplasm divides to form two daughter cells.

cytokinin A plant hormone involved in various aspects of plant growth and development.

cytoplasm The general cellular contents, exclusive of the nucleus.

cytoskeleton The internal structure, made up of microfilaments and microtubules, that gives shape and mechanical strength to cells.

day-neutral plant A plant that initiates flowering not in response to seasonal changes in the amounts of daylight and darkness but in response to some other type of environmental or internal stimulus. *Compare with* long-day, short-day, and intermediate-day plants.

deciduous Falling off at a certain season, such as leaves that are shed by trees in autumn. *Compare with* evergreen.

decomposer *See* saprotroph.

deductive reasoning Reasoning that operates from generalities to specifics and can make relationships among data more apparent. *Compare with* inductive reasoning.

deforestation The removal of forest without sufficient replanting; results in the replacement of forest with agricultural lands or other nonforested lands.

deoxyribonucleic acid (DNA) A nucleic acid, the molecule of inheritance for plants and other organisms, present in a cell's chromosomes; contains genetic information coded in specific sequences of its nucleotides.

dermal tissue system The tissue system that provides a covering for the plant body; consists of epidermis or periderm.

desert A temperate or tropical biome in which lack of precipitation limits plant growth.

desertification Degradation of once-fertile rangeland (or tropical dry forest) to nonproductive desert; caused partly by soil erosion, deforestation, and overgrazing.

determinate growth   Growth that stops after a certain size is reached; examples include the growth of flowers and leaves. *Compare with* indeterminate growth.

deuteromycetes   An artificial grouping of fungi characterized by the absence of sexual reproduction but usually having other traits similar to ascomycetes. *Also called* imperfect fungi.

development   The orderly sequence of progressive changes in the life of an organism.

diatom   One of a phylum of usually unicellular algae that are covered by ornate, siliceous shells, each consisting of two overlapping halves; an important component of plankton in both marine and fresh waters.

dichotomous   A type of branching in which a stem always branches into two approximately equal parts.

dichotomous key   A special guide that aids in the identification of an unknown plant; consists of a series of paired, contrasting statements.

dicot   *See* dicotyledon.

dicotyledon   One of the two classes of flowering plants, with each embryo in a seed having two seed leaves, or cotyledons. Also known as a dicot. *Compare with* monocotyledon.

dictyosome   *See* Golgi apparatus.

diffusion   The net movement of molecules from a region of high concentration to one of lower concentration, brought about by their tendency to disperse randomly.

dihybrid cross   A genetic cross that involves individuals differing in *two* (nonallelic) genes.

dinoflagellate   One of a phylum of algae that are unicellular, biflagellate, usually photosynthetic, and typically marine; an important component of plankton.

dioecious   Having male and female reproductive structures on separate plants. *Compare with* monoecious.

diploid (2n)   Having two sets of chromosomes per cell. *Compare with* haploid. *Also see* polyploid.

disaccharide   A sugar such as sucrose that consists of two monosaccharide subunits.

division   *See* phylum.

DNA   *See* deoxyribonucleic acid.

DNA polymerase   An enzyme complex that catalyzes DNA replication by adding nucleotides to a growing strand of DNA.

DNA replication   *See* semiconservative replication.

DNA sequencing   Determining the order of nucleotide bases in a strand of DNA.

dominant   An allele that is always expressed when it is present, regardless of whether an individual is homozygous or heterozygous for the trait. *Also see* principle of dominance. *Compare with* recessive.

double fertilization   A process in the flowering plant life cycle in which there are two fertilizations. The first fertilization results in the formation of a zygote that develops into a young plant, and the second results in the formation of endosperm (nutritive tissue).

double helix   The structure of DNA, which consists of two polynucleotide strands joined together by hydrogen bonds between complementary base pairs, with the entire molecule twisted to form a double helix.

drupe   A simple fleshy fruit in which the inner wall of the fruit is hard and stony. Peaches and cherries are drupes.

ecological niche   *See* niche.

ecological succession   *See* succession.

ecology   A discipline of biology that studies the interrelationships among organisms and between organisms and their environments.

economic botany   The branch of botany that deals with important food crops, products obtained from plant parts, and ornamental plants.

ecosphere   The interactions among and between all of the Earth's living organisms and the air, land, and water that they occupy.

ecosystem   The interacting system that encompasses a community of organisms and its nonliving physical environment. *Compare with* community.

electron acceptor molecule   A molecule in an electron transport chain that alternately accepts and releases electrons.

electron transport   A series of chemical reactions during which electrons are passed along from one electron acceptor molecule to another. During this passage, the electrons give up some of their energy, which is used to establish a proton gradient (a form of potential energy).

embryo sac   The female gametophyte generation in flowering plants.

endodermis   The innermost layer of the cortex in the root. Endodermis cells have a Casparian strip running around radial and transverse walls.

endoplasmic reticulum (ER)   An organelle composed of numerous internal membranes within eukaryotic cells; has a variety of functions, including the production of membranous organelles such as the Golgi apparatus.

endosperm   The triploid (3n) nutritive tissue that occurs at some point in the development of all angiosperm seeds.

endospore   A thick-walled, highly resistant structure formed within certain bacterial cells to withstand adverse environmental conditions.

endosymbiont theory   The theory that certain organelles, such as mitochondria and chloroplasts, originated as symbiotic prokaryotes living inside larger prokaryotic cells.

enzyme   An organic catalyst, produced within an organism, that accelerates specific chemical reactions.

epidermis   An outer tissue layer, usually one cell thick, that covers the plant body; functions primarily for protection.

ethnobotany   The branch of botany that deals with the uses of plants by indigenous peoples. This knowledge is disappearing rapidly as native people adopt modern lifestyles.

ethylene   A plant hormone involved in senescence.

eubacteria   *See* kingdom Eubacteria.

euglenoid One of a phylum of mostly freshwater algae that move by means of an anterior flagellum and are usually photosynthetic.

eukaryote An organism whose cells possess organelles surrounded by membranes. *Compare with* prokaryote.

evergreen Shedding leaves over a long time period, so that some leaves are always present. *Compare with* deciduous.

evolution Cumulative genetic change in a population of organisms through multiple generations. Evolution leads to differences among populations and explains the origins of all organisms that exist today or have ever existed.

extinction The elimination of a species from Earth; occurs when the last individual member of a species dies.

facultative anaerobe An organism that uses oxygen if it is available but that can respire anaerobically when oxygen is absent. *Compare with* obligate anaerobe.

family A taxon composed of related, similar genera.

fatty acid An organic acid composed of a long, unbranched chain of carbon and hydrogen atoms with a —COOH group at one end. Fatty acids linked to glycerol form a neutral fat or oil.

fecal coliform test A water quality test for the presence of *Escherichia coli* (fecal bacteria that are common in the intestinal tracts of humans and animals). The presence of fecal bacteria in a water supply indicates the possible presence of pathogens as well.

fermentation An anaerobic pathway in which organic fuel molecules such as glucose are incompletely broken down to yield energy for biological work. Fermentation does not release as much energy per glucose molecule as aerobic respiration.

fern One of a phylum of seedless vascular plants that reproduce by spores produced in sporangia, which are usually borne in sori on leaves. Ferns have an alternation of generations between the dominant sporophyte (a fern plant) and the independent gametophyte (a prothallus).

fertilization Fusion of male and female gametes.

fibrous root system A root system consisting of several adventitious roots of approximately equal size that arise from the base of the stem. *Compare with* taproot system.

first law of thermodynamics Energy cannot be created or destroyed, although it can be transformed from one form to another. *Compare with* second law of thermodynamics.

florigen A hypothetical hormone thought to promote flowering.

flowering plant One of a very large (about 135,000 species), very diverse phylum of seed plants that form flowers for sexual reproduction and produce seeds enclosed in fruits. Also called angiosperm. *See* dicot and monocot.

fluid mosaic model The current model for the structure of the plasma membrane and other cell membranes in which protein molecules "float" in a phospholipid bilayer.

follicle A simple dry fruit that splits open along one seam to liberate the seeds.

food chain The successive series of organisms through which energy flows in an ecosystem. Each organism in the series eats or decomposes the preceding organism in the chain. *See* food web.

food web A complex interconnection of all the food chains in an ecosystem.

forest decline A gradual deterioration (and often death) of many trees in a forest. The cause of forest decline is unclear and may involve a combination of factors.

form phylum A phylum whose members have structural similarities but do not necessarily share a common ancestry. The imperfect fungi are a form phylum.

fossil A part or trace of an ancient organism, usually preserved in rock.

frond The leaf of a fern.

fruit In angiosperms, a mature, ripened ovary. A fruit contains seeds and usually provides protection and dispersal for the seeds.

fruiting body A complex reproductive structure in which the sexual spores of certain fungi are produced; refers to the ascocarp of an ascomycete and the basidiocarp of a basidiomycete.

gametangium A unicellular or multicellular structure of plants, protists, and fungi in which gametes form.

gamete A haploid reproductive cell—that is, an egg or a sperm cell. In sexual reproduction, the union of gametes results in the formation of a zygote.

gametophyte generation The haploid or gamete-producing stage in the life cycle of a plant. *Compare with* sporophyte generation.

gemma (pl. *gemmae*) In certain liverworts, a group of reproductive cells that detaches from the thallus and develops into a new gametophyte asexually.

gene A discrete unit of hereditary information that usually specifies a polypeptide (protein); consists of part of the DNA molecule that makes up a chromosome.

gene flow The movement of alleles between local populations due to migration and subsequent interbreeding.

gene gun A nonbiological vector used to introduce recombinant DNA "bullets" into a plant cell.

gene pool All the genes that are present in a freely interbreeding population.

genetic code A code, consisting of groups of three nucleotide bases in mRNA, that specifies an amino acid or a translation start or stop signal.

genetic drift A random change in allele frequency in a small, isolated population.

genetic engineering The process of taking a specific gene from one cell and placing it into another cell, where it is expressed. Also called biotechnology.

genetic probe A radioactively labeled segment of RNA or single-stranded DNA that can be used to identify cells which contain a particular gene.

genetic recombination *See* recombination.

genotype The genetic makeup of an individual. *Compare with* phenotype.

genus (pl. *genera*) A taxon composed of related, similar species.

gibberellin A plant hormone involved in plant growth and development.

ginkgo The sole surviving species (*Ginkgo biloba*) of an ancient phylum of gymnosperms that bears exposed seeds— that is, seeds not surrounded by a cone. Ginkgoes are hardy deciduous trees with broad, fan-shaped leaves and, on the females, exposed, fleshy seeds.

glycolysis The anaerobic conversion of glucose to pyruvic acid with the production of a small amount of ATP.

gnetophyte Any of the small phylum of unusual gymnosperms that possess many features similar to flowering plants, such as vessels in xylem.

Golgi apparatus An organelle composed of stacks of flattened membranous sacs (Golgi bodies, or dictyosomes); mainly responsible for modifying, packaging, and sorting proteins.

gradualism A model of evolution in which the evolutionary changes in a species are due to a slow, steady transformation over time. *Compare with* punctuated equilibrium.

grain A simple dry fruit in which the fruit wall is fused to the seed coat, making it impossible to separate the fruit from the seed. Corn kernels are grains. Also called caryopsis.

Gram staining procedure The application of a stain that allows bacteria to be grouped as either gram-positive or gram-negative; used to help identify unknown bacteria.

gravitropism The growth of a plant in response to gravity.

green alga One of a diverse phylum of algae that contain the same pigments (chlorophylls *a* and *b* and carotenoids) and storage product (starch) as plants.

greenhouse effect Warming produced by the buildup of carbon dioxide and other greenhouse gases, which trap the sun's radiation in much the same way glass does in a greenhouse.

ground tissue The tissue in monocot stems in which the vascular tissues are embedded; performs the same functions as cortex and pith in herbaceous dicot stems.

ground tissue system All of the tissues of the plant body other than the vascular tissue system and the dermal tissue system; consists of parenchyma, collenchyma, and sclerenchyma.

growth Increases in the size and mass of an organism.

guard cell A cell in the epidermis of a stem or leaf. Two guard cells form a pore for gas exchange, called a stoma.

guttation Exudation of water droplets from leaves; caused by root pressure.

gymnosperm Any of a group of seed plants in which the seeds are not enclosed in an ovary; most gymnosperms bear their seeds in cones. Includes four phyla: conifers, cycads, ginkgoes, and gnetophytes.

haploid (*n*) Having one set of chromosomes per cell. *Compare with* diploid. *Also see* polyploid.

Hardy-Weinberg law The conditions required for allele frequencies to remain unchanged from generation to generation.

heartwood The older wood in the centers of many tree trunks that no longer functions in conduction; contains substances (pigments, tannins, gums, and resins) that color it a brownish-red. *Compare with* sapwood.

herbal A book that describes plants, published during the 15th, 16th, and 17th centuries.

herbarium A collection of dried, pressed, and carefully labeled plant specimens, used for scientific study.

heterospory Production in plants of two types of spores, microspores and megaspores, each of which gives rise to a unisexual gametophyte. *Compare with* homospory.

heterotroph An organism that cannot synthesize organic molecules from inorganic materials and therefore must obtain them from other organisms, either living or dead. *Compare with* autotroph.

heterozygous Possessing a pair of unlike alleles for a particular trait. *Compare with* homozygous.

holdfast A basal structure for attachment to solid surfaces, found in multicellular algae.

homologous chromosomes Chromosomes that are similar in structure and genetic constitution. Homologous chromosomes occur in pairs, with one member originating from the sperm cell and the other from the egg.

homologous organs Organs similar in structure because of a common evolutionary ancestry. *Compare with* analogous organs.

homospory Production of one type of spore in plants; each spore gives rise to a bisexual gametophyte. *Compare with* heterospory.

homozygous Possessing a pair of identical alleles for a particular trait. *Compare with* heterozygous.

hormone An organic chemical compound that is produced in one part of a plant and transported to another part, where it affects some aspect of metabolism, growth, or reproduction.

hornwort A member of a phylum of spore-producing, nonvascular, thalloid plants with a life cycle similar to that of mosses.

horsetail One of a phylum of seedless vascular plants with a life cycle similar to that of ferns.

humus Black or dark brown decomposed organic material.

hydathode A special pore that exudes water from a leaf by guttation.

hydrogen bond An attraction between a slightly positive hydrogen atom in one molecule and a slightly negative atom (usually oxygen) in another molecule.

hydrophilic Attracted to water. *Compare with* hydrophobic.

hydrophobic Repelled by water. *Compare with* hydrophilic.

hydroponics Cultivation of plants in an aerated solution of dissolved inorganic minerals—that is, without soil.

hypertonic   Having a solute concentration greater than that of another solution. *Compare with* isotonic and hypotonic.

hypha (pl. *hyphae*)   One of the threadlike filaments that form the mycelium of a fungus or a water mold.

hypothesis   An educated guess (based on previous observations) that may be true and is testable by observation and experimentation.

hypotonic   Having a solute concentration less than that of another solution. *Compare with* hypertonic and isotonic.

imbibition   Absorption of water by, and subsequent swelling of, a dry seed prior to germination.

imperfect fungi   *See* deuteromycetes.

incomplete dominance   A condition in which neither member of a pair of contrasting alleles is completely expressed when the other is present.

independent assortment   *See* principle of independent assortment.

indeterminate growth   Unrestricted growth. Examples include the growth of stems and roots, both of which arise from apical meristems. *Compare with* determinate growth.

inductive reasoning   Reasoning that uses specific examples to draw a general conclusion. *Compare with* deductive reasoning.

inflorescence   A cluster of flowers on a common floral stalk.

intermediate-day plant   A plant that flowers when it is exposed to days and nights of intermediate length but does not flower when daylength is too long or too short. *Compare with* long-day, short-day, and day-neutral plants.

interphase   The period in the cell cycle in which there are no visible chromosomes; the period between mitotic divisions.

interspecific hybrid   The offspring of two species.

ionic bond   An electrostatic attraction between oppositely charged ions.

isotonic   Having concentrations of solute and solvent molecules identical to those of another solution. *Compare with* hypertonic and hypotonic.

kingdom   A broad taxonomic category made up of related phyla. Many biologists currently recognize six kingdoms.

kingdom Archaebacteria   Prokaryotic organisms with certain biochemical features that set them apart from the rest of the bacteria (the eubacteria).

kingdom Eubacteria   Prokaryotic organisms other than the archaebacteria.

Krebs cycle   *See* citric acid cycle.

lateral meristem   An area of cell division on the side of a plant. The two lateral meristems, the vascular cambium and the cork cambium, give rise to secondary tissues.

leaching   The process by which dissolved materials are washed away or carried with water down through various layers of the soil.

legume   A simple dry fruit that splits open along two seams to release its seeds. Pea pods and green beans are legumes.

lenticel   A porous swelling in a woody stem that develops when the epidermis is replaced by periderm; facilitates the exchange of gases between the stem's interior and the atmosphere.

lichen   A compound organism composed of a symbiotic alga (or cyanobacterium) and fungus.

ligase   An enzyme used in genetic engineering to join pieces of DNA with sticky ends.

limiting factor   Whatever environmental variable (such as water, sunlight, or some essential mineral) tends to restrict the growth, distribution, or abundance of a particular plant population.

linkage   The tendency for a group of genes on a single chromosome to be inherited together in successive generations.

lipid   Any of a group of organic compounds that are insoluble in water but soluble in fat solvents; serves as a storage form of fuel and as an important component of cell membranes.

liverwort   A member of a phylum of spore-producing, nonvascular thalloid or leafy plants with a life cycle similar to that of mosses.

loam   A soil that has approximately 40 percent each of sand and silt and about 20 percent of clay. A loamy soil that also contains organic material makes an excellent agricultural soil.

locus (pl. *loci*)   The location of a particular gene on a chromosome.

long-day plant   A plant that flowers when the night length is equal to or less than some critical length. *Also called* short-night plant. *Compare with* short-day, intermediate-day, and day-neutral plants.

long-night plant   *See* short-day plant.

macronutrient   An essential element that is required in a fairly large amount for normal plant growth. *Compare with* micronutrient.

megaphyll   A type of leaf, found in ferns, horsetails, gymnosperms, and flowering plants, that contains multiple vascular strands (that is, complex venation). *Compare with* microphyll.

megaspore   The haploid spore in a heterosporous plant that gives rise to a female gametophyte. *Compare with* microspore.

meiosis   A process in which a diploid cell undergoes two successive divisions to produce four haploid cells; generates gametes in animals and spores in plants.

mesophyll   The photosynthetic tissue in the interior of a leaf (between the upper and lower epidermises). Sometimes differentiated into palisade and spongy mesophyll.

messenger RNA (mRNA)   RNA, transcribed from DNA, that specifies the amino acid sequence of a polypeptide (protein). *Compare with* transfer RNA and ribosomal RNA.

metabolism   The sum of all of the chemical processes that occur within a cell or organism; includes both anabolic and catabolic reactions.

microevolution Changes in allele frequencies that occur within a population over successive generations.

micronutrient An essential element that is required in trace amounts for normal plant growth. *Compare with* macronutrient.

microphyll A type of leaf, found in club mosses, that contains one vascular strand (that is, simple venation). *Compare with* megaphyll.

microspore The haploid spore in heterosporous plants that gives rise to a male gametophyte. *Compare with* megaspore.

mimicry An adaptation for survival in which an organism resembles another organism or a nonliving object.

mitochondria Spherical or elongated intracellular organelles that contain the electron transport system and enzymes for the citric acid cycle; sometimes referred to as the "power plants" of the cell.

mitosis Division of the cell's nucleus into two daughter nuclei, each with the same number of chromosomes as were in the parental nucleus.

mold The vegetative bodies of most fungi, consisting of long, branched threads called hyphae.

monocot *See* monocotyledon.

monocotyledon One of the two classes of flowering plants, with each embryo in a seed having a single seed leaf, or cotyledon. Also known as a monocot. *Compare with* dicotyledon.

monoecious Having male and female reproductive parts in separate flowers or cones on the same individual plant. *Compare with* dioecious.

monohybrid cross A genetic cross involving individuals that differ in a single pair of alleles.

monophyletic Evolved from a common ancestor; applies to a group of organisms. *Compare with* polyphyletic.

monosaccharide A simple sugar such as glucose or fructose.

moss A member of a phylum of nonvascular plants with an alternation of generations in which the dominant *n* gametophyte (a leafy moss plant) alternates with a 2*n* sporophyte that remains attached to the gametophyte.

mulch Material placed on the surface of soil around the bases of plants to help maintain soil moisture and reduce erosion. Organic mulches have the advantage of decomposing over time, thereby enriching the soil.

multiple fruit A fruit that develops from many ovaries of many separate flowers. A pineapple is a multiple fruit.

mutation A change in the DNA (a gene) of an organism; in reproductive cells, may be passed on to the next generation, where it can result in birth defects or genetic disease.

mutualism A symbiotic relationship from which both partners benefit. *Compare with* parasitism and commensalism.

mycelium The vegetative body of a fungus or water mold; consists of a branched network of hyphae.

mycorrhiza A mutualistic association between a fungus and a root that helps the plant absorb essential minerals from the soil.

natural selection The process in nature that causes evolution; the tendency of organisms that possess favorable adaptations to their environment to survive and become the parents of the next generation.

needle The leaf of a conifer, such as pine.

neo-Darwinism *See* synthetic theory of evolution.

niche The totality of an organism's adaptations, its use of resources, and the lifestyle to which it is fitted; encompasses how an organism uses materials in its environment as well as how it interacts with other organisms. Also called ecological niche.

nitrogen-fixing bacteria Gram-negative bacteria that can convert atmospheric nitrogen (N_2) to ammonia (NH_3), a form plants can use.

nodules Small swellings on the roots of leguminous plants (for example, peas, beans, and clover) that house nitrogen-fixing bacteria (*Rhizobium*).

nonpolar covalent bond A chemical bond in which electrons are shared equally among the participating atoms; does not produce any electrical charge within the molecule. *Compare with* polar covalent bond.

nonvascular plant A plant that lacks conducting tissues. *Compare with* vascular plant.

nucleic acid DNA or RNA—a polymer composed of nucleotides.

nucleolus A specialized structure in the nucleus, formed from regions of several chromosomes; the site of ribosome synthesis.

nucleotide A molecule composed of a phosphate group, a five-carbon sugar (ribose or deoxyribose), and a nitrogenous base; one of the subunits of nucleic acids.

nucleus (1) The portion of an atom that contains the protons and neutrons. (2) A cellular organelle that contains DNA and serves as the control center of the cell.

nut A simple dry fruit that has a stony wall, is usually large, and does not split open at maturity. Chestnuts, acorns, and hazelnuts are nuts.

obligate anaerobe An organism that is unable to grow or metabolize in the presence of oxygen. *Compare with* facultative anaerobe.

order A taxon composed of related, similar families.

organic compound A compound that contains carbon and usually hydrogen.

osmosis Diffusion of water, the principal solvent in biological systems, through a selectively permeable membrane from a region of higher concentration (of water) to a region of lower concentration (of water).

ovary In flowering plants, the ovule-containing base of the pistil, which consists of one or more carpels; develops into a fruit after fertilization.

ovule The structure (megasporangium) in the ovary that develops into a seed after fertilization.

oxidation The loss of electrons (or of hydrogen) from a compound. *Compare with* reduction.

oxidation-reduction reaction   A chemical reaction in which electrons are transferred from one electron acceptor molecule to another.

oxidative phosphorylation   Production of ATP using energy derived from the transfer of electrons in the electron transport system of aerobic respiration.

ozone   A blue gas, O_3, with a distinctive odor. Ozone is a human-made pollutant in the lower atmosphere but a natural and essential component in the stratosphere.

paleobotany   The branch of botany that deals with plant evolution, including the groups of extinct plants.

palisade mesophyll   Columnar leaf mesophyll cells that occur in one or more vertical stacks adjacent to the upper epidermis.

parasitism   A symbiotic relationship in which one member (the parasite) benefits and the other (the host) is adversely affected. *Compare with* mutualism and commensalism.

parenchyma   Plant cells that are relatively unspecialized, are thin-walled, may contain chlorophyll, and are typically rather loosely packed; function in photosynthesis and in the storage of nutrients.

peduncle   The stalk of a flower or inflorescence.

peptide bond   A distinctive covalent carbon-to-nitrogen bond that links the amino acids in proteins.

peptidoglycan   A component of the bacterial cell wall; consists of a modified protein or peptide molecule with an attached carbohydrate.

perennial   A plant that lives for more than 2 years. *Compare with* annual and biennial.

pericycle   A layer of meristematic cells just inside the endodermis in a root; gives rise to branch roots.

periderm   A layer of cells covering the surface of a woody stem or root—that is, the outer bark. Anatomically, the periderm is composed of cork cells, cork cambium, and cork parenchyma (phelloderm) along with traces of primary tissues.

permafrost   Permanently frozen subsoil that is characteristic of frigid areas such as the tundra.

petal   One of the often conspicuously colored modified leaves that constitute the next-to-outermost whorl of a flower, and are collectively called a corolla.

petiole   A leaf stalk; the part of a leaf that attaches the blade to the stem.

phelloderm   *See* cork parenchyma.

phenotype   The physical expression of an individual organism's genes. *Compare with* genotype.

phloem   Vascular tissue that conducts food materials (carbohydrates).

photon   A particle of radiant energy with a wavelength in the visible range of the electromagnetic spectrum.

photoperiodism   The physiological response of plants to variations in the relative lengths of day and night.

photophosphorylation   Production of ATP, using light energy absorbed in photosynthesis.

photosynthesis   The biological process that includes the capture of light energy and its transformation into chemical energy of organic molecules (such as glucose), which are manufactured from carbon dioxide and water. Photosynthesis is practiced by plants, algae, and several kinds of bacteria.

phototropism   The growth of a plant in response to the direction of light.

pH scale   A number from 0 to 14 that indicates the degree of acidity or alkalinity of a substance.

phylogeny   The complete evolutionary history of a group of organisms.

phylum   A taxon composed of related, similar classes; a category beneath the kingdom and above the class. Some botanists use the equivalent term *division* when classifying plants.

phytochrome   A blue-green proteinaceous pigment that is the photoreceptor for a wide variety of physiological responses, including initiation of flowering in certain plants.

pilus (pl. *pili*)   An external hairlike structure that occurs in great numbers on the surfaces of many bacteria; functions in attachment and conjugation.

pistil   The female part of a flower; consists of either a single carpel (a simple pistil) or two or more fused carpels (a compound pistil).

pith   Ground tissue found in the centers of many stems and roots; composed of parenchyma cells.

plankton   Free-floating, mainly microscopic aquatic organisms found in the upper layers of the ocean or bodies of fresh water.

plant biology   *See* botany.

plasma membrane   The living surface membrane of a cell; acts as a selective barrier to passage of materials into and out of the cell.

plasmid   A small circular DNA molecule that carries genes, separate from the main bacterial chromosome.

plasmodial slime mold   One of a phylum of fungus-like protists whose feeding stage consists of a plasmodium.

plasmodium   A multinucleate, amoeboid mass of living matter that constitutes the feeding phase of plasmodial slime molds.

pneumatophore   A specialized aerial root produced by certain trees living in swampy habitats (for example, black mangrove); contains a well-developed system of air spaces that may facilitate gas exchange between the atmosphere and submerged roots.

polar covalent bond   A chemical bond in which electrons are shared unequally among the participating atoms, resulting in some difference in charge at different areas of the molecule. *Compare with* nonpolar covalent bond.

polar nucleus   In flowering plants, one of two haploid cells in the embryo sac that fuse with a sperm cell during double fertilization to form the triploid endosperm.

pollen grain   The immature male gametophyte of a seed plant (that is, of all gymnosperms and angiosperms); produces sperm cells capable of fertilization.

pollen tube A tube or extension that forms after germination of the pollen grain and through which male gametes (sperm cells) pass into the ovule.

pollination In seed plants, the transfer of pollen from the male part to the female part.

pollinium In orchids, a mass of pollen grains transported as a unit during pollination.

polygenes Two or more pairs of related genes that affect the same trait in an additive fashion.

polymerase chain reaction (PCR) A method of quickly and easily increasing small amounts of DNA into millions of copies.

polypeptide A chain of many amino acids linked by peptide bonds. A protein molecule consists of one or more chains of polypeptides.

polyphyletic Evolved from two or more different ancestors; applies to a group of organisms. *Compare with* monophyletic.

polyploid Having more than two sets of chromosomes per nucleus.

polysaccharide A carbohydrate consisting of many monosaccharide subunits. Examples are starches and cellulose.

population A group of organisms of the same species that live in the same geographical area at the same time.

potassium (K$^+$) ion mechanism The mechanism by which plants open and close their stomata: the influx of potassium ions into the guard cells causes water to move in by osmosis, changing the shape of the guard cells and opening the pore.

predation An interaction between two organisms in which one (the predator) consumes the other (the prey). The consumption of plants by herbivores is considered a type of predation.

pressure-flow hypothesis The mechanism by which dissolved sugar is thought to be transported in phloem.

primary growth An increase in a plant's length. This growth occurs at the tips of the stems and roots due to the activity of apical meristems. *Compare with* secondary growth.

primary succession An ecological succession that occurs on land not previously inhabited by plants; no soil is present initially. *Compare with* secondary succession.

principle of dominance When two different alleles are present in an individual, often only one—the dominant allele—is expressed.

principle of independent assortment The chance distribution of alleles to the gametes: the distribution of the alleles governing one trait have no influence on the distribution of a different set of alleles.

principle of segregation The two alleles of each gene separate into different gametes; each gamete receives only one member of each pair.

producer An organism (such as a chlorophyll-containing plant) that manufactures complex organic molecules from simple inorganic substances. In most ecosystems, producers are photosynthetic organisms. *Compare with* consumer.

progymnosperm One of an extinct group of plants thought to have been the ancestors of gymnosperms.

prokaryote Organisms that lack membrane-bounded nuclei and other membrane-bounded organelles; the archaebacteria and eubacteria. *Compare with* eukaryote.

prop root An adventitious root that arises on the main stem and functions to support the plant.

protective coloration The coloring of an organism so that it blends into its surroundings in such a way that it is difficult to see.

protein A large, complex organic compound composed of amino acid subunits; the principal structural component of cells. *See* polypeptide.

prothallus The free-living, haploid gametophyte plant in ferns and other seedless vascular plants.

protonema A branching filament of haploid cells that grows from a moss spore; each protonema develops buds that grow into leafy gametophyte plants.

pseudobulb A thickened aboveground stem found in certain orchids.

pseudoplasmodium In cellular slime molds, an aggregation of amoeboid cells to form a spore-producing fruiting body during reproduction.

punctuated equilibrium An evolutionary model in which evolution proceeds with long periods of inactivity followed by very active phases, so that major adaptations appear suddenly in the fossil record. *Compare with* gradualism.

rain shadow An area on the downwind side of a mountain range that receives very little precipitation. Deserts often occur in rain shadows.

ray A chain of parenchyma cells that extends radially in the stems and roots of woody plants and functions for lateral transport of carbohydrates, water, and minerals.

receptacle The part of a flower stalk where the floral parts are attached.

recessive An allele not expressed in the heterozygous condition. *Compare with* dominant.

recombinant DNA Any DNA molecule made by combining genes from different organisms.

recombinant DNA technology The methods used by genetic engineers to combine genes from different organisms and transfer them into cells where they may be expressed.

recombination The formation, in gametes (and in offspring that arise from gametes), of allele combinations that are not found in either parent; the result of chromosome shuffling and crossing over in meiosis and of mutation. Also called genetic recombination.

red alga One of a diverse phylum of algae that contain the pigments chlorophyll *a*, carotenoids, phycocyanin, and phycoerythrin.

red tide A red or brown coloration of ocean water caused by a population explosion, or bloom, of dinoflagellates.

reduction The gain of electrons (or of hydrogen) by a compound. *Compare with* oxidation.

reproductive isolation The reproductive barriers that prevent a species from interbreeding with another species.

resin A clear, or translucent, viscous organic material that certain plants produce and secrete into specialized ducts. Resin may play a role in deterring plant-eating insects or disease-causing organisms. *See* resin duct.

resin duct One of a series of tubelike canals that occur in the xylem, phloem, and bark of many conifers.

restriction enzyme One of a class of bacterial enzymes that recognize specific DNA base sequences and cleave the DNA molecule at that site.

rhizoid Hairlike absorptive filaments that form part of the gametophytes of bryophytes and some vascular plants.

rhizome A horizontal underground stem that bears leaves and buds and often serves as a storage organ and a means of asexual reproduction—for example, in the iris.

ribonucleic acid (RNA) A single-stranded nucleic acid molecule that is necessary for protein synthesis. There are three forms of RNA: messenger RNA, transfer RNA, and ribosomal RNA.

ribosomal RNA (rRNA) A type of RNA that joins with various proteins to form a ribosome. *Compare with* transfer RNA and messenger RNA.

ribosome A cellular organelle that is the site of protein synthesis and consists of a larger subunit and a smaller subunit, each composed of ribosomal RNA (rRNA) and ribosomal proteins.

RNA *See* ribonucleic acid.

RNA polymerase An enzyme complex that catalyzes the synthesis of RNA molecules from DNA templates.

root cap A covering of cells over the root tip that protects the delicate meristematic tissue directly behind it.

root graft The union of roots of two different plants of either the same or different species.

root hair An extension of an epidermal cell in a root. Root hairs increase the absorptive capacity of roots.

root pressure The pressure in root xylem that occurs as a result of water moving into roots from the soil.

root system The part of the vascular plant body that is normally found belowground; anchors the plant and absorbs water and minerals from the soil. *Compare with* shoot system.

rubisco The enzyme in photosynthesis (the Calvin cycle) that catalyzes the addition of carbon dioxide to the five-carbon sugar, ribulose bisphosphate.

sand Inorganic soil particles that are larger than clay or silt.

saprotroph A microbial heterotroph that breaks down dead organic material and uses the decomposition products as a source of energy; many bacteria and fungi are saprotrophs. *Also called* decomposer.

sapwood The outermost, youngest wood in many tree trunks that conducts water and dissolved minerals. *Compare with* heartwood.

saturated fatty acid A fatty acid that has no carbon-to-carbon double bonds. *Compare with* unsaturated fatty acid.

savanna A tropical grassland with widely scattered trees that is found in areas of low rainfall and areas of seasonal rainfall with prolonged dry periods.

scientific method The steps a scientist uses to approach a problem: making observations, stating the problem, developing a hypothesis, performing an experiment, and using the results to support or refute the hypothesis or to generate other hypotheses.

sclerenchyma Cells that provide strength and support in the plant body. Sclerenchyma cells have extremely thick walls.

sclerophyllous leaf A hard, small, leathery leaf that resists water loss; characteristic of perennial plants adapted to extremely dry habitats.

secondary growth An increase in a plant's girth due to the activity of lateral meristems (vascular cambium and cork cambium). *Compare with* primary growth.

secondary succession An ecological succession that takes place after some disturbance destroys the existing vegetation; soil is already present. *Compare with* primary succession.

second law of thermodynamics When energy is converted from one form to another, some of it is degraded into a lower-quality, less useful form. Thus, with each successive energy transformation, less energy is available to do work. *Compare with* first law of thermodynamics.

seed A ripened ovule that is composed of a young multicellular plant and nutritive tissue (food reserves), enclosed by a seed coat.

seed fern One of an extinct group of seed-bearing woody plants with fernlike leaves. Seed ferns probably descended from progymnosperms and gave rise to cycads and possibly ginkgoes.

segregation *See* principle of segregation.

selective breeding The selection by humans of traits that are desired in plants, and the breeding of only those individuals that possess the desired traits. Also called artificial selection.

semiconservative replication The type of replication that is characteristic of DNA, in which each new double-stranded molecule of DNA consists of one strand from the original DNA molecule and one strand of newly synthesized DNA.

senescence The aging process.

sepal One of the outermost parts of a flower, usually leaflike in appearance, that protect the flower as a bud and are collectively called a calyx.

sexual reproduction A type of reproduction in which two gametes (usually, but not necessarily, contributed by two different parents) fuse to form a zygote. *Compare with* asexual reproduction.

shoot system The part of the vascular plant body that is normally found aboveground; consists of the stem and leaves. *Compare with* root system.

short-day plant A plant that flowers when the night length is equal to or greater than some critical length. *Also called* long-night plant. *Compare with* long-day, intermediate-day, and day-neutral plants.

short-night plant *See* long-day plant.

silt Medium-sized inorganic particles. *Compare with* clay and sand.

simple fruit   A fruit that develops from a single pistil of a single flower.

simple leaf   A leaf with a single blade. *Compare with* compound leaf.

solar tracking   The ability of certain plant parts, such as cotton leaves, to follow the movement of the sun across the sky; caused by turgor movements.

somatic cell   A cell of the body not involved in reproduction; body cell.

sorus (pl. *sori*)   In ferns, a cluster of spore-producing sporangia.

sp   The singular abbreviation for *species. Compare* spp.

species   A group of organisms with similar structural and functional characteristics that in nature breed only with one another and have a close common ancestry; a group of organisms with a common gene pool.

spindle   The apparatus, composed of microtubules, that separates chromosomes in eukaryotic cell division.

spine   A leaf modified for protection, such as a cactus spine.

spirillum (pl. *spirilla*)   A corkscrew-shaped bacterium.

spongy mesophyll   Leaf mesophyll cells that are loosely arranged and contain many air spaces. In leaves where it occurs with palisade mesophyll, spongy mesophyll is adjacent to the lower epidermis.

sporangium (pl. *sporangia*)   A spore case; the structure within which spores are formed in plants, certain protists, and fungi.

spore   A haploid reproductive cell that gives rise directly to an individual offspring in plants, fungi, and certain algae.

spore mother cell   A diploid cell that undergoes meiosis to form haploid spores in plants.

sporophyll   a leaflike structure that bears spores.

sporophyte generation   The diploid or spore-producing stage in the life cycle of a plant. *Compare with* gametophyte generation.

spp   The plural abbreviation for *species. Compare* sp.

stamen   The male part of a flower; produces pollen.

stele   The central core or cylinder of vascular tissues (xylem and phloem) and associated ground tissues (pith and pericycle) in a root or stem. The structure of the stele varies among different groups of plants.

sticky end   A single-stranded stub that extends from each end of a fragment of DNA treated with a restriction enzyme.

stigma   The portion of the pistil where the pollen lands prior to fertilization.

stimulus   A physical or chemical change in the internal or external environment of an organism, potentially capable of provoking a response.

stipule   A leaflike outgrowth at the base of a petiole.

stolon   An aboveground horizontal stem with long internodes. Stolons often form buds that develop into separate plants—for example, in the strawberry.

stoma (pl. *stomata*)   A small pore flanked by guard cells in the epidermis; allows the gas exchange that is necessary for photosynthesis.

strobilus (pl. *strobili*)   In certain plants, a conelike structure that bears spore-producing sporangia.

style   The neck connecting the stigma to the ovary of a pistil.

subspecies   A taxon below that of species; considered by many botanists to be the equivalent of the variety.

succession   The sequence of vegetation changes in a community over time. Includes primary succession and secondary succession. Also called ecological succession.

sucker   A shoot that develops adventitiously from a root; a type of asexual reproduction.

sympatric speciation   The evolution of a new species within the parent species' geographical region.

synapsis   The pairing of homologous chromosomes during prophase I of meiosis.

syngamy   The union of the gametes in sexual reproduction.

synthetic theory of evolution   The synthesis of previous theories, especially of Mendelian genetics with Darwin's theory, to formulate a comprehensive explanation of evolution. Also called neo-Darwinism.

systematics   Modern taxonomy that is based on evolutionary relationships.

taiga   A region of coniferous forests just south of the tundra in the Northern Hemisphere. Also called boreal forest.

taproot system   A root system consisting of a prominent main root with smaller lateral roots branching from it; a taproot develops directly from the embryonic radicle. *Compare with* fibrous root system.

taxon   A taxonomic grouping of any rank—for example, phylum or genus.

temperate deciduous forest   A forest biome that occurs in temperate areas where annual precipitation ranges from about 75 to 125 centimeters.

temperate grassland   A grassland characterized by hot summers, cold winters, and less rainfall than is found in a temperate deciduous forest biome.

temperate rain forest   A coniferous biome characterized by cool weather, dense fog, and high precipitation.

tendril   A leaf or stem that is modified for holding or attaching onto objects.

tension-cohesion theory   The mechanism by which water and dissolved minerals are thought to be transported in xylem: water is pulled upward under tension due to transpiration, while maintaining an unbroken column in xylem due to cohesion.

test cross   A cross between an individual with a dominant phenotype (but an unknown genotype) and a homozygous recessive individual. The test cross helps determine the genotype of the dominant individual.

thallus   The body form of certain bryophytes, consisting of flattened, ribbon-like lobes.

thigmotropism   Growth of a plant in response to contact with a solid object, as exhibited by tendrils.

tissue   A group of closely associated, similar cells that work together to carry out specific functions.

totipotency The ability of a cell (or nucleus) to provide information for the development of an entire organism.

transcription Synthesis of RNA from a DNA template.

transfer RNA (tRNA) A form of RNA that serves in the synthesis of proteins. An amino acid is bound to a specific tRNA molecule and then arranged in correct order by the complementary pairing between the codon of mRNA and the anticodon of tRNA. *Compare with* messenger RNA and ribosomal RNA.

transformed Changed by the incorporation of a piece of foreign DNA. A bacterium that takes in a plasmid carrying recombinant DNA is said to be transformed.

transgenic Said of plants and other organisms that have been genetically engineered.

translation Conversion of information provided by mRNA to a specific sequence of amino acids in the production of a polypeptide chain.

translocation The movement of materials such as dissolved sugar through phloem.

transpiration Evaporation of water vapor from a plant's aerial parts (leaves and stems).

trichome A hair or other appendage growing out from the epidermis.

triplet A sequence of three nucleotides in DNA that codes for a codon in messenger RNA.

triploid The condition of having three sets of chromosomes, as in endosperm.

trophic level Each level in a food web. All producers belong to the first trophic level, all herbivores belong to the second trophic level, and so on.

tropical rain forest A lush, species-rich forest biome that occurs in tropical areas where the climate is very moist throughout the year. Tropical rain forests are also characterized by old, infertile soils.

tuber The thickened end of a rhizome that is fleshy and enlarged for food storage. White potato tubers are an example.

tundra The treeless biome in the far north that consists of boggy plains covered by lichens and small plants such as mosses; characterized by harsh, very cold winters and extremely short summers.

turgor *See* turgor pressure.

turgor movement A temporary movement of a plant part caused by a sudden change in turgor in certain cells.

turgor pressure Pressure that develops against a cell wall when water moves into the cell by osmosis. Also called turgor.

unsaturated fatty acid A fatty acid with one or more carbon-to-carbon double bonds. *Compare with* saturated fatty acid.

vacuole A fluid-filled, membrane-bounded sac within the cytoplasm; contains a solution of salts, ions, pigments, and other materials.

variety A taxon below that of species; considered by many botanists to be the equivalent of the subspecies.

vascular bundle *See* vein.

vascular cambium A lateral meristem that produces secondary xylem (wood) and secondary phloem (inner bark).

vascular plant Any plant that possesses conducting tissues—that is, xylem and phloem. *Compare with* nonvascular plant.

vascular tissue system The tissue system that conducts materials throughout the plant body; consists of xylem and phloem.

vector In genetic engineering, an agent, such as a plasmid or virus, used to transfer DNA into a cell.

vegetative Nonreproductive.

vein A strand of vascular tissues (xylem and phloem) that conducts materials throughout the plant body. *Also called* vascular bundle.

vernalization Promotion of flowering in certain plants by exposure to a cold period.

vestigial organ An evolutionary remnant of a formerly functional organ.

virus A tiny infectious "molecule," composed of a core of nucleic acid usually encased in a coat of protein; capable of infecting living cells.

water mold One of a phylum of fungus-like protists whose bodies consist of coenocytic mycelia and that reproduce asexually by forming motile zoospores and sexually by forming oospores.

weathering process A chemical or physical process that helps to form soil from rock; the parent material is gradually broken into smaller and smaller particles.

whisk fern One of a phylum of seedless vascular plants with a life cycle similar to that of ferns.

xylem Vascular tissue that conducts water and dissolved minerals.

yeast A unicellular fungus (ascomycete) that reproduces asexually by budding or fission and sexually by ascospores.

zoospore A flagellated spore produced asexually by certain algae, water molds, and other protists.

zooxanthellae Endosymbiotic, photosynthetic dinoflagellates found in certain marine invertebrates; their relationship with corals enhances the corals' reef-building ability.

zygomycete A member of a phylum of fungi characterized by the production of nonmotile asexual spores and sexual zygospores.

zygospore A thick-walled sexual spore produced by a zygomycete.

zygote The diploid cell that results from the union of gametes in sexual reproduction.

INDEX

Boldface numbers indicate pages where terms are defined; *il* indicates an illustration on that page.